VOLUME FOUR HUNDRED AND THIRTY

Methods in
ENZYMOLOGY

Translation Initiation:
Reconstituted Systems and
Biophysical Methods

METHODS IN ENZYMOLOGY

Editors-in-Chief

JOHN N. ABELSON AND MELVIN I. SIMON

Division of Biology
California Institute of Technology
Pasadena, California

Founding Editors

SIDNEY P. COLOWICK AND NATHAN O. KAPLAN

VOLUME FOUR HUNDRED AND THIRTY

METHODS IN
ENZYMOLOGY

Translation Initiation: Reconstituted Systems and Biophysical Methods

EDITED BY

JON LORSCH
Department of Biophysics and Biophysical Chemistry
Johns Hopkins University School of Medicine
Baltimore, Maryland

AMSTERDAM • BOSTON • HEIDELBERG • LONDON
NEW YORK • OXFORD • PARIS • SAN DIEGO
SAN FRANCISCO • SINGAPORE • SYDNEY • TOKYO
Academic Press is an imprint of Elsevier

ELSEVIER

Academic Press is an imprint of Elsevier
525 B Street, Suite 1900, San Diego, California 92101-4495, USA
84 Theobald's Road, London WC1X 8RR, UK

This book is printed on acid-free paper. ∞

For information on all Elsevier Academic Press publications
visit our Web site at www.books.elsevier.com

ISBN: 978-0-12-373969-8

PRINTED IN THE UNITED STATES OF AMERICA
07   08   09   10   9   8   7   6   5   4   3   2   1

Contents

CONTRIBUTORS

Michael G. Acker
Department of Biophysics and Biophysical Chemistry, Johns Hopkins University School of Medicine, Baltimore, Maryland

M. Leah Allen
Department of Chemistry and Biochemistry and the Institute for Cell and Molecular Biology, University of Texas at Austin, Austin, Texas

Jan M. Antosiewicz
Division of Biophysics, Institute of Experimental Physics, Warsaw University, Warszawa, Poland

Dario Benelli
Dipartimento Biotecnologie Cellulari ed Ematologia, Università di Roma Sapienza, Roma, Italy

Sylvain Blanquet
Laboratoire de Biochimie, CNRS Ecole Polytechnique, Palaiseau Cedex, France

Sylvain de Breyne
Department of Microbiology and Immunology, State University of New York Downstate Medical Center, Brooklyn, New York

Karen S. Browning
Department of Chemistry and Biochemistry and the Institute for Cell and Molecular Biology, University of Texas at Austin, Austin, Texas

Lara Campbell
Department of Chemistry and Biochemistry and the Institute for Cell and Molecular Biology, University of Texas at Austin, Austin, Texas

Jayanta Chaudhuri
Immunology Program, Memorial Sloan Kettering Cancer Center, New York, New York

Brooke E. Christian
Department of Chemistry, University of North Carolina at Chapel Hill, Chapel Hill, North Carolina

David A. Costantino
Department of Biochemistry and Molecular Genetics, University of Colorado at Denver and Health Sciences Center, Aurora, Colorado

Edward Darzynkiewicz
Division of Biophysics, Institute of Experimental Physics, Warsaw University, Warszawa, Poland

Michael D. Dennis
Department of Chemistry and Biochemistry and the Institute for Cell and Molecular Biology, University of Texas at Austin, Austin, Texas

Attilio Fabbretti
Laboratory of Genetics, Department of Biology MCA, University of Camerino, Camerino, Italy

Megan E. Filbin
Department of Biochemistry and Molecular Genetics, University of Colorado at Denver and Health Sciences Center, Aurora, Colorado

Dominique Frueh
Department of Biological Chemistry and Molecular Pharmacology, Harvard Medical School, Boston, Massachusetts

Anna Maria Giuliodori
Department of Biology MCA, University of Camerino, Camerino, Italy

Domenick G. Grasso
Department of Chemistry, University of North Carolina at Chapel Hill, Chapel Hill, North Carolina

Claudio O. Gualerzi
Laboratory of Genetics, Department of Biology MCA, University of Camerino, Camerino, Italy

Laurent Guillon
Laboratoire de Biochimie, CNRS Ecole Polytechnique, Palaiseau Cedex, France

Tobias von der Haar
Protein Science Group, Department of Biosciences, University of Kent, Canterbury, United Kingdom

John Hammond
Department of Biochemistry and Molecular Genetics, University of Colorado at Denver and Health Sciences Center, Aurora, Colorado

Christopher U. T. Hellen
Department of Microbiology and Immunology, State University of New York Downstate Medical Center, Brooklyn, New York

Simpson Joseph
Department of Chemistry and Biochemistry, University of California, San Diego, California

Jeffrey S. Kieft
Department of Biochemistry and Molecular Genetics, University of Colorado at Denver and Health Sciences Center, Aurora, Colorado

Sarah E. Kolitz
Department of Biophysics and Biophysical Chemistry, Johns Hopkins University School of Medicine, Baltimore, Maryland

Victoria G. Kolupaeva
Department of Microbiology and Immunology, State University of New York Downstate Medical Center, Brooklyn, New York

Andrey L. Konevega
Institute of Physical Biochemistry, University of Witten/Herdecke, Witten, Germany, and Petersburg Nuclear Physics Institute, Russian Academy of Science, Gatchina, Russia

Darrin A. Lindhout
Department of Structural Biology, Stanford University School of Medicine, Stanford, California

Paola Londei
Dipartimento Biotecnologie Cellulari ed Ematologia, Università di Roma Sapienza, Roma, Italy

Jon R. Lorsch
Department of Biophysics and Biophysical Chemistry, Johns Hopkins University School of Medicine, Baltimore, Maryland

Umadas Maitra
Department of Developmental and Molecular Biology, Albert Einstein College of Medicine of Yeshiva University, Jack and Pearl Resnick Campus, Bronx, New York

Romit Majumdar
Department of Cell Biology, Albert Einstein College of Medicine of Yeshiva University, Jack and Pearl Resnick Campus, Bronx, New York

Assen Marintchev
Department of Biological Chemistry and Molecular Pharmacology, Harvard Medical School, Boston, Massachusetts

Steven Marsden
Manchester Interdisciplinary Biocentre, University of Manchester, Manchester, United Kingdom

Laura K. Mayberry
Department of Chemistry and Biochemistry and the Institute for Cell and Molecular Biology, University of Texas at Austin, Austin, Texas

John E. G. McCarthy
Manchester Interdisciplinary Biocentre, University of Manchester, Manchester, United Kingdom

Sean A. McKenna
Department of Structural Biology, Stanford University School of Medicine, Stanford, California

Yves Mechulam
Laboratoire de Biochimie, CNRS Ecole Polytechnique, Palaiseau Cedex, France

Pohl Milon
Laboratory of Genetics, Department of Biology MCA, University of Camerino, Camerino, Italy

Sarah F. Mitchell
Department of Biophysics and Biophysical Chemistry, Johns Hopkins University School of Medicine, Baltimore, Maryland

Patricia A. Murphy
Department of Chemistry and Biochemistry and the Institute for Cell and Molecular Biology, University of Texas at Austin, Austin, Texas

Jagpreet S. Nanda
Department of Biophysics and Biophysical Chemistry, Johns Hopkins University School of Medicine, Baltimore, Maryland

Anna Niedzwiecka
Division of Biophysics, Institute of Experimental Physics, Warsaw University, Warszawa, Poland, and Biological Physics Group, Institute of Physics, Polish Academy of Sciences, Warszawa, Poland

Kelley Ruud Nitka
Department of Chemistry and Biochemistry and the Institute for Cell and Molecular Biology, University of Texas at Austin, Austin, Texas

Frank Peske
Institute of Molecular Biology, University of Witten/Herdecke, Witten, Germany

Tatyana V. Pestova
Department of Microbiology and Immunology, State University of New York Downstate Medical Center, Brooklyn, New York, and A. N. Belozersky Institute of Physicochemical Biology, Moscow State University, Moscow, Russia

Jennifer S. Pfingsten
Department of Biochemistry and Molecular Genetics, University of Colorado at Denver and Health Sciences Center, Aurora, Colorado

Andrey V. Pisarev
Department of Microbiology and Immunology, State University of New York Downstate Medical Center, Brooklyn, New York

Cynthia L. Pon
Department of Biology MCA, University of Camerino, Camerino, Italy

Joseph D. Puglisi
Department of Structural Biology and Stanford Magnetic Resonance Laboratory, Stanford University School of Medicine, Stanford, California

Marina V. Rodnina
Institute of Physical Biochemistry, University of Witten/Herdecke, Witten, Germany

Emmanuelle Schmitt
Laboratoire de Biochimie, CNRS Ecole Polytechnique, Palaiseau Cedex, France

Takashi Shimoike
Department of Structural Biology, Stanford University School of Medicine, Stanford, California, and Department of Virology II, National Institute of Infectious Diseases, Musashi-Murayama, Tokyo, Japan

Angela Spencer
Department of Chemistry, University of North Carolina at Chapel Hill, Chapel Hill, North Carolina

Linda L. Spremulli
Department of Chemistry, University of North Carolina at Chapel Hill, Chapel Hill, North Carolina

Janusz Stepinski
Division of Biophysics, Institute of Experimental Physics, Warsaw University, Warszawa, Poland

Ryszard Stolarski
Division of Biophysics, Institute of Experimental Physics, Warsaw University, Warszawa, Poland

Sean M. Studer
Department of Chemistry and Biochemistry, University of California, San Diego, California

Anett Unbehaun
Department of Microbiology and Immunology, State University of New York Downstate Medical Center, Brooklyn, New York

Gerhard Wagner
Department of Biological Chemistry and Molecular Pharmacology, Harvard Medical School, Boston, Massachusetts

Laure Yatime
Laboratoire de Biochimie, CNRS Ecole Polytechnique, Palaiseau Cedex, France

PREFACE

Over the past 15 years, it has become clear that translation initiation is a key regulatory point in the control of gene expression. Loss-of-control of protein synthesis has been implicated in a variety of diseases ranging from cancer to viral infection, and there is increasing interest in the development of new drugs that target translation initiation. Despite the profound biological and medical importance of this key step in gene expression, we are only beginning to understand the molecular mechanics that underlie translation initiation and its control, and much work remains to be done.

These MIE volumes (429, 430, and 431) are a compilation of current approaches used to dissect the basic mechanisms by which bacterial, archaeal, and eukaryotic cells assemble, and control the assembly of, ribosomal complexes at the initiation codon. A wide range of methods is presented from cell biology to biophysics to chemical biology. It is clear that no one approach can answer all of the important questions about translation initiation, and that major advances will require collaborative efforts that bring together various disciplines. I hope that these volumes will facilitate cross-disciplinary thinking and enable researchers from a wide variety of fields to explore aspects of translation initiation throughout biology.

Initially, we had planned to publish a single volume on this subject. However, the remarkable response to my requests for chapters allowed us to scale up to three volumes. I would like to express my sincerest appreciation and admiration for the contributors to this endeavor. I am impressed with the outstanding quality of the work produced by the authors, all of whom are leaders in the field. I am especially grateful to John Abelson for giving me the opportunity to edit this publication and for his support and advice throughout the project. Finally, I am indebted to Cindy Minor and the staff at Elsevier for their help and wisdom along the way.

JON LORSCH

Methods in Enzymology

VOLUME 71. Lipids (Part C)
Edited by JOHN M. LOWENSTEIN

VOLUME 72. Lipids (Part D)
Edited by JOHN M. LOWENSTEIN

VOLUME 73. Immunochemical Techniques (Part B)
Edited by JOHN J. LANGONE AND HELEN VAN VUNAKIS

VOLUME 74. Immunochemical Techniques (Part C)
Edited by JOHN J. LANGONE AND HELEN VAN VUNAKIS

VOLUME 75. Cumulative Subject Index Volumes XXXI, XXXII, XXXIV–LX
Edited by EDWARD A. DENNIS AND MARTHA G. DENNIS

VOLUME 76. Hemoglobins
Edited by ERALDO ANTONINI, LUIGI ROSSI-BERNARDI, AND EMILIA CHIANCONE

VOLUME 77. Detoxication and Drug Metabolism
Edited by WILLIAM B. JAKOBY

VOLUME 78. Interferons (Part A)
Edited by SIDNEY PESTKA

VOLUME 79. Interferons (Part B)
Edited by SIDNEY PESTKA

VOLUME 80. Proteolytic Enzymes (Part C)
Edited by LASZLO LORAND

VOLUME 81. Biomembranes (Part H: Visual Pigments and Purple Membranes, I)
Edited by LESTER PACKER

VOLUME 82. Structural and Contractile Proteins (Part A: Extracellular Matrix)
Edited by LEON W. CUNNINGHAM AND DIXIE W. FREDERIKSEN

VOLUME 83. Complex Carbohydrates (Part D)
Edited by VICTOR GINSBURG

VOLUME 84. Immunochemical Techniques (Part D: Selected Immunoassays)
Edited by JOHN J. LANGONE AND HELEN VAN VUNAKIS

VOLUME 85. Structural and Contractile Proteins (Part B: The Contractile Apparatus and the Cytoskeleton)
Edited by DIXIE W. FREDERIKSEN AND LEON W. CUNNINGHAM

VOLUME 86. Prostaglandins and Arachidonate Metabolites
Edited by WILLIAM E. M. LANDS AND WILLIAM L. SMITH

VOLUME 87. Enzyme Kinetics and Mechanism (Part C: Intermediates, Stereo-chemistry, and Rate Studies)
Edited by DANIEL L. PURICH

VOLUME 88. Biomembranes (Part I: Visual Pigments and Purple Membranes, II)
Edited by LESTER PACKER

VOLUME 190. Retinoids (Part B: Cell Differentiation and Clinical Applications)
Edited by LESTER PACKER

VOLUME 191. Biomembranes (Part V: Cellular and Subcellular Transport: Epithelial Cells)
Edited by SIDNEY FLEISCHER AND BECCA FLEISCHER

VOLUME 192. Biomembranes (Part W: Cellular and Subcellular Transport: Epithelial Cells)
Edited by SIDNEY FLEISCHER AND BECCA FLEISCHER

VOLUME 193. Mass Spectrometry
Edited by JAMES A. MCCLOSKEY

VOLUME 194. Guide to Yeast Genetics and Molecular Biology
Edited by CHRISTINE GUTHRIE AND GERALD R. FINK

VOLUME 195. Adenylyl Cyclase, G Proteins, and Guanylyl Cyclase
Edited by ROGER A. JOHNSON AND JACKIE D. CORBIN

VOLUME 196. Molecular Motors and the Cytoskeleton
Edited by RICHARD B. VALLEE

VOLUME 197. Phospholipases
Edited by EDWARD A. DENNIS

VOLUME 198. Peptide Growth Factors (Part C)
Edited by DAVID BARNES, J. P. MATHER, AND GORDON H. SATO

VOLUME 199. Cumulative Subject Index Volumes 168–174, 176–194

VOLUME 200. Protein Phosphorylation (Part A: Protein Kinases: Assays, Purification, Antibodies, Functional Analysis, Cloning, and Expression)
Edited by TONY HUNTER AND BARTHOLOMEW M. SEFTON

VOLUME 201. Protein Phosphorylation (Part B: Analysis of Protein Phosphorylation, Protein Kinase Inhibitors, and Protein Phosphatases)
Edited by TONY HUNTER AND BARTHOLOMEW M. SEFTON

VOLUME 202. Molecular Design and Modeling: Concepts and Applications (Part A: Proteins, Peptides, and Enzymes)
Edited by JOHN J. LANGONE

VOLUME 203. Molecular Design and Modeling: Concepts and Applications (Part B: Antibodies and Antigens, Nucleic Acids, Polysaccharides, and Drugs)
Edited by JOHN J. LANGONE

VOLUME 204. Bacterial Genetic Systems
Edited by JEFFREY H. MILLER

VOLUME 205. Metallobiochemistry (Part B: Metallothionein and Related Molecules)
Edited by JAMES F. RIORDAN AND BERT L. VALLEE

VOLUME 257. Small GTPases and Their Regulators (Part C: Proteins Involved in Transport)
Edited by W. E. BALCH, CHANNING J. DER, AND ALAN HALL

VOLUME 258. Redox-Active Amino Acids in Biology
Edited by JUDITH P. KLINMAN

VOLUME 259. Energetics of Biological Macromolecules
Edited by MICHAEL L. JOHNSON AND GARY K. ACKERS

VOLUME 260. Mitochondrial Biogenesis and Genetics (Part A)
Edited by GIUSEPPE M. ATTARDI AND ANNE CHOMYN

VOLUME 261. Nuclear Magnetic Resonance and Nucleic Acids
Edited by THOMAS L. JAMES

VOLUME 262. DNA Replication
Edited by JUDITH L. CAMPBELL

VOLUME 263. Plasma Lipoproteins (Part C: Quantitation)
Edited by WILLIAM A. BRADLEY, SANDRA H. GIANTURCO, AND JERE P. SEGREST

VOLUME 264. Mitochondrial Biogenesis and Genetics (Part B)
Edited by GIUSEPPE M. ATTARDI AND ANNE CHOMYN

VOLUME 265. Cumulative Subject Index Volumes 228, 230–262

VOLUME 266. Computer Methods for Macromolecular Sequence Analysis
Edited by RUSSELL F. DOOLITTLE

VOLUME 267. Combinatorial Chemistry
Edited by JOHN N. ABELSON

VOLUME 268. Nitric Oxide (Part A: Sources and Detection of NO; NO Synthase)
Edited by LESTER PACKER

VOLUME 269. Nitric Oxide (Part B: Physiological and Pathological Processes)
Edited by LESTER PACKER

VOLUME 270. High Resolution Separation and Analysis of Biological Macromolecules (Part A: Fundamentals)
Edited by BARRY L. KARGER AND WILLIAM S. HANCOCK

VOLUME 271. High Resolution Separation and Analysis of Biological Macromolecules (Part B: Applications)
Edited by BARRY L. KARGER AND WILLIAM S. HANCOCK

VOLUME 272. Cytochrome P450 (Part B)
Edited by ERIC F. JOHNSON AND MICHAEL R. WATERMAN

VOLUME 273. RNA Polymerase and Associated Factors (Part A)
Edited by SANKAR ADHYA

VOLUME 274. RNA Polymerase and Associated Factors (Part B)
Edited by SANKAR ADHYA

Transient Kinetics, Fluorescence, and FRET in Studies of Initiation of Translation in Bacteria

Pohl Milon,* Andrey L. Konevega,[†,‡] Frank Peske,[§]
Attilio Fabbretti,* Claudio O. Gualerzi,* and Marina V. Rodnina[†]

Contents

* Laboratory of Genetics, Department of Biology MCA, University of Camerino, Camerino, Italy
† Institute of Physical Biochemistry, University of Witten/Herdecke, Witten, Germany
‡ Petersburg Nuclear Physics Institute, Russian Academy of Science, Gatchina, Russia
§ Institute of Molecular Biology, University of Witten/Herdecke, Witten, Germany

Methods in Enzymology, Volume 430
ISSN 0076-6879, DOI: 10.1016/S0076-6879(07)30001-3

Abstract

Initiation of mRNA translation in prokaryotes requires the small ribosomal subunit (30S), initiator fMet-tRNA^{fMet}, three initiation factors, IF1, IF2, and IF3, and the large ribosomal subunit (50S). During initiation, the 30S subunit, in a complex with IF3, binds mRNA, IF1, IF2·GTP, and fMet-tRNA^{fMet} to form a 30S initiation complex which then recruits the 50S subunit to yield a 70S initiation complex, while the initiation factors are released. Here we describe a transient kinetic approach to study the timing of elemental steps of 30S initiation complex formation, 50S subunit joining, and the dissociation of the initiation factors from the 70S initiation complex. Labeling of ribosomal subunits, fMet-tRNA^{fMet}, mRNA, and initiation factors with fluorescent reporter groups allows for the direct observation of the formation or dissociation of complexes by monitoring changes in the fluorescence of single dyes or fluorescence resonance energy transfer (FRET) between two fluorophores. Subunit joining was monitored by light scattering or by FRET between dyes attached to the ribosomal subunits. The kinetics of chemical steps, that is, GTP hydrolysis by IF2 and peptide bond formation following the binding of aminoacyl-tRNA to the 70S initiation complex, were measured by the quench-flow technique. The methods described here are based on results obtained with initiation components from *Escherichia coli* but can be adopted for mechanistic studies of initiation in other prokaryotic or eukaryotic systems.

1. INTRODUCTION

Biochemical analyses of the individual steps of translation initiation in bacteria suggested the pathway shown in Fig. 1.1 (reviewed in Gualerzi *et al.*, 2001; Laursen *et al.*, 2005). The 30S ribosomal subunit binds the three initiation factors, and the 30S-initiation factor complex interacts in a random manner with its two ligands, fMet-tRNA^{fMet} and mRNA. The initiation efficiency of a given mRNA is determined by its translation initiation region (TIR), namely by its secondary and tertiary structure, by the Shine-Dalgarno (SD) sequence, the distance between the SD sequence and the initiation codon, the type of initiation codon, and possibly, by the presence or absence of specific enhancer sequences. The initially formed unstable "30S preinitiation complex" is a kinetic intermediate of the bona fide "30S initiation complex" which is formed upon base pairing of the anticodon of fMet-tRNA^{fMet} with the AUG codon of mRNA in the P site. Joining of the 50S subunit with the 30S initiation complex initiates a number of reactions,

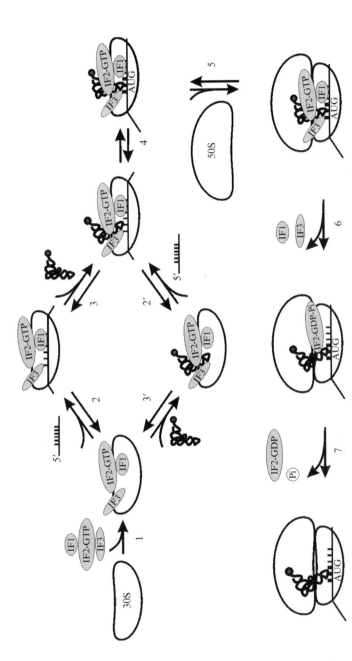

Figure 1.1   Scheme of the translation initiation events. Step 1: Initiation factors bind to the 30S subunit. The order and timing of the association of each factor is not known, although IF3 is likely to bind first. Step 2: mRNA binding. Step 3: fMet-tRNA^{fMet} binding. Binding of mRNA and fMet-tRNA^{fMet} occurs in a random fashion as indicated by steps 2' and 3'. Step 4: Conformational rearrangement of the complex, resulting in the recognition of the initiation AUG codon and formation of a stable 30S initiation complex. Step 5: 50S subunit joining. Step 6: Dissociation of IF1 and IF3 and hydrolysis of GTP by IF2. Step 7: Dissociation of IF2 and Pi, resulting in the 70S initiation complex formation.

including GTP hydrolysis by IF2, the dissociation of the initiation factors, and the stabilization of fMet-tRNA^{fMet} in the P site. In turn, each of these major reactions entails numerous elemental steps, the sequence and timing of which is largely unknown.

The structures of IF1 and IF3 are known (Biou et al., 1995; Garcia et al., 1995a,b; Kycia et al., 1995; Sette et al., 1997). Binding of IF3 to the 30S subunit, following subunit dissociation during ribosome recycling, prevents the reassociation of the subunits and couples ribosome recycling and translation initiation (Karimi et al., 1999; Peske et al., 2005; Petrelli et al., 2001). IF3 stabilizes the interaction of fMet-tRNA^{fMet} with the P site and has a role in the selection of the initiator tRNA (Gualerzi et al., 1977). IF1 binds to the A site of the 30S subunit and stimulates the activities of IF3 and IF2 (Gualerzi and Pon, 1990); it is thought to direct the initiator tRNA to the ribosomal P site by blocking the A site (Carter et al., 2001; Dahlquist and Puglisi, 2000).

IF2 is the largest of the three bacterial initiation factors. The crystal structure of bacterial IF2 has not been solved so far. Homologues of IF2 have been found in archaea and eukaryotes, where the factor is referred to as aIF5B and eIF5B, respectively, and the structure of aIF5B from *Methanobacterium thermoautotrophicum* has been solved (Roll-Mecak et al., 2000). The bacterial factor can be divided into three large domains (N-terminal, central G-, and C-terminal regions) each of them constituted by two or more subdomains. Although the N-terminal modules (N1, N2, and G1) do not have a structural equivalent in archaeal or eukaryotic factor, G2, G3, C1, and C2 are structurally conserved and correspond to domains I to VI of the aIF5B homologue (Caserta et al., 2006). IF2 is a GTP binding protein with unusual properties concerning the role of its GTPase function (Rodnina et al., 2000); in the cell it seems to act as a metabolic sensor that oscillates between an active GTP-bound form, which is present when conditions allow active protein synthesis, and an inactive ppGpp-bound form, which forms upon shortage of nutrients (Milon et al., 2006). Although the structural characterization of bacterial initiation is far from being complete, a wealth of information about the positions and orientation of mRNA, fMet-tRNA^{fMet}, and initiation factors on the ribosome has been provided by crystallography (Carter et al., 2001; Jenner et al., 2005; Pioletti et al., 2001; Yusupova et al., 2001, 2006), and cryo-EM reconstructions (Allen et al., 2005; McCutcheon et al., 1999; Myasnikov et al., 2005), as well as by cross-linking and footprinting methods (Caserta et al., 2006; Dallas and Noller, 2001; Marzi et al., 2003; Moazed et al., 1995).

In recent years, significant progress has been made in studying the kinetics of initiation. Experimental approaches were designed to study the association of mRNA with the 30S subunit by stopped-flow technique monitoring fluorescence or fluorescence resonance energy transfer (FRET) (Studer and Joseph, 2006). These studies suggested that the rate constant for the mRNA binding to the 30S subunit does not depend on the strength of the SD–anti-SD interaction and requires a single-stranded mRNA region unrelated to the SD sequence, in agreement with earlier proposals (de Smit and van Duin, 2003). The observation that all three initiation factors are required for maximal efficiency and accuracy of initiation (Canonaco *et al.*, 1986; Gualerzi and Wintermeyer, 1986; Pon and Gualerzi, 1974, 1984; Risuleo *et al.*, 1976; Wintermeyer and Gualerzi, 1983) has been confirmed and rationalized by measuring the rates of fMet-tRNA^{fMet} binding to the ribosome using rapid nitrocellulose filtration (Antoun *et al.*, 2003, 2006a). The kinetics of subunit association was studied by stopped flow monitoring light scattering (Wishnia *et al.*, 1975) and by time-resolved footprinting (Hennelly *et al.*, 2005). Guanine nucleotide binding properties of IF2 were determined by fluorescence stopped flow using fluorescent derivatives of GTP/GDP (Milon *et al.*, 2006). The kinetics of the chemical steps of initiation, namely GTP hydrolysis by IF2 and the formation of the first peptide bond, were measured using quench-flow technique (Tomsic *et al.*, 2000). Nevertheless, many fundamental questions are still open. The kinetic model of 30S initiation complex formation was incomplete because observables for studying the association and dissociation of initiation factors were lacking. The transitions between intermediates that limit the rate of initiation were not understood. The timing of the dissociation of the initiation factors after subunit joining has not been determined directly and remains a challenging issue (Celano *et al.*, 1988; Pon and Gualerzi, 1986; Tomsic *et al.*, 2000). Finally, the dynamics of the ribosome during melting of mRNA secondary structures (Studer and Joseph, 2006), the mechanism of the mRNA rearrangement from "stand-by" to "P-site decoding" on the 30S subunit (La Teana *et al.*, 1995), or the transition of the ribosome from initiation to elongation (Tomsic *et al.*, 2000) remain to be elucidated. Because the initiation of translation is a rapid process, solving these questions requires the use of transient kinetic techniques, such as stopped flow, quench flow, rapid filtration, and time-resolved footprinting and cross-linking. This review is focused on dissecting the interactions among the initiation components using stopped-flow and quench-flow techniques.

 ## 2. EXPERIMENTAL OUTLINE

Transient kinetic measurements allow for the detection of reaction intermediates and the determination of rate constants of their formation and consumption in subsequent reactions. In stopped-flow and quench-flow machines, reactions can be followed directly and in real time by using fluorescence or radioactive labels as observables. Dissecting a reaction mechanism by means of transient kinetics entails several steps. First, fluorescent dyes and components to be labeled should be chosen carefully. Because the availability of cloned, overexpressed proteins has largely eliminated the concerns regarding the large quantities of reactants required for stopped-flow measurements, it is advantageous to try different dyes and various positions in proteins/RNA for labeling. For proteins, the most common approach is to introduce single cysteine residues, preferably at nonconserved surface positions, and use thiol-reactive derivatives of fluorophores to attach fluorescent groups at the desired positions. The intrinsic cysteines in the protein to be labeled should be removed, unless they are buried and inaccessible to the labeling reagents. Another approach, which avoids mutagenesis, is to introduce fluorescence labels in a random way by labeling lysine residues at the surface of the protein. Usually, there are many potentially reactive lysine residues; to avoid detrimental effects of extensive modification on the activity of the protein, labeling should be limited to only a few residues per molecule. One potential disadvantage of the approach is that the labeling is likely to result in a heterogeneous pool of molecules with dyes attached at different positions which may report different rearrangements, resulting in multiple, often poorly defined kinetic phases. Finally, RNA can be labeled in a number of ways, at the 5′ or 3′ ends or at internal positions using fluorescent nucleotide derivatives. Labeling becomes slightly more complicated for tRNA when native, fully modified tRNA prepared from cell extracts is used, rather than chemically synthesized RNA fragments. However, natural tRNA modifications provide a number of reactive groups useful for labeling, two of which, thioU at position 8 and dihydroU in the D loop of tRNA, are commonly found in bacterial tRNAs. The methods of fluorescence labeling are well established and often can be downloaded from the web sites of the companies from which the dye is purchased (Invitrogen, Atto-Tec, or Jena Bioscience). However, an optimization of labeling conditions for each particular protein/RNA-dye pair is recommended, as it may improve both yield and purity of labeled product. Because labeling may alter properties of molecules, the functional activity of the labeled components should be tested and compared to the unlabeled species; only those fluorescence derivatives should be used which have properties that are sufficiently similar to the unmodified molecules. Practically every component of the initiation machinery can be labeled with fluorescence dyes, and we have prepared and tested fluorescence derivatives of the ribosomal subunits, fMet-tRNA^{fMet}, mRNA, initiation factors, and GTP (Table 1.1 and Fig. 1.2).

The second step is planning and performing transient-kinetics experiments. Several observables can be used when working with a stopped-flow apparatus. Fluorescence changes due to binding or conformational rearrangements of fluorescence-labeled component can be monitored. Alternatively, two fluorescence-labeled components can be used which serve as fluorescence donor and acceptor in a FRET measurement. The FRET pair should be chosen in such a way that the emission spectrum of the donor

Table 1.1 Summary of observables used to study translational initiation in bacteria

Fluorescent compound	Position in molecule	Dye	λ_{ex}	λ_{em}	Reaction monitored
30S or 50S	Lys residues	Flu	490	520	FRET upon binding of IFs, mRNA, tRNA; Subunit association/dissociation (quenching by QSY9)
30S or 50S	Lys residues	QSY9	562	—	Quencher of fluorescence upon subunit association/dissociation or upon binding of other fluorescence-labeled components
30S and 50S	Light scattering				Subunit association/dissociation
fMet-tRNA^{fMet}	Attached to thio-U^8	Flu	490	520	30S initiation complex formation (FRET)
fMet-tRNA^{fMet}	Attached to thio-U^8	Bpy	507	535	30S initiation complex formation (FRET)
fMet-tRNA^{fMet}	Replacing D20	Prf	470	505	70S initiation complex formation (FI and FRET)
fMet-tRNA^{fMet}	Radioactive amino acid				30S and 70S initiation complex formation, fMetPhe or fMetPmn synthesis
mRNA	3'end	Flu	490	520	30S initiation complex formation (FI and FRET)
IF1	Cys 4	Alx555	555	570	30S and 70S initiation complex formation (FRET)
IF3	Cys 65	Alx488	500	520	30S initiation complex formation (FI)
IF3	Cys 166	Alx488	500	520	30S initiation complex formation (FI and FRET)
IF3	Cys 166	Alx555	555	570	30S initiation complex formation (FI and FRET)
PBP	Cys 195	MDCC	425	465	Dissociation of Pi from IF2 after GTP hydrolysis
GTP/GDP	2'/3' OH	mant	355	448	Guanine nucleotide binding and dissociation
GTP	γ	[^{32}P]			GTP hydrolysis by IF2

λ_{ex} and λ_{em}, excitation and emission wavelengths of dyes used in the experiments. Note that in some cases the maximum excitation and emission wavelengths may be different.

FI, fluorescence intensity.

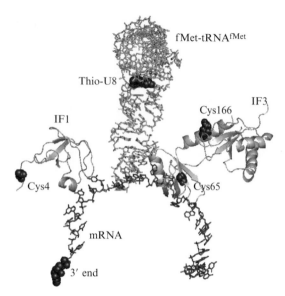

Figure 1.2 Positions of fluorescence labels in IF1, IF3, mRNA, and Met–tRNA^{fMet} in a structural model of the 30S initiation complex. The structural model was built using PyMOL software (DeLano Scientific). The 30S subunit, mRNA, and tRNA were taken from the crystal structure of the *T. thermophilus* 70S initiation complex (PDB 1YL4; Jenner *et al.*, 2005). The position of IF1 was taken from the crystal structure of the *T.thermophilus* 30S·IF1 complex (PDB 1HR0; Carter *et al.*, 2001). IF3 was manually docked according to chemical probing data (Fabbretti *et al.*, 2007).

overlaps with the excitation spectrum of the acceptor, and the expected distance between the dyes in the complex is close to the R_0 of the dye pair (Lakowicz, 1999). The donor is excited at a wavelength optimal for maximum donor excitation and minimum acceptor excitation, and fluorescence emission is measured at the maximum emission of the acceptor in most cases, although donor emission can also be measured. The FRET method is particularly suitable to follow the kinetics of ligand binding, although subsequent conformational rearrangements are often observed as well. The association of ribosomal subunits results in a change of light scattering, which is another useful observable in studies of translation initiation. The chemistry steps (i.e., GTP hydrolysis and the formation of the first peptide bond), are measured in a quench–flow apparatus using radioactively labeled GTP ($[\gamma-^{32}P]GTP$) or an amino acid attached to tRNA^{fMet} (^{3}H-, ^{14}C-, or ^{35}S-methionine). Reactions are initiated by rapid mixing of reactants and stopped by the rapid addition of an appropriate quencher. Subsequently, the reaction products are analyzed by HPLC, TLC, or other analytical techniques and quantified by radioactivity counting.

The design of a transient kinetics experiment for a multicomponent reaction such as translation initiation may be quite complex. Usually, the experiment should be planned in such a way that pseudo–first-order reaction conditions are fulfilled, which greatly simplifies the calculation of rate constants from observed rates. This is accomplished by taking limiting concentrations of the labeled component and adding all other reaction components in large excess, such that the change in concentration of unlabeled components during the reaction remains negligible. Next, the apparent rate constants of reaction are measured at constant concentration of the labeled component and increasing excess concentrations of its ligands, and the rate constants of reaction are derived from the analysis of the concentration dependence of the apparent rate constants. In recent years, the complexities of data analysis have been overcome by advances in computational methods to analyze reaction time courses by numerical integration. This allows using equimolar concentrations of reagents, and determining rate constants by global fitting; measuring many time courses at different concentrations of reagents is even more crucial in this case. Data from different approaches should be combined and the kinetic mechanism of reaction solved.

3. MATERIALS

3.1. Stock solutions

Buffer A: 50-mM Tris-HCl (pH 7.5), 70-mM NH$_4$Cl, 30-mM KCl, 7-mM MgCl$_2$. Buffer A was used for activity tests and as a common storage buffer for ribosomes.

Buffer B: 20-mM Tris-HCl (pH 7.1), 1-mM EDTA, 10% glycerol, 6-mM 2-mercaptoethanol, 0.2-mM PMSF and 0.2-mM benzamidine. 2-mercaptoethanol, PMSF, and benzamidine were added shortly before use. Buffer B was used for purification and storage of initiation factors.

All solutions were filter-sterilized (0.2 μm) and cooled to 4° before use.

3.2. Reagents

Purified tRNA^{fMet} (from E. coli) could be charged to about 1000 pmol [^{14}C] Met/A$_{260}$ U of tRNA (for charging protocol, see following). tRNA^{fMet} of this quality can be purchased from Sigma or Chemical Block (Russia), or purified from total tRNA as described by Kothe et al. (2006). Synthetic mRNAs were purchased from Curevac (Germany) or Dharmacon (USA).

3.3. Bacterial strains

E. coli MRE600: *F-, rna.*

E. coli UT5600: *ara-14, leuB6, azi-6, lacY1, proC14, tsx-67,* Δ *(ompT-fepC)* 266, *entA403, trpE38, rfbD1, rpsL109, xyl-5, mtl-1, thi-1.*

4. Experimental Procedures

4.1. Preparation and fluorescence labeling of ribosomes and ribosomal subunits

1. Preparation of 70S ribosomes

Stock solutions. All solutions were filter-sterilized (0.45 μm) and cooled to 4°. 2-mercaptoethanol was added to buffers immediately before use.

Cell opening buffer (buffer 1, 200 ml for 50 g cells): 20-mM Tris-HCl (pH 7.6), 100-mM NH$_4$Cl, 10-mM magnesium acetate, 3-mM 2-mercaptoethanol.

Sucrose cushion (buffer 2, 200 ml): 20-mM Tris-HCl (pH 7.6), 500-mM NH$_4$Cl, 10-mM magnesium acetate, 1.1-M sucrose, 3-mM 2-mercaptoethanol.

Washing buffer (buffer 3, 500 ml): 20-mM Tris-HCl (pH 7.6), 500-mM NH$_4$Cl, 10-mM magnesium acetate, 7-mM 2-mercaptoethanol.

Zonal buffer (buffer 4, 1 liter): 20-mM Tris-HCl (pH 7.6), 60-mM NH$_4$Cl, 5-mM magnesium acetate, 3-mM 2-mercaptoethanol.

Zonal gradient, 10% sucrose (buffer 5, 750 ml): 20-mM Tris-HCl (pH 7.6), 60-mM NH$_4$Cl, 5-mM magnesium acetate, 10% sucrose, 3-mM 2-mercaptoethanol.

Zonal gradient, 40% sucrose (buffer 6, 1.5 liter): 20-mM Tris-HCl (pH 7.1), 60-mM NH$_4$Cl, 5-mM magnesium acetate, 40% sucrose, 3-mM 2-mercaptoethanol.

Buffer with 50% sucrose (buffer 7, 1.5 liter): 20-mM Tris-HCl (pH 7.6), 60-mM NH$_4$Cl, 5-mM magnesium acetate, 50% sucrose, 3-mM 2-mercaptoethanol.

All procedures were carried out at 4°. 50 g of frozen cells (*E. coli* MRE600) were crushed in a cold mortar (20-cm diam.) with 100 g of cooled alumina and ground in the cold room for ~30 min. Alternatively, cells may be opened by the french press; other methods are not recommended. DNAse I was added (few crystals), ground for additional 10 min, and cell opening buffer (buffer 1) was added. To remove alumina and cell debris, the slurry was centrifuged in a Beckman JA-14 rotor for 30 min at 9000 rpm (12,500×*g*) at 4°. The supernatant was carefully removed and centrifuged in a Beckman 50.2 Ti rotor for 30 min, 16,000 rpm (31,000×*g*) at 4°. From the resulting supernatant (S-30 fraction) 14 ml were layered on 9 ml sucrose cushions (buffer 2) and centrifuged for 16 h at 33,000 rpm in the 50.2 Ti rotor (132,000×g) at 4°. After centrifugation, the supernatant was discarded, and the pellets were rinsed rapidly with washing buffer (buffer 3). The pellets were dissolved in washing buffer with the help of a glass rod and gentle stirring for 30 min on ice. The ribosomes were pooled and the total volume adjusted to 92 ml with buffer 3. 23 ml of the solution were loaded on top of 1-ml sucrose cushion (buffer 2) and centrifuged in the 50.2 Ti rotor for 6 h at 50,000 rpm (302,000×g) at 4°. The pellets were again rinsed rapidly with washing buffer (buffer 3) and dissolved in washing buffer for 30 min on ice. The vol of the solution was adjusted to 90 ml and 30 ml of the ribosome solution were loaded on three 1.5-ml sucrose cushions in a Beckman SW28 rotor, and centrifuged 13 h at 28,000 rpm (141,000×g) at 4°. The resulting ribosome pellets were rinsed and dissolved in 50 ml of the zonal buffer (buffer 4). The concentration of ribosomes was determined by measuring the absorbance at 260 nm (A_{260}). At this stage, aliquots of ribosome solutions can be nitrogen-frozen and stored at −80°. Immediately before the zonal centrifugation, 2.5 ml of 50% sucrose buffer (buffer 7) was added to 22.5 ml of ribosome solution.

4.2. Zonal centrifugation

For general instructions concerning the operation of the zonal rotor, see the manufacturer's manual. Here, we describe the procedure for the Beckmann Ti-15 rotor which has a vol of 1675 ml. Load 400 ml of the zonal buffer (buffer 4) first. Slowly load 25 ml (containing ∼10,000 to 15,000 A_{260} U) of ribosome solution. Load 1370 ml of 10 to 40% sucrose gradient (buffers 5 and 6, respectively) until 150 ml of the zonal buffer comes out of the exit tubing. Finally, load about 150 ml of 50% sucrose buffer (buffer 7). Centrifuge for 19 h at 28,000 rpm (74,000×g), at 4°. Unload the rotor with 50% sucrose (buffer 7) and collect 50-ml fractions. Measure A_{260} to identify the fractions containing 30S and 50S subunits and 70S ribosomes. Combine fractions containing 70S ribosomes and centrifuge for 24 h, 4°, at 50,000 rpm in a Beckmann 50.2 Ti rotor. Ribosome pellets are dissolved in the final storage buffer (e.g., buffer A or any other appropriate buffer), shock-frozen in liquid nitrogen in small portions, and stored at −80°. The concentration of ribosomes was determined by measuring the absorbance at 260 nm (A_{260}), using an extinction coefficient of 23 pmol/A_{260} U.

2. Preparation of 30S and 50S subunits

All buffers are as described previously for the ribosome preparation, except that the magnesium acetate concentration is 1.5 mM in buffers 4 to 7. The additional buffer required is dissociation buffer (buffer 8): 20-mM Tris-HCl (pH 7.6), 60-mM NH$_4$Cl, 1.5-mM magnesium acetate, 3-mM 2-mercaptoethanol.

All preparation steps up to the zonal centrifugation are the same as for the 70S ribosome preparation. Before the zonal centrifugation, the ribosome solution is dialyzed against subunit dissociation buffer (buffer 8) for 2 to 5 h at 4°. Zonal centrifugation is carried out in the same way as described earlier. Prior to use, 30S subunits should be activated by incubation with 20-mM MgCl$_2$ in buffer A for 1 h at 37°.

3. Fluorescence labeling of ribosomal subunits

Labeling buffer (buffer 9): 50-mM HEPES (pH 7.5), 100-mM KCl, 15-mM MgCl$_2$.

Sucrose cushion (buffer 10): 50-mM Tris-HCl (pH 7.5), 70-mM NH$_4$Cl, 30-mM KCl, 20-mM MgCl$_2$, 1.1-M sucrose.

70S ribosomes (4.5 μM) were labeled at surface lysine residues by reacting with either fluorescein succinimidyl ester (0.1 mM) or the nonfluorescent acceptor QSY9 (Invitrogen) succinimidyl ester (0.1 mM) in labeling buffer (buffer 9) for 30 min at 37°. To remove unreacted dye, ribosomes were centrifuged through 400 μl of buffer 10 for 2 h at 259,000×g in a Sorvall M120GX centrifuge. To dissociate the ribosomes into subunits, the pellets were resuspended and dialyzed against dissociation buffer (buffer 8) for 4 h at 4°. Subunits were separated by centrifugation through a 10 to 38% sucrose gradient in dissociation buffer (buffer 8) in a Beckman SW28 rotor at 19,000 rpm at 4° for 19 h. Gradients were fractionated and subunits were collected according to the A$_{260}$ profile and pelleted in a Beckman 50.2 Ti rotor at 50,000 rpm for 19 h at 4°. Pellets were resuspended in buffer A. Subunits (260 nm), fluorescein (495 nm), and QSY 9 (562 nm) were quantitated photometrically. Labeled 30S and 50S subunits both contained 3 to 5 dye molecules per subunit.

4.3. Preparation and fluorescence labeling of initiator fMet-tRNA^{fMet}

1. Fluorescence labeling of tRNA^{fMet}

Two different types of labeling were utilized for tRNA^{fMet}, at a thiouridine at position 8 of tRNA^{fMet} with iodoacetamide derivatives of fluorescein (5'-IAF) and BODIPY® 507/545 dyes (Invitrogen) or with proflavin at position 20 of tRNA^{fMet}. Labeling at thioU8 was carried out similarly to a previously published protocol (Johnson *et al.*, 1982). tRNA^{fMet} (65 A$_{260}$ units/ml) were incubated in 12-mM Hepes-KOH (pH 8.2) containing 80% DMSO and 3.5 mM of respective fluorescent dye for 2 h at 50° in the dark. Every 30 min aliquots were removed and the extent of modification was analyzed by reverse phase HPLC (LiChrospher® WP 300 RP-18 column, 250 × 4, 5 μm, Merck). With both fluorescent dyes, labeling was 100% after 2 h of incubation. The reaction was stopped by adding 0.3-M potassium acetate (pH 5.0), and labeled tRNA^{fMet} was precipitated with 2.5 vol of cold ethanol.

The preparation of proflavin-labeled tRNA^{fMet} proceeds in two steps: reduction of the dihydroU base at position 20 in the D loop of tRNA by borohydride treatment followed by the attachment of proflavin at that position (Wintermeyer and Zachau, 1974). Sodium borohydride solution is prepared by dissolving 100 mg of NaBH$_4$ in 1 ml of 10-mM KOH. tRNA^{fMet} (10 A$_{260}$ U/ml) was dissolved in 0.2-M Tris-HCl (pH 7.5) and NaBH$_4$ solution added to the final concentration of 10 mg/ml. After incubation for 30 min at 0° in the dark, the reaction was stopped by the addition of acetic acid to pH 4 to 5, and the tRNA was precipitated with cold ethanol after the addition of 0.3-M potassium acetate (pH 4.5). Ethanol precipitation was repeated 3 to 4 times to remove traces of borohydride. To prepare fMet-tRNA^{fMet}(Prf), NaBH$_4$-reduced tRNA^{fMet} was aminoacylated and formylated as described later, and then proflavin was incorporated. Proflavin labeling was carried out by adding tRNA^{fMet} (4 to 10 A$_{260}$ U/ml) to 3-mM proflavin in 0.1-M sodium acetate (pH 4.3). After incubation for 2 h at 37° in the dark, the reaction was stopped by increasing the pH value of the reaction mixture to 7.5 by addition of 1-M Tris-HCl (pH 9). Free dye was removed by phenol extraction which was repeated until the phenol phase was colorless (typically 6 extractions). fMet-tRNA^{fMet}(Prf) (or deacylated tRNA^{fMet}(Prf) when aminoacylation was omitted) was three times ethanol precipitated after addition of 0.3-M potassium acetate (pH 6.8). Incorporation of proflavin was quantified photometrically by measuring optical density for tRNA at 260 nm and 460 nm. For fully labeled tRNA, the A$_{460}$:A$_{260}$ ratio is 0.055.

2. Aminoacylation and formylation of tRNA^{fMet}

Buffer for aminoacylation (buffer 11): 25-mM Tris-acetate (pH 7.5), 8-mM magnesium acetate, 100-mM NH$_4$Cl, 30-mM KCl, 1-mM DTT. N^{10}-formyltetrahydrofolate was prepared as described (Rodnina et al., 1994b). S-100 extract from E. coli purified additionally as described later was used as a source of both methionyl-tRNA synthetase and N^{10}-formyltetrahydrofolate:Met-tRNA transformylase. S100 fraction left after the first sucrose purification step of the ribosome preparation was purified from RNA by anion-exchange chromatography on DE-52 (elution by 20-mM Tris-HCl [pH 7.5], 10-mM MgCl$_2$, 0.3-M NaCl) and dialyzed against buffer A to remove low-molecular-weight components. For preparative aminoacylation and formylation the following protocol was used: 200 A$_{260}$ U of tRNA^{fMet} were incubated for 40 min at 37° in 10 ml of aminoacylation buffer (buffer 11) containing 3-mM ATP, 60-μM [^{14}C]Met (Moravek Biochemicals, USA) or [^{3}H]Met (MP Biomedicals), 0.5-mM N^{10}-formyltetrahydrofolate, and 1 mg of S-100 extract (calculated from A$_{280}$, assuming 1 mg of protein/A$_{280}$ U). The reaction was stopped by adding 0.3-M potassium acetate (pH 5.0). The solution was extracted with water-saturated phenol, and tRNA was precipitated from the aqueous phase with 2.5 vol of cold ethanol.

3. Purification of fMet-tRNA^{fMet} and fluorescence-labeled tRNA^{fMet} derivatives

HPLC buffer A (buffer 12): 20-mM ammonium acetate (pH 5.0), 10-mM MgCl$_2$, 400-mM NaCl, 5% ethanol.

HPLC buffer B (buffer 13): 20-mM ammonium acetate (pH 5.0), 10-mM MgCl$_2$, 400-mM NaCl, 15% ethanol.

A procedure for fMet-tRNA^{fMet} purification by FPLC on a Phenyl-Superose column was described previously (Rodnina *et al.*, 1994b). Although the procedure yielded fMet-tRNA^{fMet} preparations of very high homogeneity, it had several disadvantages: (1) the removal of ammonium sulfate from the fMet-tRNA^{fMet} fractions after the Phenyl Superose chromatography is laborious and requires an additional purification step; (2) the binding capacity of the standard Phenyl-Superose column is limited to 50 to 60 A$_{260}$ U of tRNA; (3) most importantly, Phenyl-Superose is not commercially available anymore. Therefore, we have designed an alternative procedure to purify large amounts of fMet-tRNA^{fMet} by reverse-phase chromatography on HPLC.

The pellet of aminoacylated and formylated fMet-tRNA^{fMet} (up to 300 A$_{260}$ U), charged with ^{14}C or ^{3}H-labeled Met, was dissolved in 2 ml of HPLC buffer A (buffer 12) and applied to a LiChrospher® WP 300 RP-18 HPLC column (250 × 10, 5 μm, Merck), equilibrated with the same buffer. tRNA was eluted by a linear gradient of 0 to 40% HPLC buffer B (buffer 13) at a flow rate of 3 ml/min. For the purification of fluorescence-labeled tRNAs, a linear gradient from 0 to 100% of buffer 13 was used, because the retention of those tRNAs on the column is increased due to the higher hydrophobicity of modified tRNAs. 6-ml fractions were collected and 20-μl aliquots were counted in 2 ml of LumaSafe scintillation cocktail. Fractions containing radioactive fMet-tRNA^{fMet} were collected and ethanol-precipitated. fMet-tRNA^{fMet} was dissolved in H$_2$O, and stored in small aliquots at −80°. To determine the extent of formylation, an aliquot of f[^{14}C]Met-tRNA^{fMet} was deacylated by incubation with 0.5-M KOH for 30 min at 37°, and the ratio f[^{14}C]Met:[^{14}C]Met analyzed by HPLC on a LiChrospher 100 RP-8 column (250 × 4, 5 μm, Merck) using a gradient from 0 to 65% acetonitrile in 0.1% TFA. The single-step tRNA purification procedure described previously allows for the efficient separation of f[^{14}C]Met-tRNA^{fMet} from both deacylated tRNA^{fMet} and nonformylated [^{14}C]Met-tRNA^{fMet}, as well as the separation of fluorescence-labeled from nonlabeled tRNA^{fMet}. Final preparations of f[^{14}C]Met-tRNA^{fMet} (or its fluorescent derivatives) had 1400 to 1800 pmol [^{14}C]Met/A$_{260}$ U and were fully formylated. All preparations of fMet-tRNA^{fMet}, fMet-tRNA^{fMet}(Flu), fMet-tRNA^{fMet}(Bpy), and fMet-tRNA^{fMet}(Prf) were fully active in 70S initiation complex formation and peptide bond formation.

4.4. Preparation of fluorescence-labeled mRNA

Synthetic mRNAs harboring canonical (AUG) or noncanonical (AUU) initiation codons had the following sequences:

5′ G GCA **AGG AGG UA**A AUA *AUG* UUC ACG AUU 3′
5′ G GCA **AGG AGG UA**A AUA *AUU* UUC ACG AUU 3′

where the Shine-Dalgarno sequence is indicated in bold, and the initiation codon is italicized. Alternatively, a longer transcript harboring a weaker SD sequence, 022 mRNA (La Teana *et al.*, 1993), was produced by T7 RNA-polymerase *in vitro* transcription: 5′ GGG AAU UCA AAA AUU UAA AAG UUA AC**A GGU A**UA CAU ACU *AUG* UUU ACG AUU ACU ACG AUC UUC UUC ACU UAA CGC GUC UGC AGG CAU GCA AGC U 3′ ◊ 022 AUG. For 3′ labeling, mRNA (60 μM) was oxidized at the 3′-terminal ribose by incubation in 0.1-M sodium acetate (pH 5.3) and 5-mM KIO$_4$ for 30 min at 0°. The reaction was stopped by adding ethylene glycol to a concentration of 10 mM and incubating further for 10 min at 0°. After three ethanol precipitations from 0.3-M sodium acetate (pH 5.3), the RNA was dissolved in water (to 100-μM concentration) and reacted with 10 mM of fluorescein-5-thiosemicarbazide (100-mM stock in DMSO) (Invitrogen) for 12 h at 4° in the dark. Free dye was removed by phenol extraction which was repeated until the organic phase was colorless. Fluorescein-labeled mRNA(Flu) was ethanol-precipitated from the aqueous phase. Absorbance measurements of mRNA at 260 nm (extinction coefficient was calculated based on the RNA sequence) and fluorescein at 492 nm ($\varepsilon = 85,000 \ M^{-1}cm^{-1}$ at pH 9), showed that the extent of labeling was >95%.

4.5. Initiation factors

1. Expression of IF1, IF2, and IF3 and preparation of cell lysates

Lysis buffer (buffer 14): 10-mM Tris-HCl (pH 7.7), 60-mM NH$_4$Cl, 10-mM magnesium acetate.

Initiation factors were overexpressed in *E. coli* UT5600 (*ompT* $^-$) harboring two plasmids: pCI857 which encodes for the thermosensitive λ-repressor cI857 and for kanamycin resistance, and pPLC2883 (Remaut *et al.*, 1983) which confers ampicillin resistance and in which the gene encoding IF1, IF2, or IF3 was placed under the control of the lambda PL promoter. The respective constructs were pXR201 *infA** (Calogero *et al.*, 1987), pXP101 *infB* (Spurio *et al.*, 1993), or pIM302 *infC* (Brombach and Pon, 1987). It should be noted that the use of an *ompT*$^-$ strain for the overproduction of the initiation factors is very important because this strain lacks an outer membrane protease which during protein purification readily degrades the

N-terminal portions of IF2 (Lassen et al., 1992) and IF3 (C. O. Gualerzi, unpublished observations). The functional properties of N-terminally degraded factors are quite different from those of the native factors which, as previously pointed out (Caserta et al., 2006; Lammi et al., 1987), may produce misleading results, probably because the interaction with ribosomes involves the N-terminal portion of both factors (Caserta et al., 2006; Petrelli et al., 2001). For the same reason, the use of His-tagged factors is not recommended. All three initiation factor genes were engineered to make their expression more reliable and efficient. The infB (Spurio et al., 1993) and infC (Brombach and Pon, 1987) genes were separated from the neighboring genes in their transcriptional units that repress initiation factor expression. The translationally autorepressed AUU triplet of infC was substituted with the canonical AUG (Brombach and Pon, 1987). Finally, infA* is a synthetic gene that codes for the same amino acid sequence as IF1 but has an optimized usage of synonymous codons which accounts for higher expression levels of the protein (Brombach and Pon, 1987; Calogero et al., 1987).

Cells were grown in a 5 liter fermenter in LB medium with 100 μg/ml Amp and 25 μg/ml Kan at 30° and overexpression of the factors was induced at mid-log phase by temperature shift to 42°. After 30 min of induction, the incubation temperature was decreased to 37° and the cells allowed to grow for another 2 h. Cells were collected by 10 min centrifugation at 5000 rpm in a Beckman JA-10 (2700×g) rotor and washed twice by resuspension and centrifugation in lysis buffer (buffer 14). The final cell suspension (~2 ml lysis buffer/g cells) was frozen in liquid nitrogen and stored at −80°.

Before opening, 6-mM 2-mercaptoethanol, 0.2-mM PMSF, and 0.2-mM benzamidine were added to the ice-cold suspension of unfrozen cells. Cells were opened in a Misonix 3000 sonicator (Misonix, Inc.) applying 15 cycles (30-s sonication and 30-s pause) at 6 W. The lysate was centrifuged in a Sorvall SA-600 rotor for 1 h at 15,000 rpm (33,000×g). To dissociate initiation factors from ribosomes, the concentration of NH_4Cl in the supernatant (S-30) was adjusted to 1 M. Ribosomal subunits were pelleted by 16 to 18 h centrifugation of S-30 in a Beckmann 45 Ti rotor at 35,000 rpm (143,000×g), leaving the supernatant containing the initiation factors (S-150 extract).

2. Purification of IF1 and IF3

Buffer 15: Buffer B (see stock solutions) without NH_4Cl.
Buffer 16: Buffer B with 0.1 M NH_4Cl.
Buffer 17: Buffer B with 0.7 M NH_4Cl.
Buffer 18: Buffer B with 0.05 M NH_4Cl and 6 M urea.
Buffer 19: Buffer B with 0.4 M NH_4Cl and 6 M urea.
Buffer 20: Buffer B with 1 M NH_4Cl.

S-150 extract containing overexpressed IF1 was diluted with 9 volumes of buffer B without NH_4Cl (buffer 15) to a final NH_4Cl concentration of 0.1 M, loaded onto a Whatman P11 phosphocellulose (PC) column (30 mg of protein/g of resin) equilibrated with buffer 16. After washing the column with 5 bed volumes of buffer 16, IF1 was eluted with a linear gradient of 0.1- to 0.7-M NH_4Cl in buffer B (buffers 16 and 17); the total volume of the gradient amounted to 20 bed volumes of the column. Fractions containing IF1 were identified on SDS-PAGE, pooled and dialyzed against buffer B with 6-M urea and 0.05-M NH_4Cl (buffer 18) and loaded on a PC column equilibrated in the same buffer (buffer 18). The column was washed with buffer 18 and IF1 eluted by a gradient of 0.05 M- to 0.4-M NH_4Cl in 6-M urea in buffer B (buffers 18 and 19) (gradient volume 20 column bed vol). Fractions containing IF1 were pooled, diluted with buffer 15 (buffer B without NH_4Cl or urea) to a final NH_4Cl concentration of 0.05 M and loaded to a small PC column (1 ml bed vol) for concentration. The protein was eluted with buffer B containing 1 M NH_4Cl (buffer 20) and purified by gel filtration on a Superdex 75 FPLC column (Pharmacia; HiLoad 26/60). Purified IF1 was stored in buffer B containing 200-mM NH_4Cl at $-80°$.

IF3 was purified essentially as described for IF1, except for a different elution gradient 0.1 to 1-M NH_4Cl in buffer B (buffers 16 and 20) was used in the first purification step on the PC column. A gel filtration step on a Superdex 75 column (HiLoad 26/60; Pharmacia) was used only when the purity of IF3 after the second PC column was below 95%.

3. Purification of IF2

Buffer 21: Buffer B (see stock solutions) with 0.6-M NH_4Cl
Buffer 22: 20-mM Tris-HCl (pH 7.9), 1-mM EDTA, 10% glycerol, 6-mM 2-mercaptoethanol, 0.2-mM PMSF, and 0.2-mM benzamidine
Buffer 23: 20-mM sodium phosphate buffer (pH 7.1)
Buffer 24: 350-mM sodium phosphate buffer (pH 7.1)
S-150 extract containing overexpressed IF2 was diluted with 9 vol of buffer 15 to final NH_4Cl concentration of 0.1 M and loaded on a PC column preequilibrated with buffer 16. After washing the column with 5 bed vol of buffer 16, IF2 was eluted with a 20 bed-vol linear gradient 0.1- to 0.6-M NH_4Cl in buffer B (buffers 16 and 21). Fractions containing IF2 were pooled and dialyzed against buffer 22 and loaded onto a DEAE-cellulose column preequilibrated in the same buffer. The column was washed and protein eluted as described previously with a 20 bed-vol gradient (0.1- to 0.6-M NH_4Cl in buffer 22). The fractions containing IF2 were pooled, dialyzed against 20-mM phosphate buffer (pH 7.1) (buffer 23) and loaded on a hydroxylapatite column (160-ml bed vol). After washing the column with 5 bed vol of the same buffer, the protein was eluted with linear gradient (10 bed vol) of 20- to 350-mM phosphate buffer (pH 7.1) (buffers 23 and 24). Fractions containing IF2 were collected, dialyzed against buffer

B containing 200-mM NH$_4$Cl, and stored at $-80°$.

An alternative method for the rapid purification of all three initiation factors by FPLC can also be applied (Rodnina *et al.*, 1999).

4. Fluorescence labeling and purification of IF1 and IF3

IF1 and IF3 labeling buffer (buffer 25): 50-mM Tris-HCl (pH 7.1), 100-mM NH$_4$Cl, 0.1-mM EDTA.

Purification buffer (buffer 26): 10-mM Tris-HCl (pH 7.1), 200-mM NH$_4$Cl, 0.1-mM EDTA, 10% glycerol.

Elution buffer (buffer 27): 10-mM Tris-HCl (pH 7.1), 1-M NH$_4$Cl, 0.1 mM EDTA, 10% glycerol.

Single cysteine IF3(C65S/E166C) and IF1(D4C) mutants were obtained by mega-primer PCR (Quikchange, Stratagene) using vectors pIM302 *infC* and pXR201 *infA**. Mutant proteins were purified in the same way as described previously for the wild-type proteins.

For fluorescence labeling, solutions of IF1 or IF3 proteins containing single cysteines were diluted to 100 to 200 μM and the reducing agent (2-mercaptoethanol or DTT) was removed by extensive dialysis at 4° against labeling buffer (buffer 25). Disulfide bond formation was suppressed by treatment for 10 min at 37° with a 10-fold molar excess of Tris (2-carboxyethyl) phosphine hydrochloride (TCEP, Sigma), a reducing agent which does not compete with thiol modifications. Labeling was carried out with a 20-fold molar excess of dyes over the protein using Alexa 488 or 555 (Invitrogen) or CPM (Sigma) maleimide derivatives. Fluorescent dyes were dissolved in 100% DMSO at concentrations of 10 mM. The dye solution was added with a dropper to the protein solution upon gentle stirring. The kinetics and efficiency of thiol modification were assessed by SDS-PAGE (18% acrylamide), in which the labeled proteins exhibit lower mobility compared to the unmodified ones. Maximum thiol modification (>90%) was reached after 1.5 h incubation at room temperature. The reaction was stopped by the addition of a 10-fold molar excess of 2-mercaptoethanol. Stock solutions and reaction mixtures where protected from light during all labeling and purification steps.

Fluorescence-labeled IF3 was purified by chromatography on a small anion-exchange PC column (bed vol 500 μl). Reaction mixtures were directly loaded to the column and excess dye was removed by washing the column with purification buffer (buffer 26). Elution was carried out with the same buffer containing 1-M NH$_4$Cl (buffer 27) and fractions containing labeled IF3 were pooled, dialyzed against buffer B, and stored at $-80°$.

Labeled IF1 was purified by size-exclusion centrifugation (Centricon, MWCO 10 kDa). Reaction mixtures were directly loaded onto the filter and excess dye was removed by extensive washing with buffer B. Protein concentrations were determined colorimetrically (Bradford, 1976), using commercial reagent (Sigma), and densitometrically by comparing the intensities of protein bands on SDS-PAGE gels (18% acrylamide). Protein recovery after labeling and purification was typically >75% for IF3 and 70% for IF1.

5. Activity tests for initiation factors

4.5.1. IF3-dependent dissociation of pseudo-initiation complexes

N-Acetyl[^{14}C]Phe-tRNA (AcPhe-tRNA^{Phe}) dissociation from poly(U)-programmed 30S pseudo-initiation complexes was measured in the presence of increasing concentrations of wild-type IF3, IF3 containing single cysteine replacements, or fluorescence-labeled IF3 essentially as described (Pon and Gualerzi, 1979). Pseudo-initiation complexes were formed with 30S subunits (0.5 μM), IF1 (0.6 μM), IF2 (0.5 μM), poly(U) (1 mg/ml), Ac[^{14}C]Phe-tRNA^{Phe} (1 μM), and GTP (1 mM) in 30 μl of buffer A (see *Stock solutions*). After 30 min incubation at 37°, the reactions were diluted with 3 ml of ice-cold buffer A and increasing amounts of IF3 were added to induce the dissociation of Ac[^{14}C]Phe-tRNA^{Phe} from the ribosome. The residual 30S-associated radioactivity was measured by nitrocellulose filtration and liquid scintillation counting. For fully active IF3, the maximum effect is reached at a 1:1 molar ratio of IF3 to 30S subunits.

4.5.2. IF3-dependent translation *in vitro*

In vitro translation was carried out essentially as described (Brandi *et al.*, 1996; Petrelli *et al.*, 2003) using *cspA* mRNA encoding the CspA protein the synthesis of which is strongly dependent on IF3 (Giuliodori *et al.*, 2004). The reaction mixtures contained 70S ribosomes (0.6 μM) (see Preparation of 70S Ribosomes), IF1 and IF2 (1-μM each), *cspA* mRNA (0.6 μM), ^{35}S-labeled methionine (0.2 μmol, 1000 Ci/mol, Amersham), and increasing amounts of IF3 in 30 μl of buffer A. Reactions were incubated for 30 min at 37° and trichloroacetic acid-insoluble radioactivity was determined by liquid scintillation counting.

4.5.3. IF1-dependent stimulation of fMet-tRNA^{fMet} binding to the 30S subunit

IF1 activity was tested by monitoring the stimulation of f[^{35}S]Met-tRNA^{fMet} binding to the 30S subunit. Reaction mixtures contained 30S subunits (0.5 μM), IF2 (0.25 μM), IF3 (0.5 μM), AUG mRNA (1 μM), and f[^{35}S]Met-tRNA^{fMet} (1 μM), and increasing amounts of IF1 in 30 μl of buffer A. After 30 min incubation at 37° the reactions were diluted 100-fold with ice-cold buffer A and filtered through nitrocellulose membranes, then 30S-associated radioactivity was measured by liquid scintillation counting (Pon and Gualerzi, 1984).

5. RAPID KINETIC MEASUREMENTS

5.1. Measuring fluorescence intensities and FRET changes in stopped flow

Fluorescence stopped-flow measurements were performed using a SX-18MV stopped-flow apparatus (Applied Photophysics, Leatherhead, UK). Experiments were performed by rapidly mixing equal volumes (60-μl each) of reactants. The dead-time was about 1 ms and reaction rates up to 500 s^{-1} could be measured. Excitation wavelengths of fluorophores are given in Table 1.1. The emission was measured after passing appropriate cut-off filters (Schott), for example KV408, KV500, or KV590 for mant, fluorescein, or Alx555 fluorescence, respectively. In most cases, 1000 points were acquired in logarithmic sampling mode; this yields reliable data due to collecting an appropriate number of points both in the initial part of the curve where the signal changes rapidly and at the end of the reaction where the signal changes slowly or becomes constant. Notably, obtaining a reliable reaction endpoint is very important for calculating the rate constants, particularly when more than one exponential phase is observed. Several (5 to 10) individual transients were recorded and averaged. Data were evaluated by fitting the exponential function $F = F_{\infty} + A \times \exp(-k_{app} \times t)$, with a characteristic time constant, k_{app}, the amplitude of the signal change, A, the final signal, F_{∞}, and the fluorescence at time t, F. If necessary, additional exponential terms were included. Calculations were performed using TableCurve software (Jandel Scientific), Prism (Graphpad Software), or any other appropriate software.

The association of 30S subunits, 30S·IF3 or 30S initiation complexes with 50S subunits was monitored by light scattering. The method was first suggested in 1967 to study eukaryotic ribosome dissociation (Page *et al.*, 1967) and was successfully used to measure subunit association and dissociation as well as interaction with initiation factors (Antoun *et al.*, 2004; Debey *et al.*, 1975; Grunberg-Manago *et al.*, 1975; Wishnia *et al.*, 1975; Zitomer and Flaks, 1972). The principle of the method is described in detail elsewhere; in short, because the scattering intensity of a particle is proportional to the square of its molecular mass (and is independent of the shape of the molecule), the association of two particles to a single larger particle results in an increase of the intensity of light scattering. To measure light scattering in the stopped-flow apparatus, the excitation wavelength was set to 430 nm (the wavelength at which absorbance is negligible), and the emitted signal was recorded at 90° to the incident beam without any filter. The experiments were carried out either at pseudo-first order conditions, that is, at an excess of either 50S subunits or 30S subunits or at comparable concentrations of the 30S and 50S subunits. In the former case, the time courses were evaluated by exponential fitting as described previously. In the latter case, numerical integration was used.

In addition to following the time courses of reactions directly, in some instances product formation can be monitored more conveniently by using an indicator reaction. The appearance of free phosphate (Pi) in solution following GTP hydrolysis by IF2 was monitored by the fluorescence of a coumarin-labeled phosphate-binding protein (PBP) from *E. coli*, PBP-MDCC (Brune *et al.*, 1994). Binding of P_i to MDCC-labeled PBP is rapid ($k_{on} = 10^8$ $M^{-1}s^{-1}$) and tight ($K_d = 0.1$ μM), and the formation of the complex strongly increases the fluorescence of MDCC. Expression, purification, and fluorescence labeling are very efficient and were described previously (Brune *et al.*, 1994). To minimize phosphate contamination, all solutions and the stopped-flow apparatus were preincubated with 600-μM 7-methylguanosine and 0.3-U/ml purine nucleoside phosphorylase ("Pi mop") (Brune *et al.*, 1994). The concentration of MDCC-PBP (2.5 μM) was chosen such that the rate of P_i binding to MDCC-PBP ($k_1' = k_1[\text{MDCC-PBP}] = 250$ s^{-1}) was much faster than the rate of GTP hydrolysis in IF2 (30 s^{-1}) (Tomsic *et al.*, 2000) and thus the uptake of P_i liberated from the factor by MDCC-PBP was practically instantaneous. The fluorescence of MDCC was excited at 425 nm and monitored after passing a KV450 filter (Schott). Rate constants were determined by fitting a function that included the delay before the onset of P_i release, an exponential term for P_i release, and a linear slope for turnover P_i release following the initial burst; alternatively, the reaction rate constant can be calculated more precisely from global fitting of an ensemble of traces (e.g., recorded at different concentrations of ligands) using numerical integration (Savelsbergh *et al.*, 2003). In the latter case, the rate constants of Pi association with and dissociation from MDCC-PBP and the rate of Pi consumption by the Pi mop have to be determined independently and with precision.

6. QUENCH-FLOW MEASUREMENTS

Chemical reactions can be studied by rapidly mixing the components and stopping the reaction by mixing with a suitable quencher. The reaction products are analyzed by chromatographic or any other techniques, quantified, and the time course of substrate consumption or product formation analyzed in the same way as the stopped-flow transients. Rapid mixing is achieved in a quench-flow apparatus, and in all experiments described here a KinTek RQF-3 apparatus was used. The dead-time was about 3.5 ms and reaction rates of up to 200 s^{-1} could be measured. Experiments were performed by rapidly mixing equal volumes (14-μl each) of reactants and quenching them with a large volume of quencher (see following). If necessary, the solutions of reactants in the sample syringes were cooled by wrapping the syringes with ice packets to prevent sample degradation before mixing.

7. GTPase Activity

30S initiation complexes were prepared in buffer A using 30S subunits (0.6 μM), 022 mRNA (1.8 μM), IF1, IF2, and IF3 (0.9-μM each), f[^{3}H] Met-tRNA^{fMet} (0.9 μM), and [γ-^{32}P]GTP (72 μM, ~1000 dpm/pmol). After incubation at 37° for 15 min, the complexes were stored on ice. One sample syringe was filled with the solution containing 30S initiation complexes (final concentration 0.3 μM) and the other with solutions containing varying concentrations of 50S subunits. After rapid mixing, reactions were quenched by 1-M HClO$_4$ with 3-mM KH$_2$PO$_4$, and [^{32}P] phosphate was determined by molybdate extraction into ethyl acetate as described (Rodnina *et al.*, 1999). It should be noted that the extraction method cannot be used when working in buffers containing polyamines; in such a case, thin-layer chromatography (Gromadski and Rodnina, 2004) or any other suitable method for separating GTP and Pi can be utilized. The initial part of the time courses of GTP hydrolysis usually can be evaluated by fitting a single exponential function. Upon prolonged incubation, an additional linear phase accounting for the multiple-turnover GTP hydrolysis by IF2 was observed (Gualerzi *et al.*, 2001).

8. Dipeptide Formation

30S initiation complexes were prepared as described previously. One sample syringe was loaded with preformed 30S initiation complex (0.3 μM) in the presence of GTP (0.5 mM) and purified ternary complex EF-Tu·GTP·[^{14}C]Phe-tRNA^{Phe} (0.6 μM) (for the preparation of the ternary complex see Knudsen *et al.*, 2001; Rodnina *et al.*, 1994a), and the other sample syringe was loaded with 50S subunits at different concentrations. After rapid mixing, reactions were quenched with 0.8-M KOH. The quenched reaction mixtures were incubated for 30 min at 37° to hydrolyze RNA, neutralized with acetic acid, and peptides were analyzed by HPLC using a convex 0 to 65% acetonitrile gradient in 0.1% TFA. Because dipeptide formation had a significant delay at the beginning of time courses, the data were analyzed by numerical integration.

9. APPLICATIONS OF THE METHOD

A number of applications of the aforementioned methods have been described already (Antoun *et al.*, 2006b; Milon *et al.*, 2006; Tomsic *et al.*, 2000). In the following we will present a few examples of FRET and light-scattering measurements.

10. BINDING OF FMET-TRNA TO THE 30S SUBUNIT

IF2-dependent binding of fMet-tRNA^{fMet} to the 30S subunits containing mRNA, IF1, and IF3 was monitored by FRET from the donor fluorescein, attached to fMet-tRNA^{fMet}, to the acceptor Alexa555, attached to IF1 or IF3. A fluorescence increase due to FRET was observed when both fluorescence donor and acceptor were bound in the 30S initiation complex (Fig. 1.3A); in the absence of either donor or acceptor, or in the absence of the 30S subunits, no signal change was observed. FRET efficiency was higher when the acceptor was attached to IF3 (position 166) than to IF1 (position 4), which is consistent with the shorter distance between the respective positions in tRNA^{fMet} and IF3 compared to those in tRNA^{fMet} and IF1 in the structural model of the 30S initiation complex (Fig. 1.2). Single-exponential fitting of the time course (1) in Fig. 1.3A did not yield a satisfactory fit, because similar concentrations of reactants were used for mixing and thus the pseudo–first-order conditions are not satisfied. Best fitting was achieved by numerical integration using the model A + B → C, where A and B are reagents in syringes 1 and 2, respectively, and C is the reaction product with high FRET. The reaction is assumed to be irreversible, because the dissociation constant of fMet-tRNA^{fMet} from the complete 30S initiation complex is very low (La Teana *et al.*, 1993; Risuleo *et al.*, 1976). Numerical integration of time courses 1 and 2 gave similar rate constants of ~10 or 6 $\mu M^{-1}s^{-1}$ which represent the rate constant of fMet-tRNA^{fMet} association to the 30S preinitiation complex. A similar value was determined by rapid nitrocellulose filtration (12.5 $\mu M^{-1}s^{-1}$ at 37° in polymix buffer [Antoun *et al.*, 2006a]).

11. FRET TO IF3

Association of IF3 with the 30S initiation complex in the presence of all other components was monitored by FRET from fluorescein to Alexa 555. In this case, fMet-tRNA^{fMet}(Flu) or mRNA(Flu) were used as fluorescence donors and IF3(Alx) labeled at position 166 as acceptor. FRET was

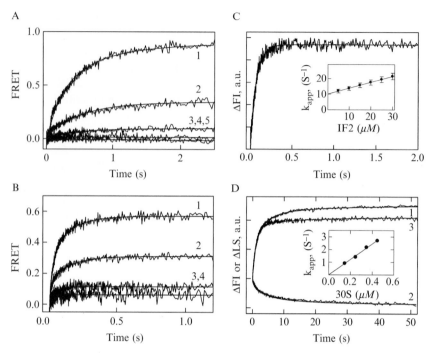

Figure 1.3 Examples of rapid kinetics measurements. (A) IF2-dependent binding of fMet-tRNA^{fMet} to the 30S subunit containing IF1 and IF3. Time courses were measured in the presence of both fluorescence donor, fMet-tRNA^{fMet}(Flu), and one of the acceptors, IF3(Alx) (trace 1) or IF1(Alx) (trace 2); in the absence of the donor with an acceptor alone (trace 3); in the presence of donor but without an acceptor (trace 4), or in the absence of the 30S subunit (trace 5). In (3) and (4) nonfluorescent fMet-tRNA^{fMet} and IFs were used, respectively, instead of their fluorescent derivatives. Syringe 1: fMet-tRNA^{fMet} (0.3 μM), IF2 (0.3 μM), and GTP (0.5 mM) in buffer A. Syringe 2: 30S subunits (0.2 μM), IF1 (0.3 μM), IF3 (0.2 μM), mRNA (0.4 μM), and GTP (0.5 mM) in buffer A. Fluorescence was measured upon excitation at 465 nm after passing a 590-nm cut-off filter. (B) Binding of IF3 to the 30S preinitiation complex. Time courses were measured in the presence of either fluorescence donors, fMet-tRNA^{fMet}(Flu) (trace 1) or mRNA (Flu) (trace 2), and the acceptor, IF3(Alx) (trace 1, 2); in the absence of the acceptor (trace 3) or donor (trace 4); in the latter two cases, nonfluorescent IF3 or fMet-tRNA^{fMet} were used, respectively. Syringe 1: 30S subunits (0.2 μM), IF1 (0.3 μM), IF2 (0.3 μM), mRNA (0.4 μM), fMet-tRNA^{fMet} (0.4 μM), and GTP (0.5 mM) in buffer A. Syringe 2: IF3 (0.2 μM) and GTP (0.5 mM) in buffer A. Fluorescence was measured upon excitation at 465 nm after passing a 590-nm cut-off filter. (C) FRET from Trp in IF2 to mant-GTP. Syringe 1: IF2 (1 μM) in buffer A. Syringe 2: mant-GTP (10 μM) in buffer A. Fluorescence was measured upon excitation at 290 nm after passing a 408-nm cut-off filter. ΔFI, fluorescence intensity change; a.u., arbitrary units. Inset: Concentration dependence of k_{app} of mant-GTP binding to IF2. (D) Subunit joining. Association of the 30S with 50S subunits measured at equal concentrations of subunits (0.25-μM each) by light scattering (trace 1), or fluorescence quenching (0.15-μM each) (trace 2), or by light scattering using pseudo-first order conditions with 30S subunits (0.15 μM) in excess of 50S subunits (0.05 μM) (trace 3). ΔLS, change in light scattering. Inset: Concentration dependence of k_{app} of subunit association.

observed only in the case when both donor and acceptor were present, whereas only very small fluorescence changes of single fluorophores (either donor or acceptor) were caused by the IF3 binding to 30S subunits (Fig. 3B). Again, the difference in FRET efficiencies is consistent with the different distances between the labeled positions in Met-tRNA^{fMet} and IF3 or mRNA and IF3 (Fig. 1.2). Because equal amounts of IF3 and the 30S preinitiation complex were used, the calculations were made by numerical integration; the rate of IF3 binding was 130 $\mu M^{-1}s^{-1}$, assuming slow dissociation of IF3 from the 30S initiation complex. This value is measured directly and is therefore much more reliable than indirect estimations reported earlier, 80 to 180 $\mu M^{-1}s^{-1}$ (Antoun et al., 2006b). Similar measurement can be performed with fluorescence-labeled IF1 instead of IF3; together with the data on fMet-tRNA^{fMet} or mRNA association (not shown) these results provide a detailed kinetic picture of 30S initiation complex assembly.

12. NUCLEOTIDE BINDING TO IF2

To study the interaction of IF2 with guanine nucleotides, the fluorescent mant-derivatives of GTP and GDP (JenaBioscience) were used. Mant-derivatives of GTP and GDP can be obtained pure without contamination by GTP or GDP (Pisareva et al., 2006; Wilden et al., 2006) and have binding properties very similar to those of unmodified GTP/GDP, as shown for many GTPases. In principle, the interaction of IF2 with mant-GTP/GDP can be monitored by changes in mant-GTP fluorescence upon direct excitation of mant fluorescence or by indirect FRET excitation from a tryptophan residue in IF2 (Gromadski et al., 2002). An example of a FRET experiment with IF2 and mant-GTP is shown in Fig. 1.3C, in which the fluorescence of mant-GTP bound to IF2 was excited at 280 nm via FRET from Trp in IF2 and measured after passing KV408 filters (Schott). The k_{app} values obtained by single exponential fitting were plotted against IF2 concentration. From the slope of the linear concentration dependence (Fig. 1.3C, inset), the association rate constant $k_1 = 0.38 \pm 0.01 \ \mu M^{-1}s^{-1}$ was obtained. The Y-axis intercept yielded the dissociation rate constant, $k_{-1} = 10 \pm 1 \ s^{-1}$. Importantly, even though rate constants can be determined with high precision in a single experiment, experiments have to be repeated with different IF2 preparations to obtain a reliable estimate of the rate constants (Milon et al., 2006).

13. SUBUNIT JOINING

The last example illustrates the use of numerical integration versus pseudo-first order approach and exponential fitting to determine rate constants of subunit association from light scattering experiments (Fig. 1.3D).

Time course 1 was obtained at equal concentrations of 30S and 50S subunits. The rate constant of subunit association $(3\ \mu M^{-1}s^{-1})$ was calculated by numerical integration, assuming the model $A + B \rightarrow C$, where A, B, and C represent 30S, 50S, and 70S ribosomes, respectively. The reaction is assumed to be practically irreversible, because 70S ribosomes are very stable in the absence of initiation factors. Furthermore, when subunit association was monitored using 50S(Flu) and 30S(QSY) subunits at equal concentrations (Fig. 1.3D, time course 2) (Peske *et al.*, 2005) and numerical integration was applied, a very similar association rate constant, about $2\ \mu M^{-1}s^{-1}$, was obtained, indicating that introducing the fluorescence labels did not affect the functional activity of the ribosomal subunits. Time course 3 was obtained at pseudo-first order conditions in excess of the 30S over the 50S subunits and apparent rate constants of subunit association were calculated by exponential fitting. The concentration dependence of the apparent rate constants (Fig. 1.3D, inset) yielded the rate constants of subunit association, $6\ \mu M^{-1}s^{-1}$ (slope). The dissociation rate constant (Y-axis intercept) is too close to zero to be determined with precision from the plot. The activity of the 30S subunit used for this experiment was >95% tested by their ability to bind fMet-tRNA^{fMet} in an initiation assay. The high activity is important for correct determination of the association rate constant, because partial activity of preparations may lead to incorrect values for the slope.

ACKNOWLEDGMENTS

We thank W. Wintermeyer (University of Witten/Herdecke) for advice on labeling of tRNA and mRNA and critical reading the manuscript; Y. P. Semenkov, V. Katunin, and V. I. Makhno (Petersburg Nuclear Physics Institute, Russia) for purified tRNA^{fMet}; and P. Striebeck, C. Schillings, A. Böhm, and S. Möbitz for expert technical assistance. The work was supported by the Deutsche Forschungsgemeinschaft, the Alfried Krupp von Bohlen und Halbach-Stiftung, and the Fonds der Chemischen Industrie (MVR), the European Union, MIUR PRIN 2005 (COG), and by the DAAD/Vigoni program (COG and MVR).

REFERENCES

Allen, G. S., Zavialov, A., Gursky, R., Ehrenberg, M., and Frank, J. (2005). The cryo-EM structure of a translation initiation complex from *Escherichia coli. Cell* **121,** 703–712.

Antoun, A., Pavlov, M. Y., Andersson, K., Tenson, T., and Ehrenberg, M. (2003). The roles of initiation factor 2 and guanosine triphosphate in initiation of protein synthesis. *EMBO J.* **22,** 5593–5601.

Antoun, A., Pavlov, M. Y., Lovmar, M., and Ehrenberg, M. (2006a). How initiation factors maximize the accuracy of tRNA selection in initiation of bacterial protein synthesis. *Mol. Cell* **23,** 183–193.

Antoun, A., Pavlov, M. Y., Lovmar, M., and Ehrenberg, M. (2006b). How initiation factors tune the rate of initiation of protein synthesis in bacteria. *EMBO J.* **25,** 2539–2550.

Antoun, A., Pavlov, M. Y., Tenson, T., and Ehrenberg, M. (2004). Ribosome formation from subunits studied by stopped-flow and Rayleigh light scattering. *Biol. Proc. Online* **6,** 35–54.

Biou, V., Shu, F., and Ramakrishnan, V. (1995). X-ray crystallography shows that translational initiation factor IF3 consists of two compact alpha/beta domains linked by an alpha-helix. *EMBO J.* **14,** 4056–4064.

Bradford, M. (1976). A rapid and sensitive method for the quantitation of microgram quantities of protein utilizing the principle of protein-dye binding. *Anal. Biochem.* **72,** 248–254.

Brandi, A., Pietroni, P., Gualerzi, C. O., and Pon, C. L. (1996). Post-transcriptional regulation of CspA expression in *Escherichia coli. Mol. Microbiol.* **19,** 231–240.

Brombach, M., and Pon, C. L. (1987). The unusual translational initiation codon AUU limits the expression of the infC (initiation factor IF3) gene of *Escherichia coli. Mol. Gen. Genet.* **208,** 94–100.

Brune, M., Hunter, J. L., Corrie, J. E., and Webb, M. R. (1994). Direct, real-time measurement of rapid inorganic phosphate release using a novel fluorescent probe and its application to actomyosin subfragment 1 ATPase. *Biochemistry* **33,** 8262–8271.

Calogero, R. A., Pon, C. L., and Gualerzi, C. O. (1987). Chemical synthesis and *in vivo* hyperexpression of a modular gene coding for *Escherichia coli* translational initiation factor IF1. *Mol. Gen. Genet.* **208,** 63–69.

Canonaco, M. A., Calogero, R. A., and Gualerzi, C. O. (1986). Mechanism of translational initiation in prokaryotes. Evidence for a direct effect of IF2 on the activity of the 30 S ribosomal subunit. *FEBS Lett.* **207,** 198–204.

Carter, A. P., Clemons, W. M., Jr., Brodersen, D. E., Morgan-Warren, R. J., Hartsch, T., Wimberly, B. T., and Ramakrishnan, V. (2001). Crystal structure of an initiation factor bound to the 30S ribosomal subunit. *Science* **291,** 498–501.

Caserta, E., Tomsic, J., Spurio, R., La Teana, A., Pon, C. L., and Gualerzi, C. O. (2006). Translation initiation factor IF2 interacts with the 30S ribosomal subunit via two separate binding sites. *J. Mol. Biol.* **362,** 787–799.

Celano, B., Pawlik, R. T., and Gualerzi, C. O. (1988). Interaction of *Escherichia coli* translation-initiation factor IF-1 with ribosomes. *Eur. J. Biochem.* **178,** 351–355.

Dahlquist, K. D., and Puglisi, J. D. (2000). Interaction of translation initiation factor IF1 with the *E. coli* ribosomal A site. *J. Mol. Biol.* **299,** 1–15.

Dallas, A., and Noller, H. F. (2001). Interaction of translation initiation factor 3 with the 30S ribosomal subunit. *Mol. Cell* **8,** 855–864.

de Smit, M. H., and van Duin, J. (2003). Translational standby sites: How ribosomes may deal with the rapid folding kinetics of mRNA. *J. Mol. Biol.* **331,** 737–743.

Debey, P., Hui Bon Hoa, G., Douzou, P., Godefroy-Colburn, T., Graffe, M., and Grunberg-Manago, M. (1975). Ribosomal subunit interaction as studied by light scattering. Evidence of different classes of ribosome preparations. *Biochemistry* **14,** 1553–1559.

Fabbretti, A., Pon, C. L., Hennelly, S. P., Hill, W. E., Lodmell, J. S., and Gualerzi, C. O. (2007). The real time path of translation factor IF3 onto and off the ribosome. *Mol. Cell* **25,** 285–296.

Garcia, C., Fortier, P. L., Blanquet, S., Lallemand, J. Y., and Dardel, F. (1995a). 1H and 15N resonance assignments and structure of the N-terminal domain of *Escherichia coli* initiation factor 3. *Eur. J. Biochem.* **228,** 395–402.

Garcia, C., Fortier, P. L., Blanquet, S., Lallemand, J. Y., and Dardel, F. (1995b). Solution structure of the ribosome-binding domain of *E. coli* translation initiation factor IF3. Homology with the U1A protein of the eukaryotic spliceosome. *J. Mol. Biol.* **254,** 247–259.

Giuliodori, A. M., Brandi, A., Gualerzi, C. O., and Pon, C. L. (2004). Preferential translation of cold-shock mRNAs during cold adaptation. *RNA* **10,** 265–276.

Gromadski, K. B., and Rodnina, M. V. (2004). Kinetic determinants of high-fidelity tRNA discrimination on the ribosome. *Mol. Cell* **13**, 191–200.

Gromadski, K. B., Wieden, H. J., and Rodnina, M. V. (2002). Kinetic mechanism of elongation factor Ts-catalyzed nucleotide exchange in elongation factor Tu. *Biochemistry* **41**, 162–169.

Grunberg-Manago, M., Dessen, P., Pantaloni, D., Godefroy-Colburn, T., Wolfe, A. D., and Dondon, J. (1975). Light-scattering studies showing the effect of initiation factors on the reversible dissociation of *Escherichia coli* ribosomes. *J. Mol. Biol.* **94**, 461–478.

Gualerzi, C., Risuleo, G., and Pon, C. L. (1977). Initial rate kinetic analysis of the mechanism of initiation complex formation and the role of initiation factor IF-3. *Biochemistry* **16**, 1684–1689.

Gualerzi, C. O., Brandi, L., Caserta, E., Garofalo, C., Lammi, M., La Teana, A., Petrelli, D., Spurio, R., Tomsic, J., and Pon, C. L. (2001). Initiation factors in the early events of mRNA translation in bacteria. *Cold Spring Harb. Symp. Quant. Biol.* **66**, 363–376.

Gualerzi, C. O., and Pon, C. L. (1990). Initiation of mRNA translation in prokaryotes. *Biochemistry* **29**, 5881–5889.

Gualerzi, C. O., and Wintermeyer, W. (1986). Prokaryotic initiation factor 2 acts at the level of the 30S ribosomal subunit. A fluorescence stopped-flow study. *FEBS Lett.* **202**, 1–6.

Hennelly, S. P., Antoun, A., Ehrenberg, M., Gualerzi, C. O., Knight, W., Lodmell, J. S., and Hill, W. E. (2005). A time-resolved investigation of ribosomal subunit association. *J. Mol. Biol.* **346**, 1243–1258.

Jenner, L., Romby, P., Rees, B., Schulze-Briese, C., Springer, M., Ehresmann, C., Ehresmann, B., Moras, D., Yusupova, G., and Yusupov, M. (2005). Translational operator of mRNA on the ribosome: How repressor proteins exclude ribosome binding. *Science* **308**, 120–123.

Johnson, A. E., Adkins, H. J., Matthews, E. A., and Cantor, C. R. (1982). Distance moved by transfer RNA during translocation from the A site to the P site on the ribosome. *J. Mol. Biol.* **156**, 113–140.

Karimi, R., Pavlov, M. Y., Buckingham, R. H., and Ehrenberg, M. (1999). Novel roles for classical factors at the interface between translation termination and initiation. *Mol. Cell* **3**, 601–609.

Knudsen, C., Wieden, H. J., and Rodnina, M. V. (2001). The importance of structural transitions of the switch II region for the functions of elongation factor Tu on the ribosome. *J. Biol. Chem.* **276**, 22183–22190.

Kothe, U., Paleskava, A., Konevega, A. L., and Rodnina, M. V. (2006). Single-step purification of specific tRNAs by hydrophobic tagging. *Anal. Biochem.* **356**, 148–150.

Kycia, J. H., Biou, V., Shu, F., Gerchman, S. E., Graziano, V., and Ramakrishnan, V. (1995). Prokaryotic translation initiation factor IF3 is an elongated protein consisting of two crystallizable domains. *Biochemistry* **34**, 6183–6187.

La Teana, A., Gualerzi, C. O., and Brimacombe, R. (1995). From stand-by to decoding site. Adjustment of the mRNA on the 30S ribosomal subunit under the influence of the initiation factors. *RNA* **1**, 772–782.

La Teana, A., Pon, C. L., and Gualerzi, C. O. (1993). Translation of mRNAs with degenerate initiation triplet AUU displays high initiation factor 2 dependence and is subject to initiation factor 3 repression. *Proc. Natl. Acad. Sci. USA* **90**, 4161–4165.

Lakowicz, J. R. (1999). "Principles of Fluorescence Spectroscopy," pp. 305–309. Kluwer Academic/Plenum Publishers, New York.

Lammi, M., Pon, C. L., and Gualerzi, C. O. (1987). The NH_2-terminal cleavage of *Escherichia coli* translational initiation factor IF3. A mechanism to control the intracellular level of the factor?. *FEBS Lett.* **215**, 115–121.

Lassen, S. F., Mortensen, K. K., and Sperling-Petersen, H. U. (1992). OmpT proteolysis of *E. coli* initiation factor IF2. Elimination of a cleavage site by site-directed mutagenesis. *Biochem. Int.* **27,** 601–611.

Laursen, B. S., Sorensen, H. P., Mortensen, K. K., and Sperling-Petersen, H. U. (2005). Initiation of protein synthesis in bacteria. *Microbiol. Mol. Biol. Rev.* **69,** 101–123.

Marzi, S., Knight, W., Brandi, L., Caserta, E., Soboleva, N., Hill, W. E., Gualerzi, C. O., and Lodmell, J. S. (2003). Ribosomal localization of translation initiation factor IF2. *RNA* **9,** 958–969.

McCutcheon, J. P., Agrawal, R. K., Philips, S. M., Grassucci, R. A., Gerchman, S. E., Clemons, W. M., Jr., Ramakrishnan, V., and Frank, J. (1999). Location of translational initiation factor IF3 on the small ribosomal subunit. *Proc. Natl. Acad. Sci. USA* **96,** 4301–4306.

Milon, P., Tischenko, E., Tomsic, J., Caserta, E., Folkers, G., La Teana, A., Rodnina, M. V., Pon, C. L., Boelens, R., and Gualerzi, C. O. (2006). The nucleotide-binding site of bacterial translation initiation factor 2 (IF2) as a metabolic sensor. *Proc. Natl. Acad. Sci. USA* **103,** 13962–13967.

Moazed, D., Samaha, R. R., Gualerzi, C., and Noller, H. F. (1995). Specific protection of 16 S rRNA by translational initiation factors. *J. Mol. Biol.* **248,** 207–210.

Myasnikov, A. G., Marzi, S., Simonetti, A., Giuliodori, A. M., Gualerzi, C. O., Yusupova, G., Yusupov, M., and Klaholz, B. P. (2005). Conformational transition of initiation factor 2 from the GTP- to GDP-bound state visualized on the ribosome. *Nat. Struct. Mol. Biol.* **12,** 1145–1149.

Page, L. A., Englander, S. W., and Simpson, M. V. (1967). Hydrogen exchange studies on ribosomes. *Biochemistry* **6,** 968–977.

Peske, F., Rodnina, M. V., and Wintermeyer, W. (2005). Sequence of steps in ribosome recycling as defined by kinetic analysis. *Mol. Cell* **18,** 403–412.

Petrelli, D., Garofalo, C., Lammi, M., Spurio, R., Pon, C. L., Gualerzi, C. O., and La Teana, A. (2003). Mapping the active sites of bacterial translation initiation factor IF3. *J. Mol. Biol.* **331,** 541–556.

Petrelli, D., LaTeana, A., Garofalo, C., Spurio, R., Pon, C. L., and Gualerzi, C. O. (2001). Translation initiation factor IF3: Two domains, five functions, one mechanism? *EMBO J.* **20,** 4560–4569.

Pioletti, M., Schlunzen, F., Harms, J., Zarivach, R., Gluhmann, M., Avila, H., Bashan, A., Bartels, H., Auerbach, T., Jacobi, C., Hartsch, T., Yonath, A., and Franceschi, F. (2001). Crystal structures of complexes of the small ribosomal subunit with tetracycline, edeine and IF3. *EMBO J.* **20,** 1829–1839.

Pisareva, V. P., Pisarev, A. V., Hellen, C. U., Rodnina, M. V., and Pestova, T. V. (2006). Kinetic analysis of interaction of eukaryotic release factor 3 with guanine nucleotides. *J. Biol. Chem.* **281,** 40224–40235.

Pon, C. L., and Gualerzi, C. (1974). Effect of initiation factor 3 binding on the 30S ribosomal subunits of *Escherichia coli*. *Proc. Natl. Acad. Sci. USA* **71,** 4950–4954.

Pon, C. L., and Gualerzi, C. (1979). Qualitative and semiquantitative assay of *Escherichia coli* translational initiation factor IF-3. *Methods Enzymol.* **60,** 230–239.

Pon, C. L., and Gualerzi, C. O. (1984). Mechanism of protein biosynthesis in prokaryotic cells. Effect of initiation factor IF1 on the initial rate of 30 S initiation complex formation. *FEBS Lett.* **175,** 203–207.

Pon, C. L., and Gualerzi, C. O. (1986). Mechanism of translational initiation in prokaryotes. IF3 is released from ribosomes during and not before 70 S initiation complex formation. *FEBS Lett.* **195,** 215–219.

Remaut, E., Tsao, H., and Fiers, W. (1983). Improved plasmid vectors with a thermo-inducible expression and temperature-regulated runaway replication. *Gene* **22,** 103–113.

Risuleo, G., Gualerzi, C., and Pon, C. (1976). Specificity and properties of the destabilization, induced by initiation factor IF-3, of ternary complexes of the 30-S ribosomal subunit, aminoacyl-tRNA and polynucleotides. *Eur. J. Biochem.* **67,** 603–613.

Rodnina, M. V., Fricke, R., and Wintermeyer, W. (1994a). Transient conformational states of aminoacyl-tRNA during ribosome binding catalyzed by elongation factor Tu. *Biochemistry* **33,** 12267–12275.

Rodnina, M. V., Savelsbergh, A., Matassova, N. B., Katunin, V. I., Semenkov, Y. P., and Wintermeyer, W. (1999). Thiostrepton inhibits the turnover but not the GTPase of elongation factor G on the ribosome. *Proc. Natl. Acad. Sci. USA* **96,** 9586–9590.

Rodnina, M. V., Semenkov, Y. P., and Wintermeyer, W. (1994b). Purification of fMet-tRNA^{fMet} by fast protein liquid chromatography. *Anal. Biochem.* **219,** 380–381.

Rodnina, M. V., Stark, H., Savelsbergh, A., Wieden, H. J., Mohr, D., Matassova, N. B., Peske, F., Daviter, T., Gualerzi, C. O., and Wintermeyer, W. (2000). GTPases mechanisms and functions of translation factors on the ribosome. *Biol. Chem.* **381,** 377–387.

Roll-Mecak, A., Cao, C., Dever, T. E., and Burley, S. K. (2000). X-Ray structures of the universal translation initiation factor IF2/eIF5B: Conformational changes on GDP and GTP binding. *Cell* **103,** 781–792.

Savelsbergh, A., Katunin, V. I., Mohr, D., Peske, F., Rodnina, M. V., and Wintermeyer, W. (2003). An elongation factor G-induced ribosome rearrangement precedes tRNA-mRNA translocation. *Mol. Cell* **11,** 1517–1523.

Sette, M., van Tilborg, P., Spurio, R., Kaptein, R., Paci, M., Gualerzi, C. O., and Boelens, R. (1997). The structure of the translational initiation factor IF1 from *E. coli* contains an oligomer-binding motif. *EMBO J.* **16,** 1436–1443.

Spurio, R., Severini, M., La Teana, A., Canonaco, M. A., Pawlik, R. T., Gualerzi, C., and Pon, C. (1993). Novel structural and functional aspects of translation initiation factor IF2. *In* "The Translational Apparatus" (K. H. Nierhaus, ed.), pp. 433–444. Plenum Press, New York.

Studer, S. M., and Joseph, S. (2006). Unfolding of mRNA secondary structure by the bacterial translation initiation complex. *Mol. Cell* **22,** 105–115.

Tomsic, J., Vitali, L. A., Daviter, T., Savelsbergh, A., Spurio, R., Striebeck, P., Wintermeyer, W., Rodnina, M. V., and Gualerzi, C. O. (2000). Late events of translation initiation in bacteria: A kinetic analysis. *EMBO J.* **19,** 2127–2136.

Wilden, B., Savelsbergh, A., Rodnina, M. V., and Wintermeyer, W. (2006). Role and timing of GTP binding and hydrolysis during EF-G-dependent tRNA translocation on the ribosome. *Proc. Natl. Acad. Sci. USA* **103,** 13670–13675.

Wintermeyer, W., and Gualerzi, C. (1983). Effect of *Escherichia coli* initiation factors on the kinetics of N-AcPhe-tRNA^{Phe} binding to 30S ribosomal subunits. A fluorescence stopped-flow study. *Biochemistry* **22,** 690–694.

Wintermeyer, W., and Zachau, H. G. (1974). Replacement of odd bases in tRNA by fluorescent dyes. *Methods Enzymol.* **29,** 667–673.

Wishnia, A., Boussert, A., Graffe, M., Dessen, P. H., and Grunberg-Manago, M. (1975). Kinetics of the reversible association of ribosomal subunits: Stopped-flow studies of the rate law and of the effect of Mg^{2+}. *J. Mol. Biol.* **93,** 499–515.

Yusupova, G., Jenner, L., Rees, B., Moras, D., and Yusupov, M. (2006). Structural basis for messenger RNA movement on the ribosome. *Nature* **444,** 391–394.

Yusupova, G. Z., Yusupov, M. M., Cate, J. H., and Noller, H. F. (2001). The path of messenger RNA through the ribosome. *Cell* **106,** 233–241.

Zitomer, R. S., and Flaks, J. G. (1972). Magnesium dependence and equilibrium of the *Escherichia coli* ribosomal subunit association. *J. Mol. Biol.* **71,** 263–279.

CHAPTER TWO

BINDING OF MRNA TO THE BACTERIAL TRANSLATION INITIATION COMPLEX

Sean M. Studer *and* Simpson Joseph

Contents

Abstract

Translation initiation is a key step for regulating the synthesis of several proteins. In bacteria, translation initiation involves the interaction of the mRNA with the ribosomal small subunit. Additionally, translation initiation factors 1, 2, and 3, and the initiator tRNA, also assemble on the ribosomal small subunit and are essential for efficiently recruiting an mRNA for protein biosynthesis. In the following chapter, we describe fluorescence-based methods for studying the interaction of mRNA with the bacterial initiation complex. Model mRNAs with a covalently attached fluorescent probe showed an increase in fluorescence intensity when bound to the bacterial initiation complex. Utilizing the increase

Department of Chemistry and Biochemistry, University of California, San Diego, California

Methods in Enzymology, Volume 430
ISSN 0076-6879, DOI: 10.1016/S0076-6879(07)30002-5

in fluorescence intensity upon mRNA binding to the bacterial initiation complex, we determined the equilibrium binding constants and the association and dissociation rate constants. These methods are important for quantitatively analyzing the effects of mRNA secondary structure and the role of the initiation factors in recruitment of mRNA by the bacterial initiation complex.

1. INTRODUCTION

Studying the dynamics of mRNA interaction with the 30S subunit is becoming ever more essential for understanding gene regulation at the translational initiation level. Several studies examining mRNA thermosensors (Johansson et al., 2002) and mRNA riboswitches (Mandal and Breaker, 2004; Mandal et al., 2003, 2004; Winkler and Breaker, 2003; Winkler et al., 2002a,b) have convincingly demonstrated how gene expression can be regulated at this level. Specifically, these studies examined how alterations in the mRNAs secondary structure, subsequently exposing the translation initiation region (TIR), resulted in higher levels of gene expression. These articles reasoned that removal of the secondary structure in the TIR, resulting in the exposure of the Shine-Dalgarno (SD) sequence, caused a more stable mRNA·30S complex. It was therefore this stable complex that ultimately led to greater levels of gene expression. The reasons why exposure of the TIR caused higher levels of gene expression can be better understood if the composition of the TIR is examined.

The TIR generally consists of three elements, all of which have been shown to affect gene expression. The three elements are the SD sequence, a start codon, and a series of adenosine or uracil adjacent to the SD sequence and the start codon.

The SD sequence, which typically ranges from 4 to 8 nucleotides in length, is a purine-rich sequence and is located 5 to 9 nucleotides from the start codon (Schneider et al., 1986; Steitz, 1980). This sequence promotes mRNA binding to the 30S subunit through its ability to hydrogen bond with a complementary region in the 3′ end of the 16S rRNA known as the Anti-Shine-Dalgarno sequence (ASD). Not only does the SD sequence promote interaction of the mRNA with the 30S subunit, but it also properly positions the mRNA on the 30S subunit. Because the SD sequence allows for the specific positioning of the mRNA and the start codon is usually a set number of nucleotides downstream, it follows that the start codon therefore must be correctly positioned as a result of the SD-ASD.

The start codon is generally composed of the bases AUG from 5′ to 3′, however, there is often a small degree of variability. The other prevalent start codons are GUG and AUU (Gualerzi and Pon, 1990). The start codon interacts specifically with the initiator tRNA.

The third factor that makes up the TIR is the unstructured regions positioned adjacent to the SD sequence and the start codon (Scherer *et al.*, 1980). These unstructured regions are often referred to as standby sites (Brandt and Gualerzi, 1991). These standby sites generally consist of a large percentage of adenosines or uracils and are believed to promote mRNA interaction with the 30S subunit. One hypothesis is that these standby sites allow for a low-affinity interaction of the mRNA with the 30S subunit. The binding of the 30S subunit to the standby site promotes melting of the secondary structure in the TIR. Melting of the TIR and subsequent unmasking of the SD sequence results in a more stable mRNA 30S complex (Studer and Joseph, 2006). The following text describes the experimental procedures employed to study mRNA interaction with the 30S subunit.

2. Experimental Procedures

2.1. Design of model mRNAs

When designing short model mRNAs representative of the TIR, one needs to consider the presence of secondary structure in the mRNA. mRNA secondary structure, which can sequester the TIR, often dramatically changes the affinity of an mRNA for the 30S subunit. These extreme changes in affinity of the mRNA for the ribosome are most evident when secondary structure masks the SD sequence.

To examine the extent of secondary structure present in a model mRNA, a helpful program to employ is mFOLD (Zuker, 2003) (http://www.bioinfo.rpi.edu/applications/mfold/rna/form1.cgi.) This program involves entering the primary sequence of an RNA and selecting from a series of parameters and constraints to give a rapid *in silico* prediction of RNA secondary structure and stability. The predicted secondary structure and its corresponding stability are partly based on the extent of hydrogen bonding that occurs between complementary nucleotides.

Although mFOLD is an excellent *in silico* program for secondary structure prediction, it does not properly account for stacking of nucleotides (Holder and Lingrel, 1975). Therefore, one needs to keep in mind that base stacking can alter the secondary structure of an mRNA and possibly affect the interaction of an mRNA with the 30S subunit.

2.2. Synthesis of model mRNAs

Once an mRNA sequence has been decided, the model mRNA can either be obtained through transcription with the use of T7 RNA polymerase and a DNA template (Milligan *et al.*, 1987) or through solid-phase synthesis.

Because solid-phase synthesis is the preferred method in our lab, we will go into more detail regarding the advantages of this technique.

The advantage of solid-phase synthesis is that it is relatively inexpensive while yielding large quantities of full-length RNA. Additionally, there are several suppliers of synthetic RNA. These manufacturers include, but are not limited to, Dharmacon, Qiagen, Operon, and Ambion. We have found that ordering from the suppliers is more efficient and sometimes less costly than purchasing phosphoramidites and synthesizing the RNA ourselves.

Another advantage of solid-phase synthesis is that it allows for the flexibility to incorporate unnatural modifications at specific locations on an RNA. For our assays, unnatural modifications on the mRNA are often essential because we incorporate fluorescent probes. The unnatural modification most commonly used is that of an amine linker on the 3' end of the mRNA. The amine linker is composed of a chain of three carbon atoms followed by a reactive amine group on the 3' end. The chain of three carbon atoms allows for free rotation of the conjugated probe, while the amine serves as the reactive moiety essential for the conjugation of a probe to the RNA.

The reason for choosing the amine linker is that a large number of probes contain amine reactive conjugates, therefore this amine linker provides a large degree of flexibility in the choice of fluorescent probes that can be linked to an mRNA. Additionally, the substitution reaction between the amine linker and n-hydroxyl succinimide ester is extremely efficient (Silverman and Cech, 1999).

2.3. Deprotection

In general, we order our RNA from Dharmacon and use the scheme they provide to remove the protecting groups acid labile orthoester (ACE) from the RNA's 2' hydroxyl. This protection group is acid labile and is efficiently removed by resuspending the synthesized, lyophilized RNA with 400 μl of the supplied Deprotection Buffer (100-mM acetic acid, adjusted to a pH 3.8 TEMED). The resuspended RNA is then incubated for between 30 min and 2 h at 60°. The incubation times vary depending on the modifications present on the mRNA. Following incubation, the RNA solution is then concentrated in a Savant Speed Vac until a pellet is formed ($\sim$6 h).

2.4. Thermal melting analysis

Thermal melting experiments were performed on a Beckman Coulter DU-640 spectrophotometer with traveling cuvette accessory that was equipped with a High-Performance Temperature Controller. mRNA was placed in a cuvette at a concentration of 1.5 μM and $\sim$5 μl of mineral oil was added to the top of the solution to prevent evaporation. A nitrogen purge was used

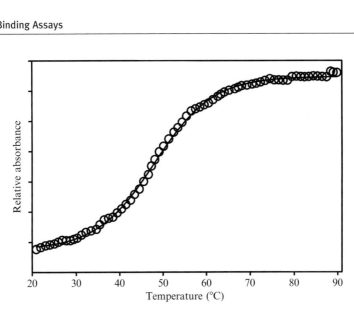

Figure 2.1 Determination of the thermal melting point for mRNA 6B: The absorbance of the mRNA was measured at every 1° from 20° to 90°. The melting curve was analyzed using MeltWin software and the T_m of this particular mRNA was determined to be 49°.

for melting points below ambient temperatures (the nitrogen purge prevented condensation from forming on the side of the cuvette). The buffer used contained 1-mM sodium phosphate (pH 7.0), 10-mM sodium cacodylate (pH 7.2), 1-mM EDTA (pH 8.0), and 150-mM sodium perchlorate as described by Cole et al. (1972). The melting curve, used for analysis, was obtained by subtracting the absorbance in the blank from the net absorbance (Fig. 2.1). Melting profiles were analyzed using MeltWin Software (http://www.meltwin.com) to determine the Tm, ΔS, ΔH, and ΔG of each mRNA.

2.5. Labeling and purification of mRNA

The fluorescence probe typically used in our lab is pyrene succinimide. The advantage of pyrene succinimide is that it is inexpensive, relative to many other fluorescent probes, and it is sensitive to changes in the microenvironment (Silverman and Cech, 1999).

Conjugating a fluorescent label to an mRNA containing an amine linker is generally very efficient. The following protocol is used to covalently attach a fluorescent label to an amine linker.

75 μl of 0.1-M sodium tetraborate (Borax) (pH 8.5)
14 μl of a Pyrene succinimide solution made by dissolving 5 mg of 1-Pyrenebutyric acid n-hydroxyl succinimide ester (Aldrich) in 560 μl of DMSO.
~20,000 pmol of mRNA
Addition of H_2O to bring the reaction vol to 100 μl

Note: Borax is made by dissolving 1.9 g of Borax in 40 ml of water. The pH of this solution is adjusted to pH 8.5 using HCl. It is then stored as 1-ml aliquots at $-80°$. The aliquots are discarded after a single use due to changes in their pH through the absorption of CO_2.

This mixture is incubated in the dark on a shaker at room temperature for $\sim$6 h. Following the incubation, at least 500 μl of ethanol is added to the mixture. The mixture is briefly vortexed and then placed at $-80°$ for at least 30 min. Following ethanol precipitation, the mixture is then centrifuged at 16,100 RCF. Centrifugation results in a chalky pellet on the bottom of the tube. This pellet is then resuspended in an appropriate vol of water and mixed with 2$\times$ Standard Loading Buffer (95% Formamide, 20-mM EDTA, 0.05% of 2.5-mM Xylene Cyanol solution, and 0.05% of 2.5-mM BromoPhenylBlue solution). This mixture is purified on a denaturing polyacrylamide gel.

Gel purification serves two purposes: (1) it separates full-length mRNA from the incompletely synthesized mRNA; and (2) it serves as a way of separating mRNA conjugated with the fluorescent probe from mRNA lacking the probe.

Following gel purification, one can identify the desired product by UV shadowing the gel on a TLC plate. What is generally found is that the band migrating most slowly is full-length mRNA with a fluorescent probe conjugated to the 3' end, whereas the products running below this upper band are incompletely synthesized RNA, unconjugated full-length mRNA, or a product of both. Further purification of the putative full-length fluorescently conjugated mRNA can be performed using reverse phase HPLC with a μBondapak C_{18} column manufactured by Waters. Purification is performed in 100-mM triethylammonium acetate, with a linear elution gradient from 0 to 60% acetonitrile over 50 min, with a flow rate of 1 ml/min (Walter and Burke, 2000). Full-length mRNA containing the hydrophobic fluorescent probe will elute at higher concentrations of acetonitrile than all of the other mRNA species. The use of HPLC, in addition to gel electrophoresis, to further purify the labeled mRNAs is sometimes extraneous and unnecessary for performing certain fluorescence-based experiments, but is essential when performing FRET-based experiments. These purification methods have been thoroughly described (Walter and Burke, 2000).

2.6. Labeling schemes

A problem encountered during FRET-based studies was determining the optimal approach for incorporating two fluorescent probes on a single strand of mRNA. The approach that works best for obtaining an mRNA with a donor and acceptor pair on the 5' and 3' ends of an mRNA, irrespective of order, is to have the mRNA synthesized with the desired probe at the 5' end, while placing a reactive moiety (i.e., amine linker) on the 3' end and conjugating the mRNA essentially in the same manner as previously described. It has been previously observed that this method gives a higher yield of full-length product that incorporates both fluorophores than compared to the double-labeled product produced purely through solid-phase synthesis. The other advantage of having the reactive moieties at the 3' end, as opposed to the completely solid-phase synthesized product, is that it allows for greater flexibility in the choice of probes the researcher is able to conjugate to the 3' end of the mRNA.

2.7. Fluorophores

The choice of fluorescent probes for FRET studies is also extremely important. It has been shown by a number of researchers that certain probes (TAMARA, Alexa 488, BODIPY FL, etc.) are quenched to differing degrees by various nucleotides (AMP, GMP, and CMP) (for a list see Torimura et al., 2001). The ability of nucleotides to quench certain probes can confer a significant advantage, which can be exploited when designing a FRET-based assay involving RNA movement or structural changes.

We prefer to use Cyanin 3 (cy3) and Cyanin 5 (cy5) (both available through GE Healthcare Life Sciences) for several reasons. First, both of these probes are readily available and can be conjugated to an mRNA during solid-phase synthesis or through substitution reactions following solid-phase synthesis. Second, they are relatively photostable and have high-fluorescence quantum yields. Last, they are not quenched by nucleotides. One disadvantage encountered with these probes is that they can be prohibitively expensive and are somewhat sensitive to the microenvironment.

2.8. Isolation of the 30S subunit and protein purification

Ribosomal 30S subunit purification has previously been described in detail (Powers and Noller, 1991). The procedure we followed was essentially the same. Briefly, 1.5 liters of MRE600 cells were grown in Luria-Bertani (LB) broth to an OD of 0.5 at a wavelength of 650 nm. These cells were pelleted and then resuspended in Buffer A. The cells were then lysed using a French Press. RQI RNAse Free, Dnase I (Promega) was added to the lysed cells. The lysate was centrifuged to remove excess cell debris. The ammonium chloride concentration of the cell lysate was then adjusted to 0.5 M. The lysate was centrifuged for 4 h at 4° in a Beckman Ti 70 rotor at 165,000 rcf. Simultaneously, six sucrose gradients were prepared containing sucrose ranging from 10 to 40% in buffer B. The gradients were stored for at least 2 h at 4° before use. The ribosome pellets formed during the 4 h centrifugation were resuspended in buffer B and were loaded onto the sucrose gradients. The sucrose gradients were centrifuged at 25,000 rpm for 16 h at 4° in a Beckman SW-28 rotor. Following centrifugation, the individual gradients of sucrose were displaced by pumping 50% sucrose solution into each centrifuge tube using a fixed rate syringe pump

made by Brandel. The sedimentation of the 30S and 50S subunit were monitored at a wavelength of 254 nm using an ISCO UA-6 detector and chart
recorder. The 30S peaks were collected in polypropelene tubes. The magnesium concentration in these individual tubes was raised to 10-mM final concentration. The 30S subunits were centrifuged at 165,000 rcf for at least 13 h at 4° in
a Ti-70 rotor. The 30S subunits were then resuspended in ∼800 μl of buffer C,
aliquots were made, quick-frozen in liquid nitrogen, and stored at −80°.

Buffer A:
   50-mM Tris-HCL (pH 7.6)
   10-mM MgCl$_2$
   100-mM NH$_4$Cl
   6-mM 2-mercaptoethanol
   0.5-mM EDTA
Buffer B:
   50-mM Tris-HCL (pH 7.6)
   1-mM MgCl$_2$
   100-mM NH$_4$Cl
   6-mM 2-mercaptoethanol
Buffer C:
   50-mM Tris-HCL (pH 7.6)
   10-mM MgCl$_2$
   100-mM NH$_4$Cl
   6-mM 2-mercaptoethanol

   Initiation factor (IF1, IF2, and IF3) purification, in addition to methionyl-
tRNA synthetase (MetRS) and methionyl-tRNA transformylase (MTF), has
previously been described in detail (Shimizu et $al.$, 2001). Only minor modifications were made to the protocol. His-tagged factors were purified from BL21
(DE3) cells grown to an OD of 0.5 in 2 liters of Luria-Bertani (LB) broth.
Isopropyl-B-D-thiogalactoside (IPTG) was added to a final concentration of
0.3 mM, and the cells were grown for ∼4 more h at 37°. The cells were pelleted
in a JA 10.5 rotor for 30 min at 5000 rpm. The cell pellets were then resuspended
in lysis buffer (50-mM N-[2-hydroxyethyl]piperazine-N$'$-[2-ethenesulfonic
acid](HEPES)-KOH, (pH 7.6), 1-M NH$_4$Cl, 10-mM MgCl$_2$, 0.3-mg/ml
lysozyme, 0.1% Triton X-100, 0.5-mM phenylmethylsulfonyl fluoride
(PMSF), and 7mM β-mercaptoethanol), and the resuspended cells were lysed
through sonication. Cell debris was removed using a Beckman Ti 70 rotor and
spinning the cell lysate at ∼100,000 rcf for 1 h at 4°. The lysate was applied to
Nickel Agarose and purified through batch purification under native
conditions as described in The QIAexpressionist^{TM} (Qiagen). Fractions
containing the His-tagged protein were dialyzed using the appropriate dialysis
membrane from Spectra/Por$^®$. The His-tagged protein was dialyzed against
storage buffer (50-mM HEPES-KOH, pH 7.6, 100-mM KCl, 10-mM
MgCl, 50% glycerol, and 7-mM B-mercaptoethanol). The buffer was

changed twice. The protein was fractionated into multiple aliquots and stored at −20°.

2.9. Purification of aminoacylated initiator tRNA

Initiator tRNA purification has previously been described (Feinberg and Joseph, 2006). Briefly, tRNA^{fmet} was purchased from Sigma–Aldrich and was aminoacylated and formylated using purified MetRS and MTF. Aminoacylated tRNAs were then purified by HPLC (Fig. 2.2). Small aliquots of purified fMet-tRNA^{fmet} were frozen at −80°.

2.10. Initiation complex assembly

The 30S ribosomal subunits (0.25 μM) were heat activated at 42° for 10 min in polyamine buffer (Bartetzko and Nierhaus, 1988) [20-mM Hepes-KOH (pH 7.6), 6-mM magnesium chloride, 150-mM ammonium chloride, 4-mM 2-mercaptoethanol, 0.05-mM spermine, and 2-mM spermidine] containing 1-mM GTP (or GTP analog). The water bath, containing the 30S subunit, was then slowly cooled to 37°. The 30S subunit was then incubated for another 10 min at 37°. Depending on the experiment, initiation factors (0.5 μM) and fMet-tRNA^{fmet} (0.75 μM) were incubated together for 10 min at 37° prior to their addition to the 30S subunit. Initiation factors in the presence or absence of fMet-tRNA^{fMet} were then added to the 30S subunits and allowed to incubate at 37° for 20 min. If the experiment did not call for initiation factors or fMet-tRNA^{fmet}, buffer was added to bring the solution containing the 30S subunit to its desired volume. Simultaneously, during the time in which the 30S subunit or 30S initiation factor complex was

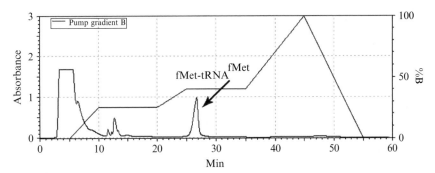

Figure 2.2 Purification of fMet-tRNA^{fMet}: Shown is an HPLC profile of fMet-tRNA^{fMet} eluting from a C18 column. The sample was eluted with a gradient of 0 to 100% buffer B over 45 min. Buffer A was 20-mM Tris-acetate (pH 5.0), 400-mM NaCl, and 10-mM magnesium acetate. Buffer B was the same as buffer A with 60% methanol. Aminoacylated tRNA elutes from the column at 40% buffer B. (See color insert.)

being incubated, the appropriate dye-labeled mRNA (0.2 μM) was then incubated separately at 37° for 30 min in 1-mM GTP (or GTP analog). Both the mRNA and the 30S subunit, regardless whether additional factors were present, were then placed at 25° for at least 5 min prior to performing mRNA-association experiments. All concentrations indicated are the concentrations before mixing. Mixing was generally performed in a 1:1 ratio. Generally, the fluorescence-based experiments were performed at 25°, instead of 37°, because of the higher signal to noise ratio at the lower temperature.

2.11. mRNA dissociation

The 30S complexes were formed as previously described for the mRNA association experiments, except that heat-activated 30S ribosomal subunits (0.25 μM) were mixed with the appropriate dye-labeled mRNA (0.2 μM) and incubated at 37° for 10 min to preform the 30S·mRNA complex. When initiation factors (0.5 μM) and fMet-tRNA^{fMet} (0.75 μM) were used, they were incubated together for 10 min at 37° in polyamine buffer with 1-mM GTP. Experiments with individual factors were performed at both 0.5-μM and 5-μM final concentration for both IF1 and IF3. Initiation factors, with or without fMet-tRNA^{fMet}, were then added to the 30S subunits and allowed to incubate at 37° for 20 min. Longer incubation times did not change the dissociation rate constants indicating that equilibrium had been achieved in the 20-min duration at 37°. The 30S initiation complexes were then placed at 25° for at least 5 min prior to measuring the dissociation kinetics. Excess unlabeled mRNA (2.5 μM) was used as a trap. The unlabeled mRNA, in polyamine buffer containing 1-mM GTP, was incubated at 37° for 30 min, then placed at 25° for at least 5 min prior to mixing with the 30S complex containing the dye-labeled mRNA. When monitoring the slow dissociation of certain mRNAs (time points longer than 60 s), manual mixing was performed on a fluorometer (Fluoromax-P, JY Horiba) sampling between every 10 and 15 s in anti-photobleaching mode. Dissociation rates were nearly identical when performed using the Fluoromax-P under manual mixing conditions and on the Stopped Flow apparatus. All dissociation curves were analyzed by least-squares fitting to a one phase exponential decay equation using GraphPad Prism software (San Diego, CA).

2.12. Steady-state fluorescence measurements

The fluorescence intensity of the 30S·mRNA complex (0.25 μM in 200-μl final vol), prepared as described, was measured at 25° with a photon-counting fluorometer (Fluoromax-P, J Y Horiba). Complexes containing pyrene-labeled mRNAs were excited at 343 nm and the emission spectrum from 360 to 420 nm wavelength was recorded. The slit width for both the emission and excitation in all cases was 1 nm, with an integration time of 1 s while sampling every 1 nm. Complexes containing cy3-/cy5-labeled mRNAs were excited at 550 nm (cy3) or 650 nm (cy5) and the emission spectrum from 560 to 720 nm wavelength were recorded. The emission maxima for cy3 and cy5 dyes are 570 nm and 670 nm, respectively.

When obtaining emission scans we do not recommend using these excitation wavelengths to excite the respective dyes. To obtain the entire cy3 and cy5 spectra, it is advisable to adjust the excitation wavelength of cy3 to ~515 nm and exciting cy5 to ~600 nm. Changing the excitation to ~515 nm seems to minimize the cross-excitation of cy5. Additionally, exciting at a shorter wavelength, and collecting over a wide emission range, yields more complete data. In changing the excitation of cy3 to a suboptimal absorbance wavelength, it then becomes necessary to open the slit widths to obtain a cleaner emission signal.

2.13. Kinetic association and dissociation experiments

Rapid kinetic experiments were performed at 25° on a stopped-flow instrument (μSFM-20, BioLogic, France). The 30S initiation complexes were prepared as described earlier and mixed at 1:1 ratio (mixing volume was 107 μl) with the mRNA. The excitation wavelength was 343 nm (band pass 10 nm), and the fluorescence emission was measured after passing a long-pass filter 361 AELP (Omega Optical, USA) installed in front of the detector. For complexes containing cy3-/cy5-labeled mRNAs, the excitation wavelength was 550 nm (cy3) or 650 nm (cy5) (band pass 10 nm) and the fluorescence emission was measured after passing a long-pass filter 3RD/570-610 (cy3 emission) or a 3RD/670LP (cy5 emission) with light blocker 3rdM/B15 (Omega Optical, USA). The association experiment for each mRNA was performed at five different mRNA concentrations (0.18, 0.36, 0.56, 0.75, and 1.2 μM before mixing) with fixed concentration of the 30S complex (0.25 μM before mixing), and each experiment was repeated at least twice.

The association data measured under second-order conditions were analyzed by nonlinear regression using the DynaFit program (Kuzmic, 1996). The first step was to create a simple text file of the association data in XY format (time versus change in fluorescence intensity). We used the text editor Windows Notepad to prepare the data file and saved the resulting file in the ASCII text format. The second step was to create a DynaFit script file for the analysis of the association data using Notepad. The script file contained the following sections: [task], [mechanism], [constants], [responses], [concentrations], [progress], [output], and [settings]. The "task" section instructs the program to fit the kinetic data to a model. The "mechanism" section describes the kinetic steps that might be involved in the association experiment.

We used a two-step reaction mechanism represented by a set of chemical equations written symbolically by the scheme: $30S + mRNA \Leftrightarrow 30S \cdot mRNA \Leftrightarrow 30S \cdot mRNA^*$ in DynaFit. Step 1 is the bimolecular association of mRNA with the 30S subunit with rate constants k_1 and k_{-1}. In step 2, the mRNA unfolds with rate constants k_2 and k_{-2}. The "constants" section was used to enter initial estimates of the adjustable parameters. Values for k_{-1} and k_{-2}, obtained independently from dissociation experiments, could be entered here. Question marks after these initial estimates indicate that these values will be optimized by the nonlinear fitting process. The "concentrations" section was used to enter the concentration of the reactants (mRNA and 30S subunit) in nM per liter. The script file was executed by DynaFit in the main program window using the menus "File", then selecting the submenu "Load", followed by "File", then selecting the submenu "Run." Statistical criteria and visual inspection of the residual distribution plots were used to obtain the best fits. More detailed description of the script files and the fitting procedure is available in the DynaFit manual.

2.14. Determination of equilibrium binding constants

The 30S subunits were heat activated as previously described. When initiation factors were present, the 30S subunit, initiation factors, and aminoacylated fMet-tRNA^{fMet} were incubated together for 30 min at 37° (Fig. 2.3). All complexes were incubated at 25° for 5 min immediately prior to obtaining titration measurements. All experiments were performed at 25° in polyamine buffer and GTP (1 mM). K_D determinations were performed by keeping pyrene derivatized mRNA's concentration constant in a 1-ml cuvette (Fisher 14-385-914B). Titration measurements were performed using a photon-counting fluorometer (Fluormax-P, JY Horiba). The 30S ribosomal subunit was titrated into the cuvette containing the pyrene-labeled mRNA. The sample was excited at 345 nm and the fluorescence emission was monitored at 379 nm (the bandwidth was adjusted for each mRNA to remain in the linear range of the instrument). Three scans were taken at each titration point with a standard deviation of $<4\%$ in all cases. A 5-min equilibration time was allowed between each titration measurement. Raw data, pertaining to the fluorescence intensity of 30S and mRNA interaction, was corrected by subtracting out the amount of fluorescence increase caused by the addition of the 30S subunit. This correction was performed by titrating corresponding amounts of 30S subunits into a polyamine buffer, containing 1-mM GTP and no mRNA, and subsequently measuring the fluorescence intensity at these various 30S concentrations. This value was then subtracted from the data containing the 30S·mRNA complex. The data was fit using a quadratic equation. $Y = m1 * ([m2 + m3 + X] - sqrt[(m2 + m3 + X)^2 - 4 * m3 * X])/(2 * m3)$

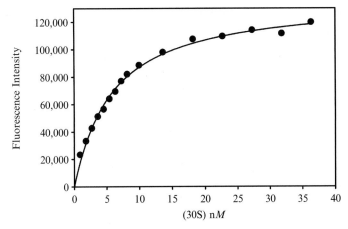

Figure 2.3 Determination of equilibrium binding constants: equilibrium binding experiments were performed by holding the concentration of mRNA constant and titrating in increasing amounts of 30S subunits. The titration experiment shown was for mRNA 4B at 1-nM concentration. 30S subunits were titrated in until the mRNA approached saturation. The K_D of the mRNA was found to be 5 nM using a quadratic fit.

Y is corrected fluorescence/maximum fluorescence
m1 is always equal to 1 (or max fluorescence if Y is not ratio)
m2 is K_D
m3 is the mRNA concentration
X is concentration of the 30S subunit being titrated

ACKNOWLEDGMENTS

This material is based upon work supported by the NSF (grant 0315780) and the NIH (grant R01 GM65265).

REFERENCES

Bartetzko, A., and Nierhaus, K. H. (1988). Mg2+/NH4+/polyamine system for polyuridine-dependent polyphenylalanine synthesis with near *in vivo* characteristics. *Methods Enzymol.* **164,** 650–658.

Brandt, R., and Gualerzi, C. O. (1991). Ribosome-mRNA contact sites at different stages of translation initiation as revealed by cross-linking of model mRNAs. *Biochimie* **73,** 1543–1549.

Cole, P. E., Yang, S. K., and Crothers, D. M. (1972). Conformational changes of transfer ribonucleic acid. Equilibrium phase diagrams. *Biochemistry* **11,** 4358–4368.

Feinberg, J. S., and Joseph, S. (2006). Ribose 2′-hydroxyl groups in the 5′ strand of the acceptor arm of P-site tRNA are not essential for EF-G catalyzed translocation. *RNA* **12,** 580–588.

Gualerzi, C. O., and Pon, C. L. (1990). Initiation of mRNA translation in prokaryotes. *Biochemistry* **29,** 5881–5889.

Holder, J. W., and Lingrel, J. B. (1975). Determination of secondary structure in rabbit globin messenger RNA by thermal denaturation. *Biochemistry* **14,** 4209–4215.

Johansson, J., Mandin, P., Renzoni, A., Chiaruttini, C., Springer, M., and Cossart, P. (2002). An RNA thermosensor controls expression of virulence genes in Listeria mono- cytogenes. *Cell* **110,** 551–561.

Kuzmic, P. (1996). Program DYNAFIT for the analysis of enzyme kinetic data: Application to HIV proteinase. *Anal. Biochem.* **237,** 260–273.

Mandal, M., Boese, B., Barrick, J. E., Winkler, W. C., and Breaker, R. R. (2003). Riboswitches control fundamental biochemical pathways in *Bacillus subtilis* and other bacteria. *Cell* **113,** 577–586.

Mandal, M., and Breaker, R. R. (2004). Adenine riboswitches and gene activation by disruption of a transcription terminator. *Nat. Struct. Mol. Biol.* **11,** 29–35.

Mandal, M., Lee, M., Barrick, J. E., Weinberg, Z., Emilsson, G. M., Ruzzo, W. L., and Breaker, R. R. (2004). A glycine-dependent riboswitch that uses cooperative binding to control gene expression. *Science* **306,** 275–279.

Milligan, J. F., Groebe, D. R., Witherell, G. W., and Uhlenbeck, O. C. (1987). Oligor- ibonucleotide synthesis using T7 RNA polymerase and synthetic DNA templates. *Nucleic Acids Res.* **15,** 8783–8798.

Powers, T., and Noller, H. F. (1991). A functional pseudoknot in 16S ribosomal RNA. *EMBO J.* **10,** 2203–2214.

Scherer, G. F., Walkinshaw, M. D., Arnott, S., and Morre, D. J. (1980). The ribosome binding sites recognized by *E. coli* ribosomes have regions with signal character in both the leader and protein coding segments. *Nucleic Acids Res.* **8,** 3895–3907.

Schneider, T. D., Stormo, G. D., Gold, L., and Ehrenfeucht, A. (1986). Information content of binding sites on nucleotide sequences. *J. Mol. Biol.* **188,** 415–431.

Shimizu, Y., Inoue, A., Tomari, Y., Suzuki, T., Yokogawa, T., Nishikawa, K., and Ueda, T. (2001). Cell-free translation reconstituted with purified components. *Nat. Biotechnol.* **19,** 751–755.

Silverman, S. K., and Cech, T. R. (1999). RNA tertiary folding monitored by fluorescence of covalently attached pyrene. *Biochemistry* **38,** 14224–14237.

Steitz, J. A. (1980). RNA-RNA interactions during polypeptide chain initiation. *In* "Ribosomes. Structure, Function and Genetics" (G. Chambliss, G. R. Craven, J. Davies, K. Davis, L. Kahan, and M. Nomura, eds.), pp. 479–496. University Park Press, Baltimore.

Studer, S. M., and Joseph, S. (2006). Unfolding of mRNA secondary structure by the bacterial translation initiation complex. *Mol. Cell* **22,** 105–115.

Torimura, M., Kurata, S., Yamada, K., Yokomaku, T., Kamagata, Y., Kanagawa, T., and Kurane, R. (2001). Fluorescence-quenching phenomenon by photoinduced electron transfer between a fluorescent dye and a nucleotide base. *Anal. Sci.* **17,** 155–160.

Walter, N. G., and Burke, J. M. (2000). Fluorescence assays to study structure, dynamics, and function of RNA and RNA-ligand complexes. *Methods Enzymol.* **317,** 409–440.

Winkler, W., Nahvi, A., and Breaker, R. R. (2002a). Thiamine derivatives bind messenger RNAs directly to regulate bacterial gene expression. *Nature* **419,** 952–956.

Winkler, W. C., and Breaker, R. R. (2003). Genetic control by metabolite-binding riboswitches. *Chembiochemistry* **4,** 1024–1032.

Winkler, W. C., Cohen-Chalamish, S., and Breaker, R. R. (2002b). An mRNA structure that controls gene expression by binding FMN. *Proc. Natl. Acad. Sci. USA* **99,** 15908–15913.

Zuker, M. (2003). Mfold web server for nucleic acid folding and hybridization prediction. *Nucleic Acids Res.* **31,** 3406–3415.

CHAPTER THREE

REAL-TIME DYNAMICS OF RIBOSOME-LIGAND INTERACTION BY TIME-RESOLVED CHEMICAL PROBING METHODS

Attilio Fabbretti, Pohl Milon, Anna Maria Giuliodori, Claudio O. Gualerzi, *and* Cynthia L. Pon

Contents

Abstract

Three protocols to perform time-resolved *in situ* probing of rRNA are described. The three methods (chemical modification with DMS and rRNA backbone cleavage by hydroxyl radicals generated by either K-peroxonitrite or Fe(II)-EDTA)

Department of Biology MCA, University of Camerino, Camerino, Italy

Methods in Enzymology, Volume 430
ISSN 0076-6879, DOI: 10.1016/S0076-6879(07)30003-7

make use of a quench-flow apparatus and exploit reactions that are faster than the interactions of ribosomal subunits with their ligands. These methods allow the investigation of the path and dynamics, in a $\cong$ 50 to 1500 ms time range, of the binding and dissociation of ribosomal ligands.

1. INTRODUCTION

Technological advancements in crystallography and analysis of diffraction patterns have resulted in an explosion of the number of high-quality, high-resolution models of important biomolecules (Berk *et al.*, 2006; Blaha, 2004 and references therein; Schuwirth *et al.*, 2005; Selmer *et al.*, 2006). However, inasmuch as these structures offer snapshots of molecules, guessing how molecules move as they carry out their functions has required assembling a flipbook of static structure.

The time-resolved chemical probing technique described enables real-time monitoring of molecular dynamics and provides a glimpse of how macromolecules move on the ribosomal surface during their function. This technique has been applied to studying the dynamics of individual steps of the translation process by monitoring the chemical reactivity of nucleotide bases in rRNA at discrete, short time-intervals following mixing the ribosome with one or more of its ligands. A standard quench-flow apparatus is used to mix the components and the chemical modification reagent as well as the quench solution to stop the modification reaction. The reactivity of the nucleotides is then determined by primer extension analysis of the rRNA.

In this article we present the successful use of this technique with three different chemical modification or cleaving reagents, namely dimethyl sulfate (DMS), K-peroxonitrite (ONOOK), and Fe(II)-EDTA. It can be envisaged that this approach may be applied to study time-dependent changes of other macromolecular structures using these or other modification reagents.

2. GENERAL STRATEGY

A standard quench-flow apparatus (BioLogic SFM-400 quench-flow apparatus) containing four syringes and two variable delay lines is employed (Fig. 3.1). The first two syringes contain the components to be mixed, the third contains the chemical modification reagent, and the fourth a quench solution to stop the reaction. The incubation time of the various components and the modification time can be varied by changing the delay lines and/or the flow parameters. As far as the choice of the reagent is concerned, this must be selected on the basis of the rate at which it can modify the target. In general the modification reaction must be no less than 10-fold faster than the interaction under study.

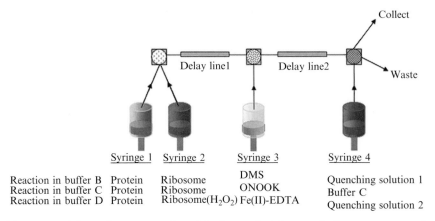

Figure 3.1 Schematic representation of the flow path of the quench–flow apparatus used in time-resolved chemical probing. The components and reagents used in the three methods of chemical probing are listed under each of the four syringes of the quench-flow apparatus.

Following the probing reactions, the rRNA is extracted from the samples and subjected to primer extension analysis to determine the region of the rRNA protected by interaction with the ligand.

To avoid damage to the quench-flow apparatus it should be extensively washed after each use with the probing reagents. Nevertheless, in our experience the O-rings are subject to rapid deterioration so that it is necessary to replace them frequently (i.e., after every 3 to 4 weeks of use).

2.1. Buffers and mixtures

Extension Mixture (for 10 reactions)

13 μl of AMV Buffer (Roche) 5×
1 μl of AMV Reverse transcriptase (Roche) 25 U/μl
6 μl of 4× NTP (1.5-mM each)

Buffers

Buffer A: 50-mM K-Hepes (pH 7.0), 100-mM KCl
Buffer B: 10-mM Mg acetate, 60-mM NH$_4$Cl, 10 mM TrisHCl (pH 7.7)
Buffer C: 7-mM Mg acetate, 60-mM NH$_4$Cl, 10 mM Hepes (pH 7.1)
Buffer D: 20-mM Na cacodylate (pH 7.4), 7-mM Mg acetate, 60-mM NH$_4$Cl
Quenching solution 1: 1:1 (v/v) solution of EtOH/β-mercaptoethanol
Quenching solution 2: EtOH, 100-mM Na acetate (pH 5.4)

2.2. Analysis of probing reactions

2.2.1. RNA extraction

The samples collected from the quench-flow apparatus after chemical modification or cleavage of the RNA are precipitated by addition of 0.1 vol of 3-M Na acetate (pH 5.4) and 3 vol of absolute EtOH. The rRNA is extracted: (1) twice with buffer-saturated phenol and 0.1% SDS, (2) twice with 1:1 (v/v) phenol/chloroform, (3) twice with chloroform, and (4) precipitated again.

2.2.2. Primer extension analysis

The extracted rRNA is stored at −80° until analysis by primer extension is performed. Sequence analysis by primer extension is carried out essentially as described (Stern *et al.*, 1988), using appropriate 5'-[^{32}P]-labeled DNA oligonucleotides complementary to sites <100 bases away from the modified base of interest. Hybridization is carried out in 4.5-μl Buffer A containing 0.6 pmol of 5'-[^{32}P]-labeled DNA oligonucleotides and ≅1 pmol of rRNA. This mixture is heated 1 min at 90° and slowly (≅7 min) cooled to 42° before addition of 2-μl extension mixture (see earlier) consisting of 1.3 μl of 5× AMV Buffer (Roche), 0.1 μl (2.5 U) AMV Reverse Transcriptase (Roche), and 0.6 μl (500-μM each) of the four dNTPs. Extension is performed at 42° for 45 min and samples are electrophoresed on a 6% (w/v) polyacrylamide sequencing gel for the appropriate period of time. The gels are exposed in a molecular imager and the band intensities quantified and normalized by reference to the intensity of the corresponding bands obtained from subunits modified in the absence of ligands. Because nonspecific stops of the reverse transcriptase may occur and discontinuities of the phosphodiester chain of the rRNA may exist, all analyses must include appropriate controls consisting of unmodified samples. The analyses are repeated at least three times and the standard deviation between the normalized intensity of the bands is used to generate error bars.

3. TIME-RESOLVED CHEMICAL PROBING WITH DMS

Dimethyl sulfate (DMS), a common reagent for chemical probing in single-stranded RNA regions, methylates adenine at N-1, guanine at N-7, and cytosine at N-3, but only the adenine and cytosine methylations inhibit the progress of reverse transcriptase while methylation of guanine at N-7 may be detected by treatment with sodium borohydride and aniline to induce strand scission. The stops of the polymerization reaction at modified A's and C's are due to the inability of the modified nucleotides of the template to form base pairs (Moazed *et al.*, 1986; Stern *et al.*, 1988).

DMS should be used with caution because it is an extremely hazardous liquid and its vapor may be fatal if inhaled. It causes severe burns to all body tissues with possible delayed injuries and burns to the lungs. In addition, based on tests with laboratory animals, it may cause cancer.

3.1. Validation of time-resolved chemical probing with DMS

The conditions for the rapid modification of the rRNA with DMS were initially defined in a time-resolved study of DMS modification of 16S rRNA bases during the formation of the inter-subunit bridges responsible for 30S to 50S subunit association (Hennelly *et al.*, 2005) and also applied to the study of the dynamics of IF3-30S (Fabbretti *et al.*, 2007) and IF1-30S (Fabbretti *et al.*, unpublished results) interaction.

Compared to "static" chemical-probing experiments, in which the reaction is allowed to proceed for an extended period of time (i.e., 30 to 120 min at 4°), a $\cong$10-fold higher concentration of DMS (i.e., 3% v/v final) must be used to ensure base modification within the short time (18 to 20 ms at 20°) allowed for the reaction. To prevent the occurrence of further modifications after this time interval which, as shown later, has been selected on the basis of empirical data (Fig. 3.2), the DMS reaction must be quenched with a 1:1 (v/v) mixture of EtOH/2-mercaptoethanol, which proved to be a satisfactory quencher having no deleterious effect toward the rRNA. In the validation tests shown in Fig. 3.2, the most suitable vol of the quench solution and the time of DMS modification for use in time-resolved studies were established. The validation tests were carried out monitoring the modification of A702, a nucleotide of 16S rRNA protected by subunit association (Hennelly *et al.*, 2005). The experimental data show that 2 vol of quench solution could efficiently stop the DMS modification and that the level of modification does not change with the use of higher vol of quench solution (Fig. 3.2A). In another experiment (Fig. 3.2B) DMS modification was carried out for different periods of time (i.e., 6 to 25 ms) with different concentrations of the reagent (3 and 4% v/v). The data obtained show that a very fast modification occurs and is complete within the first 6 ms, which represents the shortest possible delay time for delivering the large volume of the quenching solution. The level of modification remains constant between 6 and 12 ms at both DMS concentrations. However, additional modifications can be seen after 15 ms (with 4% DMS) and after 20 ms (with 3% DMS) and primer extension analyses revealed that this second, slower phase of modification probably results from nonspecific DMS-induced changes in RNA structure. Based upon the preceding data, a 3% DMS concentration, an 18- to 20-ms reaction time, and quenching with 2 vol of EtOH/2-mercaptoethanol were selected as being the most suitable conditions for time-resolved DMS modification.

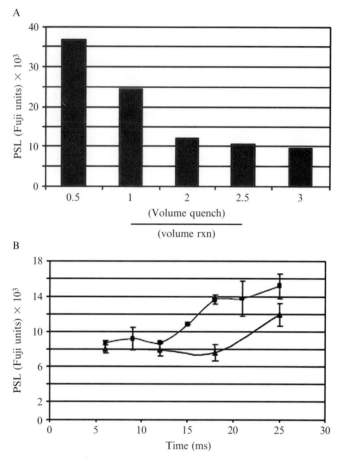

Figure 3.2 The results of the primer extension stops at A702 are expressed as the absolute intensity (photo-stimulated luminescence units) measured with a Fuji phosphor
imager as a function of (A) the ratio of the volume of the quench solution used versus the
reaction volume for a 12-ms DMS modification and (B) the times of reaction with 4%
(v/v) (■) and 3% (▲) DMS. The background of bases in unmodified control reactions is
typically < 5% of the modified value. (From Hennelly *et al.*, 2005.)

3.2. Procedure for chemical probing with DMS

Time-resolved DMS probing experiments are carried out at $20°$ in Buffer B.
Typically, syringes 1 and 2 are filled with no less than 1-ml each of
ribosomal ligand at 1.5 μM and ribosomal subunit at 1 μM, respectively.
Equal volumes (35 μl) of the two solutions are rapidly mixed and allowed to
age for 15 to 1000 ms. A 50% solution of DMS in EtOH is delivered from
syringe 3 and rapidly mixed at a 6:100 (v/v) ratio with the sample pushed
from syringes 1 and 2. The DMS reaction is finally stopped by mixing with
2 vol of quenching solution coming from syringe 4 (Fig. 3.1).

3.3. Example of time-resolved chemical probing with DMS

In the example shown in Fig. 3.3, time-resolved chemical probing with DMS was used to study the interaction between translation initiation factor IF3 and 30S ribosomal subunit (Fabbretti *et al.*, 2007). In this work, equal vol (35 *μl*) of IF3 (1.5 *μM* in syringe 1) and 30S ribosomal subunits (1 *μM* in syringe 2) were rapidly mixed and allowed to age for 60, 80, 100, 130, 280, 580, 880, or 1180 ms before being rapidly mixed with DMS as described before. After 20-ms incubation with DMS, the reaction was stopped with EtOH/2-mer-captoethanol quenching solution as described previously. Primer extension analysis of the modified 16S rRNA (Fig. 3.3A) clearly reveals specific and time-dependent effects caused by IF3 binding to the 30S ribosomal subunit in two separate regions of the ribosomal particle. The protections, which are established in a sequential order, involve initially four bases (A704, A706, C708, and A712) in the platform region near G700 and then three bases (A790, A784, and A794) located near the P-site (Fabbretti *et al.*, 2007).

4. TIME-RESOLVED PROBING WITH ONOOK

Hydroxyl radicals are generated by decomposition of peroxonitrous acid in aqueous solution at neutral pH according to the following reaction (Gotte *et al.*, 1996; King *et al.*, 1993).

$$ONOO^-K^+ + H_2O \rightarrow ONOOH + K^+ + OH^-(1);$$
$$ONOOH \rightarrow NO_2\cdot + OH\cdot(2); 2NO_2\cdot \rightarrow N_2O_4(3);$$
$$N_2O_4 + H_2O \rightarrow NO_3^- + NO_2^- + 2H^+(4)$$

Protonation of the peroxonitrite anion (1) is followed by the dissociation of its conjugate acid thereby generating hydroxyl radicals and nitrogen dioxide (2, 3) which disproportionates to form nitrite and nitrate (4). The hydroxyl radicals thus generated induce the cleavage of the RNA backbone in a way similar to that obtained by Fe(II)–EDTA (see following), as indicated by the fact that the two reagents yield almost identical cleavage patterns on the same sample of nucleic acid (King *et al.*, 1993).

4.1. Validation of time-resolved probing with ONOOK

ONOOK is quite stable as an alkaline stock solution but becomes very unstable and decays spontaneously and very rapidly upon mixing with the other reagents present in a buffered solution at pH 7.1. Therefore, under our experimental conditions ONOOK decays very rapidly (< 20 ms) and a

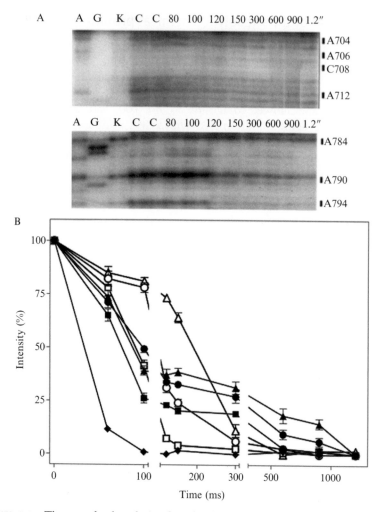

Figure 3.3 Time-resolved analysis of IF3 binding to the 30S ribosomal subunit. (A) Autoradiogram of the primer extension analysis and (B) densitometric quantification of the band intensities. Time-resolved chemical probing refers to the DMS modification in the 700 (closed symbol) and 790 (open symbol) regions. In panel B, the individual bases are indicated as A704 (▲), A706 (■), C708 (♦), A712 (●), A784 (△), A790 (○), and A794 (□). The reactivity of the individual positions is indicated as a function of time elapsed after mixing IF3 with the 30S subunit and is expressed as a percentage of the reactivity obtained before mixing. The individual lanes are indicated: 16S rRNA sequencing (A and G); 16S rRNA without DMS (K); 30S modified in the absence of factor (C); or in the presence of IF3 for the indicated times (ms or s). (From Fabbretti *et al.*, 2007.)

specific quencher for this reagent is not required. Because the duration of ONOOK-induced cleavage of rRNA depends upon the initial ONOOK concentration (Fig. 3.4), it was possible to select a reagent concentration (i.e., 6 mM) that gives no further cleavage of 16S rRNA after 20 ms of incubation.

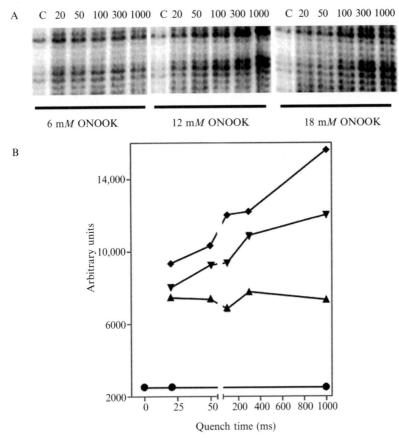

Figure 3.4 Validation of the time-resolved probing of 16S rRNA by ONOOK cleavage. Primer extension analysis of a select region (790) of 16S rRNA probed for the indicated times (expressed in ms) with ONOOK at 6 (▲), 12 (▼), and 18 mM (◆), or without reagent (●). (A) Sequencing gel. The "C" lanes contain samples incubated in the absence of ONOOK. (B) Intensity of all resolved bands shown in panel A plotted as a function of the time allowed for reagent decay (self-quenching). (From Fabbretti et al., 2007).

4.2. Preparation of ONOOK

Peroxonitrous acid is prepared essentially as described (Gotte et al., 1996; King et al., 1993) using ACS reagent grade chemicals, with the exception of the KOH, which is semiconductor grade to minimize heavy metal contaminants. Prior to use, all solutions are cooled in an ice-bath. Ten ml of 0.6-N HCl are added to 20 ml of a stirring solution of 0.6-M $NaNO_2$-0.9 M H_2O_2 followed within 2 s by 10 ml of a solution containing 1.2-N KOH and 0.4-mM diethylenetriaminepentaacetic acid (DTPA). Disproportionation of unreacted H_2O_2 is effected by immersion of a cylindrical platinum mesh electrode for 90 min at 1° with stirring. The resulting yellow ONOOK concentration normally obtained is 80 to 90 mM as determined by measuring the $A_{302 \text{ nm}}$ ($\varepsilon 302_{nm} = 1670 \ M^{-1} \ cm^{-1}$).

The ONOOK solution can be stored for up to 1 wk at $-80°$ and thawed on ice prior to the probing experiments.

4.3. Procedure for time-resolved probing with ONOOK

Time-resolved ONOOK probing experiments are carried out at $20°$ in Buffer C. Typically, syringes 1 and 2 are filled with no less than 1 ml each of ribosomal ligand at 1.5 μM and ribosomal subunit at 1 μM, respectively. Equal vol (20 μl) of the two solutions are rapidly mixed and allowed to age for 15 to 1000 ms before being rapidly mixed with 0.1 vol of a freshly prepared ONOOK solution (60 mM in syringe 3). The time allowed for the cleavage reaction (20 ms) is that required for the solution to flow through the second delay line before being pushed into a collection vessel by 10 vol of Buffer C contained in syringe 4.

4.4. Example of time-resolved probing with ONOOK

In the example shown in Fig. 3.5 equal vol (20 μl) of IF3 (1.5 μM in syringe 1) and 30S ribosomal subunits (1 μM in syringe 2) were rapidly mixed and allowed to age for 15, 30, 60, 90, 250, 500, or 1000 ms before being rapidly mixed with 0.1 vol of freshly prepared 60-mM ONOOK solution as described previously. After 20 ms (which represents the time required for the cleavage reaction), the solution was pushed into a collection vessel by 10 vol of Buffer C contained in syringe 4.

As seen in Fig. 3.5, IF3 protects from cleavage the G700 and A790 regions of 16S rRNA located in the platform and P-site region of the 30S subunit, respectively. As in the previous case (Fig. 3.3), it is clear that IF3-protection of the platform bases occurs before the protection of the bases of the A790 region. Thus, the results obtained by time-resolved probing with DMS and ONOOK are fully consistent not only for the bases that are protected but also for the order in which they become protected.

5. TIME-RESOLVED CHEMICAL PROBING WITH Fe(II)-EDTA

This method exploits hydroxyl radicals, generated by the Fenton-Haber-Weiss reaction from a free Fe(II)-EDTA complex, (i.e., Fe(II) + $H_2O_2 \rightarrow$ Fe(III) + $\bullet OH$ + OH^-) to induce cleavages of the RNA backbone, thereby probing its *in situ* accessibility (Fenton, 1894; Haber and Weiss, 1934). In fact, hydroxyl radicals attack the ribose hydrogens at the C1' and C4' positions leading to cleavage of the nucleotide chain (Latham and Cech, 1989).

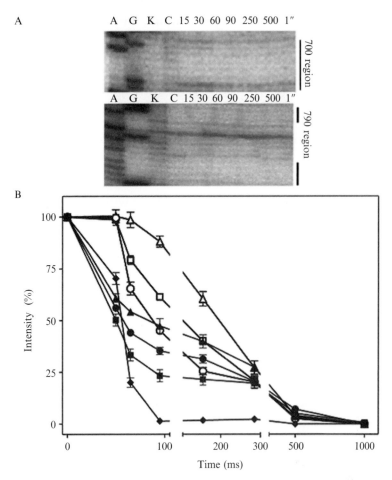

Figure 3.5 Time-resolved analysis of IF3 binding to the 30S ribosomal subunit. Time-resolved *in situ* cleavage of 16S rRNA by ONOOK as a function of time elapsed after mixing IF3 with the 30S subunit. (A) Autoradiogram of the primer extension analysis. The individual lanes correspond to 16S rRNA sequencing (A and G), 16S rRNA without ONOOK (K), 30S modified in the absence of IF3 (C) or in the presence of IF3 for the indicated times (ms or s). (B) Densitometric quantification of the A704 (▲), A706 (■), C708 (◆), A712 (●), A784 (△), A790 (○), and A794 (□) band intensities. For better comparison, the bases within the G700 and A790 regions are indicated by closed and open symbols, respectively. The intensity of each band is expressed as a percentage of that obtained before mixing 30S subunit and IF3. (From Fabbretti *et al.*, 2007.)

Because susceptibility of the RNA to attack by hydroxyl radicals is independent of its secondary structure, most of the RNA structure including double-stranded regions can be monitored, with the exception of those ribose moieties directly involved in molecular interactions (Huttenhofer and Noller, 1992; Latham and Cech, 1989).

5.1. Validation of time-resolved probing with Fe(II)-EDTA

The conditions used for the time-resolved chemical probing by Fe(II)-EDTA–induced RNA cleavage and for the subsequent quenching are the same as those defined and validated by Shcherbakova *et al.* (2006). These authors have shown that using 5-mM Fe(II)-EDTA and 0.15% H_2O_2, the hydroxyl radicals are generated within 5 ms, whereas the results shown in Fig. 3.6 demonstrate that 23S rRNA cleavage proceeds efficiently for a least 100 ms and that an incubation of 20 to 30 ms, followed by quenching with EtOH/Na acetate (see following) yields a workable level of cleavage.

5.2. Procedure for time-resolved probing with Fe(II)-EDTA

In a typical time-resolved chemical probing experiment with Fe(II)-EDTA, syringes 1 and 2 are filled with no less than 1-ml each of 1.5-μM ribosomal ligand in Buffer D and 1-μM ribosomal subunit in Buffer D containing 0.45% H_2O_2, respectively. Equal vol (20 μl) of the two solutions are rapidly mixed and allowed to age for 15 to 1000 ms before being rapidly mixed with 20 μl of a freshly prepared 15-mM Fe(II)-EDTA solution contained in syringe 3. The cleavage reaction is allowed to proceed for 20 to 30 ms before

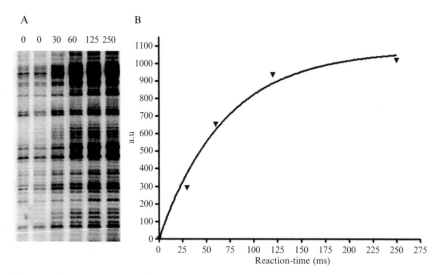

Figure 3.6 Time-resolved 23S rRNA cleavage by hydroxyl radicals generated by Fe (II)-EDTA. Time-resolved probing was carried out as described in the text by rapid mixing of 50S ribosomal subunits in Buffer D containing 0.15% H_2O_2 with 5-mM Fe(II)-EDTA. The samples were allowed to react for the indicated times before quenching with EtOH/Na acetate, sample collection, and primer extension analysis. (A) Primer extension analysis of the 2600 region of 23S rRNA as a function of the time of incubation with Fe(II)-EDTA and (B) quantification of the total intensity of the bands shown in panel A.

addition of 3 vol (180 μl) of quenching solution 2 contained in syringe 4. The entire procedure is carried out at 20°.

5.3. Example of time-resolved probing with Fe(II)-EDTA

In the example presented in Fig. 3.7, the time-resolved chemical probing technique with Fe(II)-EDTA was used to study the interaction between translation initiation factor IF2 and the 50S ribosomal subunit. The experimental details are the same as described in the previous section. Thus, equal vol (20 μl) of IF2 (1.5 μM in syringe 1) and 50S ribosomal subunits (1 μM in syringe 2) are rapidly mixed and allowed to age for 15, 60, 90, 220, 470, or 970 ms before addition of 20 μl of a freshly prepared Fe(II)-EDTA solution contained in syringe 3. After 30 ms, the cleavage reaction was stopped by mixing with 180 μl of quenching solution 2 from syringe 4 and the collected samples were analyzed by primer extension. As seen from Fig. 3.7, IF2 binding to 50S ribosomal subunits causes first protection ($\cong$50% of protection in 90 ms) in two regions of 23S rRNA (2640 to 2643 and 2665 to 2667),

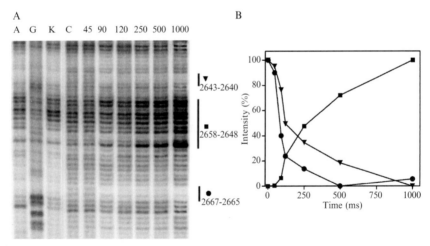

Figure 3.7 Time-resolved analysis of IF2 binding to the 50S ribosomal subunit. Time-resolved *in situ* cleavage of 23S rRNA by Fe(II)-EDTA was performed as a function of time elapsed after mixing IF2 with the 50S subunit. (A) Autoradiogram of the primer extension analysis in the 2650 region of the rRNA. The individual lanes correspond to sequencing reactions (A and G), 23S extracted from 50S incubated Fe(II)-EDTA after the indicated times following the mixing of the 50S with IF2. (K) contains 23S rRNA extracted from 50S incubated in the absence of Fe(II)-EDTA. (C) contains 23S rRNA extracted from 50S incubated in the absence of IF2. (B) Densitometric quantification of the band intensities in three different regions 2640–2643 (▲), 2648 to 2658 (■), and 2665 to 2667 (●). The intensity of each band is expressed as a percentage of that obtained before mixing 50S subunit and IF2.

and then ($\cong$ 50% of protection in 250 ms) exposure in the 2648 to 2658 region of 23S rRNA.

ACKNOWLEDGMENTS

The financial support of PRIN 2005 grants from the Italian MIUR to COG and CLP is gratefully acknowledged.

REFERENCES

Berk, V., Zhang, W., Pai, R. D., and Cate, J. H. D. (2006). Structural basis for mRNA and tRNA positioning on the ribosome. *Proc. Natl. Acad. Sci. USA* **103,** 15830–15834.

Blaha, G. (2004). Structure of the ribosome. *In* "Protein Synthesis and Ribosome Structure" (K. H. Nierhaus and D. N. Wilson, eds.), pp. 53–84. Wiley-VCH Verlag, Weinheim.

Fabbretti, A., Pon, C. L., Hennelly, S. P., Hill, W. E., Lodmell, J. S., and Gualerzi, C. O. (2007). The real time path of translation factor IF3 onto and off the ribosome. *Mol. Cell* **25,** 285–296.

Fenton, H. J. H. (1894). Oxidation of tartaric acid in presence of iron. *J. Chem. Soc.* **6,** 899–910.

Gotte, M., Parquet, R., Isel, C., Anderson, V. E., and Brenowitz, M. (1996). Probing the higher order structure of RNA with peroxonitrous acid. *FEBS Lett.* **390,** 226–228.

Haber, F., and Weiss, J. (1934). The catalytic decomposition of hydrogen peroxide by iron salts. *Proc. R. Soc. Lond. A.* **147,** 332–351.

Hennelly, S. P., Antoun, A., Ehrenberg, M., Gualerzi, C. O., Knight, W., Lodmell, J. S., and Hill, W. E. (2005). A time-resolved investigation of ribosomal subunit association. *J. Mol. Biol.* **346,** 1243–1258.

Huttenhofer, A., and Noller, H. F. (1992). Hydroxyl radical cleavage of tRNA in the ribosomal P site. *Proc. Natl. Acad. Sci. USA* **89,** 7851–7855.

King, P., Jamison, A. E., Strahs, D., Anderson, V. E., and Brenowitz, M. (1993). "Foot-printing" proteins on DNA with peroxonitrous acid. *Nucleic Acids Res.* **10,** 2473–2478.

Latham, J. A., and Cech, T. R. (1989). Defining the inside and outside of a catalytic RNA molecule. *Science* **245,** 276–282.

Moazed, D., Stern, S., and Noller, H. F. (1986). Rapid chemical probing of conformation in 16S ribosomal RNA and 30S ribosomal subunits using primer extension. *J. Mol. Biol.* **187,** 399–416.

Schuwirth, B. S., Borovinskaya, M. A., Hau, C., Zhang, W., Vila-Sanjurjo, A., Jolton, J., and Cate, J. H. D. (2005). Structures of the bacterial ribosome at 3.5 Å resolution. *Science* **310,** 827–834.

Selmer, M., Dunham, C. M., Murphy, F. V., IV, Weixlbaumer, A., Petry, S., Kelley, A. C., Weir, J. R., and Ramakrishnan, V. (2006). Structure of the 70S ribosome complexed with mRNA and tRNA. *Science* **313,** 1935–1942.

Shcherbakova, I., Mitra, S., Beer, H. R., and Brenowitz, M. (2006). Fast Fenton footprint-ing: A laboratory-based method for the time-resolved analysis of DNA, RNA and proteins. *Nucleic Acids Res.* **34,** e48.

Stern, S., Moazed, D., and Noller, H. F. (1988). Structural analysis of RNA using chemical and enzymatic probing monitored by primer extension. *Methods Enzymol.* **164,** 481–489.

CHAPTER FOUR

Overexpression and Purification of Mammalian Mitochondrial Translational Initiation Factor 2 and Initiation Factor 3

Domenick G. Grasso, Brooke E. Christian, Angela Spencer, *and* Linda L. Spremulli

Contents

Department of Chemistry, University of North Carolina at Chapel Hill, Chapel Hill, North Carolina

Methods in Enzymology, Volume 430
ISSN 0076-6879, DOI: 10.1016/S0076-6879(07)30004-9

Abstract

Two mammalian mitochondrial initiation factors have been identified. Initiation factor 2 (IF2$_{mt}$) selects the initiator tRNA (fMet-tRNA) and promotes its binding to the ribosome. Initiation factor 3 (IF3$_{mt}$) promotes the dissociation of the 55S mitochondrial ribosome into subunits and may play additional, less-well-understood, roles in initiation complex formation. Native bovine IF2$_{mt}$ was purified from liver a number of years ago. The yield of this factor is very low making biochemical studies difficult. The cDNA for bovine IF2$_{mt}$ was expressed in *Escherichia coli* under the control of the T7 polymerase promoter in a vector that provides a His$_6$-tag at the C-terminus of the expressed protein. This factor was expressed in *E. coli* and purified by chromatography on Ni-NTA resins. The expressed protein has a number of degradation products in partially purified preparations and this factor is then further purified by high-performance liquid chromatography or gravity chromatography on anion exchange resins.

IF3$_{mt}$ has never been purified from any mammalian system. However, the cDNA for this protein can be identified in the expressed sequence tag (EST) libraries. The portion of the sequence encoding the region of human IF3$_{mt}$ predicted to be present in the mitochondrially imported form of this factor was cloned and expressed in *E. coli* using a vector that provides a C-terminal His$_6$-tag. The tagged factor is partially purified on Ni-NTA resins. However, a major proteolytic fragment arising from a defined cleavage of this protein is present in these preparations. This contaminant can be removed by a single step of high-performance liquid chromatography on a cation exchange resin. Alternatively, the mature form of IF3$_{mt}$ can be purified by two sequential passes through a gravity S-Sepharose column.

1. INTRODUCTION

Mammalian mitochondria possess a translational system with a number of unusual features distinguishing it from other protein biosynthetic systems. These distinctions include protein–rich and RNA–poor ribosomes (Takemoto *et al.*, 1999) and mRNAs that lack 5′ and 3′ untranslated regions (Montoya *et al.*, 1981; Ojala *et al.*, 1981). Despite these differences, translational factors with homology to prokaryotic initiation factor 2 (IF2) and initiation factor 3 (IF3) have been detected. Furthermore, the cDNAs and/or the proteins for the three elongation factors have been characterized from bovine and human mitochondria (Bhargava *et al.*, 2004; Spremulli *et al.*, 2004; Woriax *et al.*, 1995; Xin *et al.*, 1995) as have release factor 1 and the ribosome-recycling factor (Zhang and Spremulli, 1998).

IF2$_{mt}$ promotes the binding of the initiator tRNA (fMet-tRNA^{Met}) to the small subunit of the ribosome (Liao and Spremulli, 1990, 1991). Binding is stimulated by the presence of GTP and an mRNA such as poly(A,U,G). This factor was first purified from bovine liver mitochondria using conventional chromatography followed by high-performance liquid chromatography (HPLC). It functions as a monomer with a molecular weight of 74 kDa. The sequence of mammalian IF2$_{mt}$ is easily recognized in the expressed sequence tag (EST) database because this factor is about 30 to 40% identical to the bacterial factors (Ma and Spremulli, 1995). Both bovine and human IF2$_{mt}$ have been cloned and expressed in *Escherichia coli*. The current work focuses on the human factor.

Native IF3$_{mt}$ has never been detected or purified from mammalian mitochondria. This observation probably reflects the low abundance of IF3$_{mt}$ in mitochondria. Further, detection of the cDNA for IF3$_{mt}$ in the EST databases is a challenge due to the relatively low sequence conservation of this factor (20 to 25% identity to prokaryotic IF3) (Koc and Spremulli, 2002). Despite this difficulty, the cDNA for human IF3$_{mt}$ has been cloned and the factor expressed in *E. coli* (Koc and Spremulli, 2002). This factor is expected to have a molecular weight of 29 kDa after removal of the predicted mitochondrial import sequence. The human factor is the focus of this work.

 ## 2. MATERIALS

2.1. Growth media

Cells are grown in Luria Broth (LB) containing 10 g tryptone, 5 g yeast extract, and 10 g NaCl per liter. The pH is adjusted to 7.5 with 10 N NaOH. As appropriate, the media is supplemented with 50 μg/ml kanamycin and 100 μg/ml ampicillin.

2.2. Common buffers

Buffer I: 20 mM Hepes–KOH (pH 7.6), 10 mM MgCl$_2$, 50 mM KCl, 10% glycerol, and 6 mM β-mercaptoethanol
Mg5 Wash Buffer: 50 mM Tris–HCl (pH 7.8), 5 mM MgCl$_2$, and 80 mM KCl

Mg7.5 Wash Buffer: 50 mM Tris-HCl (pH 7.8), 7.5 mM MgCl$_2$, and 35 mM KCl

2.3. Buffers for IF2$_{mt}$

Buffer 2A: 20 mM Tris-HCl (pH 7.6), 10 mM MgCl$_2$, 5 mM β-mercaptoethanol, and 0.1 mM phenylmethylsulfonyl fluoride

Buffer 2B: 20 mM Tris-HCl (pH 7.6), 10 mM MgCl$_2$, 1 M KCl, 10 mM imidazole, 5 mM β-mercaptoethanol, and 0.1 mM phenylmethylsulfonyl fluoride

Buffer 2C: 20 mM Tris-HCl (pH 7.6), 10 mM MgCl$_2$, 50 mM KCl, 150 mM imidazole, 5 mM β-mercaptoethanol, and 0.1 mM phenylmethylsulfonyl fluoride

2.4. Buffers for IF3$_{mt}$

Buffer 3A: 50 mM Tris-HCl (pH 7.8), 7 mM MgCl$_2$, 10% glycerol, 0.1 mM EDTA, 100 mM KCl, 6 mM β mercaptoethanol, 0.5 mM phenylmethylsulfonyl fluoride, and 0.8% Triton X-100

Buffer 3B: 50 mM Tris-HCl (pH 7.8), 7 mM MgCl$_2$, 10% glycerol, 1 M KCl, 10 mM imidazole, and 7 mM β-mercaptoethanol

Buffer 3C: 50 mM Tris-HCl (pH 7.8), 7 mM MgCl$_2$, 10% glycerol, 40 mM KCl, 150 mM imidazole, and 7 mM β-mercaptoethanol

2.5. Reagents

Dithiothreitol (DTT), S-Sepharose, DEAE-Sepharose, yeast tRNA (Type X Sigma number R9001), isopropyl-1-thio-β-D-galactopyranoside (IPTG), GTP lithium salt, poly(A,U,G), ampicillin sodium salt, kanamycin sulfate, lysozyme, DNase I, and folinic acid (calcium salt) are from Sigma. Triton X-100 is from Promega. Ni-NTA Agarose is obtained from Qiagen. Polyacrylamide (30%) with (0.8%) bis-acrylamide (37.5:1) is from National Diagnostics. Routine laboratory chemicals are obtained from Fisher Scientific. Oligonucleotide primers are made at the Lineberger Comprehensive Cancer Research Center at the University of North Carolina at Chapel Hill. Restriction enzymes are purchased from New England Biolabs. Nitrocellulose membrane filter paper (type HAWG, 0.45-μm pore size) is from Millipore. The TSKgel DEAE 5-PW HPLC column (7.5 cm × 7.5 mm) and the TSKgel SP-5PW column (7.5 cm × 7.5 mm) are from TosoHaas (Japan). Prior to application to the HPLC columns, protein samples are filtered through 0.45 μm HT Tuffryn membrane filters obtained from Gelman Laboratory. [^{35}S]Met is purchased from Perkin Elmer. The Benchmark^{TM} Protein Ladder is from Invitrogen.

2.6. Preparation of reagents

Yeast tRNA$_i^{Met}$ and tRNA$_m^{Met}$ are separated and partially purified from crude yeast tRNA by chromatography on BD-cellulose (Walker and RajBhandary, 1972). *E. coli* aminoacyl-tRNA synthetases free of tRNA are partially purified as described by Muench and Berg (1966). These preparations also contain the Methionyl-tRNA transformylase. [^{35}S]fMet-tRNA$_i^{Met}$ is prepared as described (Graves and Spremulli, 1983; Walker and RajBhandary, 1972) in the presence of 60 μM folinic acid. Before use, 12.5 mM folinic acid is incubated in 0.1 N HCl containing 50 mM dithiothreitol for 12 h at room temperature. This sample is then diluted 10-fold with water and stored at $-20°$ in the dark until use. *E. coli* ribosomes are prepared as described (Graves and Spremulli, 1983; Graves *et al.*, 1980). Bovine liver is obtained from a local slaughterhouse and mitochondria are prepared as described (Matthews *et al.*, 1982). Mitochondrial 55S ribosomes and 28S ribosomal subunits are prepared by sucrose density gradient centrifugation as described (Spremulli, 2007). *E. coli* initiation factors IF1, IF2, and IF3 are prepared as His-tagged constructs according to the procedure described previously (Koc and Spremulli, 2002). Antibodies against recombinant IF2$_{mt}$ were raised in rabbits (Ma and Spremulli, 1996). These antibodies were used at a 0.001 dilution in Westerns as described (Spencer and Spremulli, 2005).

2.7. The high-performance liquid chromatography (HPLC) system

The HPLC columns are attached to a two-pump system by Rainin. These pumps are connected by a mixer holding a volume of 1.2 ml. The pumps are controlled by a Macintosh computer running the MacRabbit software from Rainin. Each sample is filtered through a 0.45 μm filter prior to application to the HPLC column.

3. ASSAYS FOR IF2$_{MT}$ AND IF3$_{MT}$

3.1. IF2$_{mt}$: Principle

The activity of IF2$_{mt}$ is determined by measuring its ability to stimulate the binding of [^{35}S]fMet-tRNA$_i^{Met}$ to the P-site of either *E. coli* or mitochondrial ribosomes.

3.2. Assays of IF2$_{mt}$ on *E. coli* ribosomes

Reaction mixtures (0.1 ml) contain 50 mM Tris-HCl (pH 7.8), 5 mM MgCl$_2$, 80 mM KCl, 0.25 mM GTP, 1.25 mM phospho(enol)pyruvate, 0.37 U of pyruvate kinase, 1 mM DTT, 12.5 μg poly(A,U,G), 0.06 μM

(6 pmol) [^{35}S]fMet–tRNA [approximately 70,000 counts per min (cpm)/ pmol], 0.24 μM (24 pmol) *E. coli* IF3, 0.24 μM (58 μg) *E. coli* 70S ribosomes, and the various amounts of IF2$_{mt}$. Samples are incubated at 37° for 15 min. After incubation, each sample is rapidly diluted with 3 to 4 ml of cold Mg5 Wash Buffer and filtered through a nitrocellulose membrane that has been wetted with cold Mg5 Wash Buffer. Filtration is facilitated by use of a gentle vacuum created by a pump. The filters are washed with 3 aliquots (3 to 4 ml each) of cold Mg5 Wash Buffer with the vacuum on. Filters are dried at 100° for 5 to 7 min and counted in 5 ml toluene containing 5 g/liter 2,4 diphenyloxazole (PPO) scintillation cocktail. Background binding (<0.05 pmol) obtained in the absence of IF2$_{mt}$ is subtracted from each value. One unit is defined as the binding of 1 pmol of [^{35}S]fMet–tRNA to *E. coli* ribosomes under the assay conditions described.

3.3. Assays of IF2$_{mt}$ on *Bos taurus* (Bovine) 55S mitochondrial ribosomes

Reaction mixtures (0.1 ml) contain 50 mM Tris–HCl (pH 7.8), 7.5 mM MgCl$_2$, 35 mM KCl, 0.25 mM GTP, 1.25 mM phospho(enol)pyruvate, 0.37 U of pyruvate kinase, 1 mM DTT, 12.5 μg poly(A,U,G), 0.06 μM (6 pmol) [^{35}S]fMet–tRNA (approximately 70,000 cpm/pmol), 0.2 μM IF3$_{mt}$, 0.06 μM (6 pmol) 55S mitochondrial ribosomes (1 A$_{260}$ is 32 pmol of 55S ribosomes), and the desired amount of IF2$_{mt}$. Samples are incubated at 37° for 15 min and then treated as previously described except that Mg7.5 Wash Buffer is used.

3.4. Assays of IF2$_{mt}$ on *Bos taurus* (Bovine) 28S ribosomal subunits

Reaction mixtures (0.1 ml) contain 50 mM Tris–HCl (pH 7.6), 35 mM KCl, 0.1 mM spermine, 1 mM DTT, 7.5 mM MgCl$_2$, 0.25 mM GTP, 1.25 mM phosphoenolpyruvate, 0.37 U of pyruvate kinase, 0.06 μM (6 pmol) [^{35}S] fMet–tRNA, 0.068 μM 28S subunits (1 A$_{260}$ is 77 pmol of 28S subunits), 12.5 μg poly(A,U,G), and the desired amount of IF2$_{mt}$. Reaction mixtures are incubated at 27° for 20 min, and the amount of [^{35}S]fMet–tRNA bound to the 28S subunit is determined using the filter-binding assay described earlier and the Mg7.5 Wash Buffer.

3.5. IF3$_{mt}$: Principle

IF3$_{mt}$ activity is detected by its ability to promote the efficient binding of [^{35}S] fMet–tRNA to ribosomes in the presence of saturating amounts of IF2$_{mt}$. This assay measures primarily the ability of IF3$_{mt}$ to promote the dissociation of the ribosome into its subunits.

3.6. Assay of IF3$_{mt}$ on *E. coli ribosomes*

The standard assay (0.1 ml) contains 50 mM Tris-HCl (pH 7.8), 60 to 80 mM KCl, 5 mM MgCl$_2$, 0.25 mM GTP, 1 mM phospho(enol)pyruvate, 0.9 U of pyruvate kinase, 1 mM DTT, 12.5 μg poly(A,U,G), 0.055 μM (5.5 pmol) [^{35}S]fMet-tRNA (60,000 to 70,000 cpm/pmol), 0.4 μM *E. coli* ribosomes, a saturating amount (0.1 μM, 10 pmol) of IF2$_{mt}$, and the desired amount of IF3$_{mt}$ generally ranging from 0.01 to 0.1 μM (1 to 10 pmol). Samples are incubated at 37° for 15 min. After incubation, each sample is rapidly diluted with 3 to 4 ml of cold Mg5 Wash Buffer and filtered through a nitrocellulose membrane that has been wetted with cold Mg5 Wash Buffer. Filtration is facilitated by use of a gentle vacuum provided by a pump. The filters are washed with 3 aliquots (3 to 4 ml each) of cold Mg5 Wash Buffer with the vacuum on. Background binding (generally 0.2 to 0.5 pmol) obtained in the absence of IF3$_{mt}$ is subtracted from each value. One unit is defined as the binding of 1 pmol of [^{35}S]fMet-tRNA to ribosomes under the aforementioned assay conditions described.

3.7. Assay of IF3$_{mt}$ on bovine mitochondrial ribosomes

Reaction mixtures (0.1 ml) contain 50 mM Tris-HCl (pH 7.6), 0.1 mM spermine, 35 mM KCl, 4.5 mM MgCl$_2$, 0.25 mM GTP, 1 mM DTT, 1 mM phospho(enol)pyruvate, 0.9-U of pyruvate kinase, 12.5 μg poly(A,U,G), 0.42 μM IF2$_{mt}$, 0.06 μM [^{35}S]fMet-tRNA, 0.04 to 0.06 μM (~4 to 6 pmol) mitochondrial 55 S (1 A$_{260}$ U is 32 pmol of 55S ribosomes), a saturating amount of IF2$_{mt}$ (~0.1 μM, 10 pmol), and the desired amount of IF3$_{mt}$ (generally 0.005 to 0.1:M). These initiation complex formation assays are incubated at 37° for 15 min and analyzed as described using Mg5 Wash Buffer in which the concentration of KCl has been reduced to 40 mM. All of the values reported have been corrected for the amount of radioactivity retained on the filters in the absence of IF3$_{mt}$ (~0.1 pmol).

4. PURIFICATION OF HIS-TAGGED IF2$_{MT}$

The purification of native bovine IF2$_{mt}$ has been described previously (Liao and Spremulli, 1990, 1991). The amount of protein that can be obtained from tissues is quite small reflecting the low abundance of this factor. It is estimated that there are only 10 to 20 copies of IF2$_{mt}$ per mitochondrion. The N-terminus of bovine IF2$_{mt}$ is blocked. Hence, MitoProtII was used to predict the cleavage site for the removal of the mitochondrial import signal (Claros and Vincens, 1996). The mature form of bovine IF2$_{mt}$ (Accession Number NP_776818) is predicted to begin at amino acid residue 78.

The region of $IF2_{mt}$ encompassing the amino acids predicted to be present in the mature form of the protein (amino acids 78 to 727) has been cloned into pET-21(+) (Novagen) using the NdeI and XhoI restriction sites. This vector provides an ampicillin-resistance gene and directs the synthesis of a vector-derived His_6-tag at the C-terminus of the expressed protein (Spencer and Spremulli, 2005). This construct was transformed into an *E. coli* BL21(DE3) strain carrying the pArgU218 plasmid. This plasmid (kindly provided by Dr. Yamada, Mitsubishi Chemical Corp., Yokohama, Japan) provides the gene for the isoacceptor of $tRNA^{Arg}$ recognizing the AGA and AGG codons. It carries a kanamycin resistance marker. Use of the Stratagene strain BL21(DE3) RIL should provide the needed isoacceptor as well.

4.1. Growth of cells, induction, and preparation of cell extracts

A sample of a frozen stock of *E. coli* BL21 (DE3) carrying the coding region for mature $IF2_{mt}$ in pET21(+) is used to inoculate 20 ml of LB containing 100 μg/ml ampicillin and 50 μg/ml kanamycin. This sample (20 ml) is used to inoculate 2 liters of LB media containing 100 μg/ml ampicillin and 50 μg/ml kanamycin. These cells are grown with vigorous shaking (about 200 rpm) at 37° to an A_{595} of 0.6 to 0.8. Expression is induced by the addition of 0.1 mM IPTG and is carried out at 37° for 5 h or at 27° overnight. The expression of $IF2_{mt}$ at these two different temperatures is comparable. Cells are harvested by centrifugation in 1 liter bottles in a Sorvall RC3B centrifuge at 5000 rpm ($7300 \times g_{max}$) for 25 min at 4°. The cell pellets are resuspended in buffer containing 20 mM Tris-HCl (pH 7.6) and 10 mM $MgCl_2$ and collected by centrifugation in 50 ml falcon tubes in a RC5B Sorvall centrifuge using the SS34 rotor at 7500 rpm ($6700 \times g_{max}$) for 25 min at 4°. The supernatant is discarded and residual liquid carefully drained from the cell pellets. The pellets are then fast frozen in a dry ice-isopropanol bath and stored at −80°.

4.2. Disruption by alumina grinding

Frozen cell pellets (typically ~3 to 5 g/liter cell culture) are ground with an equal mass of alumina (Type A-5, Sigma) until completely thawed and then ground with an identical additional mass of alumina for 15 min at 4°. This grinding step is carried out in a cold mortar sitting on ice. When the cells are adequately disrupted, the mixture is viscous and makes a cracking sound when ground with the pestle. Five vol of cold Buffer 2A are added and the suspension is transferred to a 40 ml Nalgene centrifuge tube using a portion of the buffer to rinse out the mortar. The disrupted cells are subjected to centrifugation at 10,000 rpm ($12,000 \times g_{max}$) for 15 min at 4 in a Sorvall RC-5B centrifuge using a SS34 rotor. The alumina and a significant amount of cell debris are removed in this step. The supernatant is removed and its vol measured. DNase I is added to the supernatant to a final concentration of 5 $\mu g/ml$. This sample is then subjected to centrifugation at 17,000 rpm ($\sim 30,000 \times g_{max}$) for 40 min at 4° in a Sorvall SS-34 rotor. The supernatant is adjusted to 1 M NH$_4$Cl by the addition of solid NH$_4$Cl. The NH$_4$Cl is added to the sample slowly while it is stirred at 4° and this mixture is stirred for an additional 15 min at 4° before being treated with Ni-NTA resin as described later.

4.3. Disruption by sonication

In this procedure, if the cell pellets (4 to 8 g) were previously frozen, they are thawed on ice before proceeding. Five vol (5 ml/g cells) of ice-cold Buffer 2A are added to the cell pellets and they are resuspended by gentle mixing. Lysozyme (~ 1 mg) is added to the mixture followed by DNase I (to 5 $\mu g/ml$). The mixture is placed in an ice bath and sonicated in 1 s bursts with 9 s off for 7 cycles (70 s total) using a Branson SONIFER Cell Disruptor 185 with a power setting of 9 (130 W). The solution is then adjusted to a final concentration 1 M NH$_4$Cl by the addition of solid NH$_4$Cl while stirring. The sample is stirred for an additional 15 min at 4° after the salt concentration has been adjusted. The solution is subjected to centrifugation in a Beckman L8-70 ultracentrifuge using a Sorvall TFT 50.38 rotor for 1 h at 28,000 rpm ($71,000 \times g_{ave}$) at 4°. The supernatant is removed and further processed with Ni-NTA resin as described next.

4.4. Purification of IF2$_{mt}$: Ni-NTA step

The sample from alumina grinding or sonication (~ 50 ml) is mixed for 40 min at 4° with a 50% slurry of Ni-NTA agarose resin (0.2 ml slurry of resin per g original cell weight) equilibrated in Buffer 2B. Incubation is carried out on a shaking platform so that the sample is mixed gently during incubation with the resin. This mixture is transferred to a 5 ml polypropylene column (Qiagen). The resin is washed with an excess of Buffer 2B (at least 50 ml) and IF2$_{mt}$ is then eluted from the resin using ~ 10 to 15 ml Buffer 2C (1 to 2 ml/g original cell weight). This step is done by three sequential elutions of ~ 5 ml each. Each aliquot of Buffer 2C is incubated with the resin for 15 min at 4° before it is collected. These three aliquots are combined. This approach dilutes the IF2$_{mt}$ sample, which reduces its tendency to precipitate. The eluted protein is dialyzed against a 100-fold excess of Buffer I at 4° for 1.5 h with one change of buffer. In general, this sample contains some full-length IF2$_{mt}$ but a number of other protein bands are observed on SDS-polyacrylamide gel electrophoresis (Fig. 4.1, lane Ni-NTA). Many of these bands represent degradation products of

A   B
kDa   Ni-NTA   Western

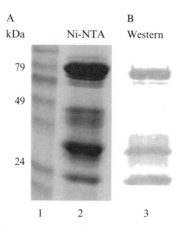

79

49

24

1   2   3

Figure 4.1 Spectrum of proteins present in Ni-NTA-purified preparations of IF2$_{mt}$. (A) A sample (10 μg) of the Ni-NTA-purified IF2$_{mt}$ was run on a 10% SDS-polyacrylamide gel and stained with Coomassie Blue (lane 2). Lane 1 contains the Benchmark^{TM} Protein Ladder. (B) Western blot (lane 3) of the partially purified IF2$_{mt}$ preparation showing that many of the contaminating bands arise from proteolytic degradation of the full-length product. (See color insert.)

IF2$_{mt}$ as indicated by Western analysis using antibodies raised against this factor (Fig. 4.1, lane 3) (Ma and Spremulli, 1996).

4.5. Purification of IF2$_{mt}$ by HPLC

The dialyzed sample from the Ni-NTA resin ($\sim$5 mg protein, about 10 ml) is first passed through a 0.45 μm filter and then injected into a 10 ml loop on the HPLC injection port. If the sample tends to precipitate at this concentration, it is diluted to $\sim$0.5 mg/ml. The sample is applied to a TSKgel DEAE 5-PW HPLC column (7.5 cm $\times$ 7.5 mm, TosoHaas Inc., Japan) equilibrated in Buffer I at a flow rate of 0.5 ml/min. This step is carried out at 4° controlled by the Macintosh computer sitting outside a cold box. The column is washed for at least 20 min with Buffer I and then developed with a linear gradient (50 ml) from 0.05 to 0.25 M KCl in Buffer I at a flow rate of 0.5 ml/min. The absorbance at 280 nm is monitored with an ISCO UA-6 UV-Visible detector and a chart recorder set to the 0.5 scale (Fig. 4.2A). The flow cell has a path length of 0.5 cm. Fractions (0.5 ml) are collected in Eppendorf tubes using an ISCO RETRIEVER 500 fraction collector. The Eppendorf tubes have the lids removed and are balanced in 13 $\times$ 100 mm glass test tubes. The fractions are fast frozen in a dry ice-isopropanol bath and stored at -70°. Fractions containing IF2$_{mt}$ are identified by 10% SDS-polyacrylamide gel electrophoresis and staining with Coomassie Blue (Fig. 4.2B). The presence of IF2$_{mt}$ is verified by Western blotting.

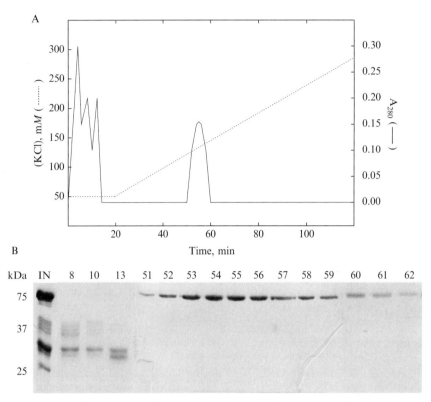

Figure 4.2 Purification of the Ni-NTA preparation of IF2$_{mt}$ on TSKgel DEAE-5PW HPLC. A sample of partially purified IF2$_{mt}$ was subjected to chromatography on HPLC as described in the text. (A) The absorbance pattern at 280 nm was monitored during chromatography with an ISCO UA-6 Absorbance detector on a 0.5 absorbance scale using a 5 mm flow cell. (B) Aliquots (16 μl) of the indicated fractions were examined by electrophoresis on 10% SDS-polyacrylamide gels for the presence of the 74-kDa mature form of the expressed IF2$_{mt}$. The designations kDa and IN indicate the Benchmark^{TM} Protein Ladder and the starting sample that had been passed through the Ni-NTA column and that was applied to the HPLC column. Fractions 52 to 59 were pooled and used as the source of purified IF2$_{mt}$. Three separate gels were required for the analysis of the full-column pattern. The figure shown is a composite of the relevant regions of these three gels. (See color insert.)

The appropriate fractions are pooled accordingly. A typical pattern is shown in Fig. 4.2. IF2$_{mt}$ elutes at a KCl concentration of approximately 0.12 M. The yield of IF2$_{mt}$ is about 0.5 mg/g cells.

4.6. Purification of IF2$_{mt}$ on a gravity DEAE-sepharose column

Because not every laboratory is equipped to purify proteins using HPLC, we developed a simple gravity-based chromatographic method for the purification of IF2$_{mt}$. The dialyzed sample from the Ni-NTA column

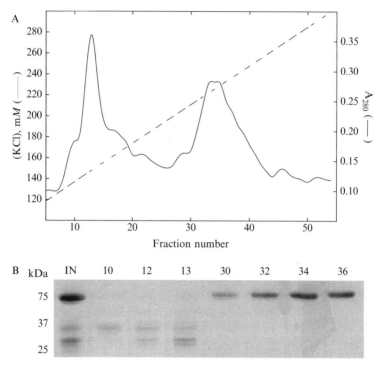

Figure 4.3 Purification of IF2$_{mt}$ on a gravity DEAE-Sepharose column. (A) The absorbance pattern at 280 nm was monitored during chromatography with an ISCO UA-6 Absorbance detector on a 0.5 absorbance scale with a 5 mm flow cell. (B) Analysis of the purity of the IF2$_{mt}$ in the column fractions. Aliquots (10 μl) of the indicated fractions were subjected to electrophoresis on 10% SDS-polyacrylamide gels and stained with Coomassie Blue. IN represents the protein pattern in the input to the DEAE-Sepharose column. (See color insert.)

(10 ml, 5 mg protein) is applied to an ~10 ml (5.7 × 1.5 cm) DEAE-Sepharose anion exchange column. The column is washed with 100 ml Buffer I containing 100 m*M* KCl until the absorbance at 280 nm is near the baseline. A linear gradient (40 ml) from 100 to 300 m*M* KCl in Buffer I is applied to the column at a flow rate of 0.75 ml/min. The absorbance at 280 nm is monitored with an ISCO UA-6 UV-Visible detector at a setting of 0.5 (Fig. 4.3A) and fractions (0.75 ml) are collected in Eppendorf tubes at a flow rate of 0.75 ml/min. Fractions containing mature IF2$_{mt}$ are identified by 10% SDS-polyacrylamide gel electrophoresis (Fig. 4.3B) and pooled. The sample is fast frozen in a dry ice-isopropanol bath and stored at −70°. IF2$_{mt}$ elutes at a KCl concentration of approximately 0.22 *M*. The yield of IF2$_{mt}$ using this approach is approximately 0.10 mg/g cells, which is equivalent to ~0.33 mg/liter of induced cell culture.

The activity of this sample is tested on both *E. coli* 70S ribosomes and mitochondrial 55S ribosomes. As indicated in Fig. 4.4, these preparations have specific activities of about 70,000 U/mg protein where 1 U is the binding of 1 pmol of [^{35}S]fMet-tRNA to *E. coli* ribosomes under the assay conditions described earlier. This value is actually slightly higher than that

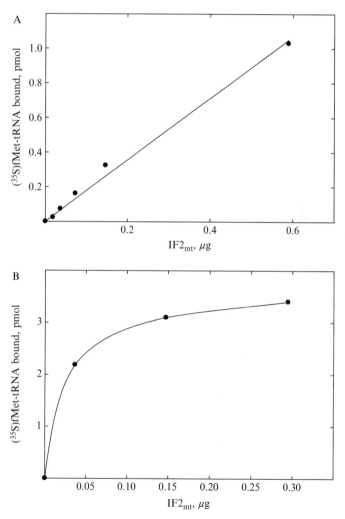

Figure 4.4 (A) Activity of the purified IF2$_{mt}$ on mitochondrial 55S ribosomes. The indicated amounts of the DEAE-Sepharose column sample of IF2$_{mt}$ were tested for activity as described in the text. A blank representing the amount of fMet-tRNA binding obtained in the absence of IF2$_{mt}$ (0.016 pmol) has been subtracted from each value. (B) Activity of the purified IF2$_{mt}$ on *E. coli* 70S ribosomes. The activity of the purified IF2$_{mt}$ was tested as described in the text. A blank representing the amount of fMet-tRNA binding obtained in the absence of IF2$_{mt}$ (0.048 pmol) has been subtracted from each value.

obtained for the native factor purified from bovine liver (Liao and Spremulli, 1991).

5. PURIFICATION OF HIS-TAGGED IF3$_{MT}$

Native IF3$_{mt}$ has never been purified from any source including mammalian mitochondria. Hence, MitoProtII was used to predict the cleavage site for the removal of the mitochondrial import signal (Claros and Vincens, 1996). This program predicted an import signal of 30 amino acids giving a 247 amino acid mature protein. An EST encoding human IF3$_{mt}$ was obtained from American Type Culture Collection and the region encompassing the mature protein (amino acids 31 to 278) was cloned into pET-21c(+) using the NdeI and XhoI restriction sites (Koc and Spremulli, 2002). This vector provides a C-terminal His$_6$-tag. This construct was transformed into *E. coli* BL21(DE3) carrying the pArgU218 plasmid to supply the isoacceptor of tRNA^{Arg} recognizing the AGA and AGG codons.

5.1. Cell growth, induction, and preparation of cell extracts

Cells from a frozen stock of *E. coli* BL21(DE3) carrying the pET21c(+) IF3$_{mt}$ construct are used to inoculate a 25 ml overnight culture of LB containing 100 μg/ml ampicillin and 50 μg/ml kanamycin. The cells are grown at 37° in a shaking incubator. Following overnight incubation, 20 ml of this culture is used to inoculate a 6 liter flask containing 2 liters of LB containing 100 μg/ml ampicillin and 50 μg/ml kanamycin. Growth of the cells is monitored until the A$_{595}$ reaches about 0.6 and the expression of IF3$_{mt}$ is induced by the addition of 50 μM IPTG. Induction is carried out for 5 h at 37° or overnight at 27°. Similar yields are obtained from cells grown under either of these conditions. The cells are harvested by centrifugation at 5000 rpm (7300×g_{max}) for 30 min at 4° in a Sorvall RC3B centrifuge using 1 liter swinging buckets in the H-6000A rotor. The cell pellets are resuspended in buffer containing 50 mM Tris-HCl (pH 7.8), 50 mM NH$_4$Cl, and 10 mM MgCl$_2$ and collected by centrifugation at 7000 rpm (5856×g_{max}) for 30 min in a Sorvall SS34 rotor. The cell pellets are carefully drained and weighed. At this point, the cells can be stored at −70° after being fast-frozen in a dry ice-isopropanol bath. The yield of cells is generally around 5 g/liter of cells.

5.2. Purification of IF3$_{mt}$: Cell lysis

Cells are resuspended in ice-cold Buffer 3A (10 ml buffer per g of cell pellet) and sonicated for 7 pulses (10 s on, 50 s off). Solid NH$_4$Cl is added to a final concentration of 0.5 M over a period of 15 min while stirring on ice. Cell debris is removed by centrifugation at 15,000 rpm ($27,000 \times g_{max}$) for 30 min in a Sorvall SS34 rotor at 4°. The supernatant is retained.

5.3. Purification of IF3$_{mt}$: Ni-NTA step

Ni–NTA agarose is equilibrated in Buffer 3B and a 50% slurry in this buffer is prepared. This slurry (1 ml/2 liter cell culture) is added to the cell lysate and the sample is mixed for 30 min at 4° using a test tube rocker. This mixture is then sequentially transferred to a 5 ml polypropylene column (Qiagen). The column is washed with 30 ml of Buffer 3B, after which the protein is eluted with 6 sequential 1 ml aliquots of Buffer 3C. Each aliquot of Buffer 3C is allowed to sit on the column for 5 min before being collected. All fractions are combined and dialyzed against a 100-fold excess of ice cold Buffer I for 2 h, with one change of Buffer I.

5.4. Purification of IF3$_{mt}$: Purification on HPLC

The IF3$_{mt}$ preparation has one major contaminant following the Ni–NTA step. The contaminant is a 19-kDa C-terminal fragment of IF3$_{mt}$ (Koc and Spremulli, 2002). This contaminant often represents 30 to 40% of the total protein in the sample. Due to the presence of this contaminant, a second step of purification is carried out using HPLC or gravity chromatography. For the HPLC method, the partially purified IF3$_{mt}$ ($\sim$3 mg in 3 ml) prepared from 2 liters of cell culture is dialyzed against a 100-fold excess of Buffer I containing 225 mM KCl for 1.5 h. The dialyzed sample is filtered through a 0.45 μm filter and applied at a flow rate of 0.5 ml/min to a TSKgel SP-5PW column (7.5 cm × 7.5 mm, TosoHaas Inc., Japan) that is equilibrated in Buffer I containing 240 mM KCl. The column is washed until the absorbance at 280 nm returns to baseline and is then developed with a linear gradient (50 ml) from 240- to 300 mM KCl in Buffer I at a flow rate 0.5 ml/min. Fractions (0.5 ml) are collected using an ISCO RETRIEVER 500 fraction collector into Eppendorf tubes from which the lids have been removed and that are balanced in 13 × 100 mm glass test tubes. Fractions containing IF3$_{mt}$ are identified by electrophoresis on 15% SDS-polyacrylamide gels. These fractions are pooled, fast-frozen in a dry ice-isopropanol bath and stored at −70°.

5.5. Purification of IF3$_{mt}$: Gravity S-sepharose column

Because HPLC equipment is not standard in all laboratories, we also developed a gravity chromatographic method for the purification of IF3$_{mt}$. In this procedure, the sample that has been dialyzed in Buffer I ($\sim$10 ml, $\sim$3 mg from 2 liter cell culture) from the Ni–NTA column is applied to a S–Sepharose column ($\sim$3 ml, 6.5 cm × 0.75 cm) equilibrated in Buffer I containing 250 mM KCl. The column is developed with a linear gradient (40 ml) from

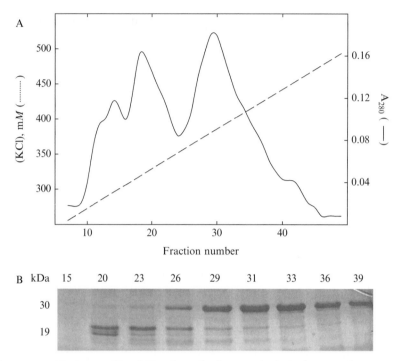

Figure 4.5 Elution profile of IF3$_{mt}$ from the first gravity S–Sepharose column using a 250 to 550 mM KCl gradient in Buffer I. (A) The absorbance pattern at 280 nm was monitored during chromatography with an ISCO UA-6 Absorbance detector and chart recorder on a 0.5 absorbance scale and a 5 mm flow cell. (B) Aliquots (5 μl) of the indicated fractions were run on a 15% SDS-polyacrylamide gel and the protein bands located by staining with Coomassie Blue. Fractions 26 to 39 were pooled and dialyzed against 100 vol of ice cold Buffer I for 2 h with a change of buffer after 1 h. This sample ($\sim$10.5 ml, $\sim$2.5 mg) was applied to a second S-Sepharose column and subjected to chromatography as shown in Fig. 6. (See color insert.)

250 to 550 mM KCl in Buffer I. Fractions (0.75 ml) are collected at a flow rate of 1 ml/min while the absorbance at 280 nm is monitored using an ISCO UA-6 UV–Visible detector at an absorbance scale setting of 0.5 and a 5 mm optical flow cell (Fig. 4.5A). Fractions containing the 29-kDa mature form of the tagged IF3$_{mt}$ are identified by SDS-polyacrylamide gel electrophoresis (Fig. 4.5B). The S-Sepharose column can be reused. Before reuse, it is washed with 10 vol of Buffer I containing 550 mM KCl, followed by 10 vol of Buffer I containing 1 M KCl. It is then equilibrated in Buffer I containing 250 mM KCl. The column is allowed to sit overnight in this buffer and then washed again briefly with approximately 50 ml of Buffer I containing 250 mM KCl before use.

The IF3$_{mt}$ eluted from the S-Sepharose column described previously still contains small amounts of the 19 kDa fragment and other minor contaminants. Fractions containing the 29 kDa band (fractions 26 to 39 in Fig. 4.5B) are combined, diluted 1.5-fold with Buffer I, and dialyzed against 100 vol of ice cold Buffer I at 4° for 2 h with a change of buffer after 1 h. The dialysate is applied to the S-Sepharose column again. The column is developed with a 40 ml linear gradient from 250 to 550 mM KCl in Buffer I and the absorbance at 280 nm is monitored as described earlier (Fig. 4.6A). Aliquots of various fractions are analyzed by SDS-polyacrylamide gel electrophoresis on 15% gels for the presence of IF3$_{mt}$ (Fig. 4.6B). The second run through the S-Sepharose column eliminates the 19 kDa fragment and has a

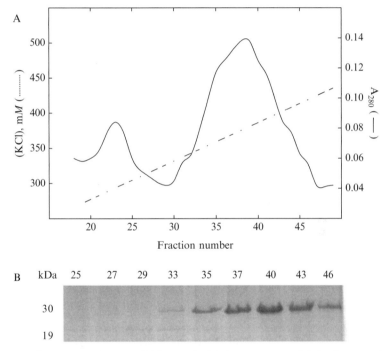

Figure 4.6 Passage of the partially purified IF3$_{mt}$ on a second S-Sepharose column using the same gradient conditions. A sample (~10.5 ml, ~2.5 mg) from the first S-Sepharose column was applied to an identical second S-Sepharose column and subjected to chromatography as described in the text. (A) The absorbance pattern at 280 nm was monitored during chromatography with an ISCO UA-6 Absorbance detector and chart recorder on a 0.5 absorbance scale. (B) Analysis of fractions for the presence of intact IF3$_{mt}$. Aliquots (10 μl) of the indicated fractions were applied to a 15% SDS-polyacrylamide gel and the protein bands were located by staining with Coomassie Blue. The material in fractions 35 to 46 was pooled and dialyzed against 100 vol of ice cold Buffer I for 2 h with a change of buffer after 1 h. (See color insert.)

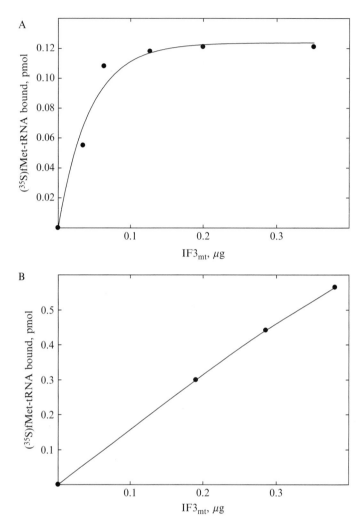

Figure 4.7 (A) Activity of the purified IF3$_{mt}$ in promoting the formation of the initiation complex on mitochondrial 55S ribosomes. The indicated amounts of IF3$_{mt}$ were tested for the ability to stimulate initiation complex formation on bovine mitochondrial ribosomes as described in the text. A blank representing the amount of fMet-tRNA binding obtained in the absence of IF3$_{mt}$ (0.1 pmol) has been subtracted from each value. (B) Activity of the purified IF3$_{mt}$ in promoting the formation of the initiation complex on *E. coli* 70S ribosomes. Reaction mixtures were prepared as described in the text and contained the indicated amount of IF3$_{mt}$. A blank representing the amount of fMet-tRNA binding obtained in the absence of IF3$_{mt}$ (0.2 pmol) has been subtracted from each value.

negligible effect on activity. Fractions containing full-length IF3$_{mt}$ are then combined, dialyzed against 100 vol of ice cold Buffer I at 4° for 2 h with a change of buffer after 1 h.

The activity of the purified IF3$_{mt}$ is tested on *E. coli* 70S ribosomes and mitochondrial 55S ribosomes (Fig. 4.7). The specific activity of the purified protein is $\sim$1700 U/mg when tested on 70S ribosomes. The total yield of purified protein is about 2 mg from 2 liters of cell culture.

REFERENCES

Bhargava, K., Templeton, P. D., and Spremulli, L. L. (2004). Expression and characterization of isoform 1 of human mitochondrial elongation factor G. *Prot. Exp. Puri.* **37,** 368–376.

Claros, M. G., and Vincens, P. (1996). Computational method to predict mitochondrially imported proteins and their targeting sequences. *Eur. J. Biochem.* **241,** 779–786.

Graves, M., and Spremulli, L. L. (1983). Activity of *Euglena gracilis* chloroplast ribosomes with prokaryotic and eukaryotic initiation factors. *Arch. Biochem. Biophys.* **222,** 192–199.

Graves, M., Breitenberger, C., and Spremulli, L. L. (1980). *Euglena gracilis* chloroplast ribosomes: Improved isolation procedure and comparison of elongation factor specificity with prokaryotic and eukaryotic ribosomes. *Arch. Biochem. Biophys.* **204,** 444–454.

Koc, E. C., and Spremulli, L. L. (2002). Identification of mammalian mitochondrial translational initiation factor 3 and examination of its role in initiation complex formation with natural mRNAs. *J. Biol. Chem.* **277,** 35541–35549.

Liao, H.-X., and Spremulli, L. L. (1990). Identification and initial characterization of translational initiation factor 2 from bovine mitochondria. *J. Biol. Chem.* **265,** 13618–13622.

Liao, H.-X., and Spremulli, L. L. (1991). Initiation of protein synthesis in animal mitochondria: Purification and characterization of translational initiation factor 2. *J. Biol. Chem.* **266,** 20714–20719.

Ma, J., and Spremulli, L. L. (1996). Expression, purification and mechanistic studies of bovine mitochondrial translational initiation factor 2. *J. Biol. Chem.* **271,** 5805–5811.

Ma, L., and Spremulli, L. L. (1995). Cloning and sequence analysis of the human mitochondrial translational initiation factor 2 cDNA. *J. Biol. Chem.* **270,** 1859–1865.

Matthews, D. E., Hessler, R. A., Denslow, N. D., Edwards, J. S., and O'Brien, T. W. (1982). Protein composition of the bovine mitochondrial ribosome. *J. Biol. Chem.* **257,** 8788–8794.

Montoya, J., Ojala, D., and Attardi, G. (1981). Distinctive features of the 5'-terminal sequences of the human mitochondrial mRNAs. *Nature* **290,** 465–470.

Muench, K., and Berg, P. (1966). Preparation of aminoacyl ribonucleic acids synthetases from *Escherichia coli*. In "Procedures in Nucleic Acid Research" (G. Cantoni and D. Davies, eds.), pp. 375–382. Harper and Row, New York.

Ojala, D., Montoya, J., and Attardi, G. (1981). tRNA punctuation model of RNA processing in human mitochondria. *Nature* **290,** 470–474.

Spencer, A. C., and Spremulli, L. L. (2005). The interaction of mitochondrial translational initiation factor 2 with the small ribosomal subunit. *Biochim. Biophys. Acta—Prot. Proteom.* **1750,** 69–81.

Spremulli, L. L., Coursey, A., Navartil, T., and Hunter, S. E. (2004). Initiation and elongation factors in mammalian mitochondrial protein synthesis. *In* "Progress in Nucleic Acids Research and Molecular Biology" (K. Moldave, ed.), pp. 211–261.

Spremulli, L. L. (2007). Large-scale isolation of mitochondrial ribosomes from mammalian tissues. *In* "Mitochondria Genomics and Proteomics Protocols" (D. Leister and J. Herrmann, eds.), pp. 265–275. Humana Press, Totowa, NJ.

Takemoto, C., Ueda, T., Miura, K., and Watanabe, K. (1999). Nucleotide sequences of animal mitochondrial tRNAs(Met) possibly recognizing both AUG and AUA codons. *Nucl. Acids Symp. Ser.* **42,** 77–78.

Walker, R. T., and RajBhandary, U. L. (1972). Studies on polynucleotides. CI. *Escherichia coli* tyrosine and formylmethionine transfer ribonucleic acids: Effect of chemical modification of 4-thiouridine to uridine on their biological properties. *J. Biol. Chem.* **247,** 4879–4892.

Woriax, V., Burkhart, W., and Spremulli, L. L. (1995). Cloning, sequence analysis and expression of mammalian mitochondrial protein synthesis elongation factor Tu. *Biochim. Biophys. Acta* **1264,** 347–356.

Xin, H., Woriax, V. L., Burkhart, W., and Spremulli, L. L. (1995). Cloning and expression of mitochondrial translational elongation factor Ts from bovine and human liver. *J. Biol. Chem.* **270,** 17243–17249.

Zhang, Y., and Spremulli, L. L. (1998). Identification and cloning of human mitochondrial translational release factor 1 and the ribosome recycling factor. *Biochim. Biophys. Acta* **1443,** 245–250.

CHAPTER FIVE

IN VITRO STUDIES OF ARCHAEAL TRANSLATIONAL INITIATION

Dario Benelli *and* Paola Londei

Contents

Dipartimento Biotecnologie Cellulari ed Ematologia, Università di Roma Sapienza, Roma, Italy

Methods in Enzymology, Volume 430
ISSN 0076-6879, DOI: 10.1016/S0076-6879(07)30005-0

Abstract

Initiation is the step of translation that has incurred the greatest evolutionary divergence. *In silico* and experimental studies have shown that archaeal translation initiation resembles neither the bacterial nor the eukaryotic paradigm, but shares features with both. The structure of mRNA in archaea is similar to the bacterial one, although the protein factors that assist translational initiation are more numerous than in bacteria and are homologous to eukaryotic proteins. This chapter describes a number of techniques that can be used for *in vitro* studies of archaeal translation and translational initiation, using as a model system the thermophilic crenarcheon *Sulfolobus solfataricus*, growing optimally at about 80° in an acidic environment.

1. INTRODUCTION

The mechanism whereby polypeptide chains are elongated is basically conserved across the primary domains of life (Bacteria, Archaea, Eukarya). Translation initiation, however, displays different complexity in the three domains and is the key control point of protein synthesis.

The mechanism and machinery for translation initiation have been studied in depth in Bacteria and Eukarya. In Archaea, however, there is still a limited amount of information about this step of protein synthesis. The most comprehensive set of data still derives from *in silico* studies, albeit several experimental reports on archaeal translation initiation have begun to appear.

Since the publication of the first complete archaeal genome (Bult *et al.*, 1996) up to the most recent database reporting the complete sequences of archaeal genomes, it has been evident that the archaeal translation initiation apparatus differs from its bacterial and eukaryotic counterparts, although sharing some features with both of them.

Genomic and biochemical (SD) analyses have revealed that the archaeal mRNAs resemble the bacterial ones in being often polycistronic, always uncapped, and lacking long poly(A) tails. In addition, archaeal mRNAs often display Shine-Dalgarno (SD) sequences located 3 to 10 nucleotides upstream from the start codon (Bell and Jackson, 1998; Dennis, 1997; Kyrpides and Woese, 1998). Moreover, computer studies revealed that in archaea, especially crenarchaeota, many genes encoded by monocistronic transcripts, as well as genes located at the $5'$-proximal end of operons, are often devoid of a SD motif and have little or no sequence upstream of the translational start codon. By contrast, internal cistrons are usually endowed with SD motifs (Sensen et al., 1996; Slupska et al., 2001; Tolstrup et al., 2000).

Although the structure of archaeal mRNAs is similar to that of bacterial ones, the situation is far different with regard to the translation initiation factors. In fact, the analysis of archaeal genomes has revealed the existence of about ten putative initiation factors homologous to the eukaryotic ones (Table 5.1).

These unexpected data raise the question of why the Archaea should possess a set of putative initiation factors similar to the eukaryotic one. Before it can be answered, it is obviously necessary to demonstrate that the archaeal proteins indeed function as translation initiation factors, and to identify the specific function of each.

To date, a limited amount of experimental data exist about the mechanism and machinery of archaeal translation, especially with regard to the initiation step, and many questions remain unanswered. To obviate this, our laboratory has undertaken a systematic investigation of the *cis-* and *trans-*acting factors involved in promoting initiation in the archaea (Londei, 2005). To this end, we have developed a series of tools for *in vitro* studies of archaeal protein synthesis, ranging from translation of natural mRNAs in cell lysates to more detailed assays for studying the function of individual initiation factors. The model organism employed in our laboratory is the thermophilic crenarcheon *Sulfolobus solfataricus*, growing optimally at ~80°.

Table 5.1 Translation initiation factors in Archaea and their homologues in bacteria and eukaryotes

Archaea	Eukaryotes	Bacteria
aIF1A	eIF1A	IF1
aIF2/5B	eIF5B	IF2
aIF1	eIF1	YCiH (only some phyla)
aIF5A	eIF5A	EFP
a/eIF2 (trimer)	eIF2 (trimer)	—
aIF6	eIF6	—
(eIF2B α, β, γ subunits)	eIF2B (pentamer)	—
(eIF4A-like)	eIF4A	—

The table lists the proteins found in all archaeal genomes having homology with known translation initiation factors in eukaryotes, bacteria, or both. The last two proteins in parentheses (eIF2B subunits and eIF4A) are listed for completeness but probably do not participate in archaeal translational initiation.

S. solfataricus has many advantages as a model archaeon: it is easy to cultivate in the laboratory, the complete genome is known, cell extracts and proteins are easy to prepare, and many molecular and biochemical data about the species are available in the literature.

This work describes in detail the array of molecular tools available for studying protein synthesis, and initiation specifically, in *S. solfataricus*. Albeit many of the data obtained with the *S. solfataricus* system are probably valid for archaea in general, it should be kept in mind that other archaeal species may present specific features that have no counterpart in *Sulfolobus*. It is hoped that more tools will be available in the near future to study the translational machinery in other species of archaea.

2. PREPARATION OF *S. SOLFATARICUS* CELLULAR EXTRACTS, RIBOSOMES, AND OTHER CELLULAR FRACTIONS

2.1. Growth of *S. solfataricus* cells

Starter cultures are prepared by inoculating pellets of *S. solfataricus* cells in 200 ml of a modified DSM 182 growth medium (in 500-ml conical flasks) at an $OD_{600} = 0.1$. The cells are grown in a $80°$ water bath on a rotary shaker at 200 rpm until an OD_{600} corresponding to the desired growth phase is reached. We usually harvest cells in the late exponential phase ($OD_{600} \cong 1.0$). The growth medium contains per liter, 1-g Difco yeast extract, 1-g Difco casamino acids, 3.1-g KH_2PO_4, 2.5-g $(NH_4)_2SO_4$, 0.2-g $MgSO_4 \bullet 7H_2O$, 0.25-g $CaCl_2 \bullet 2H_2O$, 1.8-mg $MnCl_2 \bullet 4H_2O$, 4.5-mg $Na_2B_4O_7 \bullet 10H_2O$, 0.22-mg $ZnSO_4 \bullet 7H_2O$, and 0.03-mg $Na_2MoO_4 \bullet 2H_2O$. The pH is adjusted to 3.0 with 10-N H_2SO_4 at room temperature.

The cells are harvested by centrifugation at $5000 \times g$ for 15 min at $4°$. The final yield is 1 to 2 cells (wet weights) per liter of culture. The cell pellets are stored at $-80°$, where they remain stable indefinitely. Small pieces of the frozen cell cake can be used to start new cultures even after being stored a few years.

2.2. Preparation of whole cell lysates (S-30)

An aliquot of frozen cells (usually 2 to 5 g) is placed in a precooled mortar on ice. Twice the cells' wet weight of alumina powder (Type-A5, Sigma) is added and the mixture is ground by hand with a pestle until an even paste is obtained. Then, 2 to 2.5 vol (relative to the weight of the cell pellet) of "extraction buffer" (20-mM Tris-HCl (pH 7.4), 10-mM Mg(OAc)$_2$, 40-mM NH$_4$Cl, 1-mM DTT) are gradually added to the paste. When a homogeneous mixture is obtained, RNase-free DNase I is added at the final concentration of 0.5-μg/ml^{-1} and the paste is gently stirred until all viscosity disappears. The mixture is transferred to centrifuge tubes and spun twice at 30,000g for 30 min. After each centrifugation, care should be taken to withdraw only about two thirds of the supernatant without disturbing the debris.

Portions (0.05 ml) of the last supernatant were stored at −80°, where they remained stable indefinitely. This preparation constitutes the whole cell lysate, or S-30 extract. It contains a protein concentration of ∼18 to 20 mg/ml detected by Bradford assay.

2.3. Preparation of S-100 fraction and purified ribosomes

S-30 extracts can be further processed to separate the ribosomes and a protein fraction (S-100) containing most cytoplasmic enzymes and factors.

To this end, the S-30 extract is spun in a Beckman ultracentrifuge at 100,000×g and 5° for 2 to 3 h. This procedure causes the ribosomes to sediment at the bottom of the tubes, carrying along a substantial amount of extrinsically bound proteins, including a number of translation factors. These are the crude ribosomes, which need further purification before being used in the various assays.

A first round of ribosome cleaning is carried out as follows. The pellets of crude ribosomes are resuspended in a buffer containing 20-mM Tris-HCl (pH 7.4), 500-mM NH$_4$Cl, 10-mM Mg(OAc)$_2$, 2-mM dithiothreitol and are then layered on a "cushion" of 18% (w/v) sucrose in the same buffer. The sucrose cushion should occupy half the volume of the centrifuge tube. The tubes are spun at 100,000×g for 5 to 6 h at 4°. The final pellets, which are free of most extrinsic proteins but still contain some amount of certain translation factors (e.g., aIF1 and aIF6), are resuspended in extraction buffer and stored at −20°. This preparation is termed the purified ribosomes.

Ribosome-free cytoplasmic proteins (S-100 fraction) are prepared from the 100,000×g supernatant of the "crude ribosomes" as follows. The crude S-100 is adjusted to 70% saturation with ammonium sulphate and is kept on ice for at least 1 h. The precipitated proteins are collected by centrifuging 10 min at 30,000g and are then dissolved in 0.2-fold the original S-100 volume of a buffer containing 10-mM Tris-HCl (pH 7.4), 2-mM Mg (OAc)$_2$, and 4-mM DTT. The solution is dialyzed against the same buffer, portioned, and stored at −80°.

2.4. Isolation of high-salt-purified ribosomes and "crude" initiation factors

When necessary, ribosomes essentially devoid of translation factors can be obtained by resuspending the purified ribosomes in a high salt buffer (20-mM Tris-HCl (pH 7.4), 2-M NH$_4$Cl, 10-mM Mg(OAc)$_2$, 2-mM dithiothreitol). The ribosome suspension is stirred on ice for ∼1 h to allow the release of the nonribosomal proteins. The mixture is then centrifuged on a sucrose cushion made in the same high-salt buffer following the procedure described previously. The ribosome pellets (called high-salt-washed ribosomes) are finally resuspended in "extraction buffer" and stored in small aliquots at $-20°$. The concentration of the ribosomes is determined by measuring the A_{260} and using as the extinction coefficient 1 OD_{260} 70S $\cong$ 40 pmol.

The supernatant recovered after the sedimentation of the high-salt-washed ribosomes contains most of the translation initiation factors, which can be recovered and concentrated by the following procedure. The supernatant is supplemented with 70% (final concentration) ammonium sulphate and placed on ice for $\sim$1 h to allow protein precipitation. The precipitate is collected by centrifuging 10 min at 15,000 rpm, the pellet is dissolved in resuspending buffer (20-mM Tris-HCl (pH 7.4), 2-mM dithiothreitol, 10% glycerol), dialyzed against the same buffer, portioned, and stored at $-80°$. This preparation is the high-salt ribosome wash (HSRW).

2.5. Isolation of 30S and 50S ribosomal subunits

Numerous translational assays require the use of purified 30S or 50S ribosomal subunits. To achieve subunit separation, the purified" or high-salt-washed ribosomes are centrifuged on preparative sucrose gradients. No special treatments or ionic conditions are required to promote the release of the individual subunits from 70S ribosomes. In fact, *S. solfataricus* monomeric ribosomes are quite unstable and high-speed centrifugation on density gradients is by itself sufficient to provoke their quantitative dissociation into 30S and 50S subunits.

For subunit preparation, ribosomes are resuspended in 20-mM Tris/HCl (pH 7.0), 40-mM NH_4Cl, 10-mM $Mg(OAc)_2$, and 2 mM dithiothreitol. Aliquots (1 ml) of the ribosome suspensions ($= 40$ A_{260} U) are layered onto 38-ml linear, 10 to 30% (w/v) sucrose density gradients made in the same buffer. The gradients are centrifuged in a Beckman SW 27 rotor operated at 18,000 rpm at $5°$ for 18 h. Fractions corresponding to the 30S and 50S peaks of A_{260} are separately pooled and the particles therein are precipitated by the addition of 2.5 vol of 95% ethanol. After low-speed centrifugation, the subunit pellets are resuspended in extraction buffer containing 10% (v/v) glycerol and stored at $-20°$.

2.6. Preparation of bulk tRNA

Unfractionated tRNA from *S. solfataricus* is prepared by phenol extraction from the crude S-100 fraction (i.e., before concentrating with $[NH_4]_2SO_4$). To this end, the S-100 is treated twice with an equal vol of water-saturated phenol, then once with an equal vol of a chloroform/isoamyl alcohol mixture (99:1, v/v). The RNA is precipitated from the last aqueous phase by adding 2.5 vol of 95% ethanol, and the precipitate is collected by centrifugation at 10,000 rpm for 5 min. The pellet is resuspended in 10-mM glycine (pH 9.0) and the solution is incubated 2 to 3 h at 37° to achieve alkaline deacylation of the tRNA therein contained. Lastly, the RNA is again precipitated with 2.5 vol of 95% ethanol and the resulting pellet is dissolved in 10-mM Tris-HCl (pH 7.5) at A_{260} =130 to 150. This preparation actually contains a mixture of cellular low-molecular-weight RNAs; however, most of it is tRNA and can be conveniently used as such for most purposes, except when a specific type of tRNA is absolutely required for an assay.

2.7. Isolation of initiator met-tRNA$_i$

Certain assays for translational initiation require the use of purified initiator tRNA (tRNA$_i^{Met}$). It is possible to use for this purpose *Escherichia coli* tRNA$_i^{Met}$ (unformylated), which binds to *S. solfataricus* ribosomes and factors, and is sufficiently stable even at 60 to 65°. However, employing the homologous *S. solfataricus* met-tRNAi is a better choice when possible, because it contains a number of posttranslational modifications that may be of importance in some translational reactions. Native tRNA$_i^{Met}$ can be purified from bulk *S. solfataricus* tRNA by affinity purification as follows.

A 30-mer oligonucleotide (5'-GCTTCAGGGACCAAGTTTAGG-TCCGGGGCG biotine-3') is synthesized, complementary to the 3' moiety of the sequence of *S. solfataricus* initiator tRNA$_i^{Met}$ as deduced from the published genome sequence (She *et al.*, 2001).

The biotinylated oligonucleotide (10 μg) is mixed with 0.6 mg of strep-tavidin MagneSphere paramagnetic particles (Promega), and the mixture is incubated at room temperature for 30 min in 500 μl of 10-mM Tris-HCl (pH 7.5). The particles are washed 3 times with 600 μl of 10-mM Tris-HCl (pH 7.5), then 3 times with 600 μl of 6× SSC (900-mM NaCl, 90-mM sodium citrate). The bulk tRNA, prepared as described in the previous paragraph, is diluted to a final optical density of 100 A_{260} U ml^{-1} with 6× SSC; 100 μl of this solution is mixed with 0.6-mg particles coupled with the oligonucleotide. The sample is incubated at room temperature for 30 min with continuous agitation; the particles are then separated from the supernatant using the appropriate magnetic separator and washed several times with 1 ml of 3× SSC until the A_{260} is close to zero. The particles are finally resuspended in 100 μl of 0.1× SSC and incubated at 60°for 2 min, after which time they are magnetically separated from the supernatant which is saved. The whole operation is repeated once more. Purified tRNA is finally recovered from the pooled supernatants by precipitation with 2.5 vol of 95% ethanol containing 500-mM ammonium acetate and 0.2-mg ml^{-1} (final concentration) glycogen. The pellet is resuspended in a suitable vol of sterile H$_2$O.

The problem with this procedure is that the final yield of tRNA$_i$^{Met} is very low, and it has to be repeated several times to get an amount of material sufficient for a few assays. We have also tried to obtain tRNA$_i$^{Met} by *in vitro* transcription of the cloned *S. solfataricus* gene. This technique has its problems too, the main of which is that the charging with methionine of the tRNA thus produced is very inefficient, probably because of the lack of the appropriate posttranslational modifications on the synthetic tRNA.

2.8. Charging of tRNAi with methionine

Native *S. solfataricus* tRNA$_i$^{Met} in total tRNA, or purified native tRNA$_i$^{Met}, (obtained as previously described), is charged with methionine using as the source of methionil–tRNA synthase (MetRT) either a *S. solfataricus* S100 preparation freed of nucleic acids (Benelli *et al.*, 2003) or a recombinant fragment of 547 amino acids of *E. coli* MetRT.

Aminoacylation of 4-μM tRNA$_i$^{Met} is carried out in the presence of 1.5-μM *E. coli* MetRT (or an optimal amount of RNA-free S-100) in 100 μl (final vol) of reaction mixture containing 20-mM Hepes (pH 7.5), 100-μM Na$_2$EDTA, 150-mM NH$_4$Cl, 4-mM ATP, 10-mM MgCl$_2$, and 0.1-mM unlabeled methionine or 10-μM [^{35}S]methionine (1000 Ci/mM) when it is necessary to obtain a radio-labeled product. The reaction mixture is incubated for 30 min at 37° (if *E. coli* MetRT is used) or 65° (if *S. solfataricus* S-100 is used), phenol extracted, ethanol precipitated, and resuspended in 50 μl of 10-mM KCH$_3$COOH (pH 5.5). The incorporation of [^{35}S]methionine is assayed by spotting aliquots of reactions on Whatman 3-mm paper and placing the filters in ice-cold 10% trichloroacetic acid for 30 min to precipitate the nucleic acids. The filters are then washed 3 times with 5% trichloroacetic acid, once with 95% ethanol, dried, and counted in a liquid scintillation counter.

3. *IN VITRO* TRANSLATION OF ARCHAEAL mRNAs AT HIGH TEMPERATURE

3.1. Translation in unfractionated cell lysates

Two original systems for *in vitro* translation of *S. solfataricus* natural mRNAs were developed in our laboratory. The former makes use of an unfractionated cell lysate (S-30) and is designed to preserve the natural environment for translation as much as possible (Condo *et al.*, 1999). The second (Ruggero, 1993) is a reconstituted system, employing purified ribosomes and different protein fractions, including a preparation enriched in translation initiation factors (HSRW). For the sake of completeness, both systems will be described here, with the caveat that the one based on whole cell lysates is much more active and reliable than the second and can be effectively used for most purposes. Indeed, most of the assays described here for the study of translational initiation in archaea were performed with the S-30 system. Work is still in progress to improve the performance of the reconstituted system for functional assays with individual translation initiation factors.

The standard reaction mixture with unfractionated cell lysates contains, in a final volume of 50 μl: 10-mM KCl (or NH$_4$Cl), 20-mM Tris-HCl (pH 7.4), 18- to 20-mM Mg Cl$_2$, 6-mM ATP, 3-mM GTP, 100-μg ml^{-1} *Sulfolobus* bulk tRNA, 1-μl [^{35}S]methionine (S.A. 3000 Ci mmol^{-1}), 10-μl S-30. Before use, the S-30 is preincubated at 70° for 15 min to unload the ribosomes of endogenous mRNAs and thus decrease the background expression of unspecific proteins. After mixing all components, 2 to 4 μg of the desired specific mRNA are added, and the mixture is incubated for 30 to 45 min at 65 to 73°. The mRNAs translated in this system were obtained by *in vitro* transcription of *S. solfataricus* genes cloned in an appropriate plasmid downstream of a viral T7 or T3 promoter. Before transcription, the plasmids are linearized with an appropriate restriction enzyme. The mRNAs thus prepared should include the native 5' untranslated region with the ribosome binding signals, except for the case of leaderless mRNAs (see later).

Analysis of the translation products is performed by running 10 μl of the incubation mixture (previously heated for 2 min at 100° in an equal vol of SDS-loading buffer) on polyacrilamide/SDS gels; after the run, the gels are dried and autoradiographed.

If the system works properly, the newly synthesized proteins should be visible on the autoradiographic film as sharp dark bands of the expected molecular weight on a clear background (Fig. 5.1).

3.2. RNA translation by salt-purified ribosomes

The translation system reconstituted with purified ribosomes and protein fractions operates under the same ionic and temperature conditions previously described for the unfractionated system. However, the S-30 is replaced by 10 μg of high-salt-washed ribosomes (3 to 4 pmol), and by an optimal amount of S-100 proteins and HSRW fraction containing the crude initiation factors (usually 2 to 4 μl of each). The HSRW fraction is in this case essential, because no translation is detected in its absence (Ruggero et al., 1993).

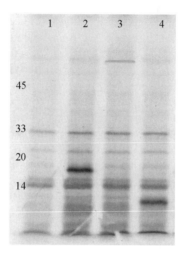

Figure 5.1 Gel electrophoretic analysis of products of *in vitro* translation of different *in vitro* transcribed mRNAs of *S. solfataricus*. Whole cell lysate programmed for translation were incubated at 73° for 40 min without added mRNA (lane 1), with 2 μg of aIF1A mRNA (lane 2), with 4 μg of mRNA aIF2/5B, (lane 3), and with 6 μg of mRNA aIF1 (lane 4). The size (in kDa) and the position of the electrophoretic markers are indicated. aIF1A and aIF2/5B mRNAs are leadered and endowed with regular SD motifs, whereas aIF1 mRNA is leaderless.

3.3. Critical parameters for *in vitro* translation

Because *S. solfataricus* is thermophilic, heat is a very important parameter for optimal *in vitro* translation. Indeed, as shown in Fig. 5.2A, the translational activity of the system is strongly dependent upon high temperature: after a 30 min incubation, amino acid incorporation is maximal at 65 to 75°, while being poorly detectable below 50°. Note that, although the optimal temperature for *S. solfataricus* cell growth is ~80°, the translational system cannot be practically operated at this temperature because evaporation of the samples inhibits translation and makes the results unreliable. For most purposes, it is better to keep the incubation temperature of the system in the range 65 to 73°.

As to the ionic milieu for translation, the most critical parameters are the concentrations of monovalent cations and of magnesium. Unlike what happens with other *in vitro* systems, *S. solfataricus* translation is extremely sensitive to monovalent cations, which are strongly inhibitory in concentrations exceeding 30 mM (KCl) or 10 mM (NH_4Cl) (Fig. 5.2B, right panel), presumably because they promote the dissociation of 70S monomers, which are very unstable in *Sulfolobus* and other crenarcheota (Londei *et al.*, 1986).

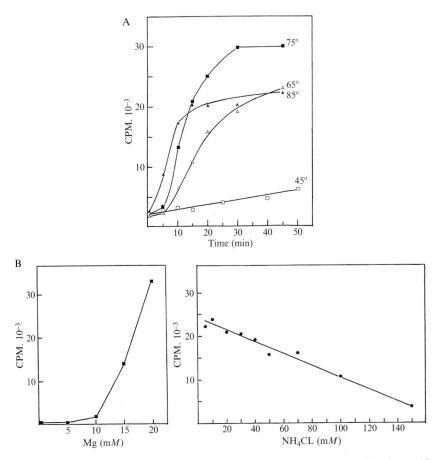

Figure 5.2 Critical parameters for *in vitro* translation. (A) Kinetics of amino acid incorporation in the cell lysate at different temperatures. (B) Ion dependence of amino acid incorporation in the cell lysate incubated at 73° for 40 min. Left panel: magnesium dependence of amino acid incorporation. Right panel: dependence of amino acid incorporation on NH_4^+ ions. (Modified from Ruggero *et al.*, 1993.)

A further relevant feature of the system is that optimal translation takes place at relatively high concentration (15 to 20 mM) of Mg^{2+} ions (Fig. 5.2B, left panel), while other described translating cell-free systems, either bacterial or eukaryotic, have Mg^{2+} optima no higher than 2 to 7 mM. Again, the role of the high Mg^{2+} ions is probably to stabilize the monomeric ribosomes (Londei *et al.*, 1986). It is interesting to note that Mg^{2+} ions cannot be replaced, even in part, by polyamines such as spermine or spermidine, which instead inhibit *in vitro* translation of natural mRNAs already at concentrations below 1 mM.

The optimal conditions for *in vitro* translation are summarized in Table 5.2.

Table 5.2 Composition of the *in vitro* cell-free systems for translation of *S. solfataricus* natural mRNAs

	Whole cell lysate (S30)	Reconstituted cell-free system
KCl (or NH$_4$Cl)	10 mM	10 mM
Tris-HCl, pH 7.4 (or TEA-HCl, pH 7.4)	20 mM	20 mM
MgCl$_2$	18–20 mM	18–20 mM
ATP	6 mM	6 mM
GTP	3 mM	3 mM
Sulfolobus bulk tRNA	100 μg ml^{-1}	100 μg ml^{-1}
Sulfolobus solfataricus crude cell lysate	3.6–4 μg/μl	—
In vitro transcribed mRNA	0.08–0.2 μg/μl	0.08–0.2 μg/μl
Salt washed ribosomes	—	10 μg
S-100 proteins	—	2–4 μl
Crude initiation factors	—	2–4 μl
Final volume	25–50 μl	25–50 μl
Temperature incubation	65–73°	65–73°
Incubation time	40–50 min	40–50 min

4. Ribosome/mRNA Interaction

4.1. Formation and detection of ribosome/mRNA complexes

A critical point in translation initiation is ribosome recognition of the translation initiation sites. This process is vastly divergent in bacteria and eukaryotes, and is regulated both by cis-elements in the mRNAs and by protein factors controlling the ribosome/mRNA interaction in trans. We have developed several assays to study the requirements for ribosome/mRNA interaction during translational initiation in *S. solfataricus*, in particular to unravel the features controlling the correct landing of the small (30S) ribosomal subunits on the translation initiation sites.

That 30S/mRNA interaction is a critical first step in *Sulfolobus* initiation was inferred by the observation that the addition of a nonhydrolyzable GTP analog, GMP-PNP, to the *in vitro* translation system inhibited translation by preventing the formation of 70S ribosomes. Under these conditions, the mRNA remained bound to the 30S ribosomal subunits (Benelli *et al.*, 2003).

The features controlling the interaction between the 30S subunits and the mRNA can be investigated using a purified assay system containing isolated 30S ribosomes and a specific mRNA species. If the mRNA includes the appropriate *cis*-elements, a binary complex 30S/mRNA is formed that can be detected by fractionating the samples on sucrose density gradients.

Specifically, binary mRNA/30S ribosome complexes are formed by mixing ∼40 pmol of purified *S. solfataricus* 30S subunits with ∼3 pmol of a specific mRNA, which can be obtained by *in vitro* transcription of a *Sulfolobus* gene cloned in an appropriate plasmid. It is not necessary to use a complete mRNA in this assay; an mRNA fragment works as well, provided that it includes the 5′ moiety with the translation initiation site. For instance, in many such experiments we have used a fragment of ∼100 nucleotides of an mRNA (mRNA104, encoding a ribosomal protein) including the 60-nucleotide-long 5′-UTR with the SD and ∼80 nucleotides following the AUG start codon (Benelli *et al.*, 2003).

To identify the binary complex on the gradient, the mRNA is uniformly radio-labeled during transcription with ^{32}P UTP (mRNA specific activity $\cong$ 30,000 cpm pmol^{-1}). The samples are incubated for 10 min in a buffer containing 10-mM KCl, 20-mM Tris-HCl (pH 7), 20-mM MgCl$_2$ (final vol 40 μl). Incubation can be performed at either low (37°) or high (65 to 70°) temperature, because mRNA/ribosome interaction is not obligatorily dependent on high temperature. However, heat (about 65°) improves the efficiency of binary complex formation, presumably because the ribosomal subunits adopt a more active conformation at temperatures closer to their physiological optimum. At the end of reaction the samples are layered on 12-ml linear, 10 to 30% sucrose gradients containing 10 mM KCl, 20-mM Tris-HCl (pH 7.5), and 20-mM MgCl$_2$; the gradients are centrifuged in a Beckman SW41 rotor at 36,000 rpm for 4 h. After centrifugation, the gradients are unloaded collecting fractions of ∼0.5 ml, which are checked for optical density at 260 nm and radioactivity content to determine the relative positions of 30S particles and mRNA.

If performed with a leadered mRNA endowed with a strong SD motif (such as mRNA 104), these experiments allow the isolation of a stable 30S/mRNA binary complex that can be recovered from the gradient and can be subjected to further analysis. The strength of the SD motif strongly influences the stability of the complex. In fact, if these assays are performed with mRNAs lacking a SD motif (we have used a mutagenized version of mRNA 104, in which the SD sequence had been destroyed) no stable binary complex is formed. The same happens with "leaderless" mRNAs, that is, mRNAs entirely, or almost so, devoid of a 5′ UTR and therefore of SD motifs.

"Leaderless" mRNAs are quite common in archaea, as revealed by *in silico* analyses of the position of the presumptive transcriptional promoters with respect to the position of the initiation codons (Sensen *et al.*, 1996; Slupska *et al.*, 2001; Tolstrup *et al.*, 2000). Because leaderless mRNAs cannot be recognized directly by the 30S ribosomal subunit by means of the SD-anti SD interaction, it has been proposed that the translation of these mRNAs is accomplished via an alternative and still uncharacterized mechanism (Benelli *et al.*, 2003; Tolstrup *et al.*, 2000). This proposal is supported by the observation that the translational block caused by mutations in the SD motif of *S. solfataricus* mRNAs can be removed by deleting the 5'-UTR altogether, namely artificially rendering the mRNA leaderless (Condò *et al.*, 1999).

Sucrose gradient analysis as described earlier using purified 30S subunits and leaderless mRNA does not evidence the formation of any binary complex, indicating that in this case no stable interaction between the ribosome and the nucleic acid takes place. The result remains the same even after addition of 50S subunits and met-tRNAi^{met} to the assays. Literature data report that in bacteria translation on leaderless mRNAs can be initiated by preformed 70S ribosomes. This assay, however, is not feasible in the case of *S. solfataricus*, in which it is impossible to isolate 70S monomers owing to their great instability.

4.2. Analysis of ribosome/mRNA interaction by "toeprinting"

A detailed analysis of the features of the mRNA/ribosome complexes, in particular, the exact determination of the position of the 30S subunits on the translation initiation site, can be performed using the technique known as "toeprinting." Briefly, toeprinting involves primer extension from an oligodeoxynucleotide, which is hybridized 3' to the initiation codon of an mRNA. Extension to the 5' end of the mRNA results in the synthesis of a "run off" cDNA fragment. If a ribosomal subunit is bound to the translation initiation site, the elongating enzyme is stopped and dislodged from the RNA, producing a shorter extension product, whose 3' terminus marks the "right"(i.e., on the 3' side of the mRNA) edge of the ribosome (Ringquist *et al.*, 1993).

Toeprinting assays on *S. solfataricus* 30S/mRNA binary complexes can be performed either on complexes previously isolated by density gradient separation as described previously, or directly on the incubation mixtures containing the desired components. In practice, the mRNA (or fragment thereof) is preannealed with a 5' end-labeled 20-mer with a sequence complementary to a region located at least 50 to 60 nucleotides downstream of the initiation codon. The annealing reaction is performed at 80° for 5 min, after which the samples are slowly cooled to room temperature and placed on ice. Binary complex formation is obtained by mixing 1 pmol of preannealed mRNA and 10 pmol of 30S subunits in 40 μl of reaction mix containing 10-mM KCl, 20 mM Tris-HCl (pH 7), and 20-mM MgCl$_2$. After incubation at 65° for 5 min, the samples are either directly supplemented with dNTP (5-mM each) and 5 U of M-MLV reverse transcriptase (RNase H minus, Promega), or previously fractionated on sucrose gradients; in the latter case, reverse transcriptase and dNTP are added to the pooled fractions containing the binary complexes. Extension is carried out at 42° for 1 h. At the end, the samples are extracted with phenol, precipitated with ethanol, and run on sequencing gels along with a sequencing reaction performed with the same oligonucleotide used for toeprinting.

If the 30S ribosome is correctly placed on the translation initiation site, as it happens in the presence of a strong SD motif in the 5' UTR, the expected toeprinting signal should be located about 15 nucleotides downstream of the first nucleotide of the initiation codon (Fig. 5.3A). Often, multiple stop signals (3 to 5) are observed, because the ribosome oscillates somewhat on the message and is "pushed" further by the collision with the reverse transcriptase (Ringquist *et al.*, 1993). Absence, or randomness, of the stop signals indicates a lack of ribosome/mRNA interaction or random positioning of the particle on the mRNA. This is usually the outcome of the experiments performed with leaderless mRNAs or with mRNAs lacking ribosome binding signals (Fig. 5.3A). However, a +12/+13 toeprint can be obtained with leaderless mRNAs if the samples are supplemented with met-tRNAi, either charged or uncharged, indicating that the binary complex is sufficiently stabilized by the codon-anticodon interaction to be detected with this technique (Fig. 5.3B). Thus, codon-anticodon recognition appears to be necessary for efficient binding of the archaeal small ribosomal subunit to a leaderless mRNA, as has already been observed in bacteria (Grill *et al.*, 2000).

5. Initiation Factors

5.1. Cloning and purification of *S. solfataricus* translation initiation factors

The analysis of over 20 complete genomes of various archaeal species has revealed that all contain some 10 proteins homologous to eukaryotic initiation factors, or to subunits thereof. A few of these factors are universal, namely they have recognizable homologues also in bacteria, although the archaeal proteins always have the greatest homology with the eukaryotic ones. Summarizing the available data, the archaeal set of (putative) translation initiation factors includes four universal proteins (aIF1, aIF2/5B, aIF1A, aIF5A) two proteins found in eukaryotes but not in bacteria (a/eIF2, aIF6), two subunits (out of five) of the eukaryotic factor eIF2B, and one or two

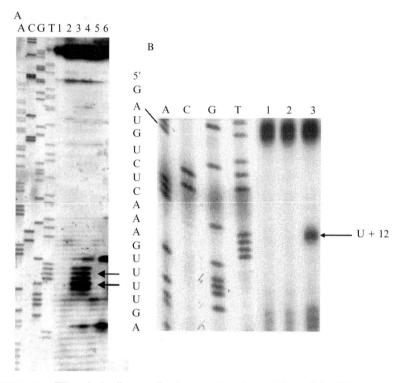

Figure 5.3 "Toeprinting" assays for determining the position of the 30S subunit on leadered and leaderless mRNAs. (A) Samples containing 30S subunits and a leadered mRNA were incubated for 15 min before primer extension. 1 and 2, wild-type mRNA 104, at 37° and 65°, without ribosomes. 3 and 4, wild-type mRNA with 30S subunits at 37° and 65°. 5 and 6, SD-less leadered mRNA with 30S subunits at 37° and 65°. (B) 1, leaderless mRNA alone; 2, mRNA and 30S subunits; 3, mRNA, 30S subunits and Met-tRNAi. Lanes ACGT, sequencing reaction. The arrows indicate the position of the toeprint signals. (Modified from Benelli *et al.*, 2003; Condo *et al.*, 1999.) (See color insert.)

homologs of a helicase (eIF4A) involved in eukaryotic initiation (Table 1). Thus archaea, despite being prokaryotes, have a more complex set of initiation factors than bacteria (which use only three, IF1, IF2, and IF3). However, the role of most of the archaeal putative factors is still unclear to a greater or lesser extent. Our laboratory has undertaken a systematic investigation of the function in translation initiation of the putative archaeal factors. This work has required the cloning and purification of most of the presumptive IFs of *S. solfataricus* and the development of various assays for determining their role in translation. In the following, the techniques currently available for the functional analysis of *S. solfataricus* initiation factors are described. Our analysis includes the proteins aIF1, aIF2/5B, aIF1A, a/eIF2, and aIF6. We have not considered the homologs of the eIF2B subunits and of eIF4A, because it is very doubtful that they have any role in translational initiation. Also, aIF5A has not yet been studied experimentally.

To clone the translational initiation factor genes, specific synthetic DNA oligomers for each gene are constructed on the sequence of S. *solfataricus*, as deduced from the published genome sequence (She *et al.*, 2001). The primers are used for PCR amplification of the desired genes; they contained specific enzymatic restriction sites, which allowed insertion of the PCR fragments into the corresponding sites of the expression plasmids pET-22b(+) (Novagen) or pRSETB (Invitrogen). The recombinant plasmids are selected in *E. coli* Top 10 (Invitrogen), and then transformed into *E. coli* BL21(DE3) (Stratagene). To ascertain the correct cloning of the gene, the purified recombinant vectors are sequenced.

The expression and purification of the recombinant proteins is achieved with standard techniques, except for the introduction of a heating step designed to enrich the cell lysates in the heat-stable factor derived from the expression of the *S. solfataricus* gene. Briefly, *E. coli* strain BL21(DE3), whose genome carries the RNA polymerase T7 gene under the control of lac UV5 promoter, are grown at 37° in LB medium containing ampicillin (100 μg/ml). Expression is induced at $OD_{600} = 0.5$ to 0.6 with 1-mM isopropyl-$_D$-thiogalactopyranoside (IPTG) and the cells grown for an additional 3 to 4 h before harvesting. The cell pellets are resuspended in Lysis buffer (50-mM NaH$_2$PO$_4$, 300-mM NaCl, 10-mM imidazole [pH 8.0]) plus 1 tablet of Protease Inhibitor Cocktail (Roche) per 10-ml lysis buffer and incubated with lysozyme on ice for 30 min. The cells are lysed by sonication and the lysates are clarified by centrifugation at 5000 rpm for 20 min at 4°. The cleared lysates are incubated 10 min at 70° to precipitate most of the heat-labile proteins of the bacterial host. The lysates enriched with the thermostable recombinant protein are clarified by centrifugation at 10,000 rpm for 30 min.

Thermostable His$_6$–proteins are purified from the lysates incubating 1 h on Ni-NTA agarose resin (Qiagen), washing with Wash buffer (50-mM NaH$_2$PO$_4$, 300-mM NaCl, 15-mM imidazole [pH 8.0]), and eluting with Elution buffer (50-mM NaH$_2$PO$_4$, 300-mM NaCl, 250-mM imidazole [pH 8.0]). These preparations are dialyzed against specific storage buffers (Table 5.3) and concentrated (when needed) with Millipore microconcentrators or by dialysis in hygroscopic environment (Sephadex G100). The concentration of the samples is determined by the Bradford Assay and the purified factors are stored in small aliquots at -80°, where they usually remain stable for several months.

It is important to note that all of the archaeal initiation factors are not tolerated equally well by the *E. coli* cells. The most relevant case is that of aIF1A, which can only be obtained by expressing the gene in the BL21 (DE3) *plys* strain, in which there is no "leakage" of the lacUV5 promoter and the transcription of the exogenous gene is effectively induced only

Table 5.3 Storage buffers for purified recombinant *S. solfataricus* initiation factors

Initiation factor	Storage buffer
a/eIF2 α, β, γ	10-mM MOPS, 200-mM KCl, 10-mM β-mercaptoethanol, 10% glycerol
aIF2/5B	20-mM Tris-HCl (pH 7.1), 60-mM KCl, 10% glycerol
aIF1	20-mM Tris-HCl (pH 7.1), 10-mM KCl, 6-mM β-mercaptoethanol, 10% glycerol
aIF6	20-mM Tris-HCl (pH 7.4), 20-mM NH$_4$Cl, 10% glycerol
aIF1A	20-mM Tris-HCl (pH 7.1), 20-mM KCl, 6-mM β-mercaptoethanol, 10% glycerol

upon addition of IPTG. This indicates that the archaeal protein is strongly deleterious for the bacterial cells, presumably because it interacts with the ribosomes without being able to participate in translational initiation. Another factor that may interfere deleteriously with bacterial translation is aIF2/5B. This factor can be expressed in the normal BL21(DE3) strain, but the cells grow slowly and the amount of protein recovered is not very high. It is interesting to observe that both aIF1A and aIF2/5B are universal factors, having homologs in all three domains of life and likely to play fundamental roles in all of them. In contrast, the factors shared by archaea and eukaryote but not present in bacteria, namely. a/eIF2 and aIF6, can be produced in *E. coli* in substantial amounts.

6. Functional Analyses of Recombinant *S. solfataricus* IFs

6.1. Reconstitution of trimeric a/eIF2

The function in translation initiation of the recombinant proteins cloned and purified as described earlier was analyzed using various *in vitro* assays. All of the recombinant proteins could be used directly after affinity purification, except for the case of a/eIF2, whose active form is a complex of three different subunits (α, β, and γ). The a/eIF2 subunits are cloned and purified independently of each other, but the analysis of the factor's function requires the reconstitution of the trimeric complex. It is also possible to analyze functionally α-γ or β-γ dimers, while the α and β subunits are unable to interact with each other. The reconstitution of the trimeric factor (or of dimers) is achieved as follows.

The standard reaction mixture for the interaction between the a/eIF2 polypeptides contains 30 pmol of α-, β- and γ-subunits (or only α-γ or β-γ)

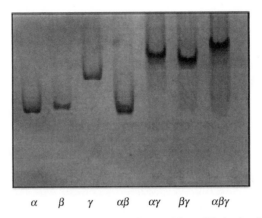

$$\alpha \quad \beta \quad \gamma \quad \alpha\beta \quad \alpha\gamma \quad \beta\gamma \quad \alpha\beta\gamma$$

Figure 5.4 Reconstitution of the trimeric a/eIF2 with purified subunits. The a-, β- and γ-polypeptides were mixed in equimolar amounts (50-pmol each) and incubated at 65° for 15 min. The individual subunits and mixtures thereof were electrophoresed on nondenaturing polyacylamide gels as described in the text.

in a buffer consisting of 30-mM KCl, 0.5-mM MgCl$_2$, 50-mM Tris-HCl (pH 7.1), and 1-mM GTP. The reaction is incubated at 65° for 15 min. The proteins and their complexes can be visualized by nondenaturing electrophoresis on 12% polyacrylamide gels prepared in acetate buffer (120-mM potassium acetate, 72-mM acetic acid [pH 4.3]). The gels included a stacking overlay of 4% polyacrylamide in acetate buffer (120-mM potassium acetate, 12-mM acetic acid [pH 6.8]). The running buffer is 133-mM acetic acid, 350-mM β-alanine (pH 4.4). After the run, the gels are stained with Coomassie brilliant blue. As shown in Fig. 5.4, under these conditions the proteins interact quantitatively to reconstitute the trimeric or dimeric complexes. We note that, although routinely performed at high temperature, the reconstitution reaction of the a/eIF2 subunits also interact readily at room temperature, indicating that they do not need heating to attain a correct functional conformation.

6.2. *In vitro* translation of archaeal mRNA in the presence of increasing amounts of recombinant initiation factors

An effective way to check whether a presumptive translation factor has indeed any function in protein synthesis is to determine whether variations in its cellular amount exert any effect on the efficiency of translation. When possible, this is best done by overexpressing the protein under analysis in living cells or trying to reduce or abolish its *in vivo* expression by various means. In practice, such *in vivo* genetic analyses are still rather difficult to perform in thermophilic Archaea such as *S. solfataricus*, because easy and reliable transformation/selection and reporter systems are not yet generally available. To circumvent this problem, we devised methods to assay the requirement for the different IFs in *S. solfataricus* by using the cell-free translational system described previously.

The most direct way to determine whether the overdosage of a putative translation initiation factor has any influence on the translational efficiency of specific (leadered and leaderless) reporter mRNAs would be to add the recombinant factor in increasing amounts to the *in vitro* translational system and measure the amount of reporter protein synthesized. Such *in vitro* experiments are similar to overexpression assays carried out *in vitro* and *in vivo* with different IFs (Grill *et al.*, 2001; Tedin *et al.*, 1999). However this approach presents several problems, the most relevant of which is that some of the recombinant factors inhibit translation unspecifically when added to the cell-free systems. This is because some factors (such as aIF2/5B) are only stable in a storage buffer containing a high concentration of monovalent cations (Table 5.3), which inhibit *in vitro* translation. Therefore, the direct approach is only feasible with factors that can be safely stored in low ionic strength buffers.

The *in vitro* translation reactions are performed according to the protocol described previously (Condò *et al.*, 1999). The samples contain in a final volume of 25 μl: 10-mM KCl, 20-mM Tris-HCl (pH 7), 20-mM MgCl$_2$, 7-mM β-mercatoethanol, 6-mM ATP, 3-mM GTP, 5 μg of bulk *S. solfataricus* tRNA, 20-μM methionine, 2 μl of [^{35}S]-methionine (1200 Ci*mmol^{-1} at 10 mCiml^{-1}), 5 μl of *S. solfataricus* S30 extract containing about 4 pmol*μl^{-1} of ribosomes (preincubated for 10 min at 73°), 0.4 pmol*μl^{-1} of *in vitro* transcribed mRNA (the reporter), and increasing amounts of the recombinant factor under study, corresponding to increasing IF:70S ratios (no aIF, 0.1:1, 0.2:1, and 0.4:1) in a final volume of 25 μl. Control samples are prepared at the same time, to which the factor's storage buffer with no protein is added. The mixtures are incubated at 73° and 5-μl aliquots are withdrawn at fixed time intervals (usually 0, 10, 15, and 20 min). The reaction products are separated on a 15% SDS-polyacrilamide gel. The quantification of the radioactivity band corresponding to the translational product of the reporter mRNA is performed using either an Instant Imager apparatus (Packard) or an X-ray film.

6.3. Cotranslation experiments

To overcome the problem of the unspecific inhibitory effect of some aIF storage buffers, we developed a different technique, consisting in expressing the putative initiation factors directly from their native mRNAs in the *in vitro* translation system.

The idea underlying these experiments is that the addition to the protein-synthesizing system of increasing amounts of an mRNA encoding a specific factor should result in the production of increasing amounts of the corresponding protein. This was usually the case in the experiments we have performed, but in any event it is something that should be checked before running the cotranslation assays. If so, the factor directly produced in the system is in turn expected to influence in trans the translation of a reporter mRNAs in a dose-dependent manner.

The success of this approach depends on the previous cloning of the gene encoding the IF under study complete of its native translation initiation region, to ensure a correct and efficient decoding of the corresponding mRNA.

In vitro translation of reporter mRNAs in the presence of the simultaneous coexpression of the mRNAs for an initiation factor is performed as described previously, except that the samples are supplemented with 0.4, 0.8, or 1.6 pmol*μl^{-1} of an aIF mRNAs instead of recombinant protein. In this manner, the system is programmed to maintain an excess of ribosomes over the exogenous mRNAs, to avoid competition between the two mRNAs for ribosome binding. At the end of reaction, the [^{35}S]-methionine-labeled proteins are separated on a SDS-polyacrilamide gel and the amount of reporter protein produced is measured by either an Instant Imager apparatus (Packard) or an X-ray film (Fig. 5.5).

The validity of this approach is demonstrated by the fact that the translation of a same reported mRNA can be either stimulated or inhibited by the coexpression of different initiation factors. For instance, the translation of both leadered and leaderless reporter mRNAs is stimulated by aIF1A (Fig. 5.5) but strongly inhibited by aIF6.

7. INTERACTION OF INITIATION FACTORS WITH RIBOSOMAL SUBUNITS

A protein acting as a translation initiation factor is expected to interact with other components of the protein synthetic apparatus at some stage of translational initiation. The components interacting with the factor can be the ribosome itself, tRNA$_i^{Met}$, other IFs, or a combination of these. In determining the function of a presumptive IF, it is therefore very important to identify its interacting partners. In particular, the study of the association of a given factor with the ribosomes can give useful clues as to the specific initiation step in which it participates. As an example, exclusive binding to the small ribosomal subunits suggests an involvement in the early stages of the initiation process, whereas preferential interaction with the large subunit or with the 70S ribosome indicates participation in late initiation events.

The association of *S. solfataricus* IFs, both native and recombinant, can be investigated by two different approaches: (1) Fractionating cell lysates or ribosomes on sucrose density gradients and then probing each fraction for

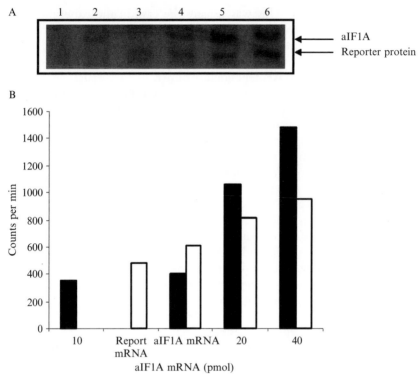

Figure 5.5 Cotranslation experiment demonstrating stimulation of protein synthesis by aIF-1A. An *in vitro* transcribed mRNA encoding initiation factor 1A was added in increasing amounts to a translation system programmed with a fixed amount of a reporter mRNA encoding a ribosomal protein. (A) Lane 1, no added mRNA; lane 2, 10 pmol of aIF-1A mRNA; lane 3, 10 pmol of reporter mRNA; lanes 4, 5 and 6, 10 pmol of 104 mRNA plus 10, 20, and 40 pmol of aIF-1A mRNA, respectively. (B) Histogram obtained by quantifying the radioactive bands corresponding to the translated proteins (10.4 and 12 kDa for 104 and aIF-1A mRNA, respectively).

the presence of the protein under study by Western blotting; or (2) running native ribosomes or subunits on a nondenaturing polyacrylamide gel and probing the position of the initiation factor in the gel by Western blotting.

7.1. Analysis of IF/ribosome association by velocity sedimentation

Using S-30 fractions, it is possible to analyze the association of the IFs with the ribosomes in different conditions of translational activity and at different stages of translation. For instance, *S. solfataricus* cell lysates that are not programmed for translation and are not incubated at the correct temperature contain only "resting" ribosomes, that is, ribosomes not engaged in protein synthesis. Lysates programmed for translation may contain different amounts of ribosomes initiating translation or involved in the elongation stage depending on the (short or long) incubation time. Moreover, *S. solfataricus* ribosomes can be observed as 70S monomers only when engaged in translation.

Therefore, the cell-free protein-synthesizing systems can be manipulated to visualize the association of the initiation factors with the ribosomes in the various phases of translation. Moreover, it is possible to detect the ribosomal localization of both native and recombinant IFs, depending on which antibody (anti-IF or anti-His) is used to perform the Western blotting.

Typical translation mixtures for monitoring the association of the IFs with the ribosomes contain 15 μl of *S. solfataricus* S-30 extract (with or without the components needed to activate an *in vitro* translation, as described) supplemented, if required, with 20 to 40 pmol of a specific initiation factor in a final volume of 75 μl. The samples are incubated at 73° for the appropriate time (about 1 to 2 min for allowing formation of the initiation complexes, up to 15 to 20 min for allowing elongation of the polypeptide chain). If 70S monomers are to be detected, prior to running the gradients, the samples should be fixed by adding formaldehyde 6% (final concentration) and keeping on ice for 30 min. In this latter case, care should be taken to prepare the S-30 fractions by substituting the customary 20-mM Tris/HCl (pH 7.0) with 20-mM Triethanolamine (TEA) (pH 7), because in the presence of Tris/HCl the addition of HCHO provokes the precipitation of the samples. Fixation with formaldehyde is essential to stabilize the 70S monomers, which would otherwise be dissociated by the hydrostatic pressure during the run.

After incubation, the samples are layered on linear 10 to 30% sucrose gradients containing 10-mM KCl, 20-mM Tris/HCl (or TEA) (pH 7.5) and 20-mM MgCl$_2$. The gradients are centrifuged in a Beckman SW41 rotor at 4° and 36,000 rpm for 4 h. After centrifugation, the gradients are unloaded with an ISCO UA-5 gradient collector and fractions of $\sim$0.5 ml are collected while monitoring the optical density at 260 nm. The individual fractions are precipitated with 4 vol of acetone for 1 h at $-20°$ and then centrifuged at 13,000 rpm. The protein pellets are resuspended in 20 μl of SDS-sample buffer, separated by SDS-PAGE, and then electroblotted to nitrocellulose membrane. The proteins of interest are visualized by probing the membrane with either anti-6-His (for recombinant factors) or polyclonal anti-aIFs antibodies (for native proteins).

Figure 5.6 shows a panel of typical results obtained with this technique, illustrating the localization of several IFs on *S. solfataricus* ribosomes.

A factor showing a markedly different behavior in resting lysates and in lysates activated for translation is aIF2/5B. In the former case, most of the protein localized in the top fractions containing low-molecular-weight

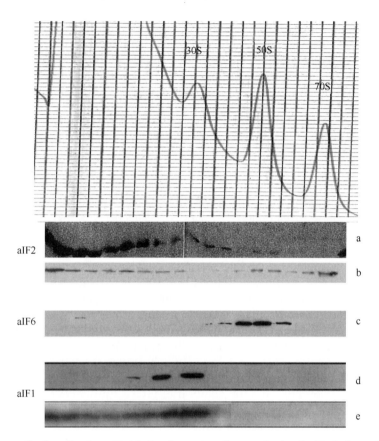

Figure 5.6 Localization of initiation factors on ribosomal subunits. Typical reactions contained 25 μl of S-30 fraction and 20 to 40 pmol of a specific initiation factor. (a) localization of endogenous aIF2/5B in resting lysates; (b) localization of endogenous aIF2/5B in lysates programmed for translation and incubated at 73° for 2 min; (c) localization of endogenous aIF6 in "resting" lysates; (d, e) localization of aIF1 in resting (d) or translating (e) lysates.

components and only a small amount of it was visible in the fractions containing the 30S and 50S ribosomal subunits (Fig. 5.6, lane a). However, in cell extracts programmed for translation and incubated for a short time (1 to 2 min) a fraction of the protein associated with the 70S peak, whereas the amount of aIF2/5B bound to the subunits, especially the 50S ones, was also increased (Fig. 5.6, lane b). These results indicate that aIF2/5B interacts with both subunits and may be involved in their association during translation to form the 70S ribosome, as it has been shown for its bacterial and eukaryotic homologs, IF2 and eIF5B (Allen *et al.*, 2005; Pestova *et al.*, 2000).

In contrast, other factors interact exclusively with only one of the ribosomal subunits, often regardless of the physiological state of the ribosomes. For instance, aIF6 (native or recombinant) associates exclusively with the 50S subunits in both resting and translating cell lysates and is never found on either the 30S or the 70S particles (Fig. 5.6, lane c). aIF1 has the opposite behavior, being associated exclusively with the 30S particles (Hasenöhrl *et al.*, 2006). In this case, the only difference between resting and translating cell extracts is that in the latter case some amount of aIF1 is also found in a free state, suggesting that this factor recycles between the 30S subunit and the soluble fraction upon entering the elongation phase (Fig. 5.6, lanes d,e; Hasenöhrl *et al.*, 2006).

7.2. Native gel assays

The interaction of the recombinant IFs with the ribosomes can be also studied using only purified ribosomes and proteins, independently of other translational components. In this case, the ribosomes, after incubation with the factors, are loaded on nondenaturing gels at low polycrylamide concentration (4%). The electrophoretic system is devised to allow the ribosomes to enter the gel and to run to the cathode, carrying along the bound factors. Free factors, in contrast, do not enter the gel and do not interfere with the final result.

The standard reaction mixture contains 50 pmol of the recombinant protein(s) and 30 pmol of ribosomes in a buffer consisting of 30-mM KCl, 0.5-mM MgCl$_2$, and 50-mM Hepes (pH 7.5). The reaction is incubated 10 min at 65°, stopped by adding an appropriate volume of 10× loading buffer (0.02% bromophenol blue, 0.02% xylene cyanol, and 50% glycerol) and immediately loaded on running nondenaturing 4% polyacrylamide gels made in 20-mM potassium acetate, 2.5-mM MgCl$_2$, and 40-mM Tris-HCl (pH 6). Electrophoresis is continued at 4° and 50 mA for 4 to 5 h. After the run, the samples are transferred to Hybond N membranes (Amersham) by electroblotting and the position of the recombinant proteins is visualized by immunostaining with anti-His tagged antibodies or with antibodies raised against the specific protein under analysis.

8. IF/tRNA$_i$^{Met} Interaction

Initiation of protein synthesis requires the correct positioning of charged initiator tRNA in the ribosomal P-site and its interaction with the start codon on mRNA. This is accomplished, in bacteria, by translation initiation factor IF2, which is able to interact directly with the fMet-tRNA^{fmet}. In eukaryotes this task is carried out by the trimeric complex eIF2, which binds Met-tRNA$_i^{Met}$ (the initiating methionine is unformylated in both eukaryotes and archaea) However, eukaryotes possess a homolog of bacterial IF2, termed eIF5B, which seems to act at a later initiation step.

As with the eukaryote, the archaea possess two proteins potentially capable of interacting with the initiator tRNA: a/eIF2, homolog to eukaryotic eIF2, and aIF2/5B, homolog to bacterial IF2 and eukaryotic eIF5B. The former protein has been shown to interact preferentially with Met-tRNA$_i^{Met}$ (Pedulla et al., 2005; Yatime et al., 2004), while the Met-tRNA$_i^{Met}$ binding capacity of aIF2/5B is still somewhat controversial. The interaction of the appropriate initiation factors with Met-tRNA$_i^{Met}$ can be assayed using several different assays, as described later.

8.1. Hydrolysis protection assay

A very sensitive test designed to detect even a rather weak interaction between Met-tRNA$_i^{Met}$ and a purified IF is the hydrolysis protection assay. It is based upon the fact that the ester bond connecting the amino acid to the tRNA is labile and is hydrolyzed spontaneously in an alkaline environment. However, binding of Met-tRNA$_i^{Met}$ by the specific IF slows down hydrolysis and hinders amino acid loss (Forster et al., 1999). The extent of protection exerted by a putative Met-tRNA$_i^{Met}$ binding factor in S. solfataricus can be measured as follows. Reaction mixtures (80 μl) containing 100-mM Tris-HCl (pH 8), 125-mM NH$_4$Cl, 6-mM MgCl$_2$, 1-mM DTT, 12 to 100 pmol of [^{35}S]Met-tRNA$_i^{Met}$ (s.a. ~500 c.p.m./pmol) and increasing quantities of the factor under analysis (10 to 200 pmol) are incubated at 65° for up to 60 min, while withdrawing samples at fixed time points. The samples are filtered under vacuum through Millipore 0.22-μm nitrocellulose disks, which are then washed extensively with incubation buffer. The nitrocellulose disks are dried and the amount of retained radioactivity is measured in a liquid scintillation counter.

8.2. Gel-retardation assay

For gel-retardation experiments, total S. solfataricus tRNA is aminoacylated with a mixture of [^{35}S]-Methionine at a specific radioactivity of 1 Ci mmol^{-1} following the protocol described previously. Complexes with the IF are formed in 50-mM Hepes/KOH (pH 7.5), 10-mM MgCl$_2$, 30-mM KCl, 1-mM GTP (when needed), adding 12 pmol of [^{35}S]Met-tRNA$_i^{Met}$ and different molar excesses (12 to 40 pmol) of the initiation factor. Incubation is carried out at 65° for 10 min, stopped by adding an appropriate volume of loading buffer consisting of 20-mM MOPS/NaOH (pH 7), 40%

glycerol, and immediately loaded on a nondenaturing gel. The gel system contains a 6 to 12% gradient of acrylamide in 20-mM MOPS/NaOH (pH 7.5) and allows optimal separation of the [^{35}S]Met-tRNA$_i$^{Met}/aIF complex from unbound tRNA. Electrophoresis is for 3 h at 100 V and room temperature in a buffer containing 20-mM MOPS/NaOH (pH 7.5).

9. INTERACTION OF [^{35}S]MET-tRNA$_i$^{MET} WITH RIBOSOMES

An essential function of the initiation factors is to stimulate the specific interaction of Met-tRNA$_i$^{Met} with the ribosomes. This task is carried out primarily by a/eIF2 in *S. solfataricus*, although aIF2/5B also seems to have a role in it. The ability of recombinant a/eIF2 to stimulate binding of Met-tRNA$_i$^{Met} to ribosomes can be tested as follows:

20 to 50 pmol of each purified a/eIF2 subunit are mixed and incubated at 65° for 5 min in incubation buffer in the presence of 1-mM GTP to allow the formation of the complete trimer. Note that GTP is not strictly required for trimer formation, but it is indispensable later to allow the interaction of the factor with Met-tRNA$_i$^{Met}. Then, ∼100 pmol of [^{35}S]Met-tRNA$_i$^{Met} are added and the incubation is continued for another 5 to 10 min to promote the formation of the Met-tRNA$_i$^{Met}/a/eIF2 complex. Finally, the samples are supplemented with 30 pmol of ribosomes and incubation is continued for another 10 min. The addition of an mRNA, or a fragment thereof containing the translation initiation site, stimulates Met-tRNA$_i$^{Met}/ribosome interaction but it is not essential.

At the end of the incubation, the samples are immediately electrophoresed on nondenaturing 4% polyacrilamide gels made in 20-mM potassium acetate, 2.5-mM MgCl$_2$, and 40-mM Tris-HCl (pH 6). After the run, the gels are stained with Coomassie brilliant blue to determine the position of the ribosomes, then dried and exposed to an X-ray film to visualize the labeled Met-tRNA$_i$^{Met}. Note that in this gel system the ribosomal subunits are separated and the labeled Met-tRNA$_i$^{Met} remains associated with the 30S subunit. The amount of ribosome-bound Met-tRNA$_i$^{Met} in the presence of increasing concentration of the factor is quantified with a PhosphoImager. Free Met-tRNA$_i$^{Met} runs much faster than the 30S/Met-tRNA$_i$^{Met} complex and can be visualized toward the bottom of the gel; however, it cannot be reliably quantified because free Met-tRNA$_i$^{Met} tends to get deacylated during the run. Figure 7 shows a typical profile of stimulation of Met-tRNA$_i$^{Met} binding to *S. solfataricus* ribosomes by trimeric a/eIF2. Note that similar results are obtained by using an α-γ dimer, whereas the isolated γ subunit does not promote Met-tRNA$_i$^{Met} binding to the ribosomes (Fig. 5.7).

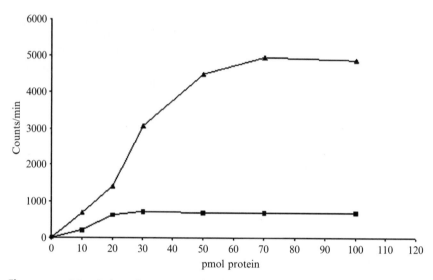

Figure 5.7   Stimulation of Met-tRNAi binding to ribosomes by a/eIF2. The ribosomes (40 pmol), radiolabeled met-tRNAi and factors were incubated at 65° for 10 min and electrophoresed on nondenaturing polyacrylamide gels. The bands corresponding to the 30S subunits/met-RNAi complexes were quantified with a PhosphoImager. Triangles: reconstituted trimeric a/eIF2; Squares: a/eIF2 γ-subunit alone.

This kind of assay can also be used to evaluate whether other IFs participate in stimulating the interaction of Met-tRNA$_i$^{Met} with the ribosome. In this case, increasing amount of the recombinant protein are added to the assay containing a fixed amount of the other components, including a/eIF2. In this way, we could ascertain aIF1 (Hasenöhrl *et al.*, 2006), aIF1A, and aIF2/5B (unpublished results) can enhance Met-tRNA$_i$^{Met}/ribosome interaction.

10. TRANSLATIONAL G-PROTEINS

A number of translation initiation factors (and both elongation factors) are G proteins that can bind and hydrolyze guanine nucleotides in a ribosome-dependent manner. Like other G proteins, translational GTPases are "molecular switch" proteins that bind and hydrolyze GTP with a cyclic mechanism which activates and inactivates the protein. This cyclic reaction may involve several other factors that either stimulate the GTP hydrolysis step or catalyze the release of bound GDP and its substitution with GTP. A core domain that is able to bind either GTP or GDP (termed G-domain) confers the characteristic switch mechanism of GTPases and its folding is nearly invariant throughout the GTPase superfamily (Sprang, 1997).

The archaea, like the eukaryote, include two G proteins, aIF2/5B and a/eIF2, in their set of translation initiation factor. These two proteins are expected to possess a ribosome-dependent GTPase activity, but its features and requirements are as yet poorly understood. In *S. solfataricus,* we have found that at least aIF2/5B appears to be able to hydrolyze GTP in the presence of both 30S and 50S ribosomal subunits. In contrast, the conditions for GTP hydrolysis by a/eIF2 remain undetermined. The GTPase assay with aIF2/5B and *S. solfataricus* ribosomes and factors is run as follows.

Increasing amounts (5 to 100 pmol) of aIF2/5B are incubated with $\sim$40 pmol of 70S ribosomes and 800-pmol γ-^{32}P-GTP and 35-μM GTP in 25-mM Tris-HCl (pH 7.1), 70-mM KCl, 7-mM MgCl$_2$, 1-mM DTT (final vol 50 μl) The samples were incubated at 65° for 20 min while withdrawing samples at fixed time points (0, 1, 5, 10, and 20 min). The reaction is stopped by adding 1-M HClO$_4$/2-M KH$_2$PO$_4$. The extent of released γ-^{32}P was determined by isopropyl acetate extraction in an acidic environment of the ^{32}P-dodecamolibdate complex: the tubes were stirred for 2 min and briefly centrifuged to separate the phases. 200 μl of the upper organic phase were counted in 2 ml of scintillation solution (adapted from Beaudry *et al.,* 1979). The rate of the hydrolysis of aIF2/5B-bound GTP is influenced by several parameters, the most important of which are the temperature and the presence of ribosomes. No GTPase activity is observed in the absence of ribosomes or at temperatures below 60°.

ACKNOWLEDGMENTS

Work in the authors' laboratory was supported by grants from the projects PRIN 2002 and PRIN 2005 of the Ministry of University and Research (MIUR). D. B. is supported by a research fellowship of the University of Bari, Italy.

REFERENCES

Allen, G. S., Zavialov, A., Gursky, R., Ehrenberg, M., and Frank, J. (2005). The cryo-EM structure of a translation initiation complex from. *Escherichia coli. Cell* **121,** 703–712.

Beaudry, P., Sander, G., Grunberg-Manago, M., and Douzou, P. (1979). Cation-induced regulatory mechanism of GTPase activity dependent on polypeptide initiation factor 2. *Biochemistry* **18,** 202–207.

Bell, S. D., and Jackson, S. P. (1998). Transcription in Archaea. *Cold Spring Harb. Symp. Quant. Biol.* **63,** 41–51.

Benelli, D., Maone, E., and Londei, P. (2003). Two different mechanisms for ribosome/mRNA interaction in archaeal translation initiation. *Mol. Microbiol.* **50,** 635–643.

Bult, C. J., White, O., Olsen, G. J., Zhou, L., Fleischmann, R. D., Sutton, G. G., Blake, J. A., FitzGerald, L. M., Clayton, R. A., Gocayne, J. D., Kerlavage, A. R.,

Dougherty, B. A., et al. (1996). Complete genome sequence of the methanogenic archaeon, Methanococcus jannaschii. *Science* **273**, 1058–1073.

Condò, I., Ciammaruconi, A., Benelli, D., Ruggero, D., and Londei, P. (1999). Cis-acting signals controlling translational initiation in the thermophilic archaeon *Sulfolobus solfataricus*. *Mol. Microbiol.* **34**, 377–384.

Dennis, P. P. (1997). Ancient ciphers: Translation in Archaea. *Cell* **89**, 1007–1010.

Forster, C., Krafft, C., Welfle, H., Gualerzi, C. O., and Heinemann, U. (1999). Preliminary characterization by X-ray diffraction and Raman spectroscopy of a crystalline complex of *Bacillus stearothermophilus* initiation factor 2 C-domain and fMet-tRNAfMet. *Acta Crystallogr. D Biol. Crystallogr.* **55**, 712–716.

Grill, S., Gualerzi, C. O., Londei, P., and Blasi, U. (2000). Selective stimulation of translation of leaderless mRNA by initiation factor 2: Evolutionary implications for translation. *EMBO J.* **19**, 4101–4110.

Grill, S., Moll, I., Hasenohrl, D., Gualerzi, C. O., and Blasi, U. (2001). Modulation of ribosomal recruitment to 5′-terminal start codons by translation initiation factors IF2 and IF3. *FEBS Lett.* **495**, 167–171.

Hasenöhrl, D., Benelli, D., Barbazza, A., Londei, P., and Blasi, U. (2006). *Sulfolobus solfataricus* translation initiation factor 1 stimulates translation initiation complex formation. *RNA* **12**, 674–682.

Kyrpides, N. C., and Woese, C. R. (1998). Universally conserved translation initiation factors. *Proc. Natl. Acad. Sci. USA* **95**, 224–228.

Londei, P. (2005). Evolution of translational initiation: New insights from the archaea. *FEMS Microbiol. Rev.* **29**, 185–200.

Londei, P., Altamura, S., Cammarano, P., and Petrucci, L. (1986). Differential features of ribosomes and of poly(U)-programmed cell-free systems derived from sulphur-dependent archaebacterial species. *Eur. J. Biochem.* **157**, 455–462.

Pedulla, N., Palermo, R., Hasenohrl, D., Blasi, U., Cammarano, P., and Londei, P. (2005). The archaeal eIF2 homologue: Functional properties of an ancient translation initiation factor. *Nucleic Acids Res.* **33**, 1804–1812.

Pestova, T. V., Lomakin, I. B., Lee, J. H., Choi, S. K., Dever, T. E., and Hellen, C. U. (2000). The joining of ribosomal subunits in eukaryotes requires eIF5B. *Nature* **403**, 332–335.

Ringquist, S., MacDonald, M., Gibson, T., and Gold, L. (1993). Nature of the ribosomal mRNA track: Analysis of ribosome-binding sites containing different sequences and secondary structures. *Biochemistry* **32**, 10254–10262.

Ruggero, D., Creti, R., and Londei, P. (1993). In vitro translation of archaeal natural mRNAs at high temperature. *FEMS Microbiol. Lett.* **107**, 89–94.

Sensen, C. W., Klenk, H. P., Singh, R. K., Allard, G., Chan, C. C., Liu, Q. Y., Penny, S. L., Young, F., Schenk, M. E., Gaasterland, T., Doolittle, W. F., Ragan, M. A., et al. (1996). Organizational characteristics and information content of an archaeal genome: 156 kb of sequence from *Sulfolobus solfataricus* P2. *Mol. Microbiol.* **22**, 175–191.

She, Q., Singh, R. K., Confalonieri, F., Zivanovic, Y., Allard, G., Awayez, M. J., Chan-Weiher, C. C., Clausen, I. G., Curtis, B. A., De Moors, A., Erauso, G., Fletcher, C., et al. (2001). The complete genome of the crenarchaeon *Sulfolobus solfataricus* P2. *Proc. Natl. Acad. Sci. USA* **98**, 7835–7840.

Slupska, M. M., King, A. G., Fitz-Gibbon, S., Besemer, J., Borodovsky, M., and Miller, J. H. (2001). Leaderless transcripts of the crenarchaeal hyperthermophile Pyrobaculum aerophilum. *J. Mol. Biol.* **309**, 347–360.

Sprang, S. R. (1997). G protein mechanisms: Insights from structural analysis. *Annu. Rev. Biochem.* **66**, 639–678.

Tedin, K., Moll, I., Grill, S., Resch, A., Graschopf, A., Gualerzi, C. O., and Blasi, U. (1999). Translation initiation factor 3 antagonizes authentic start codon selection on leaderless mRNAs. *Mol. Microbiol.* **31,** 67–77.

Tolstrup, N., Sensen, C. W., Garrett, R. A., and Clausen, I. G. (2000). Two different and highly organized mechanisms of translation initiation in the archaeon *Sulfolobus solfataricus. Extremophiles* **4,** 175–179.

Yatime, L., Schmitt, E., Blanquet, S., and Mechulam, Y. (2004). Functional molecular mapping of archaeal translation initiation factor 2. *J. Biol. Chem.* **279,** 15984–15993.

CHAPTER SIX

RECONSTITUTION OF YEAST TRANSLATION INITIATION

Michael G. Acker, Sarah E. Kolitz, Sarah F. Mitchell, Jagpreet S. Nanda, *and* Jon R. Lorsch

Contents

Department of Biophysics and Biophysical Chemistry, Johns Hopkins University School of Medicine, Baltimore, Maryland

Methods in Enzymology, Volume 430
ISSN 0076-6879, DOI: 10.1016/S0076-6879(07)30006-2

Abstract

To facilitate the mechanistic dissection of eukaryotic translation initiation we have reconstituted the steps of this process using purified *Saccharomyces cerevisiae* components. This system provides a bridge between biochemical studies *in vitro* and powerful yeast genetic techniques, and complements existing reconstituted mammalian translation systems (Benne and Hershey, 1978; Pestova and Hellen, 2000; Pestova *et al.*, 1998; Trachsel *et al.*, 1977). The following describes methods for synthesizing and purifying the components of the yeast initiation system and assays useful for its characterization.

1. Introduction

Eukaryotic translation initiation is an extremely complex process that requires at least 12 initiation factors (versus three factors in bacteria) to position an initiator methionyl-tRNA$_i^{Met}$ in the P-site of the ribosome, base-paired to the correct AUG codon of the mRNA to be translated. Decades of work have elucidated many details of this process, leading to the current model of eukaryotic initiation (Fig. 6.1; reviewed in Kapp and Lorsch, 2004b; Pestova *et al.*, 2001, 2007). Briefly, eukaryotic initiation factor (eIF) 2 forms a ternary complex (TC) with GTP and methionyl-tRNA$_i^{Met}$ that brings the methionyl-tRNA$_i^{Met}$ onto the 40S ribosomal subunit with the help of eIF1, eIF1A, and eIF3. The resulting 43S complex is thought to bind to the 5'-end of an mRNA, near the 7-methylguanosine cap, and scan in the 3' direction in search of the AUG start codon. eIF2, with the aid of the GTPase-activating protein eIF5, is able to partially hydrolyze GTP to GDP · P$_i$ prior to start codon recognition, but is unable to release the bound P$_i$. Recognition of the start codon causes a conformational change in the pre-initiation complex that results in release of eIF1

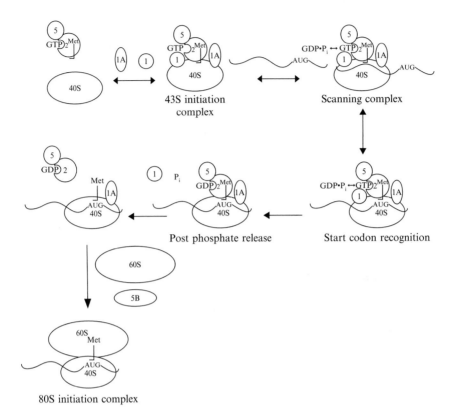

Figure 6.1 Schematic of the current model of the steps of translation initiation in the reconstituted yeast system. For clarity, factors are not shown to scale. A ternary complex (TC) formed by eIF2, methionyl-tRNA$_i^{Met}$, and GTP binds tightly to the GTPase-activating protein, eIF5. With the help of two small initiation factors, eIF1 and eIF1A, TC binds to the 40S ribosomal subunit, forming the 43S complex. This complex can then bind to and, *in vivo*, scan along mRNA in search of the start codon. During this time eIF2, with the assistance of eIF5, is able to hydrolyze GTP; P$_i$ cannot be released at this stage, however, it creates an equilibrium between GTP and GDP · P$_i$. Once the start codon has been located, a conformational change takes place, reducing the affinity of the complex for eIF1 and leading to the dissociation of this factor. After eIF1 has dissociated, phosphate is released, causing GTP hydrolysis to become irreversible, and committing the complex to initiating at the chosen codon. eIF2 · GDP is then released from the complex. The 60S ribosomal subunit joins the 40S subunit in a step facilitated by the GTPase eIF5B. This step results in an 80S initiation complex with Met-tRNA$_i^{Met}$ in the P-site, base paired to the start codon of the mRNA, ready to begin the elongation phase of translation.

from its binding site on the 40S subunit. Release of eIF1 triggers rapid P$_i$ release from eIF2 · GDP · P$_i$, making GTP hydrolysis irreversible and allowing downstream events in the pathway to take place. Recognition of the start codon is also thought to result in an additional conformational change that prevents further scanning of the mRNA. A second GTPase, eIF5B, promotes 60S ribosomal subunit joining to the 40S · mRNA · methionyl-tRNA$_i^{Met}$ complex. GTP hydrolysis by eIF5B reduces the affinity of the factor for the 80S initiation complex and dissociation of eIF5B results in a translationally competent 80S ribosome.

In vitro reconstitution has provided an invaluable tool for investigating the mechanism of translation initiation. The ability to monitor individual steps of the pathway, while at the same time controlling both the component concentrations and their structures, has enabled measurement of many of the kinetic and thermodynamic parameters at the heart of the process. Although mammalian reconstituted systems have existed for nearly 3 decades (Benne and Hershey, 1978; Pestova and Hellen, 2000; Pestova *et al.*, 1998; Trachsel *et al.*, 1977) and have provided great insight into the workings of the initiation machinery, the creation of a yeast reconstituted initiation system has greatly facilitated the coupling of the awesome power of yeast genetics (widely known as APOG) to the benefits of *in vitro* reconstitution, the awesome power of biophysical chemistry (not so widely known as APOBPC) (Algire *et al.*, 2002).

In this chapter, we describe in detail methods developed for reconstituting yeast translation initiation. Several assays useful for checking the activity of the reagents are also presented. The current minimal system uses an unstructured, model mRNA (mRNA(AUG)), obviating the need for factors involved in cap binding, mRNA recruitment, and mRNA remodeling, allowing core steps in the initiation pathway to be isolated and studied. Consequently, purification and characterization of these factors is not discussed. The incorporation of these factors and the steps that they mediate is an ongoing goal of our lab, and will hopefully be discussed in a future chapter.

Note: The compositions of all buffers, solutions, and gels referred to in the following sections are given at the end of the chapter in the Solutions section.

2. LARGE-SCALE LYSIS OF *S. CEREVISIAE* CELLS

Reconstitution of translation initiation requires the synthesis and purification of a large number of components on a relatively large scale. The only way to increase the yield of some components is to increase the culture size, often to >12 liters. This creates a problem in the speed and efficiency with which the yeast cultures can be lysed. Goode has described a protocol for large-scale lysis of *Saccharomyces cerevisiae* using liquid nitrogen and a Waring blender (see http://www.bio.brandeis.edu/goodelab/proto. html for details; Goode, 2002). This method allows relatively rapid and efficient lysis of large amounts of yeast cells, while at the same time minimizing sample degradation by stabilizing the lysate as a frozen powder.

Here, we summarize this protocol, which is used in a number of purification schemes presented in this chapter.

Pellet the washed cells in a single bottle and determine the pellet weight. Using a vol equal to 0.33 the weight of the pellet (1 ml = 1 g) resuspend the pellet in the appropriate lysis buffer or ddH$_2$O. Slowly drip the cell suspension into a bucket of liquid N$_2$ using a 25 ml pipette to create frozen cell droplets. Carefully scoop the frozen droplets into a plastic bottle, cap loosely, and store at −80° until all the liquid N$_2$ has evaporated, then cap tightly.

Dry a Waring blender canister overnight in an oven or the fume hood. Any residual water in the turning mechanism will freeze upon addition of liquid N$_2$, causing the propeller to lock in place, which can burn out the motor. For safety, set up the blender apparatus in the fume hood or behind a radiation shield to deflect liquid N$_2$ spray, and wear safety goggles and cryo gloves. Briefly cool the canister by blending a small amount of liquid N$_2$. Fill the canister no more than half-full with frozen cell pellets (30 to 150 g). Overfilling can cause yeast lysate to spray from the canister. Add liquid N$_2$ to the level of the cell pellets and, holding the canister lid in place, turn the blender on high and blend until the liquid N$_2$ evaporates, about 15 to 30 s. This will be marked by a change in the pitch of the blender sound. (Note: If yeast powder is expelled from the lid vents, stop and wait until the liquid N$_2$ level has decreased to the level of the pellets and begin again.) Tap down the powder and again add liquid N$_2$ to the canister to just above the level of the powder. Repeat the process at least 4 times for the best lysis. Transfer the powder into a 1 liter bottle using a funnel and a spatula. Add lysis buffer to the desired final volume. Shake the bottle vigorously until all the cells have been completely resuspended. You can run warm water over the bottle to help thaw the cells. (Important: Remember to open the cap occasionally until all the N$_2$ gas has escaped.)

3. Purification of 40S and 60S Ribosomal Subunits from S. cerevisiae

3.1. Culturing and storage of cells

Streak out a plate of YAS2488 cells (MATa leu2-3 112 his4-539 trp1 ura3-52 cup1::LEU2/PGK1pG/MFA2pG [Algire et al., 2002]). The stock plate should be no more than 2 wk old. Starting with cells from older stock plates results in decreased ribosome yield.

Inoculate 36 liters of YPD media in twenty-four 2800 ml baffled flasks with 5 ml from a 150 ml saturated overnight culture per flask. Incubate at 30°, shaking at 250 rpm until the culture reaches an OD$_{600}$ of 1.0 ± 0.1. This should take approximately 7 to 8 h.

3.2. Purification of 80S ribosomes

Lyse cells by the Waring blender method. The final volume of thawed lysate/buffer should be ~375 ml. Transfer the thawed lysate into 16 SS-34 tubes. Clarify the lysate by centrifugation in a Sorvall SS-34 rotor at 13,000 rpm for 30 min. Chill eight 50 ml conicals and 16 clean 26.3 ml polycarbonate tubes (with caps) (Beckman 355618) on ice. Aliquot 2.5 ml of sucrose cushion into each polycarbonate tube. When the clarification spin is finished, immediately transfer the supernatant to the chilled conicals using a 10 ml serological pipette, avoiding the lipid layer at the top. Carefully layer 22.5 ml of clarified lysate onto each of the 16 sucrose cushions. Balance the tubes and caps with lysis buffer. Centrifuge the tubes in a Beckman Type 70Ti ultracentrifuge rotor at 60,000 rpm for 106 min to pellet the 80S ribosome. During this spin, prepare the sucrose gradients; see Gradient Preparation.

Immediately upon completion of the ultracentrifuge spin, remove and discard the supernatant with a 10 ml serological pipette. Try to remove as much of the floating, white lipid layer as possible with the liquid, being very careful not to contaminate the ribosomal pellet with lipids. The pellets should be as clear as glass at this point. Add 2 ml of high salt wash buffer to each tube. Resuspend each pellet with a P1000 and pool together in a 50 ml conical. Rinse all the tubes with an additional 2 ml of high salt wash buffer and add to the 50 ml conical. Bring the vol up to 36 ml with high salt wash buffer. Place the 50 ml conical in a beaker filled with ice and stir at a moderate speed with a micro-stir bar for 1 h. During this incubation, prepare 12 TLA 100.3 tubes by placing 250 μl of sucrose cushion in the bottom of each and chilling on ice. Pellet the 80S ribosomes by layering the salt wash (3 ml per tube) onto the sucrose cushions, balancing carefully, and spinning at 100,000 rpm for 30 min at 4° in the TLA 100.3 rotor in a table-top ultracentrifuge.

Resuspend all 12 pellets in 3 ml (total) subunit separation buffer and rinse the tubes with an additional 1 ml. Transfer the ribosome solution to micro-centrifuge tubes and spin for 30 s at 14,000 rpm to remove any debris. Transfer the ribosome solution into a chilled 15 ml conical. Measure the OD_{260} of 1 μl of the ribosome solution in 1 ml of water and store the samples on ice.

3.3. Gradient preparation

Aliquot 10 ml of 5% sucrose gradient solution into each of twelve 25 × 89 mm, polyallomer tubes (Beckman 326823). Use a 25 ml serological pipette to slowly layer 20 ml of 20% sucrose gradient solution underneath the 5% layer in each tube. Top off the tubes with 5%, being careful not to mix the two layers. The meniscus should rise above the top of the tube. Cap the tubes with a rolling motion, inserting the end of the cap with the hole in it last to avoid air bubbles. Use the Gradient Master (BioComp Instruments, Inc.) to mix the gradients with the following settings: *time* = 1:21, *angle* = 76°, *speed* = 20. Be sure to start with a level platform.

3.4. Separation of 80S ribosomes into 40S and 60S subunits

Make a fresh 200 μl solution of 100 mM puromycin in ddH$_2$O. Dilute the ribosomes in subunit separation buffer to 100 to 150 ODs per ml (i.e., if you get an OD of 0.5 at 260 nm for 1 μl in 1 ml of ddH$_2$O, you currently have 500 OD per ml, dilute by approximately five-fold). Add the puromycin to the 80S solution to a final concentration of 1 mM. Incubate on ice for 15 min, then at 37° for 10 min. Gently remove 1 ml of liquid from the top of each gradient without disturbing the gradient, and layer 1 ml of ribosome solution onto the top of each. Balance the tubes with subunit separation buffer. Centrifuge the gradients at 28,000 rpm in a Beckman SW28 rotor for 7.5 h at 4°.

Prepare a gradient pump, setting the detector sensitivity to 2.0 and the recorder speed to 60 cm/h. Fill the syringe on the pump with Fluorinert F-40 (Sigma), attach an 18-gauge needle to the end of the syringe hose, and remove any bubbles. Upon completion of the spin, attach a gradient to the UV detector and insert the syringe into the bottom of the gradient tube with the pump running at a very slow flow rate. Increase the flow rate to 6 ml/min. Adjust the offset on the chart recorder as necessary.

The first peak contains bulk mRNA. Collect the entire 40S peak (the second large peak to come off) and the front 0.75 of the 60S peak (the third large peak, usually larger than the 40S peak) in separate conical tubes on ice (Fig. 6.2). Concentrate the 40S and 60S fractions with Amicon Ultra (Millipore) concentrators (100K MWCO) until the volume is <1 ml. The 60S will take longer than the 40S because they contain a higher concentration of sucrose. Exchange the buffer by diluting the concentrate to 15 ml with ribosome sucrose storage buffer and centrifuge again. Repeat the buffer exchange until the KCl concentration is lower than 20 μM. Mix the concentrate by pipetting and carefully transfer it to a clean, chilled microfuge tube. Determine the concentrations by diluting 1 μl in 1 ml ddH$_2$O and measuring the OD$_{260}$ and using extinction coefficients of 40S = 2×10^7 M^{-1}cm^{-1}; 60S = 4×10^7 M^{-1}cm^{-1}. Aliquot the subunits into convenient amounts to minimize freeze-thawing over the use of the prep. Flash-freeze in liquid N$_2$ and store at −80°.

Notes:

1. It is imperative that the yeast cell cultures be grown to an OD$_{600}$ very close to 1.0, as growth to higher ODs results in a significant reduction in ribosome yield and quality. We have not yet, however, attempted culturing yeast cells in a fermentor, which may allow for increased density of cell growth without significant effects on the ribosomes.

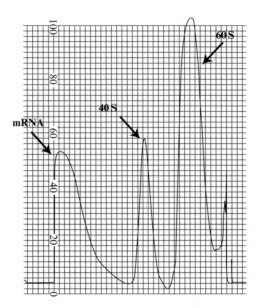

Figure 6.2   Absorbance trace of 5 to 20% sucrose gradients after ultracentrifugation of separated ribosomal subunits. The gradients are analyzed by following the absorbance at 260 nm. From the top of the gradient (left side of the graph): free mRNA is at the top of the gradient. 40S subunits are in the middle of the gradient. The transition from the 40S to the 60S peaks is discarded to maintain the purity of the samples. The 60S peak is closest to the bottom of the gradient. Only the first 0.75 of the 60S peak is collected to avoid the unseparated 80S ribosomes that pellet to the bottom of the tube.

2. Heparin is a vital component of the lysis and high salt wash buffers, serving to inhibit and remove contaminating RNases (Algire *et al.*, 2002).
3. The concentration of Mg^{2+} is critical in all steps of ribosome purification. For instance, a two-fold change in Mg^{2+} concentration in $10\times$ Ribo Buffer A can reduce the ribosome yield by 50% or more.
4. When resuspending ribosome pellets, a delicate hand is recommended. Try not to introduce bubbles.
5. Between lysing the cells and loading the gradients, all work should be performed in the cold room, if possible.

4. RIBOSOME QUALITY ANALYSIS (IDENTITY GEL)

Analyzing a sample of the rRNA on a denaturing polyacrylamide gel enables visualization of the extent of rRNA cleavage, which we have found directly correlates with the activity of the ribosomal subunits. Figure 6.3 depicts rRNA samples from two separate ribosomal subunit preps. Notice

A   B

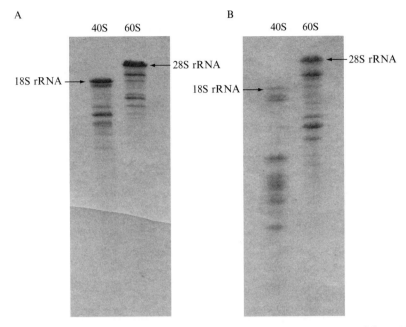

Figure 6.3 Ribosomal RNA identity gel. Ribosomal RNA was extracted from 40S and 60S ribosomal subunits and 2 μg of each sample were loaded onto a denaturing 12% polyacrylamide gel. (A) RNA extracted from active 40S and 60S ribosomal subunits. Full-length 18S and 28S ribosomal RNAs are indicated by arrows. 5S and 5.8S RNAs are not shown. (B) RNA extracted from inactive 40S and 60S ribosomal subunits. Notice the significant degradation of both 18S and 28S rRNA.

the extensive degradation of the 18S rRNA in the sample on the right, which resulted in poor 40S subunit activity.

4.1. Extracting ribosomal RNA

Add 2 μl of ribosome sample (40S, 60S, or 80S from stocks of $\sim$10 μM) to 98 μl ddH$_2$O. Add 400 μl ribosome extract buffer to the sample and extract three times with 500 μl neutral buffered phenol followed by one extraction with 500 μl chloroform. Ethanol precipitate the rRNA and resuspend the pellet in 10 μl ddH$_2$O. Quantitate the amount of rRNA extracted and store the sample on ice.

4.2. Gel analysis

Use two 20 × 30 cm glass plates (one notched) and 1.5 mm spacers. Analyze using an 8 M urea, 5% acrylamide/bisacrylamide (29:1), TBE gel. Dilute 2-μg rRNA into 10 μl ddH$_2$O and add 2 μl RNA loading dye (90% formamide,

0.02% bromophenol blue). Run the gel at 20 to 25 W for 2 h. Disassemble the apparatus, leaving the gel attached to one of the two plates. Stain the gel with methylene blue (0.04% in 0.5 M sodium acetate, pH 5.0) for 10 min and destain with water. The gel can be placed on Whatman paper and dried for permanent storage, or imaged and discarded.

Note: The gel is fragile. Minimize tearing by cutting the gel in half horizontally and discarding the bottom half, and staining and destaining while the gel is attached to the plate.

5. PURIFICATION OF HIS-TAGGED EIF2 FROM *S. CEREVISIAE*

eIF2 is overexpressed in *S. cerevisiae* strain GP3511 containing a high-copy plasmid expressing all three subunits of eIF2. eIF2γ is His-tagged at the C-terminus and serves as the sole copy of eIF2γ in this strain (Erickson and Hannig, 1996; Pavitt *et al.*, 1998).

From a glycerol stock, streak a sample of eIF2 yeast cells on a YPD plate and grow at 30° for 2 days. Inoculate twelve 1.5-l cultures of YPD media in 2800 ml baffled flasks with 750 μl from an overnight culture per flask and grow overnight (about 16 h) at 30° with shaking at 250 rpm. Spin down, store, and blend lyse cells, adding 200 ml eIF2 lysis buffer with the standard mix of protease inhibitors (see Solutions). Clarify the lysate by centrifugation in a Sorvall SS–34 rotor at 13,000 rpm for 30 min at 4°. Transfer the supernatant to a beaker and place the beaker in an ice bucket on a stir plate in the cold room. While stirring, gradually add solid ammonium sulfate to 75% saturation (48.3 g ammonium sulfate/100 ml of lysate) and stir about 1 h. Spin the slurry at 15,000 rpm for 1 h in an SS–34 rotor at 4°. During this spin set up in series two HiTrap Chelating columns (Amersham), freshly regenerated with 100 mM NiSO$_4$, on an FPLC.

Remove the supernatant and resuspend the pellet in 200 ml of eIF2 NCLB-20. Filter the solution first through 5 μm then 0.8 μm filters and load it onto the equilibrated nickel columns at a flow rate of 3 ml/min. Wash the columns with NCLB-20 and elute with NCEB-250 (the wash and elution can be done at 5 ml/min). Pool the eluted fractions and carefully dilute to 100 mM KCl (approximately five-fold) with 20 mM HEPES(pH 7.6) and 2 mM DTT by slowly adding the dilution buffer to the eIF2 solution, mixing frequently to avoid precipitation. Load the diluted sample onto an equilibrated HiTrap Heparin column (Amersham). Wash with eIF2 LSB and elute with a 60 ml linear gradient from 0 to 100% eIF2 HSB (0.1 to 1 M KCl). Analyze the fractions by SDS-PAGE using 12% polyacrylamide. The α, β, and γ subunits of eIF2 have molecular masses of 34.7, 31.6, and 57.9 kDa, respectively. eIF2 is usually the largest peak in the middle. On the example trace, eIF2 eluted in fractions 63 to 77 (Fig. 6.4).

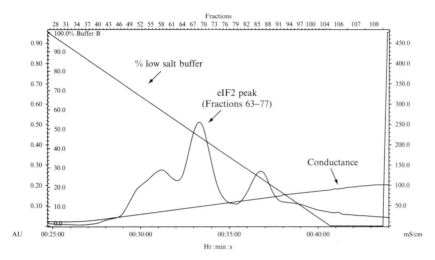

Figure 6.4 FPLC absorbance trace of eIF2 elution from heparin column. eIF2 elutes as the middle of three peaks from a HiTrap Heparin column, beginning at ~0.5 M KCl.

Pool the appropriate fractions and dilute the sample to ≤ 100 mM KCl with 20 mM Hepes (pH 7.6), and 2 mM DTT. Load the sample onto an equilibrated HiTrap Q HP column (Amersham). Wash and elute the column as for the earlier heparin step, and analyze the fractions by SDS-PAGE (Fig. 6.5). Pool the appropriate fractions and dialyze twice against 2 liters of eIF2 storage buffer. Concentrate if necessary, flash-freeze in liquid N_2, and store at $-80°$.

Notes:

1. eIF5 (45.3 kDa) binds to eIF2 and is a common contaminant in eIF2 purification. It is important to remove as much eIF5 as possible when selecting fractions from each column. Err on the side of lower eIF2 yield and higher purity.

2. A second dialysis step is used to eliminate as much GDP as possible. eIF2 binds GDP with a 20-fold higher affinity than GTP (Table 6.1; Kapp and Lorsch, 2004a), and excess GDP will interfere with ternary complex formation. We do not recommend the use of EDTA to sequester Mg^{2+} and eliminate GDP, as eIF2 contains two zinc finger domains and EDTA could sequester the Zn^{2+}, disrupting the protein's structure. Importantly, addition of Zn^{2+} to experiments measuring both ternary complex formation and binding to 40S ribosomes showed no effect on either step, suggesting the zinc-finger domains in our eIF2 are intact.

3. When analyzing eIF2 by SDS-PAGE, continue to run the gel once the dye front has migrated out of the gel. Because the α and β subunits are similar in size, they are difficult to separate during electrophoresis.

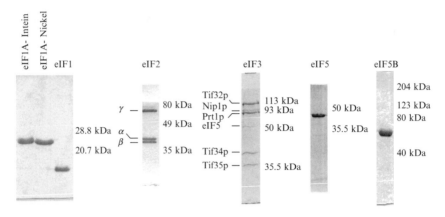

Figure 6.5   SDS-PAGE of purified yeast initiation factors. eIFs 1, 1A, 2, 3, 5, and 5B were purified as described, and analyzed by SDS-PAGE. Proteins were visualized with Coomassie blue stain. Individual subunits are identified for multisubunit factors (eIFs 2 and 3). The positions of relevant molecular weight markers are included for comparison.

4. The activity of eIF2 can be determined by measuring the factor's ability to bind GDP (Kapp and Lorsch, 2004a). In general we find our purified eIF2 to be ≥50% active, with most preps approaching 90% activity or better using GDP binding as a benchmark.

6. PURIFICATION OF HIS-TAGGED EIF3 FROM *S. CEREVISIAE*

From a glycerol stock, streak a sample of eIF3 yeast cells (LPY201 [Phan *et al.*, 1998]) on a YPD plate and grow at 30° for 2 days. Inoculate twelve 1.5 liter cultures of YPD media in 2800 ml baffled flasks with 1.5 ml of a 50 ml overnight starter culture per flask and grow overnight (~16 h) at 30° with shaking at 250 rpm. Place cultures on ice and store at 4° for 1 h.

Spin down and blend lyse cells as described, adding 200 ml zinc column loading buffer (ZCLB) to the powdered lysate. Clarify the lysate by centrifugation in a Sorvall SS-34 rotor at 13,000 rpm for 30 min at 4°. During this spin set up a HiTrap chelating column (Amersham) freshly regenerated with 100 mM ZnSO$_4$ and equilibrate on an FPLC.

Pool the supernatant and filter first through 5 μm then 0.8 μm filters. Load the lysate onto the zinc column at a flow rate of 3 ml/min. Wash the columns with ZCLB and elute with ZCEB at 5 ml/min. Analyze the fractions by SDS-PAGE using 10% polyacrylamide. Pool the appropriate fractions and concentrate to ~5 ml using an Amicon Ultra (Millipore) concentrator (10-K MWCO) if necessary. Apply the sample to an ~120 ml Superose 12 gel filtration column

Table 6.1 Summarizes the apparent equilibrium dissociation constant (K_d) values for interactions between components of the yeast translation initiation apparatus

Complex	K_d(nM)
eIF2 + [a]GDP	20 ± 5
eIF2 + GTP	1700 ± 1000
eIF2 · tRNA$_i$ + GTP	3800 ± 800
eIF2 · Met-tRNA$_i$ + GTP	200 ± 15
eIF2 + Met-tRNA$_i$	115 ± 17
eIF2 · GTP + Met-tRNA$_i$	10 ± 2
eIF2 · GDP + Met-tRNA$_i$	180 ± 20
eIF2 · GTP + unacylated tRNA$_i$	130 ± 15
eIF2 · GDP + unacylated tRNA$_i$	140 ± 15
eIF5 + eIF2 · GDP	23 ± 9
eIF5 + eIF2 · GTP · Met-tRNA$_i$	23 ± 5
eIF5 + 43S · mRNA(AUG)	≤1
eIF1 · 1A · 40S + TC[b]	54 ± 15
eIF1 · 1A · 40S · mRNA(AUG) + TC	≤1
eIF1 · 1A · 40S · TC + mRNA(AUG)	≤2
eIF1 · eIF1A · 40S + mRNA(AUG)	2200 ± 900
eIF1 + 40S	16 ± 2
eIF1 + eIF1A · 40S	1.7 ± 0.2
eIF1 + eIF1A · 40S · mRNA(AUG)	6.8 ± 0.3
eIF1 + eIF1A · 40S · TC	≤1.4 ± 0.7
eIF1 + eIF1A · 40S · mRNA(AUG) · TC	53 ± 6
eIF1A + 40S	49 ± 6
eIF1A + eIF1 · 40S	6.1 ± 0.8
eIF1A + eIF1 · 40S · mRNA(AUG) · TC	≤1

[a] '+'signifies the affinity between interacting molecules or complexes.
[b] TC: ternary complex (eIF2 · GTP · Met-tRNA$_i$).

equilibrated in enzyme storage buffer and elute with enzyme storage buffer. Analyze the fractions by SDS-PAGE as earlier (Fig. 6.5). Pool the appropriate fractions and determine the protein concentration (the molecular mass of eIF3 is 361.3 kDa). Concentrate if necessary, flash-freeze in liquid N_2, and store at $-80°$. Notes:

1. eIF3 is part of the multifactor complex (MFC) including eIF1, eIF2, and eIF5. eIF5 often copurifies with and contaminates eIF3 (Phan *et al.*, 1998). eIF5 contamination is very difficult to remove from the eIF3 prep and even small amounts will affect eIF2 GTPase assays because eIF5 is such a strong GAP.
2. Although addition of eIF3 results in only modest effects on ternary complex binding to 40S subunits in our system, these effects are

consistent with those observed *in vivo* (Jivotovskaya *et al.*, 2006; Valasek *et al.*, 2004). The purified eIF3 is active based on its ability to bind 40S ribosomal subunits, producing a supershift in a 43S native gel assay; its binding to eIF1 and eIF5; and its ability to rescue *prt1-1* mutant yeast translation extracts (Algire *et al.*, 2002).

7. OVEREXPRESSION AND PURIFICATION OF YEAST EIF1, EIF1A, AND EIF5 FROM *E. COLI*

The following protocol for purification and labeling of the single polypeptide initiation factors eIF1, eIF1A, and eIF5 makes use of the IMPACT system from New England Biolabs. The genes for each initiation factor were cloned into the expression vector pTYB2, which results in expression of a fusion protein composed of the desired eIF, a fragment of the yeast self-cleavable intein, and a chitin-binding domain protein. The chitin-binding domain is a high affinity tag that allows for stringent on-column washing of the fusion protein. The intein fragment, with the help of DTT, catalyzes on-column cleavage of the fusion protein at the point of eIF fusion, releasing the initiation factor from the column.

Freshly transform the plasmid pTYB2 (NEB) containing the desired gene into competent BL21 (DE3) CodonPlus cells (Stratagene) and grow overnight at 37° on LB plates supplemented with 50 μg/ml carbenicillin. Inoculate 1.5 liter of LB supplemented with 50 μg/ml carbenicillin and 34 μg/ml chloramphenicol to a starting OD$_{600}$ of 0.015 with a 5 ml overnight culture and grow to an OD$_{600}$ of 0.6. Induce protein expression by adding IPTG to 0.3 mM, transfer the culture to 16°, and grow overnight with shaking at 250 rpm.

Pellet the cells and resuspend in 30 ml of intein lysis buffer. Lyse the cells using a French pressure cell, passing the sample at least twice to ensure a thorough lysis. Clarify the lysate by centrifugation in a Sorvall SS-34 rotor for 30 min at 13,000 rpm at 4°. Equilibrate 1 ml of chitin resin (NEB) with 20 ml intein lysis buffer by gravity flow. Apply the clarified lysate directly to the resin and incubate at 4° with gentle shaking for 1 h. Drain the lysate by gravity flow and wash the resin with 60 ml intein wash buffer. Raise the pH and lower the salt concentration by washing the column with 20 ml of intein cleavage buffer (without DTT) and drain the column.

For unlabeled protein purification: Add DTT to a final concentration of 75 mM to the remaining intein cleavage buffer and add 3 ml to the column, allowing the buffer to flow through until about 2 ml remain above the surface of the resin. Stop the buffer flow, mix the buffer and beads to evenly distribute the DTT, and seal both ends with parafilm to prevent leaks. Incubate overnight at room temperature for optimal cleavage.

For fluorescently labeled protein purification (Maag and Lorsch, 2003; Muir *et al.*, 1998; Scheibner *et al.*, 2003): (These steps should be carried out in low light to avoid photobleaching during the labeling procedure.) To the remaining intein cleavage buffer, add sodium 2 mercaptoethanesulfonate (MESNA) to a final concentration of 200 mM. Add 0.5 ml intein cleavage buffer with MESNA to the column and allow the column to drain. Cap the column bottom to stop the buffer flow and add 0.5 ml of 1 mM Cys–Lys–fluorophore in intein cleavage buffer with 200 mM MESNA to the resin and allow to enter the column. Bring the liquid level of the column just above the resin surface by adding intein cleavage buffer with MESNA (~0.5 ml) and mix the buffer and resin by pipetting. Seal both ends of the column with parafilm to prevent leaks. Incubate overnight at room temperature for optimal cleavage/labeling. To avoid photobleaching, shield the column from all light by storing it in a dark cabinet or box or wrapping it in aluminum foil.

Uncap the column and collect the flow-through containing the cleaved protein. Rinse the column with 2 ml aliquots of cleavage buffer, collecting the flow-through each time until a 10 μl sample of the flow-through shows no significant protein in a Bradford assay.

For eIF1, eIF1A, and eIF5: Pool the appropriate fractions and dilute the sample to ≤100 mM KCl with 20 mM Hepes · KOH (pH 7.6), 10% glycerol and 2 mM DTT. Load the sample onto an equilibrated HiTrap Heparin column (Amersham). Elute with a 60 ml linear gradient from 0 to 100% HSB (0.1 to 1 M KCl). Analyze the eluted fractions by SDS-PAGE using 15% (for eIF1) or 12% (for eIF1A, eIF5) polyacrylamide (Fig. 6.5). eIF1 elutes at ~0.6 M KCl, eIF1A at ~0.35 M KCl, and eIF5 at ~0.45 M KCl.

For eIF5 only: As an additional purification step, load the sample on a HiTrap Q HP column (Amersham), using the same buffers and gradient conditions as for the heparin column. Analyze the eluted fractions by SDS-PAGE using 12% polyacrylamide (Fig. 6.5). eIF5 elutes at ~0.45 M KCl.

Pool the appropriate fractions and dialyze overnight against 2 liters of enzyme storage buffer. Determine the protein concentration (molecular masses: eIF1, 12,312 Da; eIF1A, 17,435 Da; eIF5, 45,261 Da). Concentrate if necessary, flash-freeze in liquid N$_2$, and store at −80°.

8. Overexpression and Purification of Yeast eIF1A in *E. coli*

Proteins expressed from pTYB2 and purified using the IMPACT system have an extra glycine residue on their C-termini. Likewise, fluorescent labeling of the target protein results in the addition of at least a dipeptide (Cys-Lys) coupled to a fluorophore. These additions can be detrimental if the C-terminus of a protein is important for its function. For example, the C-terminus of eIF5B interacts with the extreme C-terminus of eIF1A (Choi et al., 2000; Marintchev et al., 2003; Olsen et al., 2003), influencing the subunit joining activity of eIF5B (Acker et al., 2006). For these factors, a purification scheme resulting in a native C-terminus is required. The following purification scheme results in eIF1A containing a wild-type C-terminus and an additional N-terminal Gly-His dipeptide.

Freshly transform the plasmid pMGA-HisTEV1A (containing TIF11 [eIF1A] downstream of and in frame with a sequence encoding a His$_6$ tag and TEV protease recognition site) into competent BL21 (DE3) CodonPlus cells and grow overnight at 37° on LB plates supplemented with 30 μg/ml kanamycin. Inoculate 1.5 liters of LB supplemented with 30 μg/ml kana-mycin and 34 μg/ml chloramphenicol to a starting OD$_{600}$ of 0.015 with a 5 ml overnight culture and grow to an OD$_{600}$ of 0.6. Induce protein expression by adding IPTG to 1 mM, and grow for 4 h more at 37° with shaking at 250 rpm.

Substituting eIF1A NCLB-20 (supplemented with one complete EDTA-free protease inhibitor cocktail tablet [Roche]) for intein lysis buffer, pellet and lyse the cells, and clarify the lysate as for an intein preparation. Set up a 5 ml HiTrap Chelating column (Amersham) freshly regenerated with 100 mM NiSO$_4$ on an FPLC.

Filter the lysate first through 5 μm then 0.8 μm filters and load it onto the equilibrated nickel column at a flow rate of 3 ml/min. Wash the column with eIF1A NCLB-20 and elute with eIF1A NCEB-250 at 5 ml/min. Pool the desired fractions and dilute to 100 mM KCl (approximately five-fold) with 20 mM Hepes · KOH (pH 7.6), and 10 mM 2 mercaptoethanol. Equilibrate a HiTrap Heparin column with LSB. Load the diluted sample onto the equili-brated heparin column. Wash with LSB and elute with a 60 ml linear gradient from 0 to 100% HSB (0.1 to 1 M KCl). Analyze the fractions by SDS-PAGE using 12% polyacrylamide. eIF1A elutes as two partially overlapping peaks beginning at ~0.35 M KCl. (Control experiments show that eIF1A samples from each peak behave identically in our assays and run indistinguishably on SDS-PAGE.) Pool the desired fractions and measure the concentration of eIF1A. Add 50 μg His-tagged TEV protease per mg of target protein and incubate overnight at room temperature with gentle agitation. The TEV protease can be added directly to the pooled fractions.

The following morning, add solid imidazole to the TEV-treated sample to 20 mM. The addition of imidazole prevents nonspecific binding of protein to the column and allows the desired protein to elute entirely in the flow-through. Completely dissolve the solid imidazole and filter the sample through a 0.45 μm syringe filter. Reequilibrate the nickel-chelated HiTrap column and load the sample at 3 ml/min, being careful to collect the flow-through, as it contains the TEV-cleaved eIF1A lacking the His-tag. Wash the column with eIF1A NCLB-20 and elute with eIF1A NCEB-250. Analyze the fractions by SDS-PAGE using 12% polyacrylamide (Fig. 5). Be sure to run multiple samples of the flow-through, as well as a sample of the peak eluted with high imidazole to verify the separation of uncleaved eIF1A and His-tagged TEV protease from the cleaved eIF1A. Pool the appropriate fractions and dialyze overnight against 2 liters of enzyme storage buffer. Determine the protein concentration. Concentrate if necessary, flash-freeze in liquid N_2, and store at $-80°$.

9. OVEREXPRESSION AND PURIFICATION OF YEAST eIF5B IN *E. COLI*

As mentioned previously, the C-terminus of eIF5B plays a significant role in translation initiation. Full-length eIF5B has proven difficult to clone and overexpress, and N-terminally truncated eIF5B behaves as wild-type eIF5B *in vivo*. Based on these data, we make use of an N-terminal truncation of eIF5B, replacing the first 396 amino acids with a His_6 tag followed by a TEV protease recognition sequence.

Transform the plasmid pMGA-HisTEV5B (containing fun12 [$\Delta N396$-eIF5B] downstream of and in frame with a sequence encoding a His_6 tag and TEV protease recognition site) into competent BL21 (DE3) CodonPlus cells and grow overnight at $37°$ on LB plates supplemented with 30 $\mu g/ml$ kanamycin. Inoculate 1.5 liters of LB supplemented with 30 $\mu g/ml$ kanamycin and 34 $\mu g/ml$ chloramphenicol to a starting OD_{600} of 0.015 with a 5 ml overnight culture and grow to an OD_{600} of 0.6. Induce protein expression by adding IPTG to 1 mM, and grow for 4 h more at $37°$ with shaking at 250 rpm.

Pellet the cells and resuspend in 70 ml Buffer B-20 (Guillon *et al.*, 2005) supplemented with two complete EDTA-free protease inhibitor cocktail tablets (Roche). Divide the suspension into two 35 ml samples and lyse the cells using a French pressure cell, passing each sample at least twice to ensure a thorough lysis. Clarify the lysate by centrifugation in a Sorvall SS-34 rotor at 13,000 rpm for 30 min at $4°$. Set up a 5 ml HiTrap Chelating column (Amersham) freshly regenerated with 100 mM $NiSO_4$ on an FPLC.

Filter the lysate first through 5 μm then 0.8-μm filters and load it onto the equilibrated nickel column at a flow rate of 3 ml/min. Wash the column with Buffer B-20 and elute with Buffer B-250 at 5 ml/min. Pool the eluted fractions and dilute to 100 mM KCl (approximately five-fold) with Buffer A. Load the diluted sample onto an equilibrated HiTrap Q HP column. Wash with 100 mM NaCl Buffer A and elute with an 80 ml linear gradient from 100 mM to 500 mM NaCl in Buffer A. Analyze the eluted fractions by SDS-PAGE using 12% polyacrylamide. eIF5B elutes as one peak at ~0.2 M NaCl. Pool the desired fractions and estimate the concentration of eIF5B as for eIF1A. Add 15 μg His-tagged TEV protease per mg of target protein and incubate overnight at room temperature with gentle agitation. The TEV protease can be added directly to the pooled fractions.

Prepare the protease reaction for chromatography as in the eIF1A preparation. Separate the cleaved eIF5B peptide from the uncleaved protein and His-tagged TEV protease using a nickel column as earlier, replacing eIF1A NCLB-20 with Buffer B-20 and eIF1A NCEB-250 with Buffer B-250. Analyze the eluted fractions by SDS-PAGE as for eIF1A. Pool the appropriate fractions and concentrate to ~5 ml with an Amicon Ultra (Millipore) concentrator (10,000 MWCO). Apply the sample to an ~120 ml Superose 12 gel filtration column equilibrated in enzyme storage buffer and elute with 120 ml of enzyme storage buffer. Analyze the eluted fractions by SDS-PAGE (Fig. 6.5). Pool the appropriate fractions and measure the protein concentration (ΔN396-eIF5B molecular mass is 67.6 kDa). Concentrate if necessary, flash-freeze in liquid N_2, and store at $-80°$.

10. PURIFICATION OF YEAST METHIONYL-tRNA SYNTHETASE (YMETRS) FROM *S. CEREVISIAE*

A GST-yMetRS fusion protein is constitutively expressed in *S. cerevisiae* from a plasmid containing a TRP selectable marker (gift of Franco Fasiolo) (Kapp *et al.*, 2006). From a glycerol stock, streak out cells on a SC-Trp plate and grow at 30° for 2 days. Inoculate 18 liters of SC-Trp media in twelve 2800 ml baffled flasks with 37.5 ml per flask from a 500 ml overnight culture and grow to $OD_{600} \geq 1.5$ at 30° with shaking at 250 rpm.

Spin down and blend lyse cells, adding 220 ml PBS buffer with the standard protease inhibitors to the powdered lysate. Clarify the lysate by centrifugation in a Sorvall SS-34 rotor at 13,000 rpm for 30 min at 4°. Filter the supernatant through 5 μm then 0.8 μm filters. Pass the lysate over a 5 ml GSTrap column (Amersham) equilibrated in PBS buffer at a rate of 3 ml/min. Wash the column with PBS buffer and elute with 10 mM reduced glutathione in 50 mM Tris (pH 8.0). Analyze the fractions by SDS-PAGE using 10% polyacrylamide and pool the appropriate fractions (GST-yMetRS migrates at ~110 kDa). Dialyze the purified GST-yMetRS overnight against 2 liters of yMetRS storage buffer. Concentrate the protein to a vol of ~2 ml using an Amicon Ultra (Millipore) concentrator (10-K MWCO) and aliquot to minimize freeze-thawing. Flash-freeze in liquid N_2 and store at $-80°$. To assay the activity of the synthetase, set up small-scale (30 μl) limiting methionine tRNA charging reactions diluting the GST-yMetRS to final concentrations of 1:10, 1:20, 1:50, and 1:100 (see Charging tRNA$_i^{Met}$ with Methionine).

Notes:

1. Additional purification is not necessary, but can be performed by applying the protein to a HiTrap Q HP ion-exchange column and following the protocol for eIF5 purification.
2. Synthetase activity is not significantly altered by the GST tag.

11. RNA Synthesis and Purification

T7 polymerase run-off transcription is used to synthesize most RNAs. mRNA or tRNA can be transcribed using either a single-stranded, synthetic DNA oligonucleotide or linearized plasmid template. If a synthetic DNA template is used, it must include a double-stranded T7 polymerase binding site. This is created using a "clamp" oligonucleotide complementary to the T7 site on the template. For the poly(UC) 43mer with an AUG codon in the center (mRNA(AUG)) the template sequence is:

5'GAGAGAGAGAGAGAGAGAGAGAGCATAGAGAGAGAGAGAG-ATT**CCTATAGTGAGTCGTATTACATATGCGTGTTACC**3'.

The T7 polymerase promoter region (plus the first two encoded bases) is shown in bold and the clamp oligo sequence is complementary to this region. Plasmid templates must also include a T7 polymerase binding site but do not require use of a clamp because they are already double-stranded. We use the high copy pUC19 plasmid. Plasmids are linearized using a restriction site at the end of the template region so that the polymerase will run off the end of the DNA. Purification of the synthetic template by denaturing PAGE (as for RNA purification below) often improves results.

Transcription Reaction
1× Transcription Buffer
5 mM DTT
2.5 mM (each) NTP
1 µM clamp oligo if necessary
1 µM oligo template or 0.08 to 0.3 µg/µl plasmid template
T7 Polymerase

We purify the polymerase ourselves and change the concentration in the reaction to compensate for variations in the activity of our preparations. Incubate the reaction at 37° for at least 3 h and as long as overnight. If the reaction is working well a white precipitate should form ($Mg^{2+} \cdot PP_i$). When the reaction is complete, centrifuge the tube for several min in a tabletop, swinging bucket centrifuge at 6000 rpm, and remove the supernatant from the white solid. Extract the supernatant twice with neutral buffered phenol and once with chloroform. Bring the solution to 0.3 M sodium acetate (pH 5.0), and ethanol precipitate the RNA. Rinse the pellets with 70% ethanol and allow them to dry overnight. Resuspend in ddH$_2$O, using as little as is necessary. If it is difficult to resuspend, add a small amount of 100 mM EDTA (100 μl or less per ml). Aim to resuspend the pellet in a volume ~15% of the original reaction.

11.1. Gel purification of transcription products

To remove any truncated products, gel purify the RNA. One large gel (20 × 30 × 0.15 cm) can generally be used to purify the RNA from a 10 ml transcription reaction. Prepare and pour an 8 M urea, 12% 19:1 acrylamide: bisacrylamide RNA purification gel with 1× TBE. For transcriptions over 5 ml, insert the comb upside down, so that the teeth are pointing out of the gel. This will create one large well and not waste any space.

Add 2× RNA loading dye (90% formamide, 0.02% bromophenol blue) to the sample. Load the sample onto the gel and run at 25 W until the dye band has run ~0.75 of the way down the gel; this will take several h because of the high salt concentration in the sample. Monitor the voltage while the gel is running and be sure that the gel does not get too hot to touch. If overheating is a problem, the gel can be run in a cold room.

Disassemble the plates and wrap the gel in plastic wrap. Lay the gel over a TLC plate with fluorescent indicator wrapped in plastic wrap and shadow the RNA using a handheld, shortwave UV lamp. Remember to wear UV-absorbing glasses. The desired product is usually the major band. It is not uncommon to see a ladder of aborted products below the major product as well as n + 1 and n + 2 bands above it. The DNA template can sometimes be seen near the top of the gel. Outline the proper band with a marker on the plastic wrap. Remove the TLC plate from under the gel before excising the band with a razor. Place the gel in a 50 ml conical tube and use a 1 ml serological pipette or a glass rod to crush the gel. Cover the gel with a generous amount of 0.3 M sodium acetate (pH 5.0). For a band that crosses an entire gel, 40 ml is usually sufficient. Parafilm the cap and rotate the tube at room temperature overnight to extract the RNA.

Remove as much liquid as possible from the gel using 0.45 μm syringe filters, taking care not to push hard enough to eject or break the filter (eye protection is recommended). Ethanol precipitate the solution and resuspend the dry pellets in ddH$_2$O. If the solution appears cloudy, extract with neutral buffered phenol and then chloroform to remove the insoluble material and reethanol precipitate the RNA using 0.3 M sodium acetate (pH 5.0). For a 50 ml transcription of a 43 mer approximately 200 μl of 500 μM RNA should be obtained.

11.2. Yeast initiator tRNA

The hammerhead yeast initiator tRNA construct consists of a T7 promoter followed by the hammerhead ribozyme and initiator tRNA sequences, cloned between the SmaI and BamHI sites in the pUC19 plasmid. A BstNI site introduced at the 3′ end of the tRNA allows digestion to yield a CCA end.

5′TGCGGGCCTCTTCGCTATTACGCCAGCTGGCGAAAGGG
GGATGTGCTGCAAGGCGATTAAGTTGGGTAACGCCAGGG
TTTTCCCAGTCACGACGTTGTAAAACGACGGCCAGTGAATT
CGAGCTCGGTACCCCCAATTAAGCTTCC**TGGTAGCGCCG**
CTCGGTTTCGATCCGAGGACATCAGGGTTATGAGCCC
TGCGCGCTTCCACTGCGCCACGGCGCT*GACGGTACCGGG*
TACCGTTTCGTCCTCACGGACTCATCAGAGCG<u>*CCTATAGTG*</u>
<u>AGTCGTATT</u>AAGGGGGATCCTCTAGAGTCGACCTGCAGGC
ATGCAAGCTTGGCGTAATCATGGTCATAGCTGTTTCCTGTG
TGAAATTGTTATCCGCTCACAATTCCACACAACATACGAGCC
GGAAGCATAAAGTGTAAAGCCTGGGGTGCCTAATGAGTGAG
CTAACTCACATTAATTGCGTTGCGCTCACTGCCCGCTTTCC
AGTCGGGAAACCTGTCGTGCCAGCTGCATTAATGAATCGGC
CAACGCGCGGGGAGAGGCGGTTTGCGTATTGGGCGCTCTT3′

(The orientation is reversed. <u>Underlined</u>: T7 promoter; *Italic:* hammerhead ribozyme; **Bold**: tRNA)

This construct permits transcription beginning at a G rather than the A at which the tRNA itself begins, resulting in approximately five-fold better transcription. The hammerhead ribozyme will cleave itself from the tRNA after transcription, leaving the native tRNA (Fechter *et al.*, 1998).

Before transcription, digest the plasmid template with BstN1 for 2 h at 60°. Extract the digest twice with neutral buffered phenol and once with chloroform, and ethanol precipitate using 0.3 M sodium acetate (pH 5.0). Resuspend in ddH$_2$O. Assemble the transcription reaction as described previously with a final concentration of at least 0.08 μg/μl of digested DNA. After transcribing the RNA, incubate for 1 h at 65° to allow the hammerhead ribozyme to cleave itself from the tRNA. Extract twice with neutral buffered phenol and once with chloroform, and ethanol precipitate using 0.3 M sodium acetate (pH 5.0). Gel purify the transcript as described. Two major bands will normally be observed; the upper band is the tRNA and the lower band the hammerhead ribozyme. Extract the RNA from the gel as described.

12. Charging tRNA$_i^{Met}$ with Methionine

Translation initiation requires methionyl-tRNA$_i^{Met}$, which must be synthesized *in vitro* from purified tRNA$_i^{Met}$ and methionine by the yeast methionyl-tRNA synthetase. Two such "charging" reactions can be performed, providing two different varieties of methionyl-tRNA$_i^{Met}$ for use in different assays. The limiting charging reaction produces radiolabeled [^{35}S]methionyl-tRNA$_i^{Met}$ for use in 43S/80S gel shift assays. The stoichiometric charging reaction produces unlabeled methionyl-tRNA$_i^{Met}$ for use in many different assays. Limiting reactions are carried out on a small scale because of the short half-life of ^{35}S and the relatively small amounts required for gel shift assays. In contrast, stoichiometric charging reactions can be performed in bulk to yield the larger quantities of methionyl-tRNA$_i^{Met}$ required for fluorescence and GTP hydrolysis assays.

12.1. Limiting charging reaction

To help ensure that the tRNA is folded properly, before setting up the charging reaction heat denature the tRNA at 95° for 5 min and allow it to return to room temperature gradually over 20 to 30 min (or use a thermocycler to decrease from 95 to 4° at 0.1°/s).

Final Concentrations in Charging Reaction
40 mM Tris · Cl (pH 7.6)
10 mM Mg(OAc)$_2$
1 mM DTT
1 mM ATP · Mg^{2+}
0.3 μM ^{35}S methionine
0.02 μg/μl glycogen
5% DMSO
5 μM tRNA$_i^{Met}$
yeast methionyl aminoacyl tRNA synthetase (yMetRS)

Glycogen makes the pellet after ethanol precipitation more visible. DMSO mysteriously but reproducibly increases charging efficiency, but can be left out if desired. We have observed that some component of the commercially available ^{35}S-methionine inhibits the reaction if the methionine concentration is increased above 0.3 μM.

Set up a 100 μl reaction and, before adding the tRNA and enzyme to the mix, remove enough to do a 20 μl "no-tRNA control" reaction. This reaction will be identical in composition to the larger reaction except that it will not contain tRNA; this control allows the measurement of background for the calculation of charging efficiency. Add tRNA to the larger reaction (e.g., 4 μl to the remaining 80 μl); add ddH$_2$O instead of tRNA to the no-tRNA control (in this example, 1 μl). Add yMetRS to both reactions (e.g., 4 μl to the reaction and 1 μl to the no-tRNA control) and incubate at 30° for 30 min.

Remove 1 μl of the charging reaction and spot directly onto a Whatman GF/C filter in a scintillation vial ("input" filter). Spot 1 μl of the no-tRNA control reaction onto another filter (no-tRNA control input filter). Each input filter represents total ^{35}S counts in the reaction from which its 1 μl sample came.

Remove 5 μl of the charging reaction and add it to 50 μl of carrier RNA (1 mg/ml bulk tRNA (Roche) and 0.81 M sodium acetate, pH 5). Add 1 ml 5% TCA/95% ethanol and incubate on ice. Do the same for 5 μl of the no tRNA control reaction. Carrier RNA is used to minimize loss of the small amount of tRNA in the 5 μl sample.

While the 5 μl precipitations incubate on ice, extract the remainder of the charging reaction once with neutral buffered phenol and once with chloroform. Bring the solution to 0.3 M sodium acetate (pH 5.0), and ethanol precipitate the ^{35}S-Met-tRNA$_i^{Met}$ at $-20°$ for 1 h followed by centrifugation. Resuspend the dry pellets in ddH$_2$O to the desired volume, determined by the following method.

Once the 5 μl precipitations have been on ice for at least 15 min (and up to 1 h), spot each full precipitation reaction onto its own Whatman GF/C filter on a vacuum manifold, taking care to keep the drops near the center of the filter. Wash each filter with 3 ml of cold 5% TCA (without ethanol), then 1 ml 100% ethanol, being careful to wash the entire surface of the filter with each wash. Throughout, be careful not to touch the filter with the pipette tip. Let the filter dry for $\sim$1 min on the vacuum. Place each filter in its own scintillation vial, using tweezers. Be careful to touch only the very outer edge of the filter. Add scintillation fluid to each vial and use a scintillation counter to obtain ^{35}S counts for each of the input filters and filters from the 5 μl precipitations.

In this method ^{35}S methionine–charged tRNA$_i$ is retained on the filter and free ^{35}S-methionine washes through; thus, ^{35}S counts observed on the filter should represent charged tRNA. Because the no-tRNA control by definition could not contain any charged tRNA, any counts observed from the no-tRNA control represent a background level that should be subtracted from what is observed in the sample from the charging reaction.

Determine the charging efficiency by calculating (counts/[5 * input counts]) for the charging reaction. After correcting for the no-tRNA control reaction the resulting number is the fraction of methionine incorporated. We generally expect 20 to 50% methionine incorporated from a limiting charging. Multiply this fraction by the number of mol of methionine in the reaction to obtain the mol of methionine incorporated, and divide this number by the number of mol of tRNA in the reaction to obtain the fraction of tRNA charged. We resuspend the pellets to a final concentration of 60 nM charged tRNA.

Store charged tRNA at $-80°$ in approximately single-use aliquots. Freeze-thawing charged tRNA repeatedly leads to loss of activity. Charged tRNA will keep for months in the freezer; its useful lifespan in the absence of thawing and refreezing seems limited simply by the decay of the ^{35}S.

12.2. Stoichiometric charging reaction

The procedure for stoichiometric charging is similar to that for the limiting charge. One difference is that to calculate the level of charging, a test reaction is spiked with ^{35}S-methionine, and an equivalent amount of cold methionine is added to the actual reaction to keep concentrations consistent between the test and actual reactions. The concentrations used in a stoichiometric charging are the following:

Final Concentrations in Charging Reaction
40 mM Tris · Cl (pH 7.6)
10 mM Mg(OAc)$_2$
1 mM DTT
2 mM ATP · Mg^{++}
300 μM methionine
5 μM tRNA$_i^{Met}$
yMetRS (same dilution as used for limiting charge)

The typical reaction size for a stoichiometric charging is 5 ml. Before adding the tRNA and enzyme, remove enough mix to assemble a 30 μl no-tRNA control reaction (30 μl is a convenient size to spike with 1 μl of ^{35}S methionine). Add tRNA to the larger mix (and add ddH$_2$O to the no-tRNA control) and then remove enough mix to assemble a 30 μl "hot test" reaction. Add 1 μl of ^{35}S methionine to both the no-tRNA control reaction and the hot test reaction. Add an equivalent (1:30) volume fraction of cold 8.7 μM methionine to the larger charging reaction. Add yMetRS and incubate at 30° for 30 min.

Determine charging efficiency as described for the limiting charging. We generally expect between 20–30% tRNA charged from a stoichiometric charging reaction, and resuspend the pellets to 20 μM final charged tRNA. Repeated attempts to get 100% charging have failed, although control experiments have indicated that the uncharged tRNA does not interfere with observable steps in initiation *in vitro*.

13. FILTER BINDING ASSAY TO MONITOR TERNARY COMPLEX FORMATION

Final Concentrations in Binding Assay
25 mM Hepes · KOH (pH 7.6)
2.5 mM Mg(OAc)$_2$
80 mM KOAc (pH 7.6)
2 mM DTT
0.5 mM GDPNP · Mg^{2+}
1× eIF2 dilution
1 nM Met-tRNA$_i^{Met}$ charged with ^{35}S methionine ([^{35}S]Met-tRNA$_i^{Met}$)

Make 5× dilutions of eIF2 in eIF2 storage buffer with 0.6 μg/μl creatine kinase. The addition of creatine kinase helps to prevent eIF2 from sticking to the tube walls, a concern at low concentrations. Final concentrations from 5 to 200 nM eIF2 constitute a good range with wild-type [^{35}S]Met-tRNA$_i^{Met}$ and eIF2. Assemble the binding assay mix excluding the [^{35}S]Met-tRNA$_i^{Met}$ and the eIF2. Aliquot the mix into individual tubes, add eIF2 dilutions, and incubate 10 min at 26°. Add [^{35}S]Met-tRNA$_i^{Met}$, and incubate 10 min at 26°. Longer incubation times may be necessary according to the specifics of the experiment (e.g., if using a mutant [^{35}S]Met-tRNA$_i^{Met}$ with a kinetic defect in binding to eIF2).

On a vacuum manifold, position a Nytran Supercharge (Whatman) membrane with a nitrocellulose membrane on top. Both filters should be soaked for ~1 h in filter binding wash buffer before use. The Supercharge membrane (bottom) retains nucleic acids, whereas the nitrocellulose membrane (top) retains protein, so that [^{35}S]Met-tRNA$_i^{Met}$ bound to eIF2 will be retained on the nitrocellulose filter and free [^{35}S]Met-tRNA$_i^{Met}$ will pass through the nitrocellulose filter and be retained by the Supercharge membrane.

Turn on the vacuum just before applying the reaction to the filters, so the filters do not dry out. Filter 20 μl of the reaction, applying it to the center of the filter, and immediately wash the entire surface of the filter with 5 ml ice-cold filter binding wash buffer. Allow the filters to dry on the vacuum for a min or so.

Place each filter carefully in a separate vial, making sure to keep them pointing up (i.e., the same way in which they were filtered). Add scintillation fluid to the vials and use a scintillation counter to measure the amount of ^{35}S-methionine on each filter. Determine the fraction of [^{35}S]Met-tRNA$_i^{Met}$ bound to eIF2 by taking the counts on the top (nitrocellulose) filter and dividing by the sum of the counts on the top and bottom filters (as this sum should represent the total amount of [^{35}S]Met-tRNA$_i^{Met}$; free [^{35}S]Met will pass through both filters).

14. BENCHTOP EIF2 GTPASE ASSAY

Measuring the GTP hydrolysis activity of eIF2 in the presence and absence of eIF5 is the best way to determine the extent of eIF5 contamination in purified eIF2. These reactions are carried out on the benchtop and require significantly smaller amounts of reagents, but allow much lower kinetic detail (minimum timepoints of 2 s) than rapid quench techniques. In this experimental set-up, a structural reorganization of the pre-initiation complex is rate limiting for GTP hydrolysis (Algire et al., 2005). GTPase activity is monitored by the conversion of GTPγ[^{32}P] into GDP and ^{32}P$_i$, which are separated on a 15% polyacrylamide gel.

14.1. Gel preparation

Prepare and pour a 15% 29:1 acrylamide:bisacrylamide gel in 1× TBE (no urea) using the same plates and spacers as for the RNA identity gel. Because the signal is strong, a small loading volume is sufficient, and a 20-well comb is recommended.

14.2. Experimental setup

Aliquot 6 μl of quench/dye solution (90% formamide, 0.02% bromophenol blue, 100 mM EDTA) into a quench tube for each reaction, plus one tube for an initial background measurement. Incubate 2× ternary complex (1× reconstitution buffer [recon buffer], 1.6 μM eIF2, 1.6 μM methionyl-tRNA$_i^{Met}$, and 125 pM GTPγ[^{32}P]) at 26° for 15 min. Form 2× ribosomal complex by combining 400 nM 40S subunits, 1.6 μM each eIF1 and eIF1A, 2 μM model mRNA, and 2 mM GDP · Mg^{2+}. The addition of excess GDP in the 2× ribosomal complex prevents multiple turnovers of the reaction by saturating eIF2 with GDP once a single round of GTP hydrolysis has occurred. When interested in measuring the rate of GTP hydrolysis in the presence of eIF5, include 1.6 μM eIF5 in the 2× ribosomal complex mix. Just before beginning the experiment, add 1 μl 2× TC to 6 μl EDTA quench/dye solution; this sample represents the background level of ^{32}P$_i$ in the ternary complex. Assemble individual reactions to obtain the earliest timepoints (2 to 15 s). To do this, aliquot 2 μl of 2× ribosomal complex and initiate the reaction by adding 2 μl of 2× TC, mixing by pipetting. To stop the reaction, pipette 2 μl of the reaction into the quench tube and mix vigorously by pipetting. By mixing 2 μl of each 2× complex and quenching 2 μl total, it is possible to take relatively short timepoints without having to change pipette tips. This can be repeated for each timepoint desired up to 15 s. If the timepoints are 20 s or more apart, it is possible to set up a single reaction tube and quench aliquots over time. The quenched samples can be analyzed immediately by gel electrophoresis, or frozen at −20° indefinitely according to the activity of the [^{32}P].

14.3. Separation of GTPγ[^{32}P] and ^{32}P$_i$

Load 3 μl of each quenched sample onto the gel, using every other well. Run the gel in 1× TBE for 35 min at 25 W. Load the remaining samples in the unused wells and run for 20 min more. Place the gel between two layers of plastic wrap, and fold the ends of the plastic wrap, sealing the gel. Expose the wrapped gel to a phosphorimaging screen for 1 h. Scan the screen and quantitate the spots corresponding to GTPγ[^{32}P] and ^{32}P$_i$. GTPγ[^{32}P] will be the higher of the two bands. Calculate the fraction of GTP hydrolyzed by dividing the ^{32}P$_i$ value by the sum of the GTPγ[^{32}P] and ^{32}P$_i$ values, and subtracting from this the fraction obtained from the background sample.

Note: 2× ternary complex is relatively stable, but care should be taken to minimize the time between individual reactions for a given experiment. It may be necessary to measure the rate of GTP hydrolysis by eIF2 in ternary complex to verify that the background sample is valid throughout the experiment.

15. 43S/80S Complex Gel Shift Assay

Native gel electrophoresis can separate ternary, 43S (Fig. 6.6) and 80S complexes (Fig. 6.7), and provides a way to monitor both the binding of ternary complex to 40S subunits and the ribosomal subunit joining step of initiation (Acker *et al.*, 2006; Algire *et al.*, 2005; Maag *et al.*, 2005). We describe these assays here because they are useful for characterizing the activity of the components of the system.

15.1. Preparing an acrylamide gel for the 43S/80S gel shift assay

A picture of the gel box used is shown in Fig. 6.8 (well behind gel—26.5 cm wide, 19 cm tall, 2.5 cm deep; well in front of gel—31 cm wide, 5.5 cm tall, 6.5 cm deep). We use 18 × 30 cm plates (one notched) and 0.45 mm spacers. Because the gel is both thin and soft (4% acrylamide:bisacrylamide, 37.5:1), it is important that it stick well to one of the gel plates (we choose the unnotched plate) and slide easily off the other. It is prone to tearing, stretching, or folding otherwise, and once this happens the gel can rarely be saved. Several precautions are recommended. Prepare the notched plate by siliconizing the surface that will contact the gel (e.g., using Sigmacoat [Sigma]). Use a kimwipe dampened with ethanol to clean a 2.5 cm strip along the top of the plate where the wells will be. This will remove some of the coating and prevent the

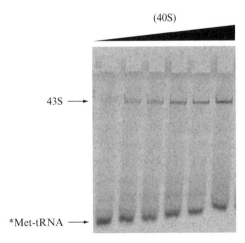

Figure 6.6 Native gel assay following 43S complex formation. Increasing amounts of 43S · mRNA complex (top band) are seen as the concentration of 40S subunits increases (moving from left to right). The lower band is free [^{35}S]Met-tRNA$_i$^{Met}. TC is generally not stable on the native gels, dissociating rapidly into free Met-tRNA$_i$^{Met}.

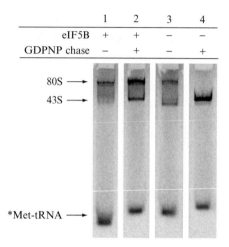

Figure 6.7 Native gel assay following 80S complex formation. 80S complexes were formed in the presence or absence of eIF5B and GDPNP (denoted by + or −) by combining 43S complexes (formed with GTP) with eIF5, 60S subunits, and the noted components. The positions of 80S and 43S complexes and Met-tRNA$_i^{Met}$ are noted. 80S complex formation is enhanced in the presence of eIF5B (compare lanes 1 and 2). GDPNP stabilizes 80S complexes formed in the presence of eIF5B (lane 2), but eliminates 80S complexes in the absence of eIF5B, suggesting complexes formed in the absence of eIF5B are not translationally competent. 43S complexes are also stabilized in the presence of GDPNP. After one round of 43S complex formation, eIF2 can (slowly) reenter the initiation pathway in a GDPNP-bound form, leading to dead-end 43S complexes (see 43S band in lanes 2 and 4).

well dividers from falling into the lanes. It is important that the coating remain on the notches themselves. To prepare the unnotched plate, scrub the surface that will contact the gel with steel wool and soap. This will roughen the plate, helping the gel to stick. Always treat the plates consistently, as once a plate has been roughened it will be extremely difficult to make it slide off the gel. Because it is difficult to insert the comb, do so before pouring the gel, pushing it only about a quarter of the way in.

The native gel is made with and run in THEM buffer. Having tested multiple buffer systems, we find that THEM yields the sharpest bands, which is important for quantitation. Insert a coil of tubing (tygon tubing can be made into a coil for this purpose), attached to a circulating water bath, into the gel box behind the gel. This coil will cool the gel as it runs, reducing the amount of gel "smiling" and minimizing the possibility of degradation of the complexes while the gel is running (numerous controls have shown the latter rarely to be a significant problem, however). Make sure the coil does not actually touch the gel plates at any point and distribute the coil as evenly throughout the box as possible. Fill both chambers with 1× THEM buffer and precool at 16° for 20 to 30 min before the gel will be

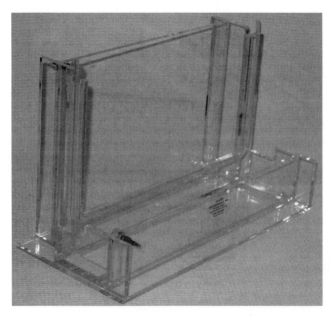

Figure 6.8   Box for native gel assays. The native gel box was specially made by Aladin Enterprises, Inc. (San Francisco, CA). Dimensions: well behind gel—26.5 cm wide, 19 cm tall, 2.5 cm deep; well in front of gel—31 cm wide, 5.5 cm tall, 6.5 cm deep.

loaded. Mark the location of the wells on the glass plate before removing the comb. Adding buffer before removing the comb will help the lane dividers remain in place. Use a syringe to gently rinse out each well, taking care not to break or bend the lane dividers. If bubbles or bits of gel remain in any of the wells, use a small piece of plastic (cutting a spacer off an old comb works well) to help clear the well.

15.2. Preparing 43S gel shift reactions

Reactions are prepared in recon buffer. Although a given experiment may require variations, our standard concentrations for testing components are 1 mM GDPNP $\cdot$ Mg^{2+}, 800 nM eIF2, 0.5 nM [^{35}S]Met-tRNA$_i^{\text{Met}}$, 1 μM eIF1 and eIF1A, 1 μM model mRNA(AUG), and 400 nM 40S ribosomal subunits. Ternary complex is formed stepwise by first incubating GDPNP $\cdot$ Mg^{2+} and eIF2 together for 10 min at 26°. [^{35}S]Met-tRNA$_i^{\text{Met}}$ is then added and the solution is incubated for an additional 5 min. Once preformed, TC can be added to 40S subunits and other factors to start the reaction, which is then incubated at 26° as necessary. With an unstructured model mRNA(AUG), 43S complex is fully formed after a few min of incubation. In the absence of mRNA, longer incubation times are required.

15.3. Preparing 80S gel shift reactions

80S complexes can be formed following a protocol similar to that for 43S complex formation (Acker *et al.*, 2006). TC is formed with 200 μM GTP $\cdot$ Mg^{2+} instead of GDPNP and added to 40S ribosomal subunits, eIF1, eIF1A, and model mRNA to form 43S complexes. An initiation mix composed of 60S ribosomal subunits, eIF5 and eIF5B is then added to the 43S complexes, initiating 80S complex formation. Typical final concentrations of the initiation mix components are 400 nM 60S ribosomal subunits, 800 nM eIF5, and 500 nM eIF5B. Alternatively, GDPNP $\cdot$ Mg^{2+} can be included with the initiation mix, resulting in a final concentration of 6 mM GDPNP in the 80S gel shift reaction. The addition of GDPNP stabilizes the 80S complexes because GDPNP-bound eIF5B has a greater affinity for the 80S complex than posthydrolysis GDP-bound eIF5B (Fig. 6.7).

15.4. Running 43S/80S gels

For thermodynamic experiments, 10 μl of each reaction is mixed with 2 μl of native gel dye (50% sucrose, 0.02% bromophenol blue, 0.02% xylene cyanol). Samples are stable in the well for the amount of time it takes to load a 24-well gel (approximately 15 min). Once all samples have been loaded, the gel is run at 25 W for ~30 min for 43S complexes. For 80S complexes, to sufficiently separate the 43S and 80S bands, the gel should be run for a total of 1 h.

For experiments measuring 43S formation kinetics, reactions can be stopped in two ways: either by loading the samples onto a running gel or by chasing with excess, cold TC. Controls have indicated that the two methods are equivalent for incubation times as short as 2 min. 80S complex reactions can only be quenched by loading the samples onto a running gel. When stopping the reaction by loading onto a gel running at 25 W, timepoints end only when the sample runs into the gel, not when they are mixed with dye. With a skilled hand the sample can be taken from the original reaction, mixed with dye, and loaded onto the gel in 20 s, allowing 3 points a min to be completed. Remember that the gel is running while you are loading and it is dangerous to put fingers in the buffer. Holding a tube in your nondominant hand or keeping your other hand behind your back can help to keep fingers away from the buffer. Accidental tests on rotation students have found that in some cases placing gloved fingers in the buffer does no actual damage to student or experiment, but erring on the side of caution is strongly advised. Once all the points have been loaded the gel should be run for an additional 30 min, but not more than 1 h total from the loading of the first sample.

Stopping reactions by quenching with cold TC can be particularly useful when taking short timepoints because there is no delay between the act of quenching and the actual end of the reaction. It is also easier to do quickly as it involves one mixing step rather than mixing and loading. To stop reactions with cold TC, add 10 μl of the reaction to 5 μl of 3× chase (3 mM GDPNP · Mg^{2+}, 1 μM eIF2, 450 nM Met-tRNA$_i^{Met}$ when using 0.5 nM hot TC). The quenched reactions can be stored at room temperature and are stable for several days. To check the effectiveness of the chase, assemble a "chase first" reaction in which 5 μl of 3× chase are added to 40S subunits and other factors just before the addition of hot TC. If the chase is effective, then very little 43S complex should be observed on the gel. Samples can be loaded as in the thermodynamic experiment.

15.5. Disassembly and analysis of gels

Once one plate (usually the notched one) has been separated from the gel, the gel is transferred to Whatman paper. It is then wrapped in plastic wrap and taped, gel side up, to the bottom of a phosphorimaging cassette. After placing a blank phosphorimaging screen on top, the entire cassette is wrapped in three layers of plastic wrap, carefully sealing the sides. It is important that it be wrapped well or moisture may damage the screen when condensation forms upon its removal from the freezer. Place the cassette in a −20° freezer overnight to prevent diffusion of the sample.

After exposing the screen, remove the cassette from the freezer and allow it to warm on the bench top. Do not unwrap the cassette before it is at room temperature or condensation may form on the cassette and screen. Once the cassette comes to room temperature, wipe any moisture from the plastic and unwrap the cassette. Scan the screen or remove the gel soon after it comes to room temperature to prevent diffusion from causing a problem. Quantitate the fraction of [^{35}S]Met-tRNA$_i^{Met}$ in 43S complex for each lane.

15.6. Testing 40S ribosomes

The 43S gel shift assay is a useful test of the quality of a ribosome preparation. Our standard tests include a concentration curve of 40S subunits from 0.5 to 50 nM in the presence of model mRNA(AUG) and from 10 to 250 nM in the absence of mRNA. We also assay several 40S concentration points in the absence of eIF1 and several in the absence of eIF1A. It is important to include mRNA(AUG) in these dropout reactions to allow sufficient 43S complex formation to be observable on the gel. A K_d of <1 nM (at the limit of detection of this assay) should be observed for the case with mRNA(AUG), and a K_d of approximately 60 nM should be observed without mRNA. A binding defect will be seen in the absence of eIF1 or eIF1A. The defect without eIF1A is generally more severe. Because both of these defects have a kinetic component, incubating these reactions for 30 min rather than the longer time period usual for the no mRNA reactions will make the defects more obvious.

16. Solutions

Standard Protease Inhibitors: 1 tablet Roche complete EDTA-free protease inhibitor cocktail per 50 ml buffer, 1 μg/ml leupeptin, 1 μg/ml aprotinin, 1 μg/ml pepstatin, 1 mM benzamidine

10$\times$ Ribo Buffer A: 200 mM Hepes $\cdot$ KOH (pH 7.4), 1 M KOAc (pH 7.6), 25 mM Mg(OAc)$_2$

Ribo Lysis Buffer: 1$\times$ Ribo Buffer A, 1 mg/ml heparin, 2 mM DTT, protease inhibitor tabs, 0.5 mM AEBSF

Sucrose Cushion: 1$\times$ Ribo Buffer A, 500 mM KCl, 1 M sucrose, 2 mM DTT

High Salt Wash: 1$\times$ Ribo Buffer A, 500 mM KCl, 1 mg/ml heparin, 2 mM DTT

Subunit Separation Buffer: 50 mM Hepes $\cdot$ KOH (pH 7.4), 500 mM KCl, 2 mM MgCl$_2$, 2 mM DTT

5% and 20% Sucrose Gradient Solutions: 50 mM Hepes $\cdot$ KOH (pH 7.4), 500 mM KCl, 5 mM MgCl$_2$, 0.1 mM EDTA, 5% or 20% sucrose, 2 mM DTT

Ribosome Sucrose Storage Buffer: 1$\times$ Ribo Buffer A, 250 mM sucrose, 2 mM DTT

Ribosome Extract Buffer: 0.3 M NaOAc (pH 5.0), 12.5 mM EDTA, 0.5% SDS

eIF2 Lysis Buffer: 75 mM Hepes $\cdot$ KOH (pH 7.6), 100 mM KCl, 100 μM GDP $\cdot$ Mg^{2+}, 10 mM BME, standard protease inhibitors

eIF2 NCLB-20: 20 mM Hepes $\cdot$ KOH (pH 7.6), 500 mM KCl, 0.1 mM MgCl$_2$, 10 μM GDP $\cdot$ Mg^{2+}, 20 mM imidazole, 10% glycerol, 10 mM BME, standard protease inhibitors

eIF2 NCEB-250: eIF2 NCLB-20 with 250 mM imidazole instead of 20 mM.

eIF2 LSB: 20 mM Hepes $\cdot$ KOH (pH 7.6), 100 mM KCl, 0.1 mM MgCl$_2$, 10 μM GDP $\cdot$ Mg^{2+}, 10% glycerol, 2 mM DTT

eIF2 HSB: eIF2 LSB with 1 m KCl instead of 100 mM.

eIF2 storage buffer: 20 mM Hepes $\cdot$ KOH (pH 7.6), 100 mM KOAc (pH 7.6), 0.1 mM Mg(OAc)$_2$, 10% glycerol, 2 mM DTT

eIF2 storage buffer with creatine kinase: eIF2 storage buffer, 0.6 μg/μl creatine kinase

Intein Lysis Buffer: 20 mM Hepes $\cdot$ KOH (pH 7.4), 0.5 M KCl (pH 7.6), 0.1% Triton X100, 1 mM EDTA, standard protease inhibitors

Intein Cleavage Buffer: 20 mM Hepes $\cdot$ KOH (pH 8.0), 0.5 M KCl, 1 mM EDTA

Intein Wash Buffer: 20 mM Hepes $\cdot$ KOH (pH 7.4), 1 M KCl (pH 7.6), 0.1% Triton X100, 1 mM EDTA

LSB: eIF2 LSB without MgCl$_2$ and without GDP

HSB: eIF2 HSB without $MgCl_2$ and without GDP

eIF3 ZCLB: 20 mM Tris (pH 7.6), 350 mM KCL, 5 mM $MgCl_2$, 20 mM imidazole, 10% glycerol, 5 mM BME

eIF3 ZCEB: eIF3 ZCLB with 250 mM imidazole instead of 20 mM

eIF1A NCLB-20: 20 mM Hepes · KOH (pH 7.6), 500 mM KCl, 20 mM imidazole, 10% glycerol, 10 mM BME

eIF1A NCEB-250: eIF1A NCLB-20 with 250 mM imidazole instead of 20 mM

Buffer B-20: 10 mM MOPS (pH 6.7), 500 mM NaCl, 20 mM imidazole (pH 7.0), 3 mM BME, 0.1 mM AEBSF

Buffer B-250: Buffer B-20 with 250 mM imidazole instead of 20 mM

Buffer A: 10 mM MOPS (pH 6.7), 100 mM or 500 mM NaCl, 10 mM BME, 0.1 mM AEBSF

yMetRS storage buffer: 40 mM Tris (pH 7.4), 10 mM $MgCl_2$, 10% glycerol, 2 mM DTT

10× Transcription Buffer: 400 mM Tris (pH 8.1), 300 mM $MgCl_2$, 20 mM spermidine, 1% Triton-X-100, 0.5 mg/ml BSA

Filter Binding Wash Buffer: 25 mM Hepes · KOH (pH 7.6), 2.5 mM $Mg(OAc)_2$, 80 mM KOAc, 2 mM DTT, 2% glycerol

10X THEM Buffer: 340 mM Tris Base, 570 mM Hepes, 1 mM EDTA, 25 mM $MgCl_2$

10X Recon Buffer: 300 mM Hepes · KOH (pH 7.4), 1-M KOAc (pH 7.6), 30 mM $Mg(OAc)_2$, 20 mM DTT

Enzyme Buffer: 20 mM Hepes· KOH (pH 7.4), 1 M KOAc (pH 7.6), 10 % glycerol, 2 mM DTT

Gels:

Identity Gel: 5% 29:1 acrylamide:bisacrylamide, 1× TBE buffer, 8 M urea, 0.1% APS, 1 μl TEMED per ml gel

RNA Purification Gel: 12% 19:1 acrylamide:bisacrylamide, 1× TBE buffer, 8-M urea, 0.1% APS, 1 μl TEMED per ml gel

GTPase Gel: 15% 29:1 acrylamide:bisacrylamide, 1× TBE buffer

Native Gel: 4% 37.5:1 acrylamide:bisacrylamide, 1× THEM buffer, 0.1% APS, 1 μl TEMED per ml gel

ACKNOWLEDGMENTS

This work was supported by grants to J.R.L. from the National Institutes of Health (GM-62128), the American Cancer Society (RSG-03-156-01-GMC), and the American Heart Association (0555466U). Many of these techniques were developed during the tenure of Mikkel Algire, Drew Applefield, Lee Kapp, and David Maag in the lab. We are grateful to Alan Hinnebusch, Tom Dever, and members of their labs for very productive collaborations and help at all stages of the development of this system.

REFERENCES

Acker, M. G., Shin, B. S., Dever, T. E., and Lorsch, J. R. (2006). Interaction between eukaryotic initiation factors 1A and 5B is required for efficient ribosomal subunit joining. *J. Biol. Chem.* **281**, 8469–8475.

Algire, M. A., Maag, D., and Lorsch, J. R. (2005). Pi release from eIF2, not GTP hydrolysis, is the step controlled by start-site selection during eukaryotic translation initiation. *Mol. Cell* **20**, 251–262.

Algire, M. A., Maag, D., Savio, P., Acker, M. G., Tarun, S. Z., Sachs, A. B., Asano, K., Nielsen, K. H., Olsen, D. S., Phan, L., Hinnebusch, A. G., and Lorsch, J. R. (2002). Development and characterization of a reconstituted yeast translation initiation system. *RNA* **8**, 382–397.

Benne, R., and Hershey, J. W. B. (1978). The Mechanism of Action of Protein Synthesis Initiation Factors from Rabbit Reticulocytes. *J. Biol. Chem.* **253**, 3078–3087.

Choi, S. K., Olsen, D. S., Roll mecak, A., Martung, A., Remo, K. L., Burley, S. K., Hinnebusch, A. G., and Dever, T. E. (2000). Physical and functional interaction between the eukaryotic orthologs of prokaryotic translation initiation factors IF1 and IF2. *Mol. Cell Biol.* **20**, 7183–7191.

Erickson, F. L., and Hannig, E. M. (1996). Ligand interactions with eukaryotic translation initiation factor 2: Role of the gamma-subunit. *EMBO J.* **15**, 6311–6320.

Fechter, P., Rudinger, J., Giege, R., and Theobald-Dietrich, A. (1998). Ribozyme processed tRNA transcripts with unfriendly internal promoter for T7 RNA polymerase: Production and activity. *FEBS Lett.* **436**, 99–103.

Goode, B. L. (2002). Purification of yeast actin and actin-associated proteins. *Methods Enzymol.* **351**, 433–441.

Guillon, L., Schmitt, E., Blanquet, S., and Mechulam, Y. (2005). Initiator tRNA binding by e/aIF5B, the eukaryotic/archaeal homologue of bacterial initiation factor IF2. *Biochemistry* **44**, 15594–15601.

Jivotovskaya, A. V., Valasek, L., Hinnebusch, A. G., and Nielsen, K. H. (2006). Eukaryotic translation initiation factor 3 (eIF3) and eIF2 can promote mRNA binding to 40S subunits independently of eIF4G in yeast. *Mol. Cell Biol.* **26**, 1355–1372.

Kapp, L. D., Kolitz, S. E., and Lorsch, J. R. (2006). Yeast initiator tRNA identity elements cooperate to influence multiple steps of translation initiation. *RNA* **12**, 751–764.

Kapp, L. D., and Lorsch, J. R. (2004a). GTP-dependent recognition of the methionine moiety on initiator tRNA by translation factor eIF2. *J. Mol. Biol.* **335**, 923–936.

Kapp, L. D., and Lorsch, J. R. (2004b). The molecular mechanics of eukaryotic translation. *Annu. Rev. Biochem.* **73**, 657–704.

Maag, D., Fekete, C. A., Gryczynski, Z., and Lorsch, J. R. (2005). A conformational change in the eukaryotic translation preinitiation complex and release of eIF1 signal recognition of the start codon. *Mol. Cell* **17**, 265–275.

Maag, D., and Lorsch, J. R. (2003). Communication between eukaryotic translation initiation factors 1 and 1A on the yeast small ribosomal subunit. *J. Mol. Biol.* **330**, 917–924.

Marintchev, A., Kolupaeva, V. G., Pestova, T. V., and Wagner, G. (2003). Mapping the binding interface between human eukaryotic initiation factors 1A and 5B: A new interaction between old partners. *Proc. Natl. Acad. Sci. USA* **100**, 1535–1540.

Muir, T. W., Dolan, S., and Cole, P. A. (1998). Expressed protein ligation: A general method for protein engineering. *Proc. Natl. Acad. Sci. USA* **95**, 6705–6710.

Olsen, D. S., Savner, E. M., Mathew, A., Zhang, F., Krishnamoorthy, T., Phan, L., and Hinnebusch, A. G. (2003). Domains of eIF1A that mediate binding to eIF2, eIF3 and eIF5B and promote ternary complex recruitment *in vivo*. *EMBO J.* **22**, 193–204.

Pavitt, G. D., Ramaiah, K. V., Kimball, S. R., and Hinnebusch, A. G. (1998). eIF2 independently binds two distinct eIF2B subcomplexes that catalyze and regulate guanine-nucleotide exchange. *Genes Dev.* **12,** 514–526.

Pestova, T., Lorsch, J. R., and Hellen, C. (2007). The Mechanism of Translation Initiation in Eukaryotes. *In* "Translational Control in Biology and Medicine" (M. B. Mathews, N. Sonenberg, and J. Hershey, eds.), pp. 87–128. Cold Spring Harbor Press, Cold Spring Harbor, NY.

Pestova, T. V., Borukhov, S. I., and Hellen, C. U. T. (1998). Eukaryotic ribosomes require initiation factors 1 and 1A to locate initiation codons. *Nature* **394,** 854–859.

Pestova, T. V., and Hellen, C. U. (2000). The structure and function of initiation factors in eukaryotic protein synthesis. *Cell Mol. Life Sci.* **57,** 651–674.

Pestova, T. V., Kolupaeva, V. G., Lomakin, I. B., Pilipenko, E. V., Shatsky, I. N., Agol, V. I., and Hellen, C. U. (2001). Molecular mechanisms of translation initiation in eukaryotes. *Proc. Natl. Acad. Sci. USA* **98,** 7029–7036.

Phan, L., Zhang, X., Asano, K., Anderson, J., Vornlocher, H. P., Greenberg, J. R., Qin, J., and Hinnebusch, A. G. (1998). Identification of a translation initiation factor 3 (eIF3) core complex, conserved in yeast and mammals, that interacts with eIF5. *Mol. Cell Biol.* **18,** 4935–4946.

Scheibner, K. A., Zhang, Z., and Cole, P. A. (2003). Merging FRET and expressed protein ligation to analyze protein-protein interactions. *Anal. Biochem.* **317,** 226–232.

Trachsel, H., Erni, B., Schreier, M. H., and Staehelin, T. (1977). Initiation of mammalian protein synthesis. II. The assembly of the initiation complex with purified initiation factors. *J. Mol. Biol.* **116,** 755–767.

Valasek, L., Nielsen, K. H., Zhang, F., Fekete, C. A., and Hinnebusch, A. G. (2004). Interactions of eukaryotic translation initiation factor 3 (eIF3) subunit NIP1/c with eIF1 and eIF5 promote preinitiation complex assembly and regulate start codon selection. *Mol. Cell Biol.* **24,** 9437–9455.

CHAPTER SEVEN

ASSEMBLY AND ANALYSIS OF EUKARYOTIC TRANSLATION INITIATION COMPLEXES

Andrey V. Pisarev,* Anett Unbehaun,* Christopher U. T. Hellen,* *and* Tatyana V. Pestova*,†

Contents

* Department of Microbiology and Immunology, State University of New York Downstate Medical Center, Brooklyn, New York
† A. N. Belozersky Institute of Physicochemical Biology, Moscow State University, Moscow, Russia

Methods in Enzymology, Volume 430
ISSN 0076-6879, DOI: 10.1016/S0076-6879(07)30007-4

Abstract

The canonical initiation process is the most complex aspect of translation in eukaryotes. It involves the coordinated interactions of at least 11 eukaryotic initiation factors, 40S and 60S ribosomal subunits, mRNA, and aminoacylated initiator tRNA (Met-tRNA$_i^{Met}$), as well as binding and hydrolysis of GTP and ATP. The factor requirements for many individual steps in this process, including scanning, initiation codon recognition, and ribosomal subunit joining, have until recently been obscure. We established the factor requirements for these steps by reconstituting the initiation process *in vitro* from individual purified components of the translation apparatus and developed approaches to explain the mechanism of individual steps and the roles of individual factors and to characterize the structure of initiation complexes. Here we describe protocols for the purification of native initiation factors and for expression and purification of active recombinant forms of all single subunit initiation factors, for the reconstitution of the initiation process, and for determination of the position of ribosomal complexes on mRNA by primer extension inhibition ("toe printing"). We also describe protocols for site-directed ultraviolet (UV) cross-linking to determine the interactions of individual nucleotides in mRNA with components of the initiation complex and for directed hydroxyl radical probing to determine the position of initiation factors on the ribosome.

1. INTRODUCTION

The canonical mechanism of translation initiation in eukaryotes is a multistep process that requires the coordinated activities of at least 11 eukaryotic initiation factors (eIFs) to correctly assemble aminoacylated initiator tRNA (Met-tRNA$_i^{Met}$), a 40S and a 60S ribosomal subunit into an 80S ribosome on the initiation codon of an mRNA (Pestova *et al.*, 2006). A 43S preinitiation complex comprising a ribosomal 40S subunit, eIF1, eIF1A, eIF3, and an eIF2·GTP/Met–tRNA$_i^{Met}$ complex binds to the 5′ cap-proximal region of a mRNA in a process that involves eIF4A, eIF4B, and the heterotrimeric cap-binding complex eIF4F. The preinitiation complex then scans in a 5′ to 3′ direction on the 5′ untranslated region (5′UTR) until it locates the initiation codon, where it forms a 48S initiation complex. eIF5 induces hydrolysis of eIF2-bound GTP, eIF5B mediates displacement of eIF1, eIF1A, eIF2·GDP, and eIF3 from the interface surface of the 40S subunit and its joining to a 60S subunit to form an 80S ribosome in which the anticodon of Met–tRNA$_i^{Met}$ is base-paired to the initiation codon in the ribosomal peptidyl (P) site. Finally, hydrolysis of eIF5B-bound GTP leads to release of eIF5B·GDP from the 80S ribosome that can then begin translation of the mRNA.

Procedures for the preparation of many of these initiation factors were first described in this series almost 30 years ago (Benne *et al.*, 1979; Merrick, 1979; Staehelin *et al.*, 1979); we have modified some of these procedures and have developed protocols for purification of additional native initiation factors (Pestova *et al.*, 1998a, 2000), for expression and purification of active recombinant forms of all single subunit initiation factors (Pestova *et al.*, 1996a,b, 1998a, 2000), for the preparation of site-specifically modified factors (Kolupaeva *et al.*, 2003; Lomakin *et al.*, 2003) and for preparation of synthetic Met–tRNA$_i^{Met}$ and site-specifically derivatized mRNAs (Pestova and Hellen, 2001; Pisarev *et al.*, 2006). Here we describe these methods and approaches that we have developed to reconstitute the entire initiation process *in vitro* from individual purified components to explain the mechanism of individual steps in initiation and the roles of individual factors in this complex process and to characterize the composition and structure of initiation complexes.

2. CHEMICALS, ENZYMES, AND BIOLOGICAL MATERIALS

Standard laboratory chemicals were from Sigma Aldrich (St. Louis, MO) and Fisher Scientific (Pittsburgh, PA), unlabeled NTPs and dNTPs were from GE Healthcare Life Sciences (Piscataway, NJ), and unlabeled amino acids were from Sigma Aldrich. Isopropyl-β-D-thiogalactopyranoside (IPTG) was from Bioworld (Dublin, OH), GMP-PNP and Pefabloc were from Roche (Indianapolis, IN), protease inhibitor cocktail set III was from Calbiochem (La Jolla, CA), puromycin was from EMD Biosciences (San Diego, CA), and Calbiochem, Fe(II)-1-(p-bromoacetamidobenzyl)-EDTA (Fe-BABE) was from Dojindo Molecular Technologies (Gaithersburg, MD), Ni^{2+}-nitrilotriacetic acid (Ni-NTA) and imidazole were from Qiagen (Valencia, CA). Kanamycin and ampicillin were from American Bioanalytical (Natick, MA) and Sigma Aldrich. 4-thiouridine (4-thioU) and 6-thioguanosine (6-thioG) were from Ambion (Austin, TX) and Jena Bioscience (Jena, Germany). 7-diethylamino-3-([4-(iodacetyl)phenyl)-4-methyl coumarin (DCIA) was from Molecular Probes/Invitrogen. Four to 12% Bis-Tris NuPAGE gel, the MOPS buffer system, and SimplyBlue Safestain were from Invitrogen (Carlsbad, CA). Oligonucleotides were from Invitrogen and MWG Biotech (High Point, NC), and the diribonucleotide transcription primer rAC was from Dharmacon (Lafayette, CO). Expression vectors from the pET series were from EMD Biosciences and from the pQE series from Qiagen. The transcription vector pBluescript SK was from Stratagene (La Jolla, CA). Heparin-Sepharose, 7-Methyl-GTP Sepharose 4B and Hoefer SG gradient makers were from GE Healthcare Life Sciences. DEAE cellulose (Whatman DE52) and phosphocellulose (Whatman P11) were from Fisher Scientific. Polyprep chromatography columns were from BioRad (Hercules, CA). Centricon YM100 and Microcon YM10 centrifugal filter units were from Millipore (Billerica, MA). Sephadex G-50 Spin-50 mini-columns were from USA Scientific (Ocala, FL). Spectrum Spectra/Por 3RC 25-mm dialysis tubing was purchased from Fisher Scientific. Polyallomer centrifuge tubes were from Beckman Coulter (Fullerton, CA) and BD PrecisonGlide 20G1{1/2} syringe needles were from Becton Dickinson (Franklin Lakes, NJ). Whatman 3MM paper was from Fisher Scientific, and X-ray film was from Denville Scientific (South Plainfield, NJ). Radiochemicals [^{35}S]methionine (37 TBq/mmol) and [α-^{32}P]dATP (111 TBq/mmol), [α-^{32}P]CTP (222 Tbq/mmol), [α-^{32}P]UTP (111 Tbq/mmol), and [γ-^{32}P]ATP (111 TBq/mmol) were purchased from MP Biomedicals (Irvine, CA).

DNA restriction endonucleases, DNA modifying enzymes, and the catalytic subunit of cAMP-dependent protein kinase were from New England BioLabs (Beverley, MA), Fermentas (Hanover, MD), and Roche. RQ1 RNase-free DNase and avian myeloblastosis virus (AMV) reverse transcriptase were purchased from Promega Corp. (Madison, WI). T7 RNA polymerase and T7 RNA polymerase buffer were from Ambion and Fermentas. The ribonuclease inhibitors RNAsin and RNAse OUT were from Promega and Invitrogen, respectively. Sequenase T7 DNA polymerase and RNase H were from USB Corp. (Cleveland, OH). RNase A was from Sigma, and RNase V1 was from Pierce (Milwaukee, WI). Polymerase chain reactions (PCR) were done by use of the Expand High Fidelity (Plus) PCR System (Roche).

Escherichia coli strain MRE 600 was from the American Type Culture Collection (Manassas, VA) (ATCC No 29417). *E. coli* BL21 Star (DE3): F-*ompT hsdS*B (rB-mB-) *gal dcm rne131* (DE3) and *E. coli* DH5α-T1^R : F-f80*lacZ*?M15 ?(*lacZYA-arg*F) U169 *deo*R *rec*A1 *end*A1 *hsd*R17 (rk-, mk+) *pho*A *sup*E44 ?- *thi*-1 *gyr*A96 *rel*A1 *ton*A were from Invitrogen. Rabbit reticulocyte lysate (RRL) was from Green Hectares (Oregon, WI). Native calf liver tRNA was from EMD Biosciences, and a mixture of native α- and β-globin mRNAs was from Invitrogen; this product offering has now been discontinued.

Items of apparatus used in the procedures described in the following include an Eppendorf thermomixer and an Eppendorf 5417C microcentrifuge, a Branson S-450D digital sonifier, and a Bransonic 1510 ultrasonic cleaner (Branson, Danbury CT), Aktapurifier chromatography systems (GE Healthcare Life Sciences), a Stratalinker UV cross linker used with a 365-nm bulb (Stratagene), a GeneAmp 2400 PCR system (Applied

Biosystems, Foster City, CA). For electrophoresis, we use BioRad Mini-PROTEAN II or Mini-PROTEAN III electrophoresis cells in conjunction with an EPS 301 power supply (GE Healthcare), a BioRad PROTEAN II xi 2-D cell two-dimensional gel apparatus in conjunction with an EPS 3501 power supply (GE Healthcare), and a vertical sequencing apparatus (Owl Separation Systems, Rochester, NY) with an EPS 3501 power supply. Gels are dried with a model 583 gel dryer (Bio-Rad). Centrifugation was done with Sorvall RC-5C Plus centrifuges (Thermo Electron Corp., Asheville, NC) and Beckman L8-M ultracentrifuges (Beckman Coulter).

3. PURIFICATION OF 40S AND 60S RIBOSOMAL SUBUNITS

3.1. Buffers

Buffer A   20 mM Tris, pH 7.5, 2 mM dithiothreitol (DTT), 4 mM MgCl$_2$, 50 mM KCl, 0.25 M sucrose

Buffer B   20 mM Tris pH 7.5, 2 mM DTT, 4 mM MgCl$_2$, 0.5 M KCl

Buffer C   20 mM Tris pH 7.5, 2 mM DTT, 2 mM MgCl$_2$, 100 mM KCl, 0.25 M sucrose

    Rabbit reticulocyte lysate stored at $-80°$ is used for purification of 40S and 60S ribosomal subunits by a slightly modified version of published protocols (Pestova and Hellen, 2005; Pestova et al., 1996a, 1998b). Fifty to 100 ml RRL is sufficient for purification of ribosomal subunits, but we usually combine ribosome and factor purification, for which we use $\sim$600 ml RRL at a time (an amount that is determined by the $\sim$300 ml capacity of the Beckman 50.2 Ti rotor). Here we first describe our protocol for the purification of ribosomal subunits alone, for instance, when purification of factors is not intended. However, the procedure would be the same if the RRL is to be used for purification of ribosomal subunits and initiation factors, in which case the pellet obtained after washing polysomes with 0.5 M KCl (see the next section) is used for ribosomal subunit purification.

    Thawed RRL is mixed with Pefabloc (0.5 mg/ml), a water-soluble and irreversible inhibitor of serine proteases. Polysomes are precipitated by centrifugation of RRL at 45,000 rpm for 4.5 h in a Beckman 50.2 Ti rotor at 4° and are resuspended in buffer A (without sucrose) to a final concentration of 100 to 150 A$_{260}$ U/ml. We initially resuspend each pellet in 1 ml buffer A by gently "stroking" it with a round-ended glass rod, add 1 ml more buffer A, and continue resuspending by use of a 1 ml Eppendorf pipette, and then add another 2 ml buffer A, combine the suspended material from all tubes, and stir it on ice for 30 min with a magnetic stirrer. If necessary, the suspension is then clarified by centrifugation for 10 min at 14,000 rpm at 4° with an Eppendorf benchtop microcentrifuge. To measure the OD_{260} of the suspension, an aliquot is first diluted ~100 to 200 fold with buffer A. If necessary, the optical density is adjusted to 100 to 150 U/ml with buffer A. The final polysome suspension is then incubated with 1 mM puromycin for 10 min on ice and then for 10 min at 37°, and 4 M KCl is added in a drop-wise manner to a final concentration of 0.5 M, with continuous mixing (so as not to create a high local concentration of KCl); 4 M KCl is a saturated solution, so that it must be stored at room temperature even before use; 1.7-ml aliquots of this suspension are then loaded immediately onto 10 to 30% sucrose density gradients prepared in buffer B with a Hoefer SG gradient maker and are centrifuged at 22,000 rpm for 16 h in a Beckman SW28 rotor at 4° to resolve the ribosomal subunits; 1-ml fractions are collected across the gradient, and their optical density is measured after diluting aliquots 10-fold. To avoid cross-contamination of 40S and 60S subunits, care must be taken to choose only peak fractions and to discard "shoulders." Although ribosomal subunits can then be precipitated after twofold dilution with buffer C without sucrose by centrifugation at 31,000 rpm for 15 h in a Beckman 50.2 Ti rotor at 4° and redissolved in buffer C, this procedure leads to a partial loss of activity of the subunits. We, therefore, prefer the faster and gentler approach of concentrating ribosomal subunits and transferring them into buffer C by use of YM-100 centricons. The final concentrations of 40S and 60S ribosomal subunits that we aim for are 25 to 50 A_{260} U/ml and 60 to 120 A_{260} U/ml, respectively. We prepare small (10 to 20 μl) aliquots for storage at −80° to minimize loss of activity because of multiple freeze/thaw cycles. The molar ribosomal subunit concentrations can be calculated assuming that 1 A_{260} unit is equivalent to 65 pmol of 40S subunits and to 29 pmol of 60S subunits, respectively (Raychaudhuri *et al.*, 1986).

4. PURIFICATION OF NATIVE EIF3 AND EIF4F

4.1. Buffers

Buffer A   20 mM Tris, pH 7.5, 2 mM dithiothreitol (DTT), 4 mM
            $MgCl_2$, 50 mM KCl, 0.25 M sucrose
Buffer D   20 mM Tris, pH 7.5, 2 mM DTT, 0.1 mM EDTA, 10%
            glycerol
Buffer E   20 mM Tris, pH 7.5, 2 mM DTT, 0.1 mM EDTA, 5% glycerol
Buffer F   20 mM Tris-HCl, pH 7.5, 1 M KCl

Ribosomal pellets derived by centrifugation of 500 to 600 ml Pefabloc-treated RRL in a Beckman Ti 50.2 rotor at 45,000 rpm for 4.5 h at 4° are suspended in buffer A to a concentration of $\sim$100 A_{260} U/ml as described previously in the protocol for purification of ribosomal subunits. The final polysomal suspension (usually between 100 and 150 ml) is gently stirred on ice for 30 min, and 4 M KCl is then added in a drop-wise manner to it, with continuous stirring, to a final concentration of 0.5 M KCl. Further gentle stirring for 30 min on ice leads to clarification of the suspension. It is then centrifuged in a Beckman Ti 50.2 rotor at 45,000 rpm for 4.5 h at 4°, yielding a pellet that can be used for purification of ribosomal subunits by the procedure described in the previous section and a supernatant known as the ribosomal salt wash (RSW) that is the starting material for the purification of native initiation factors.

Factors are precipitated from the RSW by stepwise ammonium sulfate (AS) precipitation with ammonium sulfate that has been finely powdered to ensure that it dissolves rapidly to prevent the occurrence of high local AS concentrations during precipitation. To obtain the initial 0 to 40% AS fraction, powdered AS (242 g/L RSW) is added very slowly to the RSW while stirring it on ice. After stirring for 30 min, the suspension is centrifuged in a Sorvall SS34 rotor at 15,000 rpm for 20 min at 4°, yielding a pellet that contains eIF3, eIF4B, eIF4F, and small amounts of eIF2, and a supernatant containing other factors. The pellet is dissolved in 5 to 7 ml buffer D + 100 mM KCl, dialyzed against 1 L buffer D + 100 mM KCl overnight at 4° and clarified by centrifugation at 10,000 rpm for 10 min at 4°. The supernatant left after the previous centrifugation step can be used to obtain a 40 to 50% AS fraction that contains eIF2, eIF5B, and small amounts of eIF5 by adding 62 g AS/L, and the fractionation process can be repeated to obtain a 50 to 70% fraction that contains eIF1, eIF1A, and eIF4A by addition of 135 g AS/L of supernatant left after preparation of the 40 to 50% AS fraction. The dialyzed 0 to 40%, 40 to 50%, and 50 to 70% AS fractions remain active after long-term storage at $-80°$.

After dialysis, the 0 to 40% AS fraction is applied to a $\sim$12 ml DEAE (DE52) column equilibrated with buffer D + 100 mM KCl and a fraction containing eIF3, eIF4F, eIF4B, and small amounts of eIF2 is eluted with buffer D + 250 mM KCl and diluted very slowly (to avoid hypotonic shock and precipitation of proteins) with buffer D to a final concentration of 100 mM KCl. The resulting solution is applied to a $\sim$7-ml phosphocellulose (P11) column equilibrated with buffer D + 100 mM KCl. Step elution, done with buffer D + 400 mM KCl and then buffer D + 800 mM KCl, yields $\sim$5 to 8 ml fractions containing eIF3/eIF4B/eIF4F and eIF2, respectively. The 800 mM KCl elution fraction will be used for purification of eIF2 on a MonoQ column as described in the next section.

For purification of eIF3 and eIF4F, 10 to 30% sucrose density gradients are prepared in buffer D + 400 mM KCl (to prevent association of eIF3 and eIF4F) in 4 to 6 polyallomer tubes (14 × 89 mm) with a Hoefer SG gradient maker, and up to 1.5 ml of the 400 mM KCl elution fraction is loaded on top of each gradient. Care should be taken in loading because of the high concentration of glycerol; to make loading easier, elution from P11 can be done with buffer E, which contains only 5% glycerol. After centrifugation in a SW41 rotor at 40,000 rpm for 22 h at 4°, gradients are fractionated into 22 to 25 fractions. The optical density (A_{280}) of each fraction is measured to identify the pronounced eIF3 peak, which is typically in the 7 to 10 bottom fractions of the gradient (Fig. 7.1A). The presence of eIF3 in these fractions (and of eIF4F in upper fractions) can be confirmed by SDS polyacrylamide gel electrophoresis of 5- to 7-μl aliquots (Fig. 7.1B); because fractions contain 0.4 M KCl (which forms a pellet with SDS), samples should be loaded onto gels immediately after denaturing at 95° without allowing them to cool. Fractions that contain eIF3 and eIF4F (Fig. 7.1) are then dialyzed overnight against buffer E + 100 mM KCl and against buffer D + 100 mM KCl, respectively, at 4° and can be processed further or temporarily stored at $-80°$.

To complete eIF3 purification, dialyzed eIF3-containing gradient fractions are loaded on a FPLC MonoQ HR 5/5 column that has been equilibrated with buffer E + 100 mM KCl and is then eluted with a buffer E/100 to 500 mM KCl gradient. eIF3 is eluted at ~400 mM KCl with a yield of ~5 to 7 mg, and fractions can be concentrated to an optical density (A_{280}) >4 and are transferred into buffer E + 100 mM KCl with Microcon YM100 centrifugal filter units. eIF3 prepared in this way lacks the eIF3j subunit, which dissociates during sucrose density gradient centrifugation (Pestova *et al.*, 1996a); eIF3 containing eIF3j can be prepared by omitting the sucrose density gradient centrifugation step (Unbehaun *et al.*, 2004), but in this case, eIF3 might contain eIF4F and other contaminants (Fig. 7.1C). The eIF3a (p170) subunit of eIF3 in RRL is usually degraded (Grifo *et al.*, 1983), but this does not affect the activity of eIF3 in formation of 48S and 80S initiation complexes; eIF3 with an intact eIF3a subunit can be purified from HeLa cells by use of essentially the procedure described here (Fig. 7.1C; Unbehaun *et al.*, 2004). The subunit composition of eIF3 is best analyzed by use of 4 to 12% NuPAGE Bis-Tris gel (with MES or MOPS buffer systems).

For purification of eIF4F, a column containing 1 ml 7-Methyl-GTP Sepharose[TM] 4B is first washed with 10 ml buffer D + 100 mM KCl and is then loaded three times with eIF4F-containing fractions from the sucrose density gradient after they have been dialyzed overnight against buffer E + 100 mM KCl at 4°. The column is then washed with 50 ml buffer E + 100 mM KCl, and eIF4F is eluted with 10 ml buffer D + 100 mM KCl + 75 μM 7-methyl guanosine. Nonspecifically bound material is then removed from the column by washing it with 10 ml buffer F. The eluate containing eIF4F is loaded onto an FPLC MonoQ HR 5/5 column that has been equilibrated with buffer E + 100 mM KCl and is then eluted with a buffer E/100 to 500 mM KCl gradient. eIF4F containing an intact eIF4G subunit elutes at ~360 mM KCl, and eIF4F with N-terminally truncated

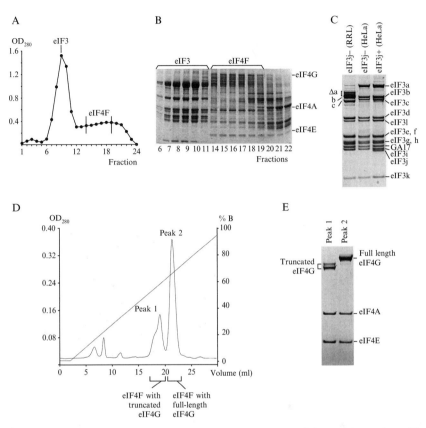

Figure 7.1 Purification of native eIF3 and eIF4F. (A) Optical density (OD_{280}) profile and (B) 10% SDS-polyacrylamide gel of proteins in fractions from a 10 to 30% sucrose density gradient after centrifugation on it of the 0 to 40% AS/250 mM KCl DE52/400 mM KCl phosphocellulose fraction of the RRL RSW. (C) Subunit composition of purified RRL eIF3j−, HeLa eIF3j+, and HeLa eIF3j− (as indicated) analyzed by gel electrophoresis. Individual subunits are indicated on the right of the panel. (D) Optical density (OD_{280}) profile and (E) subunit composition of eIF4F eluted from an FPLC MonoQ HR 5/5 column loaded with eIF4F-containing fractions that had previously been purified on 7-methyl-GTP-Sepharose. Proteins were resolved by electrophoresis on 4 to 12% Bis-Tris NuPAGE gel by use of a MES buffer system (C, E), followed by Coomassie staining. Subunits of eIF4F are indicated on panels (B) and (E).

eIF4F elutes slightly earlier at ∼330 mM KCl (Fig. 1D and E). The eluate is then concentrated by use of Microcon YM10 centrifugal filter units to an optical density (A_{280}) between 0.5 and 1. The typical yield is ∼0.5 to 0.7 mg eIF4F per 600 ml RRL.

5. Purification of Native eIF2, eIF5, and eIF5B

5.1. Buffers

Buffer D   20 mM Tris, pH 7.5, 2 mM DTT, 0.1 mM EDTA, 10%
             glycerol
Buffer E   20 mM Tris, pH 7.5, 2 mM DTT, 0.1 mM EDTA, 5% glycerol

Most native eIF2 and smaller amounts of eIF5B and eIF5 are prepared from the RSW 40 to 50% AS precipitation fraction. This fraction is dialyzed overnight against buffer D + 100 mM KCl and is then applied to a ~12 ml DEAE (DE52) column equilibrated with buffer D + 100 mM KCl. A fraction containing eIF2, eIF5, and eIF5B is eluted with buffer D + 250 mM KCl and diluted very slowly (to avoid hypotonic shock and precipitation of proteins) with buffer D to a final concentration of 100 mM KCl. The resulting solution is applied to a ~7-ml phosphocellulose (P11) column equilibrated with buffer D + 100 mM KCl. The column is then washed with buffer D + 400 mM KCl, after which eIF2, eIF5, and eIF5B are eluted with buffer D + 800 mM KCl. A ~5- to 8-ml fraction containing eIF2/eIF5/eIF5B is dialyzed overnight at 4° against buffer E + 100 mM KCl and is then loaded onto a FPLC MonoQ HR 5/5 column that has been equilibrated with buffer E + 100 mM KCl. Factors are then eluted with a buffer E/100 to 500 mM KCl gradient (Fig. 7.2A). eIF2 is eluted at ~290 mM KCl with a yield of ~2.5 to 3 mg/600 ml RRL, eIF5 at 360 mM KCl, and eIF5B at 460 mM KCl, both with yields of ~0.5 to 0.7 mg/600 ml RRL. Fractions that contain initiation factors can be concentrated and transferred into buffer D + 100 mM KCl by use of Microcon YM30 centrifugal filter units. Recombinant fully active full–length eIF5 can readily be expressed and purified, and we, therefore, no longer need to use the small amounts of native eIF5 obtained during this purification procedure in *in vitro* assembly reactions. However, although recombinant fully active N-terminally truncated eIF5B is also available, it is currently not possible to obtain full-length eIF5B in recombinant form, and the procedure described here is, therefore, the way to obtain the full-length factor. The peak that is eluted immediately before eIF2 contains eIF2 lacking the eIF2β subunit (Fig. 7.2B). eIF2 purified by this procedure also contains substoichiometric amounts of the tightly bound ATPase ABC50 (Fig. 7.2B; Tyzack *et al.*, 2000). Reasonable amounts of eIF2 (0.5 to 0.8 mg/600 ml RRL) can also be obtained from the 0 to 40% AS fraction (see the preceding section) after its FPLC chromatography on MonoQ HR 5/5 column essentially as described previously for the 40 to 50% AS fraction. eIF2 purified from the 0 to 40% AS fraction contains much less ABC50 (Fig. 7.2B). Although we have never tested directly whether ABC50 influences the activity of eIF2, no

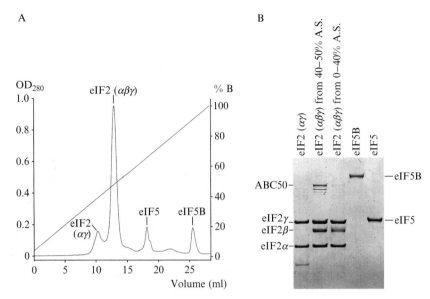

Figure 7.2 Purification of native eIF2, eIF5, and eIF5B. (A) Optical density (OD$_{280}$) profile and (B) 4 to 12% Bis-Tris NuPAGE gel of initiation factors eIF2($\alpha\gamma$), eIF2($\alpha\beta\gamma$), eIF5B, and eIF5 eluted with a Buffer E/100–500 mM KCl gradient from an FPLC MonoQ HR 5/5 column loaded with the 40 to 50% AS/250 mM KCl DE52/800 mM KCl phosphocellulose fraction of the RRL RSW. eIF2($\alpha\beta\gamma$) from the RSW 40 to 50% AS fraction was associated with ABC50. Proteins were visualized by Coomassie staining. The positions of initiation factors or their individual subunits are indicated on the right.

immediately obvious differences in the activities of eIF2 purified from 0 to 40% and 40 to 50% were observed in the assembly experiments described in the following sections.

 ## 6. EXPRESSION IN *E. COLI* AND PURIFICATION OF RECOMBINANT EIF1, EIF1A, EIF4A, EIF4B, EIF4G, EIF5, AND EIF5B

6.1. Buffers

Buffer D   20 mM Tris, pH 7.5, 2 mM DTT, 0.1 mM EDTA, 10% glycerol
Buffer E   20 mM Tris, pH 7.5, 2 mM DTT, 0.1 mM EDTA, 5% glycerol
Buffer G   20 mM Tris-HCl, pH 7.5; 10% glycerol
Buffer H   20 mM HEPES, pH 7.5, 2 mM DTT, 0.2 mM EDTA, 5% glycerol

We have prepared vectors for expression in *E. coli* of all single subunit initiation factors. This approach simplifies factor purification, eliminates the possibility of cross-contamination of native factors, and enabled us to prepare mutated forms of factors and to detect factors immunologically by N- or C-terminal tags. The expression vectors pQE31(His$_6$-eIF1) and pQE31 (His$_6$-eIF1A) were constructed by insertion into pQE31 of appropriate coding sequences from previously cloned cDNAs that were amplified by polymerase chain reaction (PCR); pET(His$_6$-eIF1) and pET(His$_6$-eIF1A) were similarly constructed by insertion of PCR fragments into pET28 (Battiste *et al.*, 2000; Fletcher *et al.*, 1999; Pestova *et al.*, 1998a). pET15b was modified to generate additional novel restriction sites in the polylinker, and PCR fragments corresponding to cloned eIF4A and eIF4B coding sequences were inserted into it to yield pET(His$_6$-eIF4A) and pET(His$_6$-eIF4B) (Pestova *et al.*, 1996a). The human eIF5 coding sequence was amplified by PCR and inserted into pET19b to yield pET19b(His$_6$-eIF5) (Pestova *et al.*, 2000) and cDNA corresponding to amino acid residues 587 to 1220 of human eIF5B ("ΔeIF5B") was cloned into pET28a, generating pET28a(eIF5B$_{587-1220}$) (Lee *et al.*, 2002). Vectors for expression of eIF4G fragments and protocols for their expression and purification are described by Kolupaeva *et al.* (2007) (this volume).

These recombinant proteins all have N-terminal Hexahistidine tags to permit purification on Ni^{2+}–NTA columns. The expression vectors are used to transform *E. coli* BL21 Star (DE3), and after overnight growth, cells on an LB/ampicillin plate (for pET15b, pET19b, and pQE31-based vectors) or on an LB/kanamycin plate (for pET28-based vectors) were washed off and suspended in 5 ml LB medium with the appropriate antibiotic, added to 1000 ml LB medium + antibiotic and incubated with shaking at 37° until OD$_{600}$ = 0.5, at which stage expression of the recombinant protein was induced by addition of 1 m*M* IPTG. The bacterial suspension is then incubated with shaking for 3 h at 37°. This protocol is appropriate for expression of all initiation factors except for ΔeIF5B, which expresses less well, and is therefore grown in 4 L LB + ampicillin and incubated with shaking at 30° until 0.3 < OD$_{600}$ < 0.4, at which stage expression of the recombinant protein is induced with 1 m*M* IPTG. The bacterial suspension is then incubated with shaking for 2 h at 30°. After centrifugation for 20 min at 5000 rpm at 4° in a Sorvall RC5B centrifuge, the supernatant is decanted and the pellet is resuspended in 20 ml or less of buffer G + 300 m*M* KCl (except for ΔeIF5B, which is resuspended in 60 ml of this buffer) and mixed with 30 *μ*l (or for ΔeIF5B, 100 *μ*l) protease inhibitor cocktail set III in 30 ml Corex (or similar heavy duty) glass tubes. The suspension is then sonicated on ice by use of a Branson S-450D sonifier for a total of 10 min, with a cycle of 10 sec sonication and 55 sec off (to minimize protein denaturation) before centrifugation in a Sorvall SS34 rotor at 15,000 rpm for 25 min at 4°. The supernatant is then added to a Ni^{2+}–NTA column that is prepared in advance with 800 *μ*l of a 50% suspension of Ni^{2+}–NTA and equilibrated with 10 ml buffer G + 300 ml KCl. After addition of the supernatant, the bound material is washed sequentially with 10-ml aliquots of buffer G + 300 mM KCl, buffer G + 800 mM KCl, buffer G + 100 mM KCl, and buffer G + 100 mM KCl + 20 mM imidazole. The washes with high salt (800 mM KCl) and low (20 mM) imidazole remove nucleic acids and proteins, respectively, that bind nonspecifically to the Ni^{2+}–NTA matrix before elution of the factor of interest with 6 to 10 aliquots of 0.5 ml buffer G + 100 mM KCl + 300 mM imidazole. DTT is then added to each 500-μl fraction to 2 mM final concentration, and the purified proteins are assayed by SDS-PAGE. Fractions containing the recombinant protein of interest are dialyzed overnight at 4° against 1 L buffer E + 100 mM KCl (for proteins that will be purified further on MonoQ) or against 1 L buffer H + 100 mM KCl (for proteins that will be purified further on MonoS). Before loading onto MonoQ, dialyzed ΔeIF5B was diluted with buffer E to a KCl concentration of 30 mM (because overnight dialysis of ΔeIF5B against buffer E + 30 mM KCl would lead to substantial precipitation of the protein).

Dialyzed fractions containing eIF1A, eIF4A, eIF4B, eIF5, and ΔeIF5B are applied to a FPLC MonoQ HR 5/5 column. Fractions are collected across a 100 to 500 mM KCl gradient, except for ΔeIF5B, which is collected across a 30 to 500 mM KCl gradient. Fractions containing apparently homogenous eIF1A elute with 290 mM KCl, eIF4A elutes with 270 mM KCl, eIF4B elutes with 360 mM KCl, eIF5 elutes with 270 mM KCl and ΔeIF5B elutes with 75 mM KCl. Dialyzed fractions containing eIF1 are applied to a FPLC MonoS HR 5/5 column. Fractions are collected across a 100 to 500 mM KCl gradient, and apparently homogenous eIF1 elutes with 250 mM KCl. Fractions that contained initiation factors (Fig. 7.3) are concentrated and transferred into buffer D + 100 mM KCl by use of Microcon YM10 centrifugal filter units, aliquoted, and stored at −80°.

7. PREPARATION AND AMINOACYLATION OF INITIATOR tRNA

7.1. Buffers

Buffer E   20 mM Tris, pH 7.5, 2 mM DTT, 0.1 mM EDTA, 5% glycerol
Buffer I   40 mM Tris, pH 7.5, 15 mM $MgCl_2$

7.2. *In vitro* transcription of initiator tRNA

Synthetic (*in vitro* transcribed) aminoacylated initiator tRNA (Met-tRNA$_i^{Met}$) is as active as native Met–tRNA$_i^{Met}$ in our reconstituted *in vitro* initiation and elongation assays (Pestova and Hellen, 2001, 2003; Unbehaun *et al.*, 2004) and has the advantages of being significantly cheaper and much easier to prepare than native tRNA$_i^{Met}$. Because the tRNA$_i^{Met}$ gene is cloned, variant tRNAs can be prepared and assayed readily (Lomakin *et al.*, 2006). We, therefore,

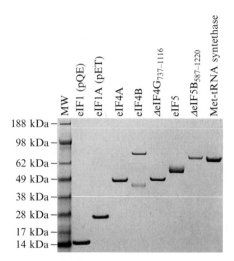

Figure 7.3 Overview of the purified recombinant proteins used in *in vitro* reconstitution of eukaryotic translation initiation. eIF1, eIF1A, eIF4A, eIF4B, ΔeIF4G$_{737-1116}$, eIF5, ΔeIF5B$_{587-1220}$, and methionyl-tRNA synthetase, as indicated, were purified as described in the text and were resolved by electrophoresis on 4 to 12% Bis-Tris NuPAGE gel by use of a MES buffer system, followed by Coomassie staining. The positions of molecular weight marker proteins are indicated in kilodaltons to the left of the panel.

routinely use synthetic Met–tRNA$_i^{Met}$ in many of our experiments. We adapted published approaches for preparation of synthetic tRNAs to prepare an initiator tRNA transcription vector by inserting long synthetic oligonucleotides (MWG Biotech) into pBR322 so that the tRNA$_i^{Met}$ gene is flanked by an upstream T7 RNA polymerase promoter and a downstream BstN1 restriction site, which permits the essential 3′-terminal –CCA sequence of tRNA to be transcribed precisely (Pestova and Hellen, 2001). Mutant derivatives were prepared by PCR and cloned under control of T7 promoter into pBluescript SK (Lomakin *et al.*, 2006). We now find that it is more cost-effective to use custom DNA synthesis services (e.g., Epoch Biolabs, Sugarland, TX; Celtek Bioscience, Nashville, TN) for preparation of new tRNA transcription vectors.

In vitro transcription is done in a 100-μl mixture that contains 20 to 40 μg linearized plasmid DNA, 10 U/μl T7 RNA polymerase, and commercially supplied buffer, 4 m*M* each of the four nucleotide triphosphates, 80 mg/ml PEG8000, 1 U/μl RNase inhibitor, 20 m*M* MgCl$_2$, and 10 m*M* DTT. This protocol, which uses high concentrations of DNA and nucleotides, allows efficient transcription of short RNAs and RNAs for which transcription is initiated at nucleotides other than G (such as tRNA$_i^{Met}$). Reaction mixtures are incubated at 37° for 1 h, after which tRNA is extracted with acidic phenol/chloroform (pH 4.7) and ethanol-precipitated. tRNA transcripts are separated from plasmid DNA restriction fragments and unincorporated nucleotides by gel filtration on a FPLC superdex G75 column, and their integrity is confirmed by gel electrophoresis. If separation from DNA fragments is not required, tRNA transcripts can be separated from unincorporated nucleotides and buffer by filtering through Sephadex G-50 Spin-50 mini-columns.

7.3. Purification of methionyl-tRNA synthetase

Aminoacyl tRNA synthetases are purified from *E. coli* strain M.R.E. 600 (which lacks detectable ribonucleases; Wade and Robinson, 1966) exactly as described (Stanley *et al.*, 1974). *E. coli* methionyl-tRNA synthetase present in this mixture is able to specifically aminoacylate eukaryotic tRNA$_i^{Met}$.

A vector for overexpression of His$_6$-tagged *E. coli* methionyl-tRNA synthetase was made by PCR amplification of the N-terminal catalytic core domain and the C-terminal tRNA-binding domain coding regions of the *metG* gene (Genbank acc. #K02671) from *E. coli* DH5α and subsequent insertion of PCR fragments into pET28a (Lomakin *et al.*, 2006). Expression of *E. coli* methionyl-tRNA synthetase in a 1 L culture of *E. coli* BL21 Star (DE3) and its subsequent purification by sequential chromatographies on Ni^{2+}-NTA matrix and MonoQ HR 5/5 column were done exactly as described previously for expression and purification of recombinant eIF1 and eIF1A. MonoQ fractions are collected across a buffer E/100 to 500 mM KCl gradient, and the His$_6$-tagged methionyl-tRNA synthetase elutes at 170 mM KCl. Fractions that contained methionyl-tRNA synthetase (Fig. 7.3) are aliquoted and stored at $-80°$.

7.4. Aminoacylation of tRNA$_i^{Met}$

Fifty to 100 μg eukaryotic tRNA$_i^{Met}$ transcripts or 250 to 500 μg total native calf liver tRNA (Novagen) are aminoacylated in a 200-μl reaction mixture also containing either unfractionated *E. coli* aminoacyl tRNA synthetases ($\sim$60 μg) or purified recombinant *E. coli* methionyl-tRNA synthetase (30 μg), 0.1 mM unlabeled methionine and 200 μCi [^{35}S] methionine (37 Tbq/mmol) or, for high activity labeling, 0.01 mM unlabeled methionine and 400 μCi [^{35}S] methionine (37 Tbq/mmol) in buffer I with 10 mM ATP, 1 mM CTP, and 1 U/μl RNAse inhibitor for 30 min at 37°. The reaction mixture was extracted with phenol-chloroform (pH 4.7), filtered with Sephadex G-50 Spin-50 mini-columns, and ethanol precipitated.

8. Assembly of 48S and 80S Initiation Complexes on β-Globin mRNA

8.1. Buffer

Buffer J   20 mM Tris, pH 7.5, 100 mM KAc, 2 mM DTT, 2.5 mM
            $MgCl_2$, 0.25 mM spermidine

48S initiation complexes can be assembled on β-globin mRNA by
incubating 2.5 to 9 pmol (depending on the purpose) native capped poly-
adenylated globin mRNA with 5 pmol Met– tRNA$_i^{Met}$ (either *in vitro*
transcribed tRNA or a constituent of a mixture of native total tRNA),
3 pmol 40S subunits, 7 pmol eIF2, 5 pmol eIF3, 7 pmol eIF4A, 7 pmol
eIF4B, 4 pmol eIF4F, 20 pmol eIF1, 20 pmol eIF1A, in a 40-µl reaction
mixture containing buffer J, 1 mM ATP and 0.2 mM GTP or its nonhydrolyzable analog GMPPNP for 10 min at 37°. If the goal is to bind most
mRNA to 40S subunits, ~90% of mRNA will be engaged into 48S
complexes if the reaction mixture contains 2.5 pmol globin mRNA,
whereas if the complete engagement of 40S subunits into 48S complexes
is required, mRNA should be present in ~2.5 to 3 times excess (~9 pmol)
over 40S subunits (3 pmol). 5′-end dependent 48S complex formation is
very sensitive to Mg^{2+} concentration; its increase to 5 mM severely reduces,
and increase to 10 mM completely abrogates, 48S complex formation.
eIF4F can be substituted by a twofold to threefold greater molar amount
of the middle domain of eIF4G (ΔeIF4G$_{737-1116}$), but in this case it is
preferable to use *in vitro* transcribed initiator tRNA, because total native
tRNA contains large amounts of short RNA contaminants that, in the
absence of the specific eIF4E-cap interaction, can efficiently compete
with globin mRNA for binding of factors and 43S complexes. Native
globin mRNA is a mixture of α- and β-globin mRNAs, and monitoring
of 48S complex formation on a specific globin mRNA (e.g., β-globin) can
be achieved by toe-printing analysis (see the relevant section later) done
with a DNA primer that is specific for the coding region of the desired
mRNA. To obtain 80S initiation complexes, 48S complexes assembled in
the presence of GTP should then be incubated with 20 pmol eIF5, 5 pmol
eIF5B, and 3 pmol 60S subunits for 10 min at 37°.
    A similar procedure can be used to assemble 48S complexes on other
mRNAs on which initiation is 5′-end dependent, but we found that those
mRNA that contain GC-rich stems in their 5′-UTRs but nevertheless can be
efficiently translated in RRL, cannot form 48S complexes in our reconstituted
system (our unpublished data). Although it has been reported that mRNAs with
G-rich secondary structure in their 5′-UTRs can form 48S complexes in the
presence of native (but not recombinant) eIF4B (Dmitriev *et al.*, 2003) we could

not assemble 48S complexes on similar mRNAs in the presence of the RSW fraction that contained native eIF4B (unpublished data).

9. RESOLUTION OF TRANSLATION INITIATION COMPLEXES BY SUCROSE DENSITY GRADIENT CENTRIFUGATION

9.1. Buffer

Buffer K   20 mM Tris, pH 8.0, 2 mM DTT, 100 mM potassium acetate

Ribosomal 43S/48S and 80S initiation complexes are stable and can be resolved from unincorporated components of the translation apparatus, from ribonucleoprotein (RNP) complexes, and from each other by sucrose density gradient centrifugation. This method can be used to assess formation of these complexes in *in vitro* reconstituted initiation reactions, as well as in cells and cell-free extracts in which their formation can be arrested at different stages in the initiation pathway by the inclusion of specific pharmacological inhibitors.

Ten to 30% linear sucrose density gradients are prepared in buffer K with 4 to 6 mM magnesium acetate (the elevated Mg^{++} concentration stabilizes ribosomal complexes during centrifugation) by use of a Hoefer SG gradient maker in polyallomer centrifuge tubes of a size that is appropriate for Beckman SW41 or SW55 rotors. Up to 10 and up to 20 40-μl reaction mixtures can be layered on top of the gradient in SW55 and SW41 centrifuge tubes, respectively. To resolve ribosomal complexes and RNP complexes, gradients are centrifuged for 3.5 h at 4° at 40,000 rpm with a Beckman SW41 rotor or for 1.5 h at 53,000 rpm with a Beckman SW55 rotor. They are then fractionated into ~25 to 30 fractions, and the optical density (A_{260}) of each is measured.

To assay the presence of components of the translation apparatus in ribosomal complexes, we use [^{35}S]Met–tRNA$_i^{Met}$ or [^{32}P]-labeled mRNA combined with detection by scintillation or Cherenkov counting of an aliquot of each fraction as appropriate and either Western blotting of factors with specific antibodies against factors or their individual subunits or the T7-tag of recombinant factors (Pestova *et al.*, 2000; Unbehaun *et al.*, 2004), or [^{32}P]-phosphorylation of a factor, as described for experiments in which eIF3's association with 40S subunits was quantified (Unbehaun *et al.*, 2004). Other investigators have labeled factors with [^{14}C] by reductive methylation for the same purpose (Benne *et al.*, 1979). It should be noted that the hydrodynamic shear forces experienced by ribosomal complexes during centrifugation lead to the loss of some factors from 48S initiation complexes, including eIF1A (Pestova *et al.*, 1998a) and eIF2·GDP (Pisarev *et al.*, 2006).

10. TOE PRINTING ANALYSIS OF RIBOSOMAL COMPLEXES

The position of 48S and 80S ribosomal complexes on mRNA templates is determined by primer extension inhibition ("toe printing") (Hartz *et al.*, 1988; Fig. 7.4A). An oligonucleotide primer annealed to an mRNA can be extended to its 5′ end by avian myeloblastosis virus reverse transcriptase (AMV RT), yielding a full-length extension cDNA product. Ribosomes that are bound stably to the mRNA will arrest primer extension, resulting in formation of a shorter than full-length cDNA extension product that can be mapped precisely by electrophoresis together with an appropriate sequence ladder on a denaturing sequencing gel. Ribosomal 40S subunits or 80S ribosomes bound to mRNA arrest RT, yielding characteristic toe prints 15 to 17 nt downstream of the first nucleotide of the P-site codon of mRNA on which they have assembled (Alkalaeva *et al.*, 2006; Anthony and Merrick, 1992; Pestova and Hellen, 2003; Pestova *et al.*, 1998a). 48S complexes and 80S ribosomes arrest RT at the same position on an mRNA, because the position of the leading edge of the 40S subunit does not change as a result of subunit joining.

Toe-printing primers are generally chosen to be 17 to 25 nt long and to hybridize to mRNA $\sim$100 to 120 nt downstream of the anticipated ribosome binding site. For example, we use the primer 5′-GCATTTGCAGAGGA-CAGG-3′ (complementary to nt 199 to 181 of rabbit β-globin mRNA) to map ribosomal complexes on its initiation codon (AUG_{54-56}) (Fig. 7.4B). After assembly of ribosomal complexes as described previously, 1 μg of primer (1 μg/μl), 1 μl magnesium acetate (320 mM), 4 μl dNTPs (5 mM dCTP, dGTP and dTTP; 1 mM dATP), 1 μl [α-^{32}P]dATP (111 TBq/mmol), and 5 U AMV-RT are added to the reaction mixture, which is then incubated for another 45 min at 37°. To increase the efficiency of phenol extraction, after incubation, reaction mixtures are supplemented with 100 μl of water, 2.5 μl 10% SDS, and 2.5 μl of 0.5 M EDTA. cDNA products are then phenol-extracted (pH 8.0), ethanol-precipitated, and resuspended in 8 μl loading buffer (0.05% bromophenol blue, 0.05% xylene cyanol FF, 20 mM EDTA [pH.8.0], 91% formamide). Samples are heated at 95° for 3 min, and aliquots (2 μl) are loaded on 7 M urea/6% polyacrylamide sequencing gels. Gels are prerun for 30 min before electrophoresis. The duration of electrophoresis required for optimal resolution of cDNA products depends on their size. cDNA products are compared with dideoxynucleotide sequence ladders obtained with the same primer and appropriate plasmid DNA in reactions done by use of Sequenase according to the manufacturer's instructions. For quantitation of ribosomal complex formation, it is, however, better to use 5′-end [^{32}P]-phosphorylated primer (which we prepare with T4

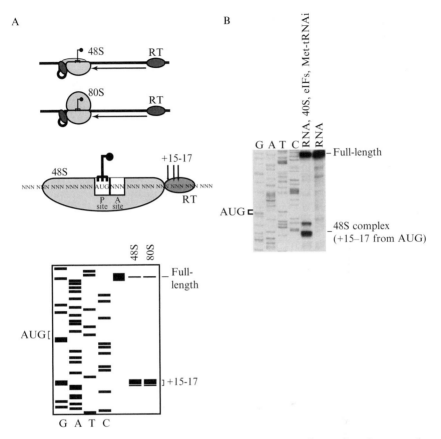

Figure 7.4 The primer extension inhibition ("toe printing") assay used to map the position of ribosomal complexes on mRNA. (A) An oligonucleotide primer annealed to the mRNA template is extended by reverse transcriptase (RT); the cDNA is radiolabeled either by use of a [^{32}P]end-labeled primer during transcription with unlabeled dNTPs or by incorporation of radiolabeled [α-^{32}P]dATP during transcription together with unlabeled dGTP, dCTP, and dTTP. On naked mRNA, RT generates a full-length cDNA product, but in 48S or 80S initiation complexes, primer extension is arrested by the leading edge of the 40S subunits, yielding cDNA products that in both instances are characteristically terminated +15 to 17 nt downstream from the first nucleotide in the P-site (by definition, +1). These radiolabeled cDNA products are resolved on denaturing sequencing gels (lower panel), and their positions can be compared with a dideoxynucleotide sequence ladder generated by use of appropriate cloned cDNA and the same primer. (B) Toe prints caused by RT arrest at the leading edge of a 48S complex assembled on native globin mRNA from individual purified components and detected with a primer specific for β-globin mRNA. cDNA products were resolved on 6% denaturing sequencing gel. The positions of toe prints +15 to 17 nt from the first nucleotide in the P-site that are attributable to the 48S complex and the full-length cDNA are indicated to the right of the gel. Lanes C, T, A, and G show cDNA sequence corresponding to β-globin mRNA derived with the same primer as for toe printing.

polynucleotide kinase) rather than including $[\alpha\text{-}^{32}\text{P}]\text{dATP}$ in the reverse transcription reaction, because each resulting cDNA product will contain a single radiolabeled nucleotide irrespective of length instead of a variable number that depends on the mRNA sequence that has been copied.

The methods previously described allow the assembly of specific initiation complexes, their position on an mRNA, and their association of initiation factors to be assayed. To characterize the interactions between initiation factors, mRNA and ribosomal subunits in initiation complexes at significantly higher resolution, we have adapted two techniques, site-directed UV cross-linking (Dontsova *et al.*, 1992; Favre *et al.*, 1998) and directed hydroxyl radical probing (Culver and Noller, 2000; Datwyler and Meares, 2000) that have proved to be valuable in analyzing bacterial ribosomal complexes for use in our eukaryotic translation system.

11. SITE-DIRECTED UV CROSS-LINKING OF MRNA IN INITIATION COMPLEXES

11.1. Buffers

Buffer J   20 mM Tris, pH 7.5, 100 mM potassium acetate, 2 mM DTT, 2.5 mM Mg acetate, 0.25 mM spermidine

Buffer L   20 mM Tris-HCl, pH 7.5, 50 mM KCl, 2 mM $MgCl_2$, 2 mM DTT, 0.1 mM EDTA

Buffer M   20 mM Tris-HCl, pH 7.5, 100 mM KCl

The interaction of mRNA with ribosome and initiation factor components of preinitiation and initiation complexes likely plays specific roles in ensuring processivity during scanning, in initiation codon selection, and in ensuring the stability of assembled complexes. To characterize these interactions at nucleotide resolution, we designed vectors for transcription of mRNAs that incorporate 4-thiouridine (4^{S}U) or 6-thioguanosine (6^{S}G) residues (Fig. 7.5A) at single unique positions (in addition to the initiation codon) that can subsequently be selectively cross-linked to adjacent proteins and nucleic acids. 4^{S}U has several properties that render it particularly appropriate for this purpose: it can be readily incorporated into mRNA transcripts by T7 RNA polymerase; it engages in conventional base-pairing interactions; it can be selectively activated at wavelengths (330 to 370 nm) that minimize side reactions involving other nucleotides in mRNA, tRNA, and rRNA; and, importantly, it is highly photoreactive with both amino acid residues and nucleic acid bases, yielding "zero-length" cross-links that represent direct contacts with 48S complex constituents and that can, therefore, be analyzed with high precision. We routinely use a Stratagene UV Stratalinker for UV-crosslinking at 365 nm.

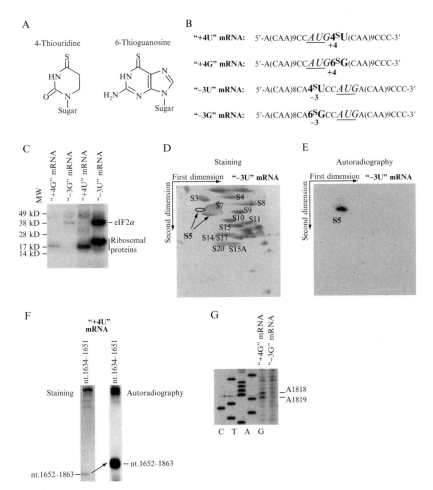

Figure 7.5 Site-directed UV cross-linking of mRNA in ribosomal initiation complexes. (A) Structural formulae of 4-thiouridine (4^SU) and 6-thioguanosine (6^SG). (B) Sequences of $(CAA)_n$–AUG–$(CAA)_m$ mRNA derivatives containing 4^SU or 6^SG at −3 or +4 positions (bold) relative to the A of the initiation codon (underlined). (C–G) Contacts of nucleotides at −3 and +4 positions of mRNA with rRNA and protein components of ribosomal 48S complexes. (C) UV cross-linking of ^{32}P-labeled $(CAA)_n$–AUG–$(CAA)_m$ mRNAs containing 4^SU or 6^SG at [−3] and [+4] nucleotides as indicated with protein components of 48S complexes assembled with a full set of initiation factors, assayed by electrophoresis on 4 to 12% Bis-Tris NuPAGE gel by use of a MES buffer system and autoradiography. The positions of molecular weight marker proteins are indicated in kilodaltons to the left of the panel. (D, E) Analysis by 2D gel electrophoresis of ribosomal proteins UV-cross-linked to ^{32}P-labeled $(CAA)_n$–AUG–$(CAA)_m$ mRNAs containing 4^SU at position [−3]. (D) A gel of 40S subunit proteins stained with Simply Blue Safe Stain and (E) an autoradiograph of the same gel. The position corresponding to the radioactive spot (E) is shown as a black oval on the stained gel (D). The positions of some ribosomal proteins on the basis of our sequencing data or according to Madjar *et al.* (1979) are indicated. (F, G) UV cross-linking of 4^SU at position [+4] of mRNA with

6^SG is incorporated less efficiently into RNA than 4^SU and has an absorption maximum of $\sim$310 nm (instead of $\sim$330 nm for 4^SU) so that its efficiency of cross-linking at 365 nm is considerably lower than that of 4^SU. Nevertheless, occasions occur when it is nevertheless particularly useful, for example as a control for the base specificity of interactions at specific positions in an mRNA (Pisarev et al., 2006). 6^SG exists in a thiol–thione equilibrium that could potentially lead to misincorporation during transcription (Favre et al., 1998; Sergiev et al. 1997), but we determined that 6^SG -containing mRNAs are functional (Pisarev et al., 2006).

We will illustrate the application of this technique by describing our studies on identification of specific functional interactions of the -3 and $+4$ context nucleotides with components of the 48S complex (Pisarev et al., 2006). Scanning 40S subunits can bypass the first AUG triplet if it is <10 nucleotides (nt) from the 5' end of mRNA or if its context deviates from the optimum sequence GCC(**A/G**)CCAUG**G**, particularly at -3 and $+4$ positions (in bold) (Kozak, 1986, 1991). eIF1 maintains the fidelity of initiation codon selection and enables 43S complexes to recognize the initiation codon context. We hypothesized that the role of -3 and $+4$ context nucleotides could be to stabilize conformational changes in ribosomal complexes that occur on codon-anticodon base pairing by interacting with components of these complexes. To investigate the interactions of the -3 and $+4$ context nucleotides with components of the 48S complex, we used mRNAs of the [CAA]n-AUG-[CAA]m series that contained a single uridine or guanosine (in addition to the AUG codon; Fig. 7.5B). The unavoidable presence of the additional U and G nucleotides in the initiation codon does not interfere with the experimental approach, because they are base paired with the anticodon of initiator tRNA and, therefore, do not cross-link with other components of 48S initiation complexes (Pisarev et al., in preparation). Inclusion of additional U or G nucleotides in these

---

18S rRNA in 48S complexes. (F) Preliminary mapping of the site of cross-linking to 18S rRNA by RNase H digestion. 18S rRNA UV cross-linked to a ^{32}P-labeled (CAA)$_n$–AUG–(CAA)$_m$ mRNA derivative containing 4^SU at the [+4] position in 48S complexes assembled with a full set of initiation factors. 18S rRNA was digested in the presence of a DNA primer complementary to nucleotides 1652 to 1683 and analyzed by electrophoresis in denaturing 12% PAGE with subsequent staining (left panel) or autoradiography (right panel). (G) Determination of the exact sites of cross-linking of (CAA)n–AUG–(CAA)m mRNA derivatives containing 6^SG at [+4] to 18S rRNA in 48S complexes by primer extension analysis. In a control reaction, UV cross-linking was done with 48S complexes assembled on (CAA)n–AUG–(CAA)m mRNA derivatives containing 6^SG at [−3]. cDNA products were resolved on 6% denaturing sequencing gel. The positions of RT stop sites at AA$_{1818-1819}$ are indicated on the right. Lanes C, T, A, and G depict 18S rRNA sequence generated with cloned rRNA and the same primer. These panels are reproduced from Pisarev et al. (2006) with permission (Copyright © Cold Spring Harbor Laboratory Press, Genes and Development 20, 624-636-2797 [2006]).

mRNAs was avoided by flanking AUG codons with multiple CAA triplets, which also minimize secondary structure and which we have previously found to yield highly efficient templates for assembly of initiation complexes (Pestova and Kolupaeva, 2002). These transcripts initiated from A to avoid the presence of a $5'$-terminal guanosine, but because transcription by T7 RNA polymerase initiates less efficiently from A than G at the $+1$ position, 100-μl transcription mixtures were supplemented with 1 mM rAC transcription primer to increase initiation efficiency. These transcription reactions also contained 1 mM 4^SU or 6^SG and unlabeled ATP (1 mM), CTP (0.1 mM) and GTP or UTP, as appropriate (1 mM), 500 U T7 RNA polymerase, 100U RNase inhibitor, and 7 μl [α^{32}P]CTP (222 Tbq/mmol). mRNA transcripts were separated from unincorporated components by use of Sephadex G-50 Spin-50 mini-columns.

For UV cross-linking, 48S complexes are assembled by incubating 100 ng ^{32}P-labeled (CAA)n-AUG-(CAA)m mRNA containing either 4^SU or 6^SG , 10 pmol Met–tRNA$_i^{Met}$, 8 pmol 40S subunits, 5 μg eIF2, 15 μg eIF3, 2.5 μg eIF4A, 0.5 μg eIF4B, 2.5 μg eIF4F, 0.2 μg eIF1A, 0.2 μg eIF1 in 100 μl buffer J containing 1 mM ATP and 0.4 mM GTPγS for 10 min at 37°. These 48S complexes are resolved from unincorporated components by centrifugation in a Beckman SW55 rotor for 1 h and 40 min at 4° and 50,000 rpm in 10 to 30% sucrose density gradients prepared in buffer J. The gradients are fractionated into $\sim$25 fractions, and the presence of [^{32}P]-labeled mRNA in ribosomal fractions is monitored by Cherenkov counting. Peak 48S fractions containing equal amounts of counts ($\sim$200,000 cpm) are irradiated at 365 nm for 30 min on ice in quartz cuvettes (1-mm path length) and used to identify cross-linked nucleotides in 18S rRNA and cross-linked proteins.

11.2. Identification of cross-linked initiation factors and ribosomal proteins

To identify cross-linked proteins, UV-irradiated ribosomal fractions are treated with RNase A. However, depending on the composition and secondary structure of the mRNA, a combination of different RNases (RNAses One, V1, T1) might be required. Most cross-linked initiation factors and their subunits can be identified by autoradiography on the basis of the electrophoretic mobility in NuPAGE 4 to 12% Bis-Tris-gel of cross-linked proteins from RNAase-digested fractions (Fig. 7.5C). On the other hand, the identification of ribosomal proteins, which are similar in molecular weight, requires two-dimensional (2D) gel electrophoresis. 2D gel electrophoresis might also be required for identification of the small eIF3k subunit of eIF3 (which comigrates with ribosomal proteins) and to distinguish the closely migrating eIF3 subunits eIF3e–eIF3h. In our case, acidic-SDS 2D gel was appropriate; however, the choice of the first dimension (acidic versus basic) depends on the nature of the cross-linked proteins and must be empirically determined if the nature of the cross-linked proteins is unknown. We did not modify the protocol for 2D gel electrophoresis and, therefore, do not describe it in detail here. Interested readers are referred to the original publication (Yusupov and Spirin, 1988). All cross-linked peak fractions (~200 μl each) were combined, transferred to buffer M, and concentrated on microcon-YM10 centrifugal filter units to 100 μl final volume, and treated with 50 μg RNase A for 30 min at 37°. These samples were combined with 100 μl 40S subunits (100 OD$_{260}$ U/ml) in buffer M. Proteins were extracted from these mixtures with 100 mM MgCl$_2$ in 67% acetic acid and precipitated with 6.5 volumes acetone (Hardy *et al.*, 1969), resuspended in 40 μl of 8 M urea, 1% 2-mercaptoethanol, 10 mM Bis-Tris acetate, pH 4.2, incubated for 15 min at 37°, and subjected to first dimension electrophoresis (Yusupov and Spirin, 1988) in 120-mm long glass tubes with a 2.4-mm inner diameter. First dimension gels were incubated for 10 min in cathode buffer and combined with second dimension gels prepared as described in detail elsewhere (Schagger and von Jagow, 1987). The separating gel (16.5% T (total monomer concentration) and 3% C (weight percentage of cross-linker) also contained 13.3% w/v glycerol. After electrophoresis for 12 h at 40 mA, gels were stained with Simply Blue Safe Stain (Fig. 7.5D) and were either dried and autoradiographed (Fig. 5E) to determine the location of cross-linked proteins or used for in-gel tryptic digestion to generate peptides for LC-nanospray tandem mass spectrometry to confirm the identity of cross-linked proteins. To attribute the cross-linked band to a particular ribosomal protein (Fig. 7.5D,E), the so-called north–west shift of a cross-linked protein caused by the presence of cross-linked nucleotides must be taken into account. It should be noted that a small proportion of 48S complexes can dissociate during the cross-linking procedure and that the released mRNA can rebind nonspecifically to RNA-binding initiation factors that are components of 48S complexes (e. g., several subunits of eIF3), which can lead to false-positive cross-links. To avoid this potential problem, another sucrose density gradient centrifugation separation step must be performed after cross-linking if interaction of mRNA with RNA-binding factors (e.g., eIF3) is being studied.

11.3. Identification of cross-linked nucleotides in 18S rRNA

Identification of sites of cross-linking of individual 4^{S}U or 6^{S}G nucleotides in ^{32}P-labeled mRNAs to 18S rRNA can be done in two stages (Dontsova *et al.*, 1992). First, cross-linked rRNA is hybridized with a panel of deoxy-oligonucleotides complementary to different regions of rRNA, digested with RNAse H, and separated by gel electrophoresis, followed by autoradiography to localize (within ~100 nt) the approximate site of cross-linking. The cross-linked nucleotide is then identified precisely by primer extension with a primer chosen on the basis of preliminary identification of the cross-linked region by RNAse H digestion. In this experimental protocol, rRNA, mRNA, and tRNA from cross-linked peak fractions are extracted with phenol/chloroform, ethanol precipitated, and resuspended in 20 μl H$_2$0, after which 2-μl aliquots were hybridized with pairs of a panel of ~20-mer DNA oligonucleotides complementary to different regions of 18S rRNA in 20 μl annealing mixture that also contained buffer M, 1 μg oligonucleotide and 4 μg unlabeled 18S rRNA to allow visualization of unlabeled 18S rRNA fragments. After hybridization, the reaction mixture is incubated with 5U RNAse H at 37° for 30 min. 18S rRNA fragments were separated by electrophoresis in a denaturing 8 M urea/12% polyacrylamide minigel. Cross-linked and uncross-linked 18S rRNA fragments generated by RNAse H treatment were visualized by autoradiography and methylene blue staining, respectively (Fig. 7.5F). Identification of cross-linked regions and their attribution to corresponding uncross-linked fragments of 18S rRNA on stained gels is done, taking into account the reduced mobility of cross-linked rRNA fragments because of the covalently bound mRNA (Fig. 7.5F). Cross-linked nucleotides are then located precisely in 18S rRNA by primer extension inhibition in a 20 μl reaction mixture containing 2-μl aliquots of resuspended cross-linked 18S rRNA, 1 μg of oligodeoxyribonucleotide complementary to a region of 18S rRNA (chosen on the basis of preliminary mapping by RNase H digestion), and 5 U AMV RT. The reaction mixture is incubated for 30 min at 37°, and cDNA products are analyzed by electrophoresis through 7 M urea/6% polyacrylamide sequencing gel (Fig. 7.5G).

12. Directed Hydroxyl Radical Probing to Localize Initiation Factors on the 40S Subunit

12.1. Buffers

Buffer N   80 mM HEPES, pH 7.5, 300 mM KCl, 10% glycerol
Buffer O   80 mM HEPES, pH 7.5, 100 mM KCl, 2.5 mM MgAc, 10% glycerol
Buffer P   80 mM HEPES, pH 7.5, 100 mM KCl, 3 mM MgAc, 1 mM GMPPNP, 10% glycerol

A complete understanding of the mechanism of initiation in eukaryotes requires detailed knowledge of the location of initiation factors on the ribosome and relative to mRNA, tRNA, and other initiation factors. Eukaryotic ribosomal initiation complexes have so far proved refractory to analysis by cryoelectron microscopy, and we have, therefore, instead applied the method of directed hydroxyl radical probing to map the locations of initiation factors in initiation complexes. This method was originally developed for and very successfully applied to the characterization of prokaryotic ribosomal complexes (Culver and Noller, 2000). In this approach, a panel of active mutant versions of a protein is generated to allow iron (II) to be tethered to different single-surface–exposed cysteine residues by way of the linker 1-(*p*-bromoacetamidobenzyl)-EDTA (BABE). The tethered Fe(II) then catalyzes Fenton chemistry, generating hydroxyl radicals that cleave nearby rRNA; correlation of the sites of cleavage on the ribosome with the positions of the Fe(II)–BABE-linked cysteine residues on the mutant factors yields a set of constraints that can be used to dock the factor onto the ribosome. The relatively short radius of action of hydroxyl radicals ($\sim$20 Å) allows quite precise localization of the protein.

The value of data obtained with this approach requires high-resolution structures or structural models of the factor in question and of the surface of the ribosome to which it binds. Fortunately, the structures of most eukaryotic initiation factors have been determined by NMR or X-ray crystallography (Pestova *et al.*, 2006) and the sequence of mammalian 18S rRNA is highly homologous to that of prokaryotic 16S rRNA (Mears *et al.*, 2002), which justifies the use of the crystal structure of the 30S subunit from *Thermus thermophilus* (Yusupov *et al.*, 2001), as well as other more recent and higher resolution ribosomal structures (Schuwirth *et al.*, 2005; Selmer *et al.*, 2006) to model the interactions of factors with elements of the 40S subunit.

The principles of this technique are discussed in great detail in Culver and Noller (2000), and here we will therefore only describe its application to studying eukaryotic ribosomal complexes using mapping of the position of eIF1 on the 40S subunit as an example (Lomakin *et al.*, 2003). The availability of an NMR structure of the complete protein (Fletcher *et al.*, 1999), and our observations that eIF1's unstructured N-terminus was not essential for its function and that the two existing cysteine residues could be replaced without loss of function (Lomakin *et al.*, 2003) enabled us to create mutants in which each cysteine residue was retained individually and, with the cysteine-less eIF1 mutant as a base, to select a panel of well-distributed, surface-exposed amino acid residues for replacement by cysteine residues (Fig. 7.6A). This was done by site-directed mutagenesis with conventional PCR approaches. If possible, phylogenetic comparisons are used to select nonconserved polar residues for substitution.

Before proceeding further, it is important to verify that single-cysteine mutant proteins have activities that are comparable to the wild-type factor, and thus that they adopt similar structures and interact with ligands in a similar manner. These assays should be repeated after derivatizing mutant factors with Fe(II)-BABE (see later). In the case of eIF1, these assays included analysis of eIF1's interaction with the 40S subunit and analysis

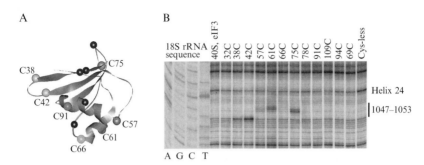

Figure 7.6 Directed hydroxyl radical cleavage of 18S rRNA in 40S/eIF3/eIF1 complexes from Fe(II) tethered to different positions on the surface of eIF1. (A) Ribbon diagram of the structured domain of human eIF1 (Fletcher *et al.*, 1999). The spheres indicate the positions of cysteines introduced on the surface of eIF1 for tethering of Fe(II)–BABE. Residues (C38, C42, C57, C61, C66, C75, and C91) from which hydroxyl radicals were able to cleave 18S rRNA are indicated. (B) Primer extension analysis of directed hydroxyl radical cleavage of helix 24 of 18S rRNA in 40S/eIF3/eIF1 complexes from Fe(II) tethered to positions on eIF1 as indicated. The reaction mixtures corresponding to the lane marked "40S, eIF3" did not contain eIF1; the lane marked "Cys-less" corresponds to a reaction mixture that contained the cysteine-less eIF1 mutant. Reference lanes G, A, C, and T depict 18S rRNA sequence generated from the same primer. The positions of cleaved nucleotides are shown as a black bar on the *right*. These panels are reproduced from Lomakin *et al.* (2003) with permission (Copyright © Cold Spring Harbor Laboratory Press, Genes and Development 17, 2786–2797 [2003]).

by toe printing of eIF1's activity in supporting 48S complex formation on native β-globin mRNA (Lomakin *et al.*, 2003).

Fe(II)–BABE derivatization of single-cysteine eIF1 mutants was done by use of an adaptation of a previously described procedure (Culver and Noller, 2000). Pure single-cysteine eIF1 mutants and (as a negative control) cysteine-less eIF1 mutant (100 μg each) were incubated with 1 mM Fe(II)-BABE at 37° for 30 min in 60 μl buffer N, and Fe(II)–BABE-derivatized eIF1 was then separated from unincorporated Fe(II)–BABE on Microcon YM10 centrifugal filter units by several steps of concentration and buffer dilution, and stored at a concentration of 0.3 to 0.4 mg/ml at −80°. The efficiency of Fe(II)–BABE derivatization of mutant eIF1 proteins was estimated essentially as described (Culver and Noller, 2000) by comparing the reactivity of the fluorescent reagent with DCIA with cysteine sulfhydryl groups before derivatization and after, when reactive groups should be blocked by bound Fe(II)–BABE as a result of derivatization. The cysteine-less mutant form of eIF1 was used as a control that should not bind DCIA and thus should not fluoresce after incubation with it.

Fe(II)–BABE derivatized factors were used for hydroxyl radical cleavage of 18S rRNA and of initiator tRNA after verification that they were active. To analyze eIF1's proximity to elements of 18S rRNA, 40S/eIF3/[Fe(II)–BABE]–eIF1 complexes were formed by incubating 7 pmol 40S subunits,

10 pmol eIF3, 3 μg poly(U) RNA, and 50 pmol derivatized eIF1 in 20 μl buffer O at 37° for 10 min and chilled on ice. To generate hydroxyl radicals, the reaction mixture was supplemented with 0.025% H_2O_2 and 5 mM ascorbic acid to initiate Fenton chemistry and incubated on ice for 10 min. Reactions were quenched by addition of 10 mM thiourea. 18S rRNA was phenol-extracted, ethanol precipitated, and analyzed by primer extension by use of AMV RT and primers complementary to different regions of 18S rRNA (Fig. 6B). The intensity of cleavage generally reflects the proximity of bound Fe(II)–BABE to sites of cleavage.

To analyze the potential cleavage of tRNA$_i^{Met}$ by [Fe(II)–BABE]-eIF1 in 43S complexes, they were assembled by incubating 12 pmol 40S subunits, 12 pmol eIF3, 10 pmol eIF2, and 2 pmol Met-tRNA$_i^{Met}$ transcript in 35 μl buffer P at 37° for 10 min and were chilled on ice. After hydroxyl radical cleavage, done as described for 18S rRNA, initiator tRNA was analyzed by primer extension by use of the primer 5′-TGGTAGCAGAG-GATGG-3′, which is complementary to the 16 nucleotides at its 3′-terminus. Analysis of initiator tRNA cleavage by primer extension does not allow inspection of the ~20 3′-terminal nucleotides and also requires unmodified *in vitro* transcribed tRNA (because modified nucleotides in native tRNA strongly arrest reverse transcription). However, these drawbacks can be overcome by assembling initiation complexes with 5′[^{32}P]-labeled Met-tRNA$_i^{Met}$, cleavage of which can be analyzed directly by gel electrophoresis and autoradiography.

In the absence of high-resolution structures of eukaryotic ribosomes, the crystal structures of prokaryotic ribosomes (e.g., Yusupov *et al.*, 2001) have instead been used to model the eIF1/40S subunit interaction. Analysis of the hydroxyl radical cleavage data in the context of the ribosome structure and manual docking were done with MOLMOL (Koradi *et al.*, 1996). The radius of action of hydroxyl radicals from tethered Fe(II) for strong cleavage of RNA is known from calibration experiments (Culver and Noller, 2000) and, accordingly, initial modeling was based on 15-Å restraints between sites of strong cleavage on rRNA and the side chains of mutated amino acids of eIF1. Slight adjustments to eIF1's position were made to avoid backbone clashes between eIF1 and tRNA bound to the P-site. According to the resulting model, eIF1 binds to the interface surface of the platform of the 40S subunit in the proximity of the ribosomal P-site (Lomakin *et al.*, 2003).

ACKNOWLEDGMENTS

This work was supported by grant R01 GM59660 from the National Institute of General Medical Sciences.

REFERENCES

Alkalaeva, E. Z., Pisarev, A. V., Frolova, L. Y., Kisselev, L. L., and Pestova, T. V. (2006). *In vitro* reconstitution of eukaryotic translation reveals cooperativity between release factors eRF1 and eRF3. *Cell* **125**, 1125–1136.

Anthony, D. D., and Merrick, W. C. (1992). Analysis of 40S and 80S complexes with mRNA as measured by sucrose density gradients and primer extension inhibition. *J. Biol. Chem.* **267**, 1554–1562.

Battiste, J. L., Pestova, T. V., Hellen, C. U. T., and Wagner, G. (2000). The eIF1A solution structure reveals a large RNA-binding surface important for scanning function. *Mol. Cell* **5**, 109–119.

Benne, R., Brown-Luedi, M. L., and Hershey, J. W. B. (1979). Protein synthesis initiation factors from rabbit reticulocytes: Purification, characterization, and radiochemical labeling. *Methods Enzymol.* **60**, 15–35.

Culver, G. M., and Noller, H. F. (2000). Directed hydroxyl radical probing of RNA from iron(II) tethered to proteins in ribonucleoprotein complexes. *Methods Enzymol.* **318**, 461–475.

Datwyler, S. A., and Meares, C. F. (2000). Protein-protein interactions mapped by artificial proteases: Where sigma factors bind to RNA polymerase. *Trends Biochem. Sci.* **25**, 408–414.

Dmitriev, S. E., Terenin, I. M., Dunaevsky, Y. E., Merrick, W. C., and Shatsky, I. N. (2003). Assembly of 48S translation initiation complexes from purified components with mRNAs that have some base pairing within their 5′ untranslated regions. *Mol. Cell. Biol.* **23**, 8925–8933.

Dontsova, O., Dokudovskaya, S., Kopylov, A., Rinke-Appel, J., Jünke, N., and Brimacombe, R. (1992). Three widely separated positions in the 16S RNA lie in or close to the ribosomal decoding region; a site-directed cross-linking study with mRNA analogues. *EMBO J.* **11**, 3105–3116.

Favre, A., Saintomé, C., Fourrey, J.-L., Clivio, P., and Laugâa, P. (1998). Thionucleobases as intrinsic photoaffinity probes of nucleic acid structure and nucleic acid-protein interactions. *J. Photochem. Photobiol. B.* **42**, 109–124.

Fletcher, C. M., Pestova, T. V., Hellen, C. U., and Wagner, G. (1999). Structure and interactions of the translation initiation factor eIF1. *EMBO J.* **18**, 2631–2637.

Grifo, J. A., Tahara, S. M., Morgan, M. A., Shatkin, A. J., and Merrick, W. C. (1983). New initiation factor activity required for globin mRNA translation. *J. Biol. Chem.* **258**, 5804–5810.

Hardy, S. J., Kurland, C. G., Voynow, P., and Mora, G. (1969). The ribosomal proteins of *Escherichia coli*. I. Purification of the 30S ribosomal proteins. *Biochemistry* **8**, 2897–2905.

Hartz, D., McPheeters, D. S., Traut, R., and Gold, L. (1988). Extension inhibition analysis of translation initiation complexes. *Methods Enzymol.* **164**, 419–425.

Kolupaeva, V. G., Lomakin, I. B., Pestova, T. V., and Hellen, C. U. T. (2003). Eukaryotic initiation factors 4G and 4A mediate conformational changes downstream of the initiation codon of the encephalomyocarditis virus internal ribosomal entry site. *Mol. Cell. Biol.* **23**, 687–698.

Kolupaeva, V., de Breyne, S., Pestova, T., and Hellen, C. (2007). *In vitro* reconstitution and biochemical characterization of translation initiation by internal ribosomal entry. *Methods Enzymol.* **430**, 409–439 (this volume).

Koradi, R., Billeter, M., and Wuthrich, K. (1996). MOLMOL: A program for display and analysis of macromolecular structures. *J. Mol. Graph.* **14**, 51–55.

Kozak, M. (1986). Point mutations define a sequence flanking the AUG initiator codon that modulates translation by eukaryotic ribosomes. *Cell* **44**, 283–292.

Kozak, M (1991). Structural features in eukaryotic mRNAs that modulate the initiation of translation. *J. Biol. Chem.* **266**, 19867–19870.

Lee, J. H., Pestova, T. V., Shin, B. S., Cao, C., Choi, S. K., and Dever, T. E. (2002). Initiation factor eIF5B catalyzes second GTP–dependent step in eukaryotic translation initiation. *Proc. Natl. Acad. Sci. USA* **99**, 16689–16694.

Lomakin, I. B., Kolupaeva, V. G., Marintchev, A., Wagner, G., and Pestova, T. V. (2003). Position of eukaryotic initiation factor eIF1 on the 40S ribosomal subunit determined by directed hydroxyl radical probing. *Genes Dev.* **17**, 2786–2797.

Lomakin, I. B., Shirokikh, N. E., Yusupov, M. M., Hellen, C. U. T., and Pestova, T. V. (2006). The fidelity of translation initiation: Reciprocal activities of eIF1, IF3 and YciH. *EMBO J.* **25**, 196–210.

Madjar, J. J., Arpin, M., Buisson, M., and Reboud, J. P. (1979). Spot position of rat liver ribosomal proteins by four different two-dimensional electrophoreses in polyacrylamide gel. *Mol. Gen. Genet.* **171**(2), 121–134.

Mears, J. A., Cannone, J. J., Stagg, S. M., Gutell, R. R., Agrawal, R. K., and Harvey, S. C. (2002). Modeling a minimal ribosome based on comparative sequence analysis. *J. Mol. Biol.* **321**, 215–234.

Merrick, W. C. (1979). Purification of protein synthesis initiation factors from rabbit reticulocytes. *Methods Enzymol.* **60**, 101–108.

Pestova, T. V., Borukhov, S. I., and Hellen, C. U. T. (1998a). Eukaryotic ribosomes require initiation factors 1 and 1A to locate initiation codons. *Nature* **394**, 854–859.

Pestova, T. V., and Hellen, C. U. T. (2001). Preparation and activity of synthetic unmodified mammalian tRNAi(Met) in initiation of translation *in vitro*. *RNA* **7**, 1496–1505.

Pestova, T. V., and Hellen, C. U. T. (2003). Translation elongation after assembly of ribosomes on the Cricket paralysis virus internal ribosomal entry site without initiation factors or initiator tRNA. *Genes Dev.* **17**, 181–186.

Pestova, T. V., and Hellen, C. U. T. (2005). Reconstitution of eukaryotic translation elongation *in vitro* following initiation by internal ribosomal entry. *Methods* **36**, 261–269.

Pestova, T. V., Hellen, C. U. T., and Shatsky, I. N. (1996a). Canonical eukaryotic initiation factors determine initiation of translation by internal ribosomal entry. *Mol. Cell. Biol.* **16**, 6859–6869.

Pestova, T. V., and Kolupaeva, V. G. (2002). The roles of individual eukaryotic translation initiation factors in ribosomal scanning and initiation codon selection. *Genes Dev.* **16**, 2906–2922.

Pestova, T. V., Lomakin, I. B., Lee, J. H., Choi, S. K., Dever, T. E., and Hellen, C. U. T. (2000). The joining of ribosomal subunits in eukaryotes requires eIF5B. *Nature* **403**, 332–335.

Pestova, T. V., Lorsch, J. R., and Hellen, C. U. T. (2006). The mechanism of translation in eukaryotes. *In* "Translational Control in Biology and Medicine" (N. Sonenberg, J. W. B. Hershey, and M. B. Mathews, eds.), pp. 87–128. Cold Spring Harbor Laboratory Press, Cold Spring Harbor, New York.

Pestova, T. V., Shatsky, I. N., Fletcher, S. P., Jackson, R. J., and Hellen, C. U. T. (1998b). A prokaryotic-like mode of cytoplasmic eukaryotic ribosome binding to the initiation codon during internal translation initiation of hepatitis C and classical swine fever virus RNAs. *Genes Dev.* **12**, 67–83.

Pestova, T. V., Shatsky, I. N., and Hellen, C. U. T. (1996b). Functional dissection of eukaryotic initiation factor 4F: The 4A subunit and the central domain of the 4G subunit are sufficient to mediate internal entry of 43S preinitiation complexes. *Mol. Cell. Biol.* **16**, 6870–6878.

Pisarev, A. V., Kolupaeva, V. G., Pisareva, V. P., Merrick, W. C., Hellen, C. U. T., and Pestova, T. V. (2006). Specific functional interactions of nucleotides at key −3 and +4 positions flanking the initiation codon with components of the mammalian 48S translation initiation complex. *Genes Dev.* **20**, 624–636.

Raychaudhuri, P., and Maitra, U. (1986). Identification of ribosome-bound eukaryotic initiation factor 2.GDP binary complex as an intermediate in polypeptide chain initiation reaction. *J. Biol. Chem.* **261,** 7723–7728.

Schagger, H., and von Jägow, G. (1987). Tricine-sodium dodecyl sulfate-polyacrylamide gel electrophoresis for the separation of proteins in the range from 1 to 100 kDa. *Anal. Biochem.* **166,** 368–379.

Schuwirth, B. S., Borovinskaya, M. A., Hau, C. W., Zhang, W., Vila-Sanjurjo, A., Holton, J. M., and Cate, J. H. (2005). Structures of the bacterial ribosome at 3.5 Å resolution. *Science* **310,** 827–834.

Selmer, M., Dunham, C. M., Murphy, F. V., 4th, Weixlbaumer, A., Petry, S., Kelley, A. C., Weir, J. R., and Ramakrishnan, V. (2006). Structure of the 70S ribosome complexed with mRNA and tRNA. *Science* **313,** 1935–1942.

Sergiev, P. V., Lavrik, I. N., Wlasoff, V. A., Dokudovskaya, S. S., Dontsova, O. A., Bogdanov, A. A., and Brimacombe, R. (1997). The path of mRNA through the bacterial ribosome: A site-directed crosslinking study using new photoreactive derivatives of guanosine and uridine. *RNA* **3,** 464–475.

Staehelin, T., Erni, B., and Schreier, M. H. (1979). Purification and characterization of seven initiation factors for mammalian protein synthesis. *Methods Enzymol.* **60,** 136–1165.

Stanley, W. M., Jr. (1974). Specific aminoacylation of the methionine-specific tRNA's of eukaryotes. *Methods Enzymol.* **29,** 530–547.

Tyzack, J. K., Wang, X., Belsham, G. J., and Proud, C. G. (2000). ABC50 interacts with eukaryotic initiation factor 2 and associates with the ribosome in an ATP-dependent manner. *J. Biol. Chem.* **275,** 34131–34139.

Unbehaun, A., Borukhov, S. I., Hellen, C. U. T., and Pestova, T. V. (2004). Release of initiation factors from 48S complexes during ribosomal subunit joining and the link between establishment of codon-anticodon base-pairing and hydrolysis of eIF2-bound GTP. *Genes Dev.* **18,** 3078–3093.

Wade, H. E., and Robinson, H. K. (1966). Magnesium ion-dependent ribonucleic acid depolymerases in bacteria. *J. Bacteriol.* **101,** 467–479.

Yusupov, M. M., and Spirin, AS (1988). Hot tritium bombardment technique for ribosome surface topography. *Methods Enzymol.* **164,** 426–439.

Yusupov, M. M., Yusupova, G. Z., Baucom, A., Lieberman, K., Earnest, T. N., Cate, J. H., and Noller, H. F. (2001). Crystal structure of the ribosome at 5.5 A resolution. *Science* **292,** 883–896.

CHAPTER EIGHT

RECONSTITUTION OF MAMMALIAN 48S RIBOSOMAL TRANSLATION INITIATION COMPLEX

Romit Majumdar,* Jayanta Chaudhuri,[†] and Umadas Maitra[‡]

Contents

* Department of Cell Biology, Albert Einstein College of Medicine of Yeshiva University, Jack and Pearl Resnick Campus, Bronx, New York
† Immunology Program, Memorial Sloan Kettering Cancer Center, New York, New York
‡ Department of Developmental and Molecular Biology, Albert Einstein College of Medicine of Yeshiva University, Jack and Pearl Resnick Campus, Bronx, New York

Methods in Enzymology, Volume 430
ISSN 0076-6879, DOI: 10.1016/S0076-6879(07)30008-6

Abstract

Initiation of translation is defined as the process by which a 40S ribosomal subunit, containing bound initiator methionyl-tRNA (Met-tRNA$_i$), is positioned at the initiation AUG codon of an mRNA to form the 48S initiation complex. Subsequently, a 60S ribosomal subunit joins the 48S initiation complex to form an elongation-competent 80S initiation complex. By use of highly purified eukaryotic translation initiation factors (eIFs), ribosomes, Met-tRNA$_i$, mRNA, GTP as an effector molecule, and ATP as a source of energy, the initiation step of translation can be efficiently reconstituted. In this chapter, we describe the detailed procedure for efficient binding of Met-tRNA$_i$ to the 40S ribosomal subunit, the subsequent binding of the resulting 43S preinitiation complex to an mRNA, and scanning and positioning of the 43S complex at the AUG start codon of the mRNA to form the 48S initiation complex.

1. INTRODUCTION

Initiation of translation is defined as the process by which a 40S ribosomal subunit, containing bound initiator methionyl-tRNA (Met-tRNA$_i$), is positioned at the initiation AUG codon of an mRNA to form the 40S initiation complex (also called the 48S initiation complex). Recognition of the AUG start codon by the 40S complex sets the reading frame for the subsequent elongation reaction. Subsequently, a 60S ribosomal subunit joins the 40S complex to form the functional 80S initiation complex (80S ·Met-tRNA$_i$·mRNA) that is competent to form the first peptide bond. The initiation process occurs by a series of partial reactions and requires the participation of a large number of eukaryotic translation initiation factors (eIFs), ATP as the source of energy, and GTP as an effector molecule (for a review, see Kapp and Lorsch [2004]).

Initiation of translation occurs in several distinct phases that involve the following partial reactions:

1. Binding of the Met-tRNA$_i$ to an active GTP-bound eIF2 protein to form the ternary complex:

$$eIF2 + GTP + Met\text{-}tRNA_i \rightarrow [GTP \cdot eIF2 \cdot Met\text{-}tRNA_i]$$
Ternary complex

2. Binding of the ternary complex to a 40S ribosomal subunit to form a 43S preinitiation complex requiring the participation of eIF1A and eIF1, which remain bound to the 43S preinitiation complex:

$$[GTP \cdot eIF2 \cdot Met\text{-}tRNA_i] + 40S \cdot eIF3 \xrightarrow{eIF1A, \ eIF1} [GTP \cdot eIF2 \cdot Met\text{-}tRNA_i \cdot 40S]$$
43S preinitiation complex

3. Binding of the 43S preinitiation complex to the 5′ end of an mRNA, which is aided by the presence of the cap-binding initiation factor eIF4F. This is then followed by scanning of the 5′ untranslated region of the mRNA by the 43S preinitiation complex and positioning of the complex at the AUG codon of the mRNA to form a 48S initiation complex (40S · Met-tRNA$_i$ · mRNA). The scanning reaction is facilitated by the presence of the factors eIF4A and eIF4B and also requires the continued participation of eIF1A and eIF1:

$$[GTP \cdot eIF2 \cdot Met\text{-}tRNA_i \cdot 40S \cdot eIF3] + mRNA \xrightarrow{eIF, \ -4F, \ -1A, \ -1, \ -4A, \ -4B, \ ATP}$$
$$[GTP \cdot eIF2 \cdot Met\text{-}tRNA_i \cdot 40S \cdot eIF3 \cdot mRNA]$$
48S initiation complex

4. eIF5-mediated hydrolysis of GTP bound to the 48S initiation complex and the release of the initiation factors and the guanine nucleotide from 40S subunit:

$$[GTP \cdot eIF2 \cdot Met\text{-}tRNA_i \cdot 40S \cdot eIF3 \cdot mRNA] \xrightarrow{eIF5}$$
$$[40S \cdot mRNA \cdot Met\text{-}tRNA_i] + [eIF2 \cdot GDP] + P_i + eIF3$$

5. Joining of a 60S ribosomal subunit to the 48S initiation complex to form the 80S initiation complex in a reaction catalyzed by eIF5B:

$$[40S \cdot mRNA \cdot Met\text{-}tRNA_i] + 60S \xrightarrow{eIF5B} [80S \cdot mRNA \cdot Met\text{-}tRNA_i]$$
80S initiation complex

6. Recycling of eIF2 from the inactive (eIF2 · GDP) form to the active (eIF2 · GTP) form, catalyzed by the guanine nucleotide exchange factor (GEF), eIF2B:

$$[eIF2 \cdot GDP] + GTP \xrightarrow{eIF2B} [eIF2 \cdot GTP] + GDP$$

$$[eIF2 \cdot GTP] + Met\text{-}tRNA_i \rightarrow [Met\text{-}tRNA_i \cdot eIF2 \cdot GTP]$$

Each of these partial reactions can be conveniently assayed by labeling Met-tRNA$_i$ with either ^{35}S or ^{3}H followed by nitrocellulose membrane filter binding (partial reaction 1) or by sucrose density gradients (partial reactions 2 to 5). Thus, reconstitution of the translation initiation process requires the purification of these eukaryotic initiation factors, along with 40S and 60S ribosomal subunits and Met-tRNA$_i$. Here we describe in detail the procedure for efficient binding of Met-tRNA$_i$ to the 43S ribosomal subunit to form the 43S preinitiation complex and its subsequent binding to the mRNA and positioning of the complex at the AUG start codon to form the 48S initiation complex.

2. REAGENTS

1. Buffers
2. Initiator Met-tRNA$_i$
3. Purified 40S and 60S ribosomal subunits
4. Synthetic trinucleotide AUG codon and mRNA
5. Purified translation initiation factors

3. BUFFERS

3.1. Met-tRNA$_i$ preparation

Buffer T1: 20 mM Tris-HCl (pH 7.8), 10 mM Mg(OAc)$_2$, 30 mM KCl, 10 mM 2-mercaptoethanol, 0.5 mM EDTA

Buffer T2: 0.25 M potassium phosphate (pH 7.5), 1 mM Mg(OAc)$_2$, 10 mM 2-mercaptoethanol, 0.1 mM EDTA, 5% glycerol

Buffer T3: 50 mM Tris-HCl (pH 7.8), 0.5 mM EDTA, 1 mM dithiothreitol, 10% glycerol

Buffer T4: 50 mM potassium acetate (pH 5.0), 2 mM MgCl$_2$, 100 mM KCl

Buffer T5: 50 mM potassium acetate (pH 5.0), 2 mM MgCl$_2$, 240 mM KCl

Buffer T6: 50 mM potassium acetate (pH 5.0), 2 mM MgCl$_2$, 1.2 M KCl

Buffer T7: 7 ml Buffer T5 + 18 ml 100% ethanol

Buffer T8: 50 mM potassium acetate (pH 4.5), 2 mM MgCl$_2$, 100 mM KCl

3.2. Ribosome preparation

Buffer R1: 20 mM Tris-HCl (pH 7.5), 70 mM KCl, 9 mM MgCl$_2$, 0.1 mM EDTA, 1 mM dithiothreitol, 5% glycerol

Buffer R2: 20 mM Tris-HCl (pH 7.5), 700 mM KCl, 4 mM MgCl$_2$, 2 mM dithiothreitol

Buffer R3: 30 mM MgCl$_2$, 2 mM dithiothreitol

Buffer R4: 20 mM Tris-HCl (pH 7.5), 100 mM KCl, 5 mM MgCl$_2$, 0.1 mM EDTA, 1 mM dithiothreitol, 55% glycerol

3.3. Purification of translation initiation factors from rabbit reticulocyte lysates

3.3.1. Assays and reconstitution of the 48S ribosomal complex

Buffer A: 20 mM Tris-HCl, pH 7.5, 100 mM KCl, 1 mM MgCl$_2$, 5 mM 2-mercaptoethanol

Buffer B: 20 mM Tris-HCl, pH 7.5, 100 mM KCl, 5 mM MgCl$_2$, 5 mM 2-mercaptoethanol

Buffer C: 50 mM Tris-HCl, pH8.3, 75 mM KCl, 6 mM MgCl$_2$

3.3.2. Purification of initiation factors

Buffer D: 0.25 M sucrose, 40 mM Tris-HCl, pH 7.5, 1 mM dithiothreitol, 0.1 mM EDTA, 0.5 mM phenylmethylsulfonyl fluoride (PMSF), 35 mM KCl

Buffer E: 20 mM Tris-HCl, pH 7.5, 1 mM dithiothreitol, 0.1 mM EDTA, 0.5 mM PMSF, 10% glycerol

Buffer F: 20 mM Tris-HCl, pH 7.5, 1 mM dithiothreitol, 0.1 mM EDTA, 650 mM KCl

Buffer G: 20 mM HEPES, pH 8.0, 0.2 mM EDTA, and 14 mM 2-mercaptoethanol

Buffer H: 20 mM HEPES, pH 7.5, 0.2 mM EDTA, 7 mM 2-mercaptoethanol, 0.5 mM PMSF, 500 mM KCl

Buffer I: 20 mM Tris-HCl, pH 8.0, 10 mM MgCl$_2$, 100 mM KCl, 1 mM EDTA, 0.5 mM PMSF

Buffer J: 20 mM Tris-HCl, pH 7.5, 0.1 mM EDTA, 5 mM 2-mercaptoethanol, 10% glycerol, 0.5 mM PMSF

Buffer K: 20 mM Tris-HCl, pH 7.5, 10 mM MgCl$_2$, 50 mM KCl, 1 mM EDTA

Buffer L: 20 mM phosphate buffer, pH 7.8, 500 mM NaCl, 5 mM potassium imidazole, 10% glycerol

4. General Methods

4.1. Preparation of initiator methionyl-tRNA (Met-tRNA$_i$)

4.1.1. Isolation of *Escherichia coli* tRNA synthetase

Frozen *E. coli* MRE600 cells (10 g) are ground with 20 g of alumina powder (Alcoa A-301) in a precooled mortar until a fine paste is obtained. The paste is extracted with 30 ml of Buffer T1. The suspension is centrifuged at 13,000 rpm for 30 min in a Sorvall SS-34 rotor at 4°. The supernatant is recentrifuged in the same manner. The resulting supernatant is treated with 9 μg of RNase-free DNase I (Roche Molecular Biochemicals) for 30 min at room temperature and then centrifuged for 2½ h at 45,000 rpm in a Beckman Ti-50.2 rotor at 4° to remove ribosomes. The postribosomal supernatant is diluted with an equal volume of 0.5 M potassium phosphate, pH 7.5, and loaded onto a DEAE cellulose column (45-ml bed volume) (Fisher Scientific) that was equilibrated with Buffer T2. The column is washed with the same buffer at a flow rate of 1 ml/min. Under these conditions, nucleic acids are retained on the column, whereas proteins are mostly unretarded. The absorbance (at 280 nm) of eluted fractions is monitored. The peak fractions are pooled and treated with solid ammonium sulfate to a saturation of 75%. The solution is stirred for 30 min at 4°, the precipitated protein was recovered by centrifugation at 25,000 rpm for 20 min in a Beckman Ti-50.2 rotor, and dissolved in minimum volume of Buffer T3. The protein solution is dialyzed overnight against 400 ml of Buffer T3, divided into small aliquots, and stored at −70° until use. Under these conditions, the synthetase preparation is active for several years.

4.1.2. Preparation of Met-tRNA$_i$

Unfractionated tRNA is isolated from rabbit liver as described by Deutscher (Deutscher, 1974), and the initiator tRNA is specifically aminoacylated with [^{35}S] methionine by use of the *E. coli* synthetase preparation, prepared as described previously. Under these conditions, only tRNA$_i^{Met}$ is amino acylated, because the *E. coli* methionine synthetase does not recognize the mammalian tRNA$_m$ species (Bose *et al.*, 1974). The procedure for aminoacylation is as follows. The reaction mixture contains 50 mM Tris-HCl (pH 7.5), 10 mM MgCl$_2$, 10 mM KCl, 2.5 mM 2-mercaptoethanol, 5 mM ATP, 1 mM [^{35}S] methionine (10,000 to 50,000 cpm/pmol) (GE Healthcare Life Sciences), 200 A$_{260}$ units of crude rabbit liver tRNA, and 0.8 mg *E.coli* synthetase preparation. After incubation at 37° for 30 min, the reaction mixture is chilled. A small aliquot (10 μl) is assayed for aminoacylation by treatment with ice-cold 5% trichloroacetic acid (TCA). After approximately 10 min at 0°, Millipore filtration is performed followed by determination of acid–insoluble radioactivity. Aminoacylation is normally complete after approximately 25 min of incubation. The bulk reaction is then treated with KOAc (pH 4.5) to a final concentration of 0.3 M and incubated on ice for 10 min. The precipitated proteins are removed by centrifugation at 15,000 rpm for 15 min at 4°. The supernatant is diluted with an equal volume of sterile water and loaded onto a DEAE–cellulose column (7.5-ml bed volume) that has been equilibrated with Buffer T4. The column is washed with 20 ml of the same buffer and then with 80 ml Buffer T5, and the labeled Met–tRNA$_i$ is finally eluted with Buffer T6. Fractions containing the radioactive peak are pooled. The charged tRNA (as well as uncharged tRNA) is precipitated with 3 volumes of ethanol at −20°. The precipitate is recovered by centrifugation at 9000 rpm for 15 min and evenly suspended in 10 ml Buffer T7. The suspension is incubated at −20° for 2 h followed by centrifugation at 9000 rpm for 15 min. The precipitate is briefly air-dried and dissolved in Buffer T8 to a final concentration of 250 to 300 A$_{260}$ units/ml. The Met–tRNA$_i$ preparation is then divided into small aliquots and stored at −70°. Such a preparation is stable for at least 6 months. Purified [^{3}H] Met–tRNA$_i$ is prepared as previously described except that [^{3}H] methionine (8000 cpm/pmol) (GE Healthcare Life Sciences) is used instead of [^{35}S] methionine.

4.2. Preparation of ribosomal subunits

A large number of investigators have used ribosomes isolated from mammalian liver, rabbit reticulocyte lysates, and also Artemia eggs (Benne et al., 1979; Merrick, 1979; Staehelin et al., 1979). We have observed that although ribosomes isolated from all sources were active, those purified from Artemia eggs are most efficient in the formation of both 43S and 48S initiation complexes. Moreover, ribosomes isolated from multicellular eukaryotes have similar RNA and protein composition. Because of the ease of preparation of highly active ribosomal subunits, we have routinely used Artemia eggs as a source of ribosomes.

All glassware and buffers used for ribosomal subunit preparation are autoclaved before use. Artemia eggs (50 g) (Brine Shrimp Eggs, Ogden, UT, USA) are suspended in 100 ml of a 20% Chlorax solution and allowed to settle for 5 min (Raychaudhuri and Maitra, 1986). The cysts are filtered through a Millipore filter screen, washed with ice-cold sterile water, and transferred to a dry beaker. The cysts are then treated with 500 ml of water; the suspension is stirred and allowed to settle for 15 min. The floating particles are drained off at the end of the incubation. This process is repeated six times until the amount of floating particles is reduced to a minimum. The cysts are once again filtered over a Millipore filter screen, transferred to a precooled mortar, and ground with 37.5 g of sea sand (Fisher Scientific) with the gradual addition of 20 ml of Buffer R1 till a fine paste is obtained. An additional 60 ml Buffer R1 is added to obtain a smooth suspension, which is filtered through four layers of cheese cloth. The solid paste is transferred back to the mortar and ground again after addition of 40 ml Buffer R1. The suspension is re-extracted through four layers of cheese cloth. The filtrate from these operations is centrifuged at 10,000 rpm for 30 min in a Sorvall SS-34 rotor. The supernatant is filtered through glass wool avoiding the orange lipid layer, and the filtrate is centrifuged at 28,000 rpm for 20 min in a Beckman Ti-50.2 rotor. The supernatant is once again filtered through glass wool and centrifuged at 40,000 rpm for 2½ h in a Beckman Ti-50.2 rotor. The supernatant is drained off and the ribosomal pellet washed slightly with Buffer R1 before being suspended in 0.5 ml Buffer R2. The ribosomal suspension (150 to 170 A_{260} units) is layered on to 15 to 30% (w/v) sucrose gradients containing 20 mM Tris-HCl, pH 7.5, 700 mM KCl, 11 mM MgCl$_2$, 2 mM dithiothreitol and centrifuged at 25,000 rpm for 13 h in a Beckman SW 28 rotor to separate the ribosomal subunits. Fractions are collected from the top by use of an ISCO gradient fractionator and monitored by absorbance at 254 nm. Separate peaks are obtained for the 40S and 60S ribosomal subunits. The individual subunits are conservatively pooled avoiding cross-contamination. Each pooled fraction is diluted with an equal volume of Buffer R3 and centrifuged at 40,000 rpm for 14 h in a Beckman Ti-45 rotor at 4°. The supernatant is carefully decanted and then discarded. Each individual ribosomal subunit pellet is resuspended in 250 μl Buffer R4 by slow shaking at 4° for approximately 10 h. After centrifugation at 15,000 rpm for 15 min in a tabletop centrifuge to remove any undissolved particles, the 40S and 60S ribosomal subunits are divided into small aliquots and stored at −135°, where they are stable for approximately 6 months.

4.3. Preparation of AUG trinucleotide codon and mRNA

4.3.1. Synthesis of AUG

The trinucleotide codon ApUpG is synthesized chemically on ribo-G columns (iPR-PAC-G-RNA-CpG, Glen Research Corporation) with A and U phosphoramidites (PAC-A-CE Phosphoramidite and U-CE Phosphoramidite, respectively, Glen Research Corporation) by use of the Expedite Oligo Synthesizer. After synthesis, the column beads are transferred to an Microfuge tube, resuspended in 1 ml of a freshly prepared solution of a 1:1 mixture of ammonium hydroxide/methylamine and incubated at 65° for 10 min. The beads are vortexed, centrifuged briefly, and then incubated at 65° for an additional 10 min. The resuspended material is cooled on ice. After centrifugation, the supernatant is transferred to a fresh tube. The beads are washed twice with 100 μl of a 1:1:1 mixture of ethanol/acetonitrile/water. The supernatant from each wash is pooled with the initial supernatant and freeze-dried overnight in a Speed-Vac. The next day, any air in the tube is driven out by flushing with argon and the dried material is resuspended in 0.5 ml of a 2:1:1.5 mixture of N-methylpyrrolidinone/triethylamine/triethylamine trihydrofluoride. Desilation is carried out by incubating the reaction mixture at 65° for 90 min. The reaction mixture is then cooled to room temperature, and the reaction is quenched with 0.2 volumes of 60 mM MgCl$_2$. The ionic strength of the reaction mixture is adjusted to 100 mM by diluting with an equal volume of 75 mM ammonium formate (pH 9.25) and water. The pH is adjusted to 9.25 by adding 100 to 250 μl of ammonium hydroxide. The reaction mixture is then loaded onto a DEAE-cellulose column (10-ml bed volume) that is previously equilibrated in 75 mM ammonium formate (pH 9.25). All subsequent operations are carried out at room temperature. The column is initially washed with 30 ml of the equilibrating buffer (75 mM ammonium formate, pH 9.25). A linear gradient of 200 ml total volume, from 75 mM ammonium formate, pH 9.25, to 1 M ammonium formate, pH 9.25, is then applied. Fractions of 3 ml are collected, and A$_{260}$ of each fraction is determined. AUG, free of unreacted components, elutes in fractions 23 through 30 under these conditions. The pooled fractions are repeatedly lyophilized over a few days to remove the ammonium formate. The final dried preparation is dissolved in sterile water and then stored at $-135°$ at a concentration $\sim$50 A$_{260}$ unit/ml.

4.3.2. Synthesis of capped mRNA

It should be noted that although the mRNA used here for the reconstitution studies is a model mRNA generated from a cloned DNA by T7 RNA polymerase catalyzed transcription, any mRNA isolated from natural sources (e.g., globin mRNA) can also be used in these reactions.

Accordingly, a 5′-capped pCON-PK mRNA, containing a 50 nucleotide-long 5′ UTR, a single AUG codon, and a 150-nucleotide–long open-reading frame (ORF) that terminated in a pseudo-knot structure at the 3′ end is generated by use of the pCON-PK plasmid (linearized with *EcoRI*) as template. The plasmid expressing this mRNA from a T7 promoter was originally described by Kozak (Kozak, 1997, 1998) and subsequently modified by Majumdar (Majumdar and Maitra, 2005). The sequence of the T7 polymerase transcribed mRNA is as follows:

5′GAAGCUAAAACAAAUCAAUCAAUCAAAACACAAGCUUCAC
UCAAGAAAAACUACACC*A*UGGCAAGGAUCCAAAGUCAGCCA
AAUCAAGAUCCGAGAUUUUCCCCCGGGCGAGCUCAGCUUG
GGGUAUCAGUCAGGCUCGGCUGGUACCCCUUGCAAAGCGA
GCCUACAGGGCAUCGUAAAG AACAUUUUGAGGAAUUC3′

The mMESSAGE mMACHINE™ T7 Ultra coupled transcription-capping kit (Ambion) is used to synthesize the capped mRNA per the manufacturer's protocol. The capped mRNA is purified from the enzymatic reaction by use of the MEGAclear Purification Kit (Ambion). The

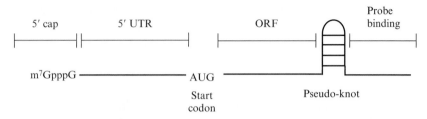

Figure 8.1 Schematic diagram of the 200 nt-long pCON-PK mRNA. The procedure for construction of the mRNA is in the text. The pseudo-knot containing a stem-loop structure is introduced to prevent the scanning 40S ribosomal complex from falling off the mRNA. "Probe binding" denotes the region where the ^{32}P-labeled oligonucleotide probe was hybridized during toe-printing analysis (see text).

mRNA solution is further purified to ensure removal of unincorporated materials by passing through NucAway Spin Columns (Ambion) according to the manufacturer's protocol. The mRNA is quantified by measuring absorbance at 260 nm. A schematic of the pCON-PK mRNA is shown in Fig. 8.1.

To generate ^{32}P-labeled pCON-PK mRNA, the 5' end is phosphorylated with T4 polynucleotide kinase (Fisher Scientific) and [γ-^{32}P] ATP (6000 Ci/mmol) (GE Healthcare Life Sciences) according to the manufacturer's protocol. The ^{32}P-labeled mRNA is purified from unincorporated nucleotides by using G-50 quick spin Sephadex columns (Roche Molecular Biochemicals).

4.4. Purification of translation initiation factors from rabbit reticulocyte lysates

4.4.1. Assays
A variety of different methods are used to assay individual initiation factors as outlined in the following.

Assay of eIF2 activity   The activity of purified eIF2 is assayed by its ability to bind Met-tRNA$_i$ and GTP to form an eIF2 · GTP · Met-tRNA$_i$ ternary complex as described by Stringer *et al.* (1979).

$$\text{eIF2} + \text{GTP} + \text{Met-tRNA}_i \rightarrow [\text{Met-tRNA}_i \cdot \text{eIF2} \cdot \text{GTP}] \rightarrow \text{Measured by}$$
$$\text{Ternary complex} \qquad \text{nitrocellulose}$$
$$\text{filter binding}$$

Reaction mixtures (50 μl) contain 20 mM Tris-HCl, pH 7.5, 100 mM KCl, 5 mM 2-mercaptoethanol, 0.4 mM GTP, 10 μg of bovine serum albumin, 0.5 A$_{260}$ unit of tRNA containing approximately 10 pmol of [^{35}S] Met-tRNA$_i$, and 0.8 to 1.2 μg of purified eIF2. After incubation at

37° for 4 min, the reaction mixtures are filtered through nitrocellulose membrane filters. The filters are washed with 5 to 6 ml of ice-cold reaction buffer, dried, and assayed for radioactivity in a liquid scintillation spectrometer. One unit of eIF2 activity is defined as the amount of protein that promoted the GTP-dependent binding of 1 pmol of Met-tRNA$_i$ to a nitrocellulose filter. Reaction mixtures containing no eIF2 or GTP are used as controls.

Initiation factor requirement for ribosomal complex formation vis-à-vis Mg^{2+} concentration   It should be noted that the requirement of initiation factors for the binding of the ternary complex (formed as described previously) to the 40S ribosomal subunit is dependent on Mg^{2+} concentration. In the presence of 5 mM Mg^{2+} and synthetic AUG trinucleotide codon, the ternary complex can be nearly quantitatively transferred to the 40S ribosomal subunit to form a stable 40S initiation complex (40S • AUG • Met-tRNA$_i$ • eIF2 • GTP) in the absence of any additional initiation factors. The resulting 40S initiation complex can be isolated by sucrose gradient centrifugation. Such an isolated 40S initiation complex serves as an efficient substrate for eIF5-mediated GTP hydrolysis (without the requirement of any additional factors), resulting in the quantitative hydrolysis of bound GTP and the release of eIF2•GDP and P$_i$ from the ribosomal complex. Furthermore, addition of the 60S subunit to the reaction results in efficient formation of the 80S complex without the requirement of the initiation factor eIF5B. In addition, Met-tRNA$_i$ bound in such an 80S initiation complex can interact with puromycin to form methionyl-puromycin indicating the formation of a functional 80S ribosomal initiation complex (Majumdar *et al.*, 2005).

In contrast, when the binding of the ternary complex to the 40S subunit is carried out at a presumed physiological Mg^{2+} concentration of 1 to 2 mM (which has been found to be optimal for translation of an mRNA in a reticulocyte lysate system) and the subsequent sucrose gradient analysis is also carried out in buffers containing 1 mM Mg^{2+}, the binding reaction depends on the presence of the initiation factors eIF1A, eIF3, and eIF1. The three factors are together required for the quantitative transfer of the ternary complex to the 40S ribosomal subunits (Fig. 8.2A). In addition, under these conditions all three factors remain stably bound to the 40S preinitiation complex (Majumdar *et al.*, 2003). It should be noted that in the formation of the 40S preinitiation complex, eIF1A plays a major role in the transfer of the ternary complex to the 40S subunits, whereas the presence of eIF1 stimulates the transfer reaction. The primary role of eIF3 is to stabilize the resulting 40S preinitiation complex. The stabilizing effect of eIF3 becomes more evident when 60S ribosomal subunits are also included in the reaction. In the absence of eIF3, addition of 60S ribosomal subunits leads to complete disruption of the 40S preinitiation complex formed in the

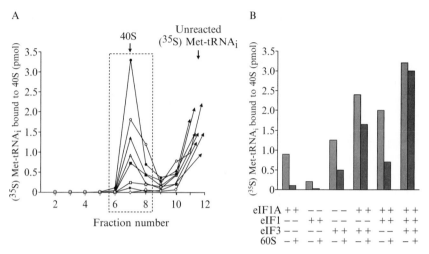

Figure 8.2   Dependence of the formation of 43S preinitiation complex on eIF1, eIF1A, and eIF3 and the effect of addition of 60S ribosomal subunits. (A) Sucrose gradient analysis of 43S preinitiation complex formation. A preformed [^{35}S] Met-tRNA$_i$ · eIF2 · GTP ternary complex was incubated with 40S ribosomal subunits at 1 mM Mg^{2+} in the presence of eIF1, eIF1A, and eIF3, either alone or in combination. After incubation at 37° for 4 min, the reaction mixture was analyzed in sucrose density gradients in buffers containing 1 mM Mg^{2+}. The gradients were fractionated and analyzed as described in the text. Initiation factor additions were as follows: ◇, no addition; ◆, eIF1; □, eIF3; △ eIF1A; ■ eIF1 + eIF3; ▲, eIF1A + eIF3; ○, eIF1A + eIF1; ●, eIF1 + eIF1A + eIF3. The ascending arrows indicate ^{35}S radioactivity recovered at the top of each gradient corresponding to unreacted free Met-tRNA$_i$ and unreacted Met-tRNA$_i$ · eIF2 · GTP ternary complex. (B) 60S-mediated disruption of the 43S preinitiation complex. Identical sets of reaction mixtures as in (A) were prepared to form the 43S preinitiation complex. Before sucrose gradient analysis, each reaction mixture was divided into two parts. The first part was subjected to sucrose density gradient centrifugation and analyzed as described under (A). On the basis of the total amount of radioactivity used, the amount (pmol) of 43S preinitiation complex formed was determined and indicated as a bar graph. In the other part, for each reaction mixture, the Mg^{2+} concentration was raised to 5 mM, and 60S ribosomal subunits were added. After incubation at 37° for 4 min, the reaction mixture was analyzed by sucrose density gradients in buffers containing 5 mM Mg^{2+}. The total radioactivity in the 40S region (as indicated) was calculated and plotted as a bar graph.

presence of eIF1A and eIF1, whereas the complex formed in the presence of all three factors (eIF1A, eIF1, and eIF3) is completely resistant to the disruptive effects of the 60S ribosomes (Fig. 8.2B). This is important in the light of the fact that *in vivo*, when both 40S and 60S ribosomal subunits are present in the same milieu, the presence of eIF3-bound 40S ribosomes is essential to generate a stable 43S preinitiation complex. Furthermore, the presence of bound eIF3 in the 43S preinitiation complex may also facilitate the positioning of the preinitiation complex at the 5′ cap structure of the mRNA (Dever, 2002; Hinnebusch, 2006).

Assay of eIF1A activity   eIF1A activity is routinely assayed for its ability to mediate the transfer of ^{35}S or ^{3}H–labeled Met-tRNA$_i$ from Met-tRNA$_i$ · eIF2 · GTP ternary complex to 40S ribosomal subunits in the presence of AUG and at 1 mM Mg^{2+} (Chaudhuri *et al.*, 1997b).

$$40S + [\text{Met-tRNA}_i \cdot \text{eIF2} \cdot \text{GTP}] + \text{AUG} \xrightarrow{\text{eIF1A}} [\text{Met-tRNA}_i \cdot \text{eIF2} \cdot \text{GTP} \cdot 40S \cdot \text{AUG}] \rightarrow \text{Measured by}$$

40S initiation complex   sucrose density
gradients

Reactions are carried out in two stages as follows: In stage 1, 0.8 to 1.2 μg of purified eIF2, 8 pmol of [^{3}H] Met-tRNA$_i$, and 0.4 mM GTP are incubated in reaction mixtures (50 μl) containing 20 mM Tris-HCl, pH 7.5, 100 mM KCl, 5 mM 2-mercaptoethanol, and 4 μg of nuclease-free bovine serum albumin for 4 min at 37° to promote formation of the [^{3}H] Met-tRNA$_i$ · eIF2 · GTP ternary complex. In stage 2, reaction mixtures (125 μl total volume) are supplemented with eIF1A (50 to 500 ng), 0.5 A$_{260}$ units of AUG codon, 0.6 A$_{260}$ unit of 40S ribosomal subunits, and MgCl$_2$ (1 mM final concentration). After incubation at 37° for 4 min, reaction mixtures are chilled in ice-water and layered onto 7.5 to 30% (w/v) sucrose gradients containing Buffer A and centrifuged at 48,000 rpm for 105 min in a Beckman SW 50.1 rotor. Fractions (200 to 300 μl) are collected from the bottom of each tube, and the radioactivity is measured in Aquasol (Packard Instrument Inc.) in a liquid scintillation spectrometer (Fig. 8.2A). Under the conditions of the assay, formation of the 40S initiation complex is directly proportional to the amount of eIF1A added.

eIF1A activity can also be assayed in the absence of AUG codon or mRNA, provided that following incubation at 1 mM Mg^{2+}, the Mg^{2+} concentration of the reaction mixture is adjusted to 5 mM, and the sucrose gradient centrifugation is also carried out in buffers containing 5 mM Mg^{2+} (Chaudhuri *et al.*, 1999). Such an assay method is particularly useful for the assessment of eIF1 and eIF3 activities as described in the following.

Assay of eIF1 activity   The assay method is based on the principle that although eIF1A alone is sufficient for the AUG-dependent binding of the ternary complex to 40S subunits at 1 mM Mg^{2+}, in the absence of AUG, eIF1 activity is essential for the eIF1A-mediated binding of the ternary complex to the 40S ribosomes, particularly at low (and presumably more physiological) concentrations of eIF1A (Majumdar *et al.*, 2003).

$$40S + [\text{Met-tRNA}_i \cdot \text{eIF2} \cdot \text{GTP}] \xrightarrow{\text{eIF1A, eIF1}} [\text{Met-tRNA}_i \cdot \text{eIF2} \cdot \text{GTP} \cdot 40S] \rightarrow \text{Measured by}$$

40S preinitiation complex   sucrose density
gradients

eIF1 activity is assayed in a reaction similar to that described above for the assay of eIF1A activity, except that in stage 2, reaction mixtures are supplemented with eIF1A (200 ng), eIF1 (250 ng), 0.6 A_{260} units of 40S subunits, and $MgCl_2$ to a final concentration of 1 mM. AUG is omitted from the reaction. After incubation at 37° for 4 min, the chilled reaction mixtures are analyzed by sucrose gradient centrifugation at 1 mM Mg^{2+} as described under assay of eIF1A activity. Under the conditions of this assay, eIF1A-mediated binding of Met-tRNA$_i$ to the 40S ribosomes depends on the presence of eIF1 in the reaction mixture (Fig. 8.2A).

Assay of eIF3 activity   Two different assay methods may be used to measure eIF3 activity:

Method A   eIF3 activity is measured by its ability to bind to 40S ribosomal subunits and stimulate the AUG-dependent binding of ternary complex to 40S subunits in the presence of 1 mM $MgCl_2$ (Chaudhuri *et al.*, 1997a).

40S + [Met-tRNA$_i$·eIF2·GTP] + AUG $\xrightarrow{\text{eIF3}}$ [Met-tRNA$_i$·eIF2·GTP·40S·AUG] → Measured by
                                                    40S initiation complex   sucrose density
                                                                                   gradients

This assay is similar to that described previously for "Assay of eIF1A Activity," except that eIF3 is added in lieu of eIF1A. However, although eIF1A stimulates the binding of the ternary complex 5-to 10-fold, eIF3 stimulates this reaction only 2-to 4-fold (Fig. 8.2A).

Method B   A more specific assay for eIF3 is based on the following observation. Although eIF1 and eIF1A together can mediate the efficient binding of the ternary complex to 40S ribosomes, the addition of 60S subunits to this reaction disrupts the 40S complex, resulting in the dissociation of Met-tRNA$_i$ from the 40S subunit. In contrast, prior incubation of 40S subunits with eIF3 prevents this 60S-mediated disruption (Chaudhuri *et al.*, 1999; Majumdar *et al.*, 2003)

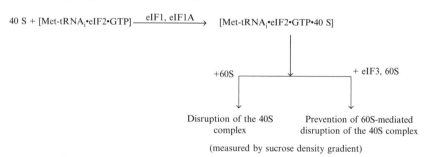

(measured by sucrose density gradient)

The assay consists of the following two stages. The first stage is similar to that described for eIF1A for the formation of the ternary complex. In stage 2, reaction mixtures (125 μl total volume) are supplemented with eIF1A

(200 ng), eIF1 (200 ng), eIF3 (3 μg), $0.6A_{260}$ unit of 40S ribosomal sub-units, and $MgCl_2$ to a final concentration of 1 mM. After incubation at 37° for 5 min to form the 40S complex, reaction mixtures (175 μl total volume) are supplemented with 1.2 A_{260} unit of 60S ribosomal subunits, and the Mg^{2+} concentration is then raised to 5 mM. After incubation at 37° for an additional 4 min, the mixtures are chilled on ice, and the amount of stable 40S complex remaining in the reaction mixture is analyzed by sucrose gradient centrifugation containing Buffer B (5 mM $MgCl_2$). Under the conditions of the assay, omission of eIF3 from the reaction leads to no detectable formation of the 40S complex, presumably because of disruption of the complex by 60S ribosomal subunits (Fig. 8.2B).

Assay of eIF5 activity   The assay for eIF5 activity is based on the principle that this initiation factor promotes the hydrolysis of GTP bound to the 40S initiation complex. When such a complex, formed in the presence of AUG codon and at 5 mM Mg^{2+}, is incubated with 60S subunits, an 80S initiation complex ($80S \cdot AUG \cdot Met\text{-}tRNA_i$) is formed in the absence of any other initiation factors, including eIF5B. This 80S complex reacts with puromy-cin to form methionyl-puromycin, indicating that the complex is fully active in peptide bond formation. Thus, eIF5 activity can be conveniently assayed either by its ability to promote hydrolysis of GTP bound to the 40S initiation complex or by its ability to mediate the joining of a 60S ribosomal subunit to a 40S initiation complex, formed with AUG, to form an 80S initiation complex.

$$[Met\text{-}tRNA_i \cdot eIF2 \cdot GTP \cdot 40S \cdot AUG] \xrightarrow{eIF5} [Met\text{-}tRNA_i \cdot 40S \cdot AUG] + [eIF2 \cdot GDP] + eIF3 \xrightarrow{60S} [Met\text{-}tRNA_i \cdot 80S \cdot AUG]$$

+Pi	80S initiation complex
↓	↓
Measured by phosphomolybdate assay	Measured by sucrose density gradient

Accordingly, either of the following two methods can be used to assay eIF5 activity:

Method A   This method for assaying eIF5 activity directly measures the ability of the protein to mediate the hydrolysis of $[\gamma\text{-}^{32}P]$ GTP bound to eIF2 on the 40S initiation complex (Chakravarti *et al.*, 1993). For this assay, first a 40S initiation complex ($40S \cdot AUG \cdot [^3H]$ Met-tRNA$_i \cdot$ eIF2 $\cdot [\gamma\text{-}^{32}P]$ GTP) is prepared and isolated, free of unreacted components, as described in the following.

Large-scale isolation of the 40S initiation complex   Reaction mixtures (150 μl) containing 20 mM Tris-HCl, pH 7.5, 100 mM KCl, 5 mM 2-mercaptoethanol, 60 μg of nuclease-free bovine serum albumin 20 μM

$[\gamma\text{-}^{32}\text{P}]$ GTP (8000 to 10,000 cpm/pmol) (GE Healthcare Life Sciences), 50 pmol of $[^3\text{H}]$ Met-tRNAi (500 to 1000 cpm/pmol), and 10 μg of eIF2 are incubated at 37° for 5 min to form a $[^3\text{H}]$ Met-tRNA$_i \cdot$ eIF2 $\cdot [\gamma\text{-}^{32}\text{P}]$ GTP ternary complex. Subsequently, $MgCl_2$ is added to a final concentration of 5 mM, followed by the addition of 0.5 A_{260} unit of AUG codon and 3.0 A_{260} units of 40S ribosomal subunits, and the reaction mixture (200 μl total volume) is incubated for an additional 4 min at 37°. Under these conditions (i.e., at 5 mM Mg^{2+} and in the presence of AUG codon), the ternary complex is quantitatively transferred to the 40S ribosomal subunit, in the absence of any other initiation factors, to form the 40S initiation complex. The reaction mixtures are chilled in an ice-water bath and then layered onto a 5 ml of 7.5 to 30% (w/v) sucrose density gradient containing Buffer B (5 mM $MgCl_2$) and centrifuged at 48,000 rpm for 105 min in a SW 50.1 rotor. Fractions (250 μl) are collected from the bottom of each gradient, and an aliquot (10 μl) from each fraction is counted in a liquid scintillation spectrometer to determine the radioactivity profile. The 40S initiation complex fractions containing bound $[^3\text{H}]$ Met-tRNA$_i$ and $[\gamma\text{-}^{32}\text{P}]$ GTP are pooled, divided into equal aliquots of 1 pmol each, and stored at $-135°$ until further use.

Assay   To assay the eIF5-mediated GTPase activity, approximately 80 μl of isolated 40S complex containing 1 pmol of bound $[\gamma\text{-}^{32}\text{P}]$ GTP is incubated with eIF5 (25 ng) at 22° for 10 min (isolated 40S complexes contained 20 mM Tris-HCl, pH 7.5, 5 mM $MgCl_2$, 5 mM 2-mercaptoethanol, 100 mM KCl, and approximately 10% sucrose). After incubation, the release of $^{32}\text{P}_i$ by the cleavage of $[\gamma\text{-}^{32}\text{P}]$ GTP bound to the 40S complex is determined by a modification of the method of Conway and Lipmann (1964) as follows. The reaction is terminated by adding 100 μl of 20 mM silicotungstic acid in 10 mM H_2SO_4, followed by centrifugation in an Eppendorf Microfuge. The supernatant is then treated with 200 μl of 2 mM KH_2PO_4 and 100 μl of 5% ammonium molybdate in 4 N H_2SO_4. The reaction tubes are incubated for 1 min at 37° followed by the addition of 500 μl of isobutanol/benzene (1:1). The tubes are vortexed vigorously for 20 sec followed by a brief (30 sec) centrifugation in an Eppendorf Microfuge to separate the phases. A 200-μl aliquot from the upper organic layer from each tube is assayed for radioactivity by counting in a liquid scintillation spectrometer. One unit of eIF5 activity is defined as the amount of protein required for hydrolysis of 1 pmol of the bound GTP.

Method B   The second assay method, developed by Raychaudhuri *et al.* (1985) from this laboratory, measures the ability of eIF5 to mediate the formation of an 80S initiation complex in an AUG-dependent system. For this purpose, a 40S initiation complex is formed as described under "Large-scale Isolation of 40S Initiation Complex," except that unlabeled GTP is used instead of [γ-^{32}P] GTP. Either [^{3}H] Met–tRNA$_i$ (8000 to 12,000 cpm/pmol) or [^{35}S] Met–tRNA$_i$ (8000 to 12,000 cpm/pmol) may be used. To assay for eIF5-mediated 80S initiation complex formation, approximately 80 μl of isolated 40S complex containing 1 pmol of bound [^{3}H] Met–tRNA$_i$ is incubated with 0.8 A$_{260}$ unit of 60S ribosomal subunits and eIF5 (25 ng) at 37° for 5 min. The MgCl$_2$ in the reaction mixtures is maintained at a final concentration of 5 mM. Reaction mixtures (125 μl total volume) are chilled and then analyzed for the formation of an 80S initiation complex by sucrose gradient centrifugation containing Buffer B (5 mM MgCl$_2$).

4.5. Purification of initiation factors

All mammalian translation initiation factors required for the efficient formation of the 48S ribosomal initiation complex could be isolated from rabbit reticulocyte lysates. In recent years, however, monomeric initiation factors (e.g., eIF1, eIF1A, eIF4A, eIF4B, and eIF5) have been conveniently purified as bacterially expressed recombinant proteins. These proteins are as active as the proteins isolated from rabbit reticulocyte lysates.

All subsequent operations are carried out between 0 and 4° unless otherwise indicated (Chevesich et al., 1993). For a scheme for the purification of the translation initiation factors, see Fig. 8.3.

4.5.1. Preparation of crude ribosomal salt-wash proteins from rabbit reticulocyte lysates

Rabbit reticulocyte lysates (300 ml) obtained from Green Hectares Co., Oregon, WI, are thawed, protease inhibitors (leupeptin 0.5 μg/ml, pepstatin 0.7 μg/ml, aprotinin 2.0 μg/ml, phenylmethylsulfonyl fluoride 0.5 mM) are added, and the lysate is centrifuged for 3½ h at 48,000 rpm in a Beckman Ti–50.2 rotor. The polysomal pellets are suspended in 26 ml of Buffer D. The suspension is homogenized in a Douce homogenizer and made up to 0.5 M KCl by the slow addition of 4 M KCl, stirred for 30 min at 4°, then centrifuged for 3 h at 48,000 rpm in a Beckman Ti–50.2 rotor. The post-ribosomal supernatant is dialyzed overnight against 1 L of Buffer E + 100 mM KCl.

4.5.2. Initial separation of the initiation factors

The dialyzed ribosomal salt wash proteins (approximately 120 mg protein) are loaded onto a DEAE-cellulose column (40-ml bed volume) that has been equilibrated in buffer E + 100 mM KCl. After washing the column with the same buffer until the A$_{280}$ of the effluent is below 0.1, the absorbed proteins are eluted by use of buffer E + 300 mM KCl. The A$_{280}$ of each fraction is monitored, the peak fractions are pooled and dialyzed against 1 L of Buffer E + 75 mM KCl for 2 h to reduce the ionic strength of the fraction to that of Buffer E + 100 mM KCl. This dialyzed fraction

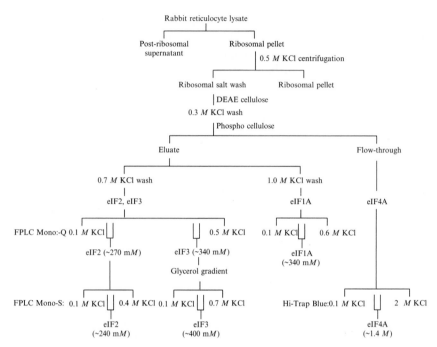

Figure 8.3 Flow scheme for the purification of translation initiation factors from rabbit reticulocyte lysates. For details of the procedure, see text.

(approximately 60 mg protein) is then applied to a phosphocellulose column (20-ml bed volume) that has been equilibrated in Buffer E + 100 mM KCl (the flow-through fraction of the phosphocellulose column is preserved for the subsequent purification of eIF4A). After washing the column with 60 ml of the same buffer, the absorbed proteins are eluted with Buffer E containing increasing concentrations of KCl as follows: (1) 700 mM KCl, (2) 1 M KCl, and assayed for eIF2 activity.

*Purification of eIF2, eIF3, and eIF5 from the 700-m*M *KCl phosphocellulose eluate*   The phosphocellulose eluate fractions are assayed for eIF2 activity. The phosphocellulose-700 mM eluate fractions, containing the bulk of eIF2 activity, are pooled, dialyzed successively against Buffer E + 250 mM KCl for 2 h and Buffer E + 75 mM KCl for 3 h to reduce the ionic strength of the fraction to that of Buffer E + 100 mM KCl. The dialyzed material is then loaded onto a Fast Protein liquid chromatography (FPLC)-Mono Q (HR 5/5) column (1-ml bed volume) that has been equilibrated with Buffer E + 100 mM KCl. The column is washed with 5 ml of the same buffer to remove any nonspecifically bound proteins. Bound proteins are eluted with a linear gradient of 3 ml (total volume) from Buffer E + 100 mM KCl to Buffer E + 225 mM KCl, followed by another linear gradient of 20 ml (total volume) from Buffer E + 225 mM KCl to Buffer E + 550 mM KCl. During the Mono Q gradient step, eIF2, eIF3, and eIF4F activities are separated from one another. eIF2 (assayed as described under "Assay of eIF2 Activity") elutes at $\sim$270 mM KCl, whereas eIF5 activity (assayed as described under "Assay of eIF5 Activity") elutes at $\sim$340 mM KCl. The eIF3 and eIF4F activities (assayed by Western Blotting by use of specific anti-eIF3 and anti-eIF4G [Santa Cruz Biotechnology] antibodies, respectively) elute at $\sim$370 mM and $\sim$450 mM KCl, respectively (Fig. 8.3).

4.5.3. Further purification of eIF2

Mono Q fractions containing eIF2 activity are pooled and further purified by FPLC-Mono S chromatography by an adaptation of the procedure of Dholakia and Wahba (Dholakia and Wahba, 1987) as follows. The pooled fractions containing eIF2 activity are dialyzed against Buffer E + 150 mM KCl to reduce the ionic strength of the protein solution to that of Buffer E + 150 mM KCl. The dialyzed material is then applied to an FPLC-Mono S (HR 5/5) column (1-ml bed volume) that has been equilibrated with Buffer E + 100 mM KCl. The column is washed with 5 ml of the same buffer, and bound proteins are eluted with two successive linear gradients: (1) a gradient of 2 ml (total volume) from Buffer E + 150 mM KCl to Buffer E + 200 mM KCl, followed by (2) a gradient of 10 ml (total volume) from Buffer E + 200 mM KCl to Buffer E + 400 mM KCl. Fractions (0.5 ml) containing eIF2 activity (eluting at $\sim$240 mM KCl) are pooled, dialyzed against Buffer E + 100 mM KCl, and stored at $-20°$. Purified eIF2 has a specific activity of approximately 2000 units/mg protein and exhibits three major polypeptide bands, corresponding to the α-, β-, and γ-subunits of eIF2 (Fig. 8.4).

4.5.4. Further purification of eIF3

For further purification of eIF3 (Chaudhuri et al., 1997a), the pooled Mono Q-eIF3 fractions are concentrated to approximately 1 ml by Centricon-30 (Amicon) filtration and then purified by centrifugation at 40,000 rpm in two 15 to 40% (w/v) glycerol gradients containing Buffer F for 20 h in a SW41 rotor at 4°. Fractions are analyzed by SDS-PAGE followed by Coomassie blue staining. Fractions (0.5 ml each) containing eIF3 are pooled and dialyzed against Buffer E + 100 mM KCl for 3 h and then fractionated on a FPLC-Mono S (HR 5/5) column (1 ml bed volume) with a linear gradient of 20 ml (total volume) from Buffer E + 100 mM KCl to Buffer E + 700 mM KCl. Fractions (0.5 ml each) are analyzed by SDS-PAGE followed by Coomassie blue staining. The activity of eIF3-containing fractions is assayed as described under "Assay of eIF3 Activity, Method B." The eIF3 activity elutes at $\sim$400 mM KCl. Active fractions are pooled, dialyzed for approximately 5 h against Buffer E containing 55% glycerol + 100 mM KCl, and then stored in small aliquots at $-70°$ (Fig. 8.4).

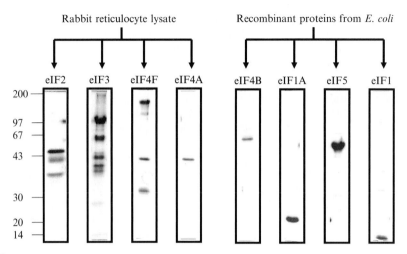

Figure 8.4 SDS-polyacrylamide gel electrophoresis of purified translation initiation factors. (See color insert.)

4.5.5. Further purification of eIF5

Although the cloning of mammalian eIF5 has allowed us to purify large amounts of active recombinant eIF5 protein from *E. coli* with relative ease as described in a later section, eIF5 may also be purified from rabbit reticulocyte lysates as described previously from our laboratory (Chevesich *et al.*, 1993).

4.6. Further purification of eIF1A from the 1 *M* KCl phosphocellulose eluate

eIF1A is purified from the phosphocellulose-1 *M* KCl eluate (Chaudhuri *et al.*, 1997b) as follows. The phosphocellulose-1 *M* KCl eluate (0.5 mg protein) is dialyzed against Buffer E + 100 m*M* KCl and then applied to a 1-ml bed volume FPLC-Mono Q (HR 5/5) column. The column is washed with 5 ml of this buffer, and bound proteins are eluted with two consecutive linear gradients in Buffer E (1 ml/min) as follows: 100 to 250 m*M* KCl (total volume 5 ml), followed by 250 to 600 m*M* KCl (total volume 25 ml). Fractions of 0.5 ml are collected and assayed for eIF1A activity. Fractions containing eIF1A activity (eluting at approximately 340 m*M* KCl) are pooled and stored in small aliquots at −70°.

4.6.1. Purification of eIF4A

The factor eIF4A is purified from the flow-through fraction of the phosphocellulose column purification step described previously under "Purification of Initiation Factors: Initial Separation of the Initiation Factors." The phosphocellulose flow-through fraction (10 ml containing approximately 2 mg protein) is loaded on to a 1-ml bed volume of a FPLC- Hi-Trap Blue column (GE Healthcare Life Sciences) that is equilibrated with Buffer E $+$ 100 mM KCl. The column is washed with 15 ml of the same buffer and eIF4A is eluted by use of a linear gradient of 10 ml (total volume) from Buffer E $+$ 100 mM KCl to Buffer E $+$ 2 M KCl. Fractions are analyzed by Western blotting by use of a specific anti-eIF4A antibody. Fractions containing eIF4A (eluting at approximately 1.4 M KCl) are pooled, dialyzed against (1) Buffer E $+$ 500 mM KCl for 2 h, (2) Buffer E $+$ 100 mM KCl for 4 h, and stored at $-70°$ in small aliquots (Fig. 8.4).

4.7. Purification of eIF4F from rabbit reticulocyte lysate

Although eIF4F can be obtained from the Mono Q fractionation of the phosphocellulose-700 mM KCl eluate (the eIF4F protein, assayed by Western Blot by use of a specific anti-eIF4G antibody eluted at approximately 450 mM KCl), this eIF4F is highly dilute and displays poor cap-binding activity. Thus, we purify eIF4F separately by use of an adaptation of the procedure described by Sonenberg and associates (Edery et al., 1983) as follows.

Crude ribosomal 0.5 M KCl-wash proteins are prepared from 750 ml rabbit reticulocyte lysates and precipitated by the addition of ammonium sulfate to 40% saturation. The pellet is resuspended in minimum volume (approximately 1 ml) of Buffer G $+$ 100 mM KCl and dialyzed against 1 L of the same buffer for 4 h. The dialyzed material is clarified by centrifuging at 12,000 rpm for 20 min and then adjusted to 0.5 M KCl by the slow addition of 4 M KCl followed by stirring for 30 min. The resulting protein fractions (10 to 20 A_{280} units/tube) is sedimented in a 10 to 35% sucrose gradient (11 ml) containing Buffer H at 40,000 rpm for 22 h in a Beckman SW41 rotor at 4°. Fractions are analyzed by SDS-PAGE followed by Western blotting by use of antibodies specific for eIF3 and eIF4F. The part of the gradient containing eIF4F (excluding the fast-sedimenting eIF3) is pooled and dialyzed against Buffer G containing 10% glycerol $+$ 100 mM KCl. The dialyzed material is incubated with m^7GTP-Sepharose (GE Healthcare Life Sciences) (equilibrated in Buffer G containing 10% glycerol and 100 mM KCl) for 1 h at 4° with gentle shaking, and the unadsorbed protein solution is removed from the m^7GTP-Sepharose beads by pouring the suspension into an empty 1-ml Econo-column Chromatography column (BioRad). The column is washed with 20 ml of the same buffer to remove unabsorbed proteins and then with 6 ml of the same buffer containing 100 μM GTP (GE Healthcare Life Sciences). Proteins that specifically bound to the 5′ cap structure are eluted with 4 ml of the same buffer containing 75 μM m^7GTP (GE Healthcare Life Sciences). Fractions are analyzed by SDS-PAGE followed by Coomassie blue staining. Fractions containing eIF4F are pooled and the excess m^7GTP nucleotide is removed from these fractions in a Centricon–30 filtration step involving repeated dilutions with Buffer E + 100 mM KCl. Complete removal of the eIF4F-bound m^{7}GTP nucleotide is indicated when the protein fraction reached a final A$_{280}$/A$_{260}$ ratio of 1.7 to 2.0. The purified eIF4F protein exhibits three major polypeptide bands corresponding to 220 kDa (eIF4G), 46 kDa (eIF4A), and 24 kDa (eIF4E) (Fig. 8.4).

4.8. Purification of recombinant translation initiation factors eIF5, eIF1, eIF1A, and eIF4B

4.8.1. Recombinant eIF5 protein

E. coli XL1-Blue cells, transformed with recombinant pGEX-KG plasmid containing mammalian eIF5 coding sequences (Chaudhuri *et al.*, 1994), are grown in 2 L of 2YT medium to an A$_{600}$ of 0.8 and induced with 0.5 mM isopropyl-D-thiogalactopyranoside (IPTG) (Sambrook *et al.*, 1989). Cells are harvested 2.5 h after induction, washed with ice-cold 0.9% NaCl, quick-frozen in a dry ice-ethanol bath, and stored at $-70°$ until use.

For isolation of recombinant eIF5, the frozen cells are suspended in 30 ml of Buffer I, treated with lysozyme (final concentration of 400 μg/ml) (Roche Molecular Biochemicals), incubated for 20 min at 4°, and then disrupted by sonication. The cell debris is removed by centrifugation at 15,000 rpm for 20 min, the supernatant is treated with a mixture of protease inhibitors, and then incubated with 30 μg of pancreatic DNase I for 30 min at 4°. After centrifugation at 45,000 rpm for 2½ h, the clear supernatant is mixed with 2 ml of a suspension of glutathione (GSH)-Sepharose beads (GE Healthcare Life Sciences) previously equilibrated in Buffer J containing 100 mM KCl. The mixture is gently incubated in a rotator at 4° for 1 h, and the liquid containing unabsorbed proteins is then removed from the beads by pouring the suspension into an empty 5-ml Econo-column Chromatography column (BioRad). The column containing GST-eIF5–bound beads is then washed with 20 ml of Buffer J + 280 mM KCl (until the A$_{280}$ was below 0.1) and then with 10 ml of phosphate-buffered saline (PBS, 10 mM potassium phosphate, pH 7.0, 150 mM NaCl). The beads are then resuspended with 3 ml of PBS and incubated with 200 U of thrombin (GE Healthcare Life Sciences) overnight at 4°. This results in the release of eIF5 from GST-eIF5 fusion protein bound to the beads into the supernatant. The released eIF5 is isolated free of beads by pouring the entire mixture onto the 5-ml Econo-column Chromatography column and collecting the flow-through liquid. The beads in the column are washed with 2 ml of PBS, and the resulting flow-through fraction is also collected. The pooled flow-through fractions are dialyzed against 600 ml of Buffer E + 80 mM KCl for approximately 2 h and then applied to a 1-ml bed-volume FPLC-MonoQ column equilibrated in Buffer E + 100 mM KCl. The column is washed with 5 ml of the same buffer, and bound proteins are eluted with a linear gradient (0.5 ml/min) of 15 ml (total volume) from Buffer

E + 100 mM KCl to Buffer E + 500 mM KCl. Fractions of 0.5 ml are collected and assayed for eIF5 by Western blotting by use of polyclonal anti-eIF5 antibodies. Fractions containing eIF5 (eluting at approximately 360 mM KCl) are pooled and dialyzed against 600 ml of Buffer J containing 55% glycerol + 100 mM KCl for approximately 8 h and then stored at $-20°$ (Fig. 8.4).

4.8.2. Recombinant eIF1 protein

The open-reading frame of eIF1 cDNA (Fields and Adams, 1994) has been cloned into a pET-5a expression plasmid (Novagen) (Majumdar *et al.*, 2003). This pET-5a-eIF1 expression vector is used to transform *E. coli* BL21 (DE3) cells (Novagen). Transformants are then grown at 37° in 1 L LB medium (Sambrook *et al.*, 1989) containing 50 μg/ml ampicillin to an A_{600} of approximately 0.8, induced with 1 mM IPTG, and grown for an additional 2 h. The cells are harvested by centrifugation, washed with 0.9% NaCl, quick-frozen in a dry ice/ethanol bath, and stored at $-70°$.

For purification of recombinant eIF1, frozen *E. coli* cells (5 g) are suspended in 15 ml of Buffer K, treated with lysozyme (final concentration of 400 μg/ml), incubated for 20 min at 4°, and are disrupted by sonication. The postribosomal supernatant is prepared as described previously for the isolation of recombinant eIF5 from overproducing *E. coli* cells (Chaudhuri *et al.*, 1994). The postribosomal supernatant (170 mg protein in 16 ml total volume) is loaded onto a DEAE-cellulose column (60-ml bed volume) equilibrated in Buffer E + 50 mM KCl and the column is washed with Buffer E + 50 mM KCl. eIF1 is virtually unretarded under these conditions and appears in the initial wash along with the unretarded protein peak. Fractions containing the protein peak are pooled (26 mg protein), adjusted to 0.1 M KCl, and loaded onto a phosphocellulose column (9-ml bed volume) equilibrated in Buffer E + 100 mM KCl. The column is washed with the same buffer until A_{280} of the effluent is below 0.1. Bound eIF1 is then eluted from the column with a linear gradient (total volume 72 ml) in buffer E from 100 to 800 mM KCl. Fractions containing eIF1 (eluting at ~450 mM KCl) are pooled, dialyzed against Buffer E + 50 mM KCl for 2 h to reduce the ionic strength to that of buffer E + 100 mM KCl, and then applied to a FPLC-Mono Q (HR 5/5) column (1-ml bed volume). The column is washed with 5 ml of Buffer E + 100 mM KCl, and bound proteins are eluted by a gradient elution (12.5 ml total volume) from Buffer E + 100 mM KCl to Buffer E + 400 mM KCl. Fractions containing eIF1 (eluting at ~150 mM KCl) are pooled, dialyzed against buffer J containing 55% glycerol + 100 mM KCl, and stored at $-20°$. The yield is approximately 1 mg of homogeneous protein (Fig. 8.3). eIF1 is monitored at different purification steps by SDS-polyacrylamide gel electrophoresis followed by Coomassie blue staining (Fig. 8.4), as well as by immunoblot analysis by use of rabbit polyclonal anti-eIF1 antibodies.

4.8.3. Recombinant eIF1A protein

The open-reading frame of eIF1A cDNA (Fields and Adams, 1994) is cloned into the *Nde I/Eco*RI sites of pET–5a plasmid (Novagen) (Chaudhuri *et al.*, 1997b). This expression vector pET-5a–eIF1A is used to transform *E. coli* BL21 (DE3) cells (Novagen). Transformants are then grown at 37° in 1 L LB medium (Sambrook *et al.*, 1989) containing 50 μg/ml ampicillin to an A_{600} of approximately 0.8 and induced with 1 mM IPTG. The cells are harvested 2 h after induction by centrifugation, washed with 0.9% NaCl, quick-frozen in a dry ice/ethanol bath, and stored at $-70°$.

For purification of eIF1A, frozen *E. coli* cells (5 g) are suspended in 15 ml of Buffer K, treated with lysozyme (final concentration of 400 μg/ml), incubated for 20 min at 4°, and are disrupted by sonication. The postribosomal supernatant is prepared as described previously for the isolation of recombinant eIF5 from overproducing *E. coli* cells (Chaudhuri *et al.*, 1994). The postribosomal supernatant (75 mg protein in 10 ml total volume) is loaded onto a DEAE-cellulose column (25-ml bed volume) equilibrated in Buffer E + 100 mM KCl and the column is washed with Buffer E + 100 mM KCl to remove unadsorbed proteins. Bound proteins are eluted with Buffer E + 300 mM KCl. Fractions containing the protein peak are pooled (25 mg) and directly applied to a phosphocellulose column (5-ml bed volume) that is equilibrated with Buffer E + 300 mM KCl. The column is washed with Buffer E + 450 mM KCl until A_{280} is below 0.1. Bound eIF1A is then eluted from the column with Buffer E + 1 M KCl. The eluate is dialyzed against (1) Buffer E + 500 mM KCl for 2 h and (2) Buffer E + 100 mM KCl for 4 h to reduce the ionic strength of the protein solution to that of Buffer E + 100 mM KCl. The dialyzed protein solution is then applied to FPLC-Mono Q (HR 5/5) column (1-ml bed volume). The column is washed with 5 ml of the same buffer, and bound proteins are eluted with a linear gradient of 5 ml (total volume) from Buffer E + 100 mM KCl to Buffer E + 250 mM KCl, followed by another linear gradient of 25 ml (total volume) from Buffer E + 250 mM KCl to Buffer E + 600 mM KCl. Fractions (0.5 ml) containing eIF1 (eluting at approximately 340 mM KCl) are pooled, dialyzed against Buffer J containing 55% glycerol + 100 mM KCl, and stored at $-70°$. The yield is approximately 1 mg of homogeneous protein. eIF1A is monitored at different purification steps by SDS-polyacrylamide gel electrophoresis followed by Coomassie blue staining (Fig. 8.4), as well as by immunoblot analysis by use of rabbit polyclonal anti-eIF1A antibodies.

4.8.4. Recombinant eIF4B protein

The pET28A-eIF4B plasmid containing the mammalian eIF4B coding sequence fused to His$_6$-tag at its N terminus is transformed into *E. coli* BL21(DE3) cells. His$_6$-eIF4B is purified by use of Ni^{2+}-nitrilotriacetic acid

(NTA) agarose column as described previously (Pause *et al.*, 1994). Expression of His$_6$-eIF4B is induced by the addition of 0.5 m*M* IPTG to an exponentially growing (A$_{600}$ of approximately 0.8) 1 L of the bacterial culture (Sambrook *et al.*, 1989). Cells are harvested 2½ h after induction and suspended in Buffer L in the ratio of 3 ml buffer/gm of cell. Lysozyme (to a final concentration of 400 *μ*g/ml) is added, and the cells are disrupted by sonication. The cell lysate is clarified by centrifugation at 15,000 rpm for 20 min. The supernatant is treated with 30 *μ*g DNase I for 30 min. A mixture of protease inhibitors is added, and the solution is centrifuged for 2½ h at 45,000 rpm in a Beckman Ti-50.2 rotor. The supernatant is incubated with 1 ml Ni^{2+}-NTA-agarose beads, pre-equilibrated in Buffer L containing 30 m*M* potassium imidazole, pH 6.0, for 1 h at 4°. All unadsorbed proteins are removed from the beads by pouring the protein suspension into an empty 1-ml column and by subsequent washing the beads with 20 ml of Buffer L + 30 m*M* potassium imidazole. The bound His$_6$-eIF4B fusion protein is eluted by Buffer L + 300 m*M* potassium imidazole. The presence of His$_6$-eIF4B in the eluted fractions is monitored by Western blotting with a specific anti-His antibody. Peak fractions are pooled, dialyzed against Buffer E + 100 m*M* KCl for 4 h, and stored in small aliquots at −70° (Fig. 8.4).

5. Reconstitution of the 48S Ribosomal Complex

5.1. mRNA binding by the 43S preinitiation complex

Reactions are carried out in two stages. In stage 1, reaction mixtures (50 *μ*l) containing 20 m*M* Tris-HCl, pH 7.5, 100 m*M* KCl, 5 m*M* 2-mercaptoethanol, 4 *μ*g of nuclease-free bovine serum albumin, 400 *μ*M GTP, 1.2 *μ*g of purified rabbit reticulocyte eIF2, and 8 pmol [^{3}H] Met-tRNA$_i$ (10,000 cpm/pmol) are incubated at 37° for 4 min to promote the formation of the [^{3}H] Met-tRNA$_i$ · eIF2 · GTP ternary complex. Another set of reaction mixtures (50 *μ*l) containing Buffer A, 3 *μ*g of eIF3, 200 ng of eIF1A, 250 ng of eIF1, and 0.6 A$_{260}$ unit of 40S ribosomal subunits are incubated at 37° for 4 min and then supplemented with 50 *μ*l of the Stage 1 reaction mixture. Mg^{2+} is added to a final concentration of 1 m*M*, and the reaction mixtures are incubated at 37° for 4 min to form a 40S preinitiation complex and then chilled in an ice-water bath. In stage 2, reaction mixtures (30 *μ*l) containing Buffer A, 4 *μ*g of eIF4F (devoid of cap analog) and [^{32}P-labeled] pCON-PK mRNA (1 *μ*g) are incubated at 25° for 3 min, chilled on ice for 2 min, and then supplemented with 100 *μ*l of the stage 1 reaction. The Mg^{2+} concentration of the reaction mixtures (now 150 *μ*l) is adjusted to 1 m*M*. The reaction mixtures are incubated at 37° for 4 min to promote binding of the 43S preinitiation complex to the eIF4F-bound pCON-PK mRNA, layered onto a 5-ml 7.5 to 30 % (w/v) sucrose density gradient containing Buffer A and centrifuged at 48,000 rpm for 105 min in a SW 50.1 rotor. Fractions (200 to 300 μl) are collected from the bottom of each gradient, and the ^{3}H and ^{32}P radioactivity in each fraction is determined in a liquid scintillation spectrometer (Fig. 8.5A).

5.2. Primer extension assays

The binding of a labeled mRNA to 43S preinitiation complex can be conveniently measured by sucrose gradient analysis as described previously. However, this does not distinguish between a 43S preinitiation complex bound at the 5′ end of the mRNA and a complex positioned at the AUG start codon. To distinguish between these two possibilities, it is necessary to perform a toe-printing analysis (inhibition of primer extension) as was originally described by Pestova and associates (1996a,b, 1998) and Kozak (1997, 1998), and subsequently used in our laboratory (Majumdar and Maitra, 2005).

The oligonucleotide 5′-GGCATCGTAAAGAACATTTTGAG-3′ serves as the primer for the reverse transcription reaction and is labeled at the 5′ end with T4 polynucleotide kinase and [γ-^{32}P] ATP (6000 Ci/mmol, Amersham). The primer is complementary to nucleotides 180 to 192 of the pCON-PK wild type mRNA (Fig. 8.1). The ^{32}P-labeled primer ($\sim$4 pmol) is preannealed to pCON-PK mRNA (0.25 μg or $\sim$3 pmol) by heating for 1 min at 65°, followed by incubation at 37° for 8 min in a buffer containing 40 mM Tris-HCl, pH 7.5, and 0.2 mM EDTA. The primer–mRNA complexes are then incubated on ice for 15 min while the 40S complexes were assembled separately as follows.

A preformed ternary complex (eIF2 $\cdot$ Met-tRNA$_i$ $\cdot$ GTP) is incubated with 40S ribosomal subunits (1.2 A$_{260}$ units) in the presence of eIF1 (200 ng), eIF1A (250 ng), eIF3 (7 μg), and eIF4F (5 μg), devoid of cap analog, at 37° for 5 min in a buffer containing 20 mM Tris-HCl, pH 7.5, 100 mM KCl, 5 mM β-mercaptoethanol, 20 μg BSA, and 1 mM MgCl$_2$ to form the 43S preinitiation complex (40S $\cdot$ eIF2 $\cdot$ Met-tRNA$_i$ $\cdot$ GTP $\cdot$ eIF1 $\cdot$ eIF1A $\cdot$ eIF3). The preinitiation complex thus formed is incubated with the pCON-PK mRNA (preannealed to the radiolabeled primer) and 1 mM ATP, eIF4A (1 μg), and eIF4B (1 μg) at 37° for 4 min in the presence of 100 U of RNaseOUT (Invitrogen).

To determine the position of the 40S complex on the mRNA, reaction mixtures (125 μl each) are then incubated with deoxynucleotide triphosphates (dATP, dTTP, dCTP, dGTP) to a final concentration of 0.5 mM each in Buffer C. Primer extension reaction is initiated by adding 2 U/μl Superscript II reverse transcriptase (Invitrogen) and incubating at 25° for 15 min. The reactions are terminated by extracting with phenol-chloroform (1:1). cDNA products are precipitated by addition of an equal

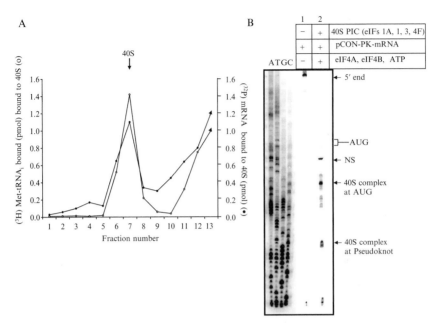

Figure 8.5 Formation of the 48S ribosomal complex. (A) Equimolar binding of mRNA and Met-tRNAi to 40S ribosomal subunits. A 43S preinitiation complex was formed as described in the legend for Fig. 8.4, except that [³H] Met-tRNA$_i$ was used. To determine mRNA binding by this 43S preinitiation complex, ³²P-labeled pCON-PK mRNA, preincubated with eIF4F, was added. After incubation at 37° for 4 min, reaction mixtures were analyzed by sucrose density gradients to determine the ³²P-radioactivity sedimenting in the 40S region. (B) Toe-printing analysis to determine the position of the 40S ribosomal complex on the mRNA. A preformed 43S preinitiation complex was incubated with a capped pCON-PK mRNA, preannealed to a ³²P-labeled primer, along with eIF4A, eIF4B, and ATP under standard conditions for primer extension assays as described in the text. In the complete system (lane 2), the different cDNA products obtained in the reaction are indicated by arrows (NS, nonspecific cDNA product). Reference lanes A, T, G, and C depict a dideoxynucleotide ladder obtained by primer extension with the pCON-PK mRNA in the presence of dideoxynucleotides. Lane 1 was similar to lane 2, except that no 40S ribosomal subunit, or initiation factor or ATP, was added. It should be noted that owing to the large size of the 40S complexes, the reverse transcriptase terminates 17 nt downstream of the point at which the ribosomal complex was arrested on the mRNA as was described originally by Pestova *et al.* (2000).

volume of isopropanol followed by overnight incubation at −80°. The cDNA products are then resuspended, mixed with formamide, heated at 90° for 1 min, and analyzed by electrophoresis through 8% polyacrylamide sequencing gels followed by autoradiography. The products are compared with a dideoxynucleotide ladder obtained by primer extension with the pCON-PK mRNA in the presence of dideoxynucleotides by use of avian myeloblastosis virus (AMV) reverse transcriptase (Promega) (Fig. 8.5B).

6. Remarks

6.1. eIF2-mediated ternary complex assay

Although initiation factors are routinely isolated from 0.5 M KCl-wash proteins of crude ribosomal pellets, we have observed that significant concentrations of many of these factors (e.g., eIF2 and eIF5) are also present in the postribosomal supernatant. In the case of eIF2, the initiation factor is present in reticulate lysates as an eIF2·GDP, as well as an eIF2·eIF2B complex in addition to free eIF2. We have observed that eIF2 isolated from ribosomal salt wash by the procedure described here contains a significant amount of eIF2·GDP. If the ternary complex assay is carried out in the presence of 1 mM Mg^{2+}, eIF2·GDP present cannot form a ternary complex under the conditions of the assay unless purified eIF2B is also present in the reaction. In contrast, in the absence of Mg^{2+} (under conditions of the assay as described here), the presence of excess GTP alone in the reaction mixture is able to displace GDP from the eIF2·GDP complex, resulting in the formation of an active eIF2·GTP complex. For this reason, during eIF2-mediated ternary conplex formation, Mg^{2+} is not included in the assay.

6.2. Assays for initiation factors

Although assays for individual initiation factors have been described here, clearly it is also possible to follow the elution of the factors during purification by use of specific antibodies against each of these initiation factors. However, such a procedure is more time consuming than the specific assays described here. More importantly, purification of the proteins on the basis of immunoreactivity does not reveal whether the factors are active.

6.3. Binding of mRNA to the 43S preinitiation complex

The procedure described here uses 5′-capped message in general and does not address the issue of translation from mRNA that uses internal ribosome entry sites. It should be noted that when a ternary complex binds to a 40S ribosomal subunit in the presence of an AUG codon at 5 mM Mg^{2+} and in the absence of all other initiation factors, the resulting 40S comlpex (40S·AUG·eIF2·GTP·Met-tRNA$_i$), in this article, is termed the 40S initiation complex. In contrast, the 40S complex formed in the presence of eIF1A, eIF1 and eIF3 (40S·eIF2·GTP·Met-tRNA$_i$·eIF1A·eIF1·eIF3) is termed a 43S preinitiation complex. The 43S preinitiation complex, positioned at the AUG start codon of an mRNA, is termed a 48S initiation complex (40S·eIF2·GTP·Met-tRNA$_i$·eIF3·mRNA).

6.4. eIF5B-/mediated 80S initiation complex formation

We have described in detail the factor requirements for the efficient formation of the 48S initiation complex. However, the subsequent joining of the 60S ribosomes to the 43S initiation complex, leading to the formation of an 80S initiation complex, has not been addressed in this chapter. Presumably, such an 80S initiation complex can be efficiently formed by the addition of purified eIF5B and 60S ribosomal subunits as described by Pestova and associates (2000).

ACKNOWLEDGMENTS

Research in the author's laboratory was supported by Grant GM 15399 from the National Institutes of Health and by Cancer Core Support Grant P30CA13330 from the National Cancer Institute.

REFERENCES

Benne, R., Brown-Luedi, M. L., and Hershey, J. W. (1979). Protein synthesis initiation factors from rabbit reticulocytes: Purification, characterization, and radiochemical labeling. *Methods Enzymol.* **60**, 15–35.

Bose, K. K., Chatterjee, N. K., and Gupta, N. K. (1974). Fractionation of rabbit liver methionyl-tRNA species. *Methods Enzymol.* **29**, 522–529.

Chakravarti, D., Maiti, T., and Maitra, U. (1993). Isolation and immunochemical characterization of eukaryotic translation initiation factor 5 from *Saccharomyces cerevisiae. J. Biol. Chem.* **268**, 5754–5762.

Chaudhuri, J., Chakrabarti, A., and Maitra, U. (1997a). Biochemical characterization of mammalian translation initiation factor 3 (eIF3). Molecular cloning reveals that p110 subunit is the mammalian homologue of *Saccharomyces cerevisiae* protein Prt1. *J. Biol. Chem.* **272**, 30975–30983.

Chaudhuri, J., Chowdhury, D., and Maitra, U. (1999). Distinct functions of eukaryotic translation initiation factors eIF1A and eIF3 in the formation of the 40S ribosomal preinitiation complex. *J. Biol. Chem.* **274**, 17975–17980.

Chaudhuri, J., Das, K., and Maitra, U. (1994). Purification and characterization of bacterially expressed mammalian translation initiation factor 5 (eIF-5): Demonstration that eIF-5 forms a specific complex with eIF-2. *Biochemistry* **33**, 4794–4799.

Chaudhuri, J., Si, K., and Maitra, U. (1997b). Function of eukaryotic translation initiation factor 1A (eIF1A) (formerly called eIF-4C) in initiation of protein synthesis. *J. Biol. Chem.* **272**, 7883–7891.

Chevesich, J., Chaudhuri, J., and Maitra, U. (1993). Characterization of mammalian translation initiation factor 5 (eIF-5). Demonstration that eIF-5 is a phosphoprotein and is present in cells as a single molecular form of apparent M(r) 58,000. *J. Biol. Chem.* **268**, 20659–20667.

Conway, T. W., and Lipmann, F. (1964). Characterization of a ribosome-linked guanosine triphosphatase in *Escherichia coli* extracts. *Proc. Natl. Acad. Sci. USA* **52**, 1462–1469.

Deutscher, M. P. (1974). Aminoacyl-tRNA synthetase complex from rat liver. *Methods Enzymol.* **29**, 577–583.

Dever, T. E. (2002). Gene-specific regulation by general translation factors. *Cell* **108**, 545–556.

Dholakia, J. N., and Wahba, A. J. (1987). The isolation and characterization from rabbit reticulocytes of two forms of eukaryotic initiation factor 2 having different beta-polypeptides. *J. Biol. Chem.* **262,** 10164–10170.

Edery, I., Humbelin, M., Darveau, A., Lee, K. A., Milburn, S., Hershey, J. W., Trachsel, H., and Sonenberg, N. (1983). Involvement of eukaryotic initiation factor 4A in the cap recognition process. *J. Biol. Chem.* **258,** 11398–11403.

Fields, C., and Adams, M. D. (1994). Expressed sequence tags identify a human isolog of the suil translation initiation factor. *Biochem. Biophys. Res. Commun.* **198,** 288–291.

Hinnebusch, A. G. (2006). eIF3: A versatile scaffold for translation initiation complexes. *Trends Biochem. Sci.* **31,** 553–562.

Kapp, L. D., and Lorsch, J. R. (2004). The molecular mechanics of eukaryotic translation. *Annu. Rev. Biochem.* **73,** 657–704.

Kozak, M. (1997). Recognition of AUG and alternative initiator codons is augmented by G in position +4 but is not generally affected by the nucleotides in positions +5 and +6. *EMBO J.* **16,** 2482–2492.

Kozak, M. (1998). Primer extension analysis of eukaryotic ribosome-mRNA complexes. *Nucleic Acids Res.* **26,** 4853–4859.

Majumdar, R., Bandyopadhyay, A., and Maitra, U. (2003). Mammalian translation initiation factor eIF1 functions with eIF1A and eIF3 in the formation of a stable 40S preinitiation complex. *J. Biol. Chem.* **278,** 6580–6587.

Majumdar, R., and Maitra, U. (2005). Regulation of GTP hydrolysis before ribosomal AUG selection during eukaryotic translation initiation. *EMBO J.* **24,** 3737–3746.

Merrick, W. C. (1979). Assays for eukaryotic protein synthesis. *Methods Enzymol.* **60,** 108–123.

Pause, A., Methot, N., Svitkin, Y., Merrick, W. C., and Sonenberg, N. (1994). Dominant negative mutants of mammalian translation initiation factor eIF-4A define a critical role for eIF-4F in cap-dependent and cap-independent initiation of translation. *EMBO J.* **13,** 1205–1215.

Pestova, T. V., Borukhov, S. I., and Hellen, C. U. (1998). Eukaryotic ribosomes require initiation factors 1 and 1A to locate initiation codons. *Nature* **394,** 854–859.

Pestova, T. V., Hellen, C. U., and Shatsky, I. N. (1996a). Canonical eukaryotic initiation factors determine initiation of translation by internal ribosomal entry. *Mol. Cell. Biol.* **16,** 6859–6869.

Pestova, T. V., Lomakin, I. B., Lee, J. H., Choi, S. K., Dever, T. E., and Hellen, C. U. (2000). The joining of ribosomal subunits in eukaryotes requires eIF5B. *Nature* **403,** 332–335.

Pestova, T. V., Shatsky, I. N., and Hellen, C. U. (1996b). Functional dissection of eukaryotic initiation factor 4F: The 4A subunit and the central domain of the 4G subunit are sufficient to mediate internal entry of 43S preinitiation complexes. *Mol. Cell. Biol.* **16,** 6870–6878.

Raychaudhuri, P., Chaudhuri, A., and Maitra, U. (1985). Eukaryotic initiation factor 5 from calf liver is a single polypeptide chain protein of Mr = 62,000. *J. Biol. Chem.* **260,** 2132–2139.

Raychaudhuri, P., and Maitra, U. (1986). Identification of ribosome-bound eukaryotic initiation factor 2. GDP binary complex as an intermediate in polypeptide chain initiation reaction. *J. Biol. Chem.* **261,** 7723–7728.

Sambrook, J., Fritsch, E. F., and Maniatis, T. (1989). "Molecular Cloning: A Laboratory Manual." Cold Spring Harbor Laboratory, Cold Spring Harbor, NY.

Staehelin, T., Erni, B., and Schreier, M. H. (1979). Purification and characterization of seven initiation factors for mammalian protein synthesis. *Methods Enzymol.* **60,** 136–165.

Stringer, E. A., Chaudhuri, A., and Maitra, U. (1979). Purified eukaryotic initiation factor 2 from calf liver consists of two polypeptide chains of 48,000 and 38,000 daltons. *J. Biol. Chem.* **254,** 6845–6848.

BIOPHYSICAL APPROACH TO STUDIES OF CAP–EIF4E INTERACTION BY SYNTHETIC CAP ANALOGS

Anna Niedzwiecka,*,† Janusz Stepinski,* Jan M. Antosiewicz,* Edward Darzynkiewicz,* *and* Ryszard Stolarski*

Contents

* Division of Biophysics, Institute of Experimental Physics, Warsaw University, Warszawa, Poland
† Biological Physics Group, Institute of Physics, Polish Academy of Sciences, Warszawa, Poland

Methods in Enzymology, Volume 430
ISSN 0076-6879, DOI: 10.1016/S0076-6879(07)30009-8

Abstract

Specific recognition of mRNA $5'$ cap by eukaryotic initiation factor eIF4E is a rate-limiting step in the translation initiation. Structural determination of the eIF4E–cap complexes, as well as complexes of eIF4E with other proteins regulating its activity, requires complementary experiments that allow for energetic and dynamic aspects of formation and stability of the complexes. Such a combined approach provides information on the binding mechanisms and, hence, may lead to mechanistic models of eIF4E functioning and regulation on the molecular level. This chapter summarizes in detail the method of experiments used to probe the cap-binding center of eIF4E, steady state and stopped-flow fluorescence, and microcalorimetry. The studies were performed with a wide class of synthetic, structurally modified cap analogs that resembles in some respect an application of site directed mutagenesis of the protein. The chapter presents a general recipe as to how to investigate protein–ligand interactions if the protein has no enzymatic activity and both the protein and the ligand absorb and emit UV/VIS radiation in the same spectral ranges.

1. INTRODUCTION

Cap-dependent translation of messenger RNA in eukaryotes begins with specific recognition of a cap structure at the mRNA $5'$ terminus by a highly conserved, tryptophan-rich, and the least abundant eukaryotic initiation factor, eIF4E (for review see Gingras *et al.*, 1999). The cap, or monomethylguanosine cap (MMG-cap, m⁷GpppN) consists of 7-methylguanosine (m⁷G) linked by a $5'$-to-$5'$ triphosphate bridge to the first transcribed nucleoside, N. In *nematodes* and chordate species, acquisition of an alternative, trimethylguanosine cap (TMG-cap, $m_3^{2,2,7}$GpppN) by mRNAs occurs in the *trans*-splicing process (Blumenthal, 1998). Binding of eIF4E to the cap is a rate-limiting step for translation initiation (Raugh *et al.*, 2000). During the formation of the 48S initiation complex, eIF4E is directly bound to eukaryotic initiation factor 4G (eIF4G), a scaffold for other factors involved in mRNA recruitment (Morley *et al.*, 1997). The eIF4E–eIF4G interface is a target for translation control by small, inhibitory 4E-binding proteins (4E-BPs) that share a common recognition motif with eIF4G (Mader *et al.*, 1995). Several other proteins were found to regulate eIF4E activity by competing with eIF4G, Cup, Bicoid, CPEB, and homeodomain proteins (Richter and Sonenberg, 2005; Topisirovic and Borden, 2005). The role of a variety of eIF4E isoforms (Hernandez and Vazquez-Pianzola, 2005) in both the cytoplasm and the nucleus makes eIF4E a focus of considerable biochemical and biophysical studies. This is one of the central regulators of cell growth, proliferation, and survival, and its overexpression leads to malignant transformation and tumorigenesis (Watkins and Norbury, 2002).

Structures of several binary complexes composed of eIF4E and a synthetic cap analog were resolved by X-ray diffraction studies in crystal (Marcotrigiano et al., 1997; Niedzwiecka et al., 2002a; Tomoo et al., 2003) and multidimensional NMR in solution (Matsuo et al., 1997). Structural changes of apo-eIF4E that accompany formation of the eIF4E-cap complex were characterized by resolving the cap-free structure of the protein in solution (Volpon et al., 2006). Among the contacts stabilizing the complexes, sandwich cation-π stacking of the 7-methylguanine moiety in between two tryptophan rings is the most important for specific recognition of the cap by eIF4E.

Investigations of the cap–eIF4E interactions by emission spectroscopy and isothermal titration calorimetry (ITC) complement the structural data by providing information on the stability of the complexes and dynamics of their formation. The cap affinity for eIF4E can be expressed in terms of the equilibrium association constant, K_{as}, and the corresponding standard Gibbs free-energy change, $\Delta G°$. Kinetic constants of the association process can be obtained by stopped-flow techniques. Analysis of these parameters allowed formulation of a two-step molecular mechanism of the eIF4E-cap association (Niedzwiecka et al., 2002a). Anchoring of the cap phosphate groups to the basic amino acids is followed by cooperative formation of the contacts stabilizing m⁷G inside the cap-binding pocket (i.e., sandwich cation-π stacking and hydrogen bonding of Watson–Crick type). The binding is accompanied by partial protonation of the cap and extensive hydration of the complex.

Thermodynamic parameters of binding (i.e., changes of standard enthalpy, $\Delta H°$, entropy $\Delta S°$, and molar heat capacity under constant pressure $\Delta C_p°$) are related to submolecular origins of the specific interactions. Nontrivial isothermal enthalpy–entropy compensation found for a large series of chemically different cap analogs was the first experimental evidence of structural instability and great energetic fluctuations of apo-eIF4E unless bound to the cap (Niedzwiecka et al., 2004).

Our approach exploited a wide class of structurally modified cap analogs in probing requirements of the binding center of eIF4E. In this regard, it may be treated as analogous to site-directed protein mutagenesis. In this chapter, we describe in detail a route, from syntheses of chemical cap analogs through running the steady-state fluorescence, ITC, and stopped-flow fluorescence experiments, up to numerical analysis of the acquired data. Potential methodological pitfalls and errors resulting from oversimplification of the data treatment are analyzed and suitable corrections described. Much attention is given to eIF4E activity during the measurements, which seems a nontrivial problem in case of nonenzymatic proteins.

2. SYNTHESES AND PURIFICATION OF CAP ANALOGS

Simple 7-substituted derivatives of guanosine, guanosine 5′-mono-, di-, and triphosphates (mononucleotide cap analogs) can be prepared by alkylation of guanine ring at N7 in mild conditions with appropriate alkyl iodide or bromide (Scheme 9.1) (Darzynkiewicz *et al.*, 1985; Jankowska *et al.*, 1993). Among the dinucleotide cap analogs, P^1-guanosine-5′ P^3-(7-methylguanosine-5′) triphosphate (m^7GpppG) is a standard compound. It can be modified in one or both bases, in the ribose ring(s), and/or in the triphosphate bridge.

Generally, it is possible to prepare dinucleoside 5′,5′-polyphosphates by means of condensation of a given nucleoside mono-, di-, or triphosphates with a second nucleoside 5′-polyphosphate activated at the terminal phosphate with a good leaving group. Imidazole derivatives, imidazolidephosphates (phosphorimidazolidates), are the most widely used (Cramer *et al.*, 1961; Hoard and Ott, 1965; Kadokura *et al.*, 1997; Sawai *et al.*, 1991, 1992; Stepinski *et al.*, 1995, 2001). Other leaving groups of choice include phenylthio- (Nakagawa *et al.*, 1980), 4-methoxyphenylthio- (Kohno *et al.*, 1985), 4-chlorophenylthio- (Fukuoka *et al.*, 1994), morpholine- (Adam and Moffat, 1966), or 5-chloro-8-quinolyl- (Fukuoka *et al.*, 1994) nucleoside 5′-phosphates derivatives. The new pyrophosphate bond is formed by means of a nucleophilic attack of the terminal phosphoryl hydroxyl group on the activated phosphate. Usually, the coupling reactions proceed in anhydrous solutions, dimethylformamide (DMF), pyridine, 1-methyl-2-pyrrolidinone, or dimethylsulfoxide (DMSO) with an appropriate catalyst. However, in the case of imidazolides as the substrates, the coupling can be performed in water buffer at pH 7 (Sawai *et al.*, 1991, 1992).

Chemical syntheses of m^7GpppG and m^7GpppA on a small scale were first accomplished with di-*n*-butylphosphinothioyl- (Hata *et al.*, 1976) or phenylthio- (Nakagawa *et al.*, 1980) activating groups. A new, two-step synthesis of m^7GpppG on a larger scale proceeded by formation of GpppG,

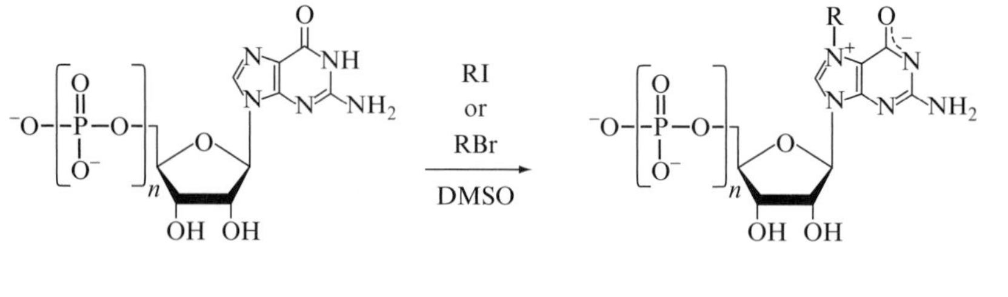

Scheme 9.1 Preparation of mononucleotide 5′ mRNA cap analogs modified at N7 of the guanine ring; $n = 1, 2,$ or 3 for mono-, di-, and triphosphates, respectively.

and its kinetically controlled methylation at N7 (Fukuoka *et al.*, 1994; Stepinski *et al.*, 1995). This approach is superior to alternative methods because of efficient isolation of GpppG and m^{7}GpppG from the reaction mixtures (Scheme 9.2). The improved procedure, which is described later in detail, exploits the use of $ZnCl_2$ as promoter during formation of the pyrophosphate bond (Kadokura *et al.*, 1997; Stepinski *et al.*, 2001, 2002). A similar synthetic procedure can be used for the cap analogs substituted at N7 by a residue other than methyl (e.g., benzyl or ethyl group).

2.1. Synthesis of P^1-guanosine-5$'$ P^3- (7-methylguanosine-5$'$) triphosphate (m^{7}GpppG)

In the first step, synthesis of P^1,P^3-bisguanosine-5$'$ triphosphate (GpppG) was carried out as follows (Scheme 9.2A). Guanosine 5$'$-monophosphate (from Sigma) converted to triethylammonium (TEA) salt (0.46 g, 1 mmol), imidazole (0.34 g, 5 mmol), and 2,2$'$-dithiodipyridine (0.44 g, 2 mmol, Aldrich) were mixed in anhydrous DMF (5 ml) and TEA (140 μl). Triphenylphosphine (0.52 g, 2 mmol) was added, and the mixture was stirred overnight at room temperature. One hundred milliliters of anhydrous acetone solution of sodium perchlorate (0.49 g, anhydrous) was added to the reaction mixture with an intensive stirring and cooled down in a refrigerator for 2 h. The precipitate was filtered off and washed (being deeply grounded each time) three times with new portions of anhydrous acetone. After drying over P_4O_{10} in a vacuum desiccator, the GMP imidazolide was dissolved in DMF (10 ml), and GDP (Sigma, converted to the TEA salt, 0.58 g, 0.9 mmol) was added. After addition of $ZnCl_2$ (0.8 g), the mixture was stirred at room temperature overnight, poured into a beaker containing a solution of EDTA (2 g) in water (250 ml), and neutralized with 1 M NaHCO$_3$. GpppG was isolated by DEAE-Sephadex ion-exchange column chromatography by use of a linear gradient of TEAB (triethylammonium bicarbonate buffer, pH 7.4), 0 to 1.4 M. Purity of the fractions of the main peak was monitored by reverse-phase HPLC. The fractions were pooled and evaporated to dryness on a rotary evaporator under diminished pressure ($\leq$20 mmHg), with addition of ethanol for TEAB decomposition, to give 0.73 g GpppG (TEA salt, yield: 82%).

GpppG (TEA salt, 495 mg, 0.5 mmol) was mixed with 6 ml of dimethylsulfoxide and 0.1 ml of methyl iodide at room temperature (Scheme 9.2B). Progress of the methylation was monitored by reverse-phase HPLC. When the peak intensity of m^{7}GpppG was equal to that of GpppG (after 1.5 h at ~25°), the reaction was quenched by adding 60 ml of cold water. The water solution was extracted three times with 10-ml portions of diethyl ether. Chromatography of the aqueous phase on DEAE-Sephadex by use of a linear gradient of TEAB, 0 to 1.2 M, and evaporation of the pooled fractions to dryness gave m^{7}GpppG (TEA salt).

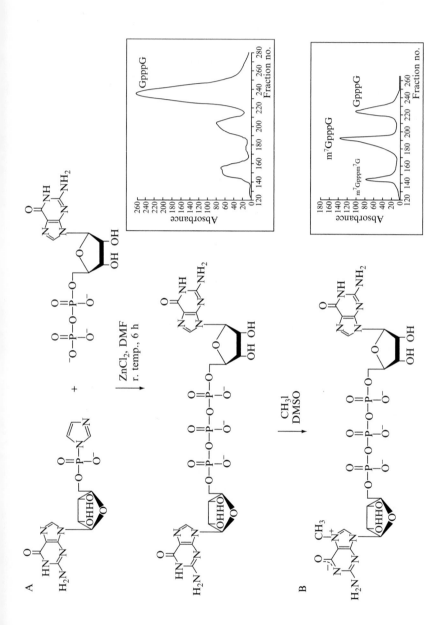

Scheme 9.2 Two-stage synthesis of m⁷GpppG: (A) coupling reaction to GpppG; (B) methylation to the final product. Inserts show clear separations of GpppG and m⁷GpppG from the reaction mixtures by ion-exchange chromatography (absorption at $\lambda = 260$ nm, arbitrary units). Conditions of the elution: DEAE-Sephadex A-25 column in HCO_3^- form, diameter, 3.6 cm; height, 90 cm; linear gradient of trietyloammonium bicarbonate as the mobile phase.

During the evaporation, the temperature of a water bath was kept below 30°, and ethanol was added to facilitate the TEAB decomposition. The final product was converted to Na^+ salt by ion exchange on a small column of Dowex 50Wx8 (Na^+ form), evaporated to a small volume, precipitated with ethanol, centrifuged, and dried in a desiccator over P_4O_{10} to give amorphous white powder (yield: 114 mg, 27%).

^{1}H NMR (D_2O), δ in ppm vs. internal TSP: 8.00 (1H, s, H8-G), 5.90 (1H, d, H1'-m^{7}G; $J_{1',2'} = 3.4$ Hz), 5.80 (1H, d, H1'-G, $J_{1',2'} = 6.3$ Hz), 4.67 (1H, t, H2'-G, $J_{1'-2'} = 6.3$ Hz, $J_{2',3'} = 5.1$ Hz), 4.54 (1H, dd, H2'- m^{7}G; $J_{1',2'} = 3.4$ Hz, $J_{2',3'} = 4.8$ Hz), 4.47 (1H, dd, H3'-G, $J_{2',3'} = 5.1$ Hz, $J_{3',4'} = 3.3$ Hz), 4.42 (1H, dd, H3'-m^{7}G; $J_{2'-3'} = 4.8$ Hz, $J_{3',4'} = 5.7$ Hz), 4.39 (1H, m, H4'-m^{7}G; $J_{4',5'} = 2.5$ Hz, $J_{4',5''} = 2.1$ Hz), 4.36 (1H, m, H5'-m^{7}G; $J_{4'',5'} = 2.5$, $J_{5',5''} = 11.5$ Hz, $J_{5',P} = 4.2$ Hz), 4.34 (1H, m, H4'-G; $J_{3',4'} = 3.3$, $J_{4',5'} = 3.6$ Hz, $J_{4',5''} = 4.5$ Hz), 4.27 (1H, m, H5''-m^{7}G; $J_{4',5''} = 2.1$ Hz, $J_{5',5''} = 11.5$ Hz, $J_{5,''P} = 5.7$ Hz), 4.26 (1H, m, H5'-G; $J_{4',5'} = 3.6$ Hz, $J_{5',5''} = 11.5$ Hz, $J_{5',P} = 5.5$ Hz), 4.24 (1H, m, H5''-G; $J_{4',5''} = 4.5$ Hz, $J_{5',5''} = 11.5$ Hz, $J_{5,''P} = 7.0$ Hz), 4.04 (3H, s, CH$_3$-m^{7}G); ^{31}P NMR (D_2O), δ in ppm vs. external H_3PO_4: -12.2 (2P, m, α and γ), -22.8 (1P, t, β, $J_{P,P} = 19.5$ Hz).

Modifications of the bases or of the ribose ring(s) require more laborious synthetic procedures. They should be introduced at the nucleoside (base) level, and the intermediates can be successively converted to suitable nucleotides and dinucleotides; for a representative example of m$_3^{2,2,7}$GpppG synthesis, see later (Scheme 9.3). First syntheses of this TMG-cap analog were accomplished by use of phenylthio groups as the activating agents (Darzynkiewicz et al., 1988, 1990; Iwase et al., 1989). An alternative procedure (Stepinski et al., 1995) exploited a coupling of N^2,N^2,7-trimethylguanosine 5'-monophosphate imidazolide with GDP in aqueous HCl/N-ethylmorpholine buffer, pH 7, in the presence of Mg(II) ions. Our improved procedure for the synthesis of m$_3^{2,2,7}$GpppG is described in the following section. It exploits catalytical properties of anhydrous zinc chloride in DMF, which seemed to be a more effective promoter of the pyrophosphate bond formation than manganese salt in aqueous conditions (Stepinski et al., 2001).

2.2. Synthesis of P^1-guanosine-5' P^3-(N^2,N^2,7-trimethyguanosine-5') triphosphate (m$_3^{2,2,7}$GpppG)

N^2,N^2-Dimethyguanosine (from Biolog, 311 mg, 1.0 mmol) was stirred overnight with trimethylphosphate (10 ml) and phosphorus oxychloride (0.370 ml) at 6°. Addition of 100 ml water and neutralization with 1 M TEAB quenched the reaction. DEAE-Sephadex chromatography with a linear gradient of TEAB, 0 to 0.7 M afforded 444 mg N^2, N^2-dimethyguanosine 5'-monophosphate (TEA salt), yield 90%.

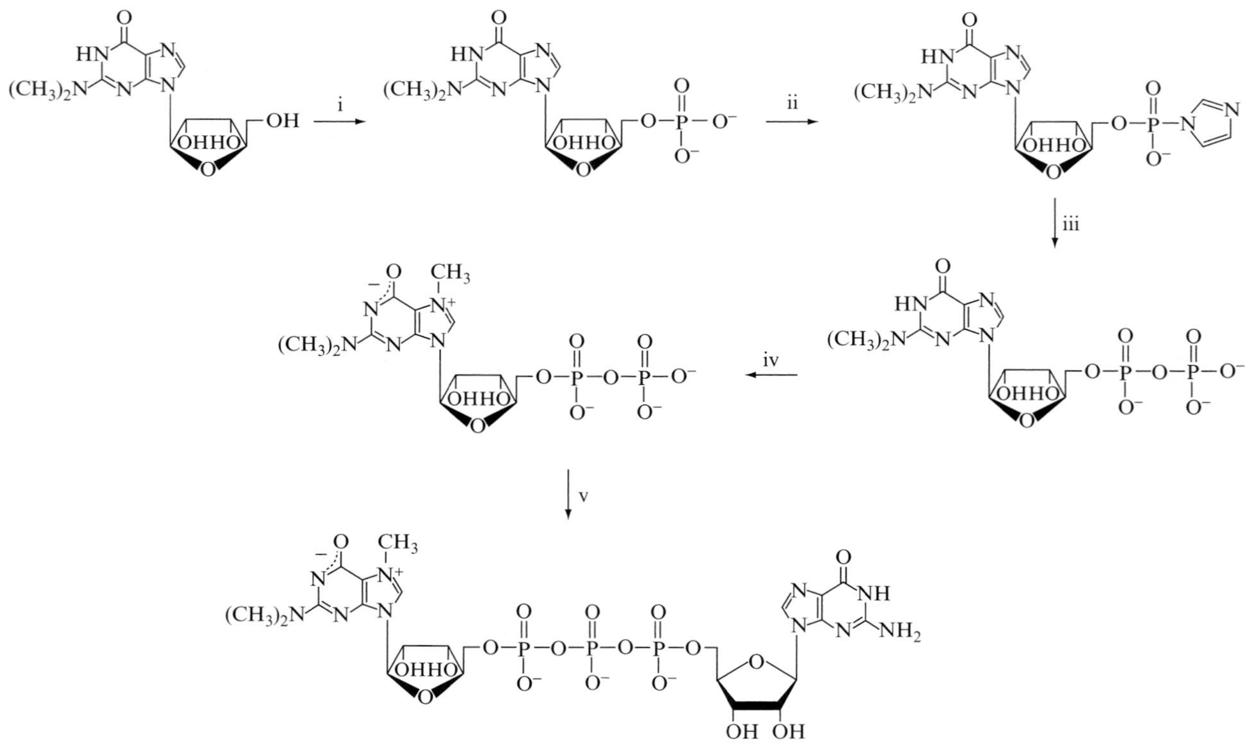

Scheme 9.3 Synthesis of m$_3^{2,2,7}$GpppG: (i), POCl$_3$, (CH$_3$O)$_3$PO; (ii), imidazole, 2,2′-dithiodipyridine, triphenylphosphine, DMF; (iii), ZnCl$_2$, triethylammonium dihydrogenphosphate, DMF; (iv), CH$_3$I, DMSO; (v), ZnCl$_2$, guanosine 5′-(P-imidazolido) monophosphate (triethylammonium salt), DMF.

N^2,N^2-Dimethyguanosine 5′-monophosphate (443 mg, TEA salt, 0.9 mmol), imidazole (306 mg, 4.5 mmol), and 2,2′-dithiodipyridine (Aldrich, 396 mg, 1.8 mmol) were mixed in anhydrous DMF (10 ml) and TEA (126 μl). Triphenylphosphine (472 mg, 1.8 mmol) was added, and the mixture was stirred for 5 h at room temperature. The mixture was placed in a centrifuge tube, and sodium perchlorate (0.45 g, anhydrous), dissolved in acetone (60 ml), was added. After cooling for 2 h in a refrigerator, the mixture was centrifuged, and the supernatant was discarded. The precipitate was ground with a new portion of acetone, cooled, and centrifuged again. The process was repeated, and the precipitate was dried in a vacuum desiccator over P_4O_{10}. The imidazolide was dissolved in 20 ml of DMF, and 2 g of tris(triethylammonium)phosphate was added. The latter was prepared from TEA and phosphoric acid and dried over P_4O_{10} in a desiccator to semicrystalline mass. Finally, 0.8 g of $ZnCl_2$ was added, and the reaction mixture was stirred at room temperature for 6.5 h, poured into a beaker containing a solution of 2.5 g EDTA in 150 ml water, and neutralized with 1 M NaHCO$_3$. Chromatographic isolation on DEAE-Sephadex with a linear gradient of TEAB, 0 to 1 M, yielded 451 mg (67%) N^2, N^2-dimethyguanosine 5′-diphosphate (TEA salt).

N^2,N^2-Dimethyguanosine 5′-diphosphate (TEA salt, 337 mg, 0.5 mmol) was mixed with 5 ml of dimethylsulfoxide and 0.5 ml of methyl iodide at room temperature. After 5 h, the reaction mixture was treated with 80 ml of cold water and extracted three times with 10-ml portions of diethyl ether. Chromatography of the aqueous phase, after neutralization with NaHCO$_3$, on DEAE-Sephadex with a linear gradient of TEAB, 0 to 0.8 M, yielded 193 mg (56%) $N^2,N,7^2$-trimethyguanosine 5′-diphosphate (TEA salt).

Guanosine 5′-monophosphate (from Sigma) converted into TEA salt (345 mg, 0.75 mmol), imidazole (255 mg, 3.75 mmol), and 2,2′-dithiodipyridine (330 mg, 1.5 mmol, from Aldrich) were mixed in anhydrous DMF (5 ml) and TEA (105 μl). Triphenylphosphine (0.4 g, 1.5 mmol) was added, and the mixture was stirred for 5 h at room temperature. Next, 100 ml anhydrous acetone solution of sodium perchlorate (375 mg, anhydrous) was added with an intensive stirring, and the reaction mixture was cooled down for 2 h in a refrigerator. The precipitate was filtered off and washed three times with new portions of anhydrous acetone being deeply grounded each time. After drying over P_4O_{10} in a vacuum desiccator, the imidazolide of GMP was dissolved in DMF (10 ml), and N^2,N^2,7-trimethyguanosine 5′-diphosphate (TEA salt, 172 mg, 0.25 mmol) was added. Next, $ZnCl_2$ (600 mg) was added, and the mixture was stirred at room temperature overnight, poured into a beaker containing a solution of 2 g of EDTA in 250 ml of water, and neutralized with 1 M NaHCO$_3$. Chromatographic isolation on DEAE-Sephadex with a linear gradient of TEAB 0 to 1 M yielded 198 mg (77%) $m_3^{2,2,7}GpppG$ (TEA salt). The final product was converted to Na$^+$ salt on a small Dowex 50Wx8 column, evaporated to a small volume, precipitated with ethanol, centrifuged, and dried in a desiccator over P_4O_{10} to amorphous white powder.

^{1}H NMR (D_2O, ppm vs. internal TSP): δ 7.94 (1H, s, H8-G), 5.91 (1H, d, H1'- $m_3^{2,2,7}$G; $J_{1',2'} = 3.4$ Hz), 5.73 (1H, d, H1'-G, $J_{1',2'} = 5.9$ Hz), 4.55 (1H, t, H2'-G, $J_{1',2'} = 5.9$ Hz, $J_{2',3'} = 5.1$ Hz), 4.54 (1H, dd, H2'-$m_3^{2,2,7}$G; $J_{1'-2'} = 3.4$ Hz, $J_{2',3'} = 4.8$ Hz), 4.42 (1H, dd, H3'-G, $J_{2',3'} = 5.1$ Hz, $J_{3',4'} = 3.7$ Hz), 4.40 (1H, dd, H3'- $m_3^{2,2,7}$G; $J_{2'-3'} = 4.8$ Hz, $J_{3',4'} = 5.8$ Hz), 4.39 (1H, m, H5'- $m_3^{2,2,7}$G; $J_{4',5'} = 2.5$ Hz, $J_{5',5'} = 11.5$ Hz, $J_{5',P} = 4.0$ Hz), 4.36 (1H, m, H4'- $m_3^{2,2,7}$G; $J_{3',4'} = 5.8$, $J_{4',5'} = 2.5$ Hz, $J_{4',5'} = 2.3$ Hz), 4.33 (1H, m, H4'-G; $J_{3',4'} = 3.7$, $J_{4',5'} = 4.1$ Hz, $J_{4',5''} = 3.9$ Hz), 4.27 (1H, m, H5''- $m_3^{2,2,7}$G; $J_{4',5''} = 2.3$ Hz, $J_{5',5''} = 11.5$ Hz, $J_{5,''P} = 5.8$ Hz), 4.26 (1H, m, H5'-G; $J_{4',5'} = 4.1$ Hz, $J_{5',5''} = 11.5$ Hz, $J_{5',P} = 5.8$ Hz), 4.23 (1H, m, H5''- G; $J_{4',5''} = 3.9$ Hz, $J_{5',5''} = 11.5$ Hz, $J_{5,''P} = 7.0$ Hz), 4.04 (3H, s, N^{7}CH$_3$- $m_3^{2,2,7}$G), 3.14 (6H, s, N^{2}CH$_3$- $m_3^{2,2,7}$G). ^{31}P NMR analogous to that of m^{7}GpppG.

Similarly, other dinucleotide 5',5'-di-, tri-, tetra-, penta-, and hexa-phosphate modified in the base(s) and/or in the ribose(s) can be synthesized (Jemiality *et al.*, 2003; Stepinski *et al.*, 2001). The synthesis of "anti-reversed" cap analogs, ARCAs, which are incorporated into RNA transcripts only in the right orientation (Kadokura *et al.*, 1997), is described in the parallel chapter (Grudzien-Nogalska *et al.*, 2007).

Modifications of the triphosphate bridge (other than the length) may include replacement of oxygen(s) of the pyrophosphate bonds by other atom or group (e.g., by methylene group: direct phosphorylation of a first nucleoside with methylenebis (phosphonic dichloride) and coupling of the product with a second nucleotide) (Kalek *et al.*, 2005, 2006). The synthetic procedures are thoroughly described in the parallel chapter (Grudzien-Nogalska *et al.*, 2007).

In general, the final products and intermediates of the syntheses can be safely stored at -20 to $-80°$ for months or even years. The structures and purity should be checked routinely by HPLC, mass spectrometry, ^{1}H and ^{31}P NMR, UV absorption, and fluorescence.

3. Binding of the Cap Analogs to eIF4E by Fluorescence Titration Experiments

3.1. Thermodynamic and apparent association constant

A thermodynamic equilibrium constant, K_t, as a function of temperature (T) and pressure only, is expressed in terms of equilibrium molar activities, a_i, of all species participating in the reaction:

$$K_t = \prod_{i=1}^{\alpha} a_i^{v_i}, \tag{1}$$

where v_i are stoichiometric coefficients of the reaction. A logarithmic form of this relationship makes it possible to analyze the dependence of K_t on the presence of individual participants of the reaction (i.e., macromolecules, ligands, cations, anions, protons, and water):

$$\log(K_t) = \sum_{i=1}^{\alpha} v_i \cdot \log(a_i) \tag{2}$$

An experimentally observed equilibrium association constant (K_{as}) is defined in terms of concentrations of selected reactants (e.g., apo-protein [P_{act}] and ligand [L]) and products (e.g., protein–ligand 1:1 complex [cx]):

$$K_{as} = \frac{[cx]}{[P_{act}]_0 \cdot [L]_0} \tag{3}$$

Index "0" indicates equilibrium concentrations of unbound reactants. This apparent constant involves hidden information on both the true thermodynamic association constant, K_t, and changes of activity coefficients on solution conditions. Therefore, K_{as} depends on the environmental variables, pH, ionic strength, and osmolality of the solution. To probe molecular origins of stability and specificity of protein–ligand interactions in solution, the binding studies must be enriched by measurements at different experimental conditions. This is a starting point to find and analyze possible intermolecular processes that can accompany binding of cap analogs to eIF4E, as protonation, and ion or water exchange.

The standard Gibbs free energy change related to the binding equals:

$$\Delta G^{\circ} = -RT \cdot \ln K_{as} \tag{4}$$

ΔG° is the most suitable parameter to compare protein–ligand affinity in a quantitative way.

Titration of a macromolecule by a ligand is a standard method of determination of specificity and stoichiometry of a binding reaction (e.g., Eftink, 1997) if a change of a detected signal is clearly related to the binding event. Titration experiments at different temperatures provide K_{as} and thermodynamic parameters of the binding, changes of standard enthalpy, ΔH°, entropy, ΔS°, and molar heat capacity, from the vant't Hoff equation in the linear or nonlinear form (e. g., Ha et al., 1989).

Fluorescence titration was widely used to characterize the association between eIF4E and various analogs of the cap or capped RNA oligomers (see Niedzwiecka *et al.*, 2002a and citations therein). Before the introduction of our method (Niedzwiecka *et al.*, 2002a,b, 2004; Niedzwiecka-Kornas *et al.*, 1999), the reported K_{as} values yielded puzzling conclusions as a result of neglecting several systematic sources of errors during the experiments and numerical analysis. The most important were as follows: lacking of control of the activity for highly unstable, nonenzymatic eIF4E, neglecting a quite efficient emission of a free ligand, the presence of unknown, unremovable contamination of eIF4E by a cap analog after affinity chromatography, by use of uncorrected values of fluorescence signal, and simplifying the data analysis by linear transformations. A proper design of the titration experiment and taking into account all the aforementioned sources of errors gave for the first time the equilibrium association constants that reflect true affinity of eIF4E for the mRNA 5′ cap. The method is of general validity for studies of any protein–ligand interactions, especially if both absorption and emission spectra of a protein that contains many fluorescent residues and of a ligand are completely overlapped (e.g., Worch *et al.*, 2005).

3.2. Preparation of the samples

The concentrations of the cap analogs were obtained from weighed amounts ($\pm5\%$) and checked by absorption on the basis of extinction coefficients from Cai *et al.* (1999). At elevated pH, the guanine moiety substituted at N(7) undergoes opening of the five-member ring, followed by a hydrolytic cleavage of the glycosidic bond (Darzynkiewicz *et al.*, 1990). As checked by NMR, the cap was stable at pH 7.2, as well as stable enough to perform brief (<1 h) experiments at pH 8 to 9.

Full-length human eIF4E and murine eIF4E(28–217) were expressed in *E. coli* (Edery *et al.*, 1988), purified from inclusion bodies pellets, and refolded by one-step dialysis from 6 *M* guanidinium hydrochloride, followed by ion exchange chromatography on a HiTrap SP or MonoS column, without any contact with cap to the purity of $>95\%$ on the SDS PAGE. If necessary, the protein solutions were buffer exchanged on Ultrafree-15-ml filters (Millipore, MA) with Biomax 5 kDa NMWL membrane. The protein sample was filtered through Millipore Ultrafree-0.5-ml Biomax 100 kDa NMWL or Millipore 0.22-μm filter, and softly degassed before the experiment. Total protein concentration was determined from absorbance, 53,900 $cm^{-1}M^{-1}$. The samples for the ITC and stopped-flow experiments were prepared in a similar manner. Full-length human eIF4E was also purified from the soluble fraction by affinity chromatography (Webb *et al.*, 1984).

3.3. Experimental conditions of fluorescence titration

Fluorescence spectra (Fig. 9.1) were recorded on a Perkin Elmer LS–50B spectrofluorometer, in 50 mM HEPES/KOH, pH 7.20, 100 mM KCl, 1 mM dithiothreitol (DTT), and 0.5 mM disodium ethylenediaminetetra-acetate (EDTA), by use of a quartz semi-micro, 4-mm × 10-mm, cuvette (119.004F QS, Hellma, Germany) with continuous, slow magnetic stirring for mixing the components and keeping a constant temperature in the entire cuvette. Temperature inside the thermostated cuvette was controlled with a thermocouple (±0.2°). The protein sample was equilibrated at least 30 min before each experiment.

Intrinsic protein fluorescence was monitored after excitation at 280 nm (excitation slit 2.5 nm, emission slit 4–5 nm, cutoff filters) in a "Time Drive" mode at a single wavelength in the range from 320 to 340 nm, selected so that both the quenching of the protein fluorescence and the increasing fluorescence from the free ligand were distinctly visible at different ranges of ligand concentrations (Fig. 9.1). Because it is impossible either to excite eIF4E and the cap selectively or to detect the fluorescence of only one species, both processes should be clearly visible and precisely measured. An integration time of 30 sec and a gap of 30 sec for adding the ligand were found to ensure quite short duration of one titration containing enough data points (>30) with effective reduction of the noise (<1% of the initial signal). Measurements of the signal at constant time intervals allowed control of a

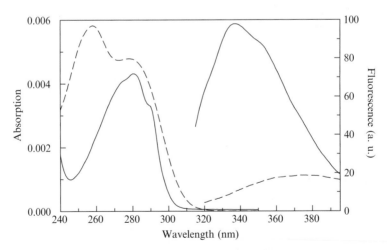

Figure 9.1 Absorption (left) and fluorescence (right, $\lambda_{ex} = 280$ nm) spectra of the protein (eIF4E, solid lines, 0.2 μM) and the ligand (m^{7}GTP, broken lines, 2 μM) at 20°, pH 7.2, in HEPES/KOH 50 mM, 100 mM KCl, 0.5 mM EDTA, 1 mM DTT.

possible fluorescence drift in time. During the gap, the UV xenon flash lamp was switched off to avoid photobleaching of the protein sample. This can be replaced by shutting down a shutter in other spectrofluorometers possessing CW lamps. The cuvette was not touched during the titration to keep the unchanged geometry. The 30-sec integration of the fluorescence signal at a single wavelength was checked to be superior to registration of the entire emission band, 310 nm to 400 nm, and its subsequent integration (noise level 3 to 5%). Our procedure minimized the error of a single titration point and maximized the number of points, which most influence the correctness and accuracy of determination of the K_{as} value, as shown by theoretical simulations (fluorescence data analysis).

3.4. Titration assay

Titrations were performed at several eIF4E concentrations, 50 nM to 1 μM, in steady-state conditions. Ligand solutions of increasing concentrations from 1 μM to 5 mM were injected manually by aliquots of 1 to 1400 μl of the eIF4E solution. Each titration consisted of more than 30 data points, with a suitable number in the range where the binding isotherm, Eqs. (11) to (12), attained an apparent plateau (Fig. 9.2). This region of the maximal curvature is the most important for proper determination of K_{as}. Both initial and final data points are of minor importance. For hypothetical, infinite K_{as}, the quenching attains a maximum at $[L] = [P_{act}]$, and remains constant with increasing $[L]$ if the ligand is nonflorescent. In case of cap analogs, the real maximal quenching is never attained. The apparent plateau results from two counteracting processes: protein fluorescence quenching and increase of free ligand fluorescence. This apparent plateau may not be interpreted as a result of protein saturation with the ligand, because both the apparent depth of the quenching and the position of the apparent plateau on the ligand concentration scale depend on the ligand fluorescence efficiency and on the selection of the excitation and emission wavelengths. Such a misinterpretation was a common source of errors in earlier works on eIF4E–cap interactions.

Subtraction of the free cap fluorescence to avoid the problem is groundless, because the equilibrium concentration of the free ligand depends on the association constant to be determined.

3.5. Corrections of fluorescence raw data

The fluorescence intensities should always be corrected for the *inner filter* effect (Parker, 1968). The correction has to be measured for each spectrofluorometer with a mixture of two noninteracting species, one fluorescent and the other absorbing (e.g., tryptophan and GMP, respectively).

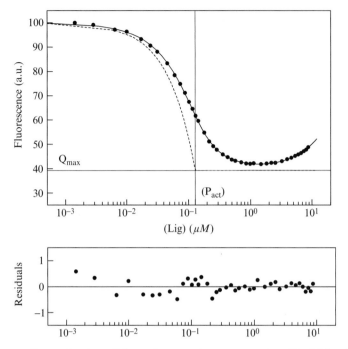

Figure 9.2 Determination of the active protein concentration, [P_{act}]. The titration curve (solid line) determines both K_{as} and [P_{act}] as free parameters of the fitting, Eqs (11) and (12). [P_{act}] is graphically represented at the x-axis by the point where the hypothetical binding curve (dotted line) for the infinite K_{as} value and a nonfluorescent ligand attain the maximal quenching, Q. Residuals show the correctness (random distribution) and the quality (<1% deviations) of the fitting.

A fourth-order polynomial dependence of the fluorescence intensity, F, on the absorbance, A, proved to be a good estimate of this correction for any cuvette:

$$\frac{F(A)}{F(0)} = 1 - a \cdot A + b \cdot A^2 - c \cdot A^3 + d \cdot A^4 \qquad (5)$$

The factor F(A)/F(0) was determined individually for each titration. The final data corrected for the absorption (F_{corrA}) were:

$$F_{corrA} = F_{obs} \div \frac{F(A)}{F(0)} \qquad (6)$$

The *inner filter* effect was almost negligible for ligands of the highest affinity but could change the K_{as} values by several-fold for weakly interacting or strongly absorbing analogs.

In some cases the protein fluorescence decreased in time at higher temperatures, $\sim 40°$, most probably because of partial denaturation of eIF4E. Therefore, the fluorescence signal was further monitored for 20 min after completing the titration. The exponential curve was fitted to these points:

$$F(t) = a \cdot \exp(-b \cdot t) + c \tag{7}$$

The fluorescence intensity, F_{obs}, was corrected to yield F_{corrT}:

$$F_{corrT} = F_{obs} \cdot \exp(b \cdot t) \tag{8}$$

The correction significantly improved the goodness of fit, R^2, and the P value (see *Statistical analysis*).

The fluorescence intensity was also corrected for dilution if it exceeded 3%.

3.6. Fluorescence data analysis

Total *apo*-eIF4E concentration is a sum of the active, $[P_{act}]$, and the inactive, $[P_{inact}]$, fractions. The initial fluorescence intensity, $F(0)$, is equal to:

$$F(0) = [P_{act}] \cdot \phi_{Pact-free} + [P_{inact}] \cdot \phi_{Pinact} \tag{9}$$

Where ϕ is the efficiency of the fluorescence. After adding a ligand with the efficiency $\phi_{lig-free}$ and of the total concentration $[L]$, the corrected fluorescence intensity (F), Eqs. (5) to (8), is:

$$F = [P_{act}]_0 \cdot \phi_{Pact-free} + [cx] \cdot \phi_{cx} + [L]_0 \cdot \phi_{lig-free} \\ + [P_{inact}] \cdot \phi_{Pinact} \tag{10}$$

No assumption regarding ϕ_{Pinact} is necessary. The fluorescence intensity as a function of the total ligand concentration is expressed by:

$$F = F(0) - [cx] \cdot (\Delta\phi + \phi_{lig-free}) + [L] \cdot \phi_{lig-free} \tag{11}$$

where $\Delta\phi = \phi Pact-free - \phi_{cx}$ is the difference between the fluorescence efficiencies of the active *apo*-protein and that of the complex, and $[cx]$ is:

$$[cx] = \frac{[L] + [P_{act}]}{2} + \frac{\sqrt{(K_{as}([L] - [P_{act}]) + 1)^2 + 4K_{as} \cdot [P_{act}]}}{2K_{as}} \tag{12}$$

The theoretical binding curve, Eqs. (11) to (12), with the parameters: K_{as}, $[P_{act}]$, $\Delta\phi$, $F(0)$, and $\phi_{lig-free}$ was fitted to the experimental data points (Fig. 9.2) by means of a nonlinear, least-squares method by use of PRISM 3.02 (GraphPad Software Inc., San Diego, CA, USA) or ORIGIN 6.0 (Microcal Software Inc., Southampton, MA, USA). The ligand fluorescence efficiency, $\phi_{lig-free}$, was verified independently in the absence of the protein with the accordance of $\pm 4\%$.

The maximal fluorescence quenching on binding, Q, can be now calculated as:

$$Q = [P_{act}] \cdot \Delta\phi \tag{13}$$

To test our titration procedure, theoretical simulations of experiments were performed for a typical set of parameters, $[P_{act}] = 0.2\ \mu M$, and $K_{as} = 100 \cdot 10^6\ M^{-1}$, with Gaussian noise imposed on the fluorescence data points, 0 to 3.3% of the initial fluorescence signal, $F(0) = 300$. The K_{as} values obtained by fitting, Eq. (11) to (12), to the simulated data points were shown to deviate from the expected value of $100 \cdot 10^6\ M^{-1}$ according to the increasing noise level (Fig. 9.4). The uncertainty, ΔK_{as}, attained 50% of K_{as} only for the noise over 2%. Hence, dumping of the noise is crucial for the correctness and accuracy of the numerical analysis and must be reduced to 1% or less. This was possible only in the continuous titration experiment, in which the fluorescence signal at each ligand concentration was integrated by 30 sec, with the same time intervals between measurements.

3.7. Statistical analysis

Quality of the fitting can be determined by goodness of the fit, R^2:

$$R^2 = 1 - \frac{\sum\limits_{i} \left(y_i^{exp} - y_i^{fit}\right)^2}{\sum\limits_{i} \left(y_i^{exp} - y^{mean}\right)^2} \tag{14}$$

Where: y_i^{exp}, y_i^{fit} are the measured and the fitted values, respectively, and y^{mean} is the mean of the measured values. A fitting was accepted if R^2 was not <0.99.

Correctness of the fitting can be checked by distribution of the residuals (Fig. 9.2) and a statistical one–parameter P value (Beyer, 1987) that denotes the probability that the distribution of the fitting residuals is random. A low one–parameter P value (<0.05) signalizes significant systematic deviations of experimental data from the assumed model.

Discrimination between models of different numbers of degrees of freedom (v_1 and v_2) is based on a statistical two–parameter Snedecor's F

test (Beyer, 1987). The test provides an assessment whether improvement of the results obtained by fitting of a greater number of parameters (e g., $[P_{act}]$ fitted) is statistically important in comparison with the simpler model ($[P_{act}]$ fixed). The F ratio quantifies the relative decrease of sum-of-squares in relation to the relative decrease of the number of degrees of freedom:

$$F = \frac{\left(\sum (y_{exp} - y_{fit_1})^2 - \sum (y_{exp} - y_{fit_2})^2\right)\bigg/\sum (y_{exp} - y_{fit_{-2}})^2}{(v_1 - v_2)/v_2}$$

(15)

The two-parameter $P(v_1,v_2)$ value is the probability that the improvement of the fit expressed by F could be obtained by chance. A low two-parameter $P(v_1,v_2)$ value (<0.05) indicates a significant improvement of the model with additional fitting parameters.

Distribution of the K_{as}^i values from several experiments performed under the same conditions is not symmetrical, with the maximum shifted to the lower values and a long "tail" toward the higher ones. A Gaussian distribution is obtained for logarithms of $\overline{K}_{as}^i$ Fig. 9.3. Therefore, the final weighed average $\overline{K}_{as}$, and its uncertainty, $\Delta\overline{K}_{as}$, should be calculated according to Eq. (16) and Eq. (17), respectively.

$$\overline{K}_{as} = \exp\left(\frac{\sum_i \ln K_{as}^i \cdot \left(\frac{K_{as}^i}{\Delta K_{as}^i}\right)^2}{\sum_i \left(\frac{K_{as}^i}{\Delta K_{as}^i}\right)^2}\right)$$

(16)

$$\Delta\overline{K}_{as} = \frac{\overline{K}_{as}}{\sqrt{\sum_i \left(\frac{K_{as}^i}{\Delta K_{as}^i}\right)^2}}$$

(17)

3.8. Activity of the nonenzymatic eIF4E protein

Quantitative control of the protein activity belongs to the fundamental state-of-the-art biochemical methods in enzymology. Because eIF4E is not an enzyme, it was impossible to determine the specific activity of each

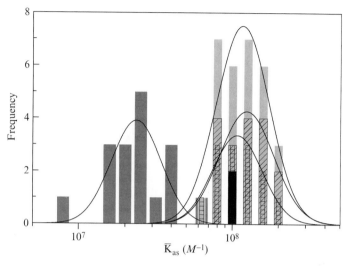

Figure 9.3 Statistics of results obtained by titration with m⁷GTP of murine eIF4E(28–217) refolded from inclusion bodies, purified without contact with cap, all samples (light grey bars, $\overline{K}_{as} = 105 \pm 3 \cdot 10^6\ M^{-1}$), murine eIF4E(28–217) from inclusion bodies, freshly prepared (diagonally striated bars, $\overline{K}_{as} = 108 \pm 4 \cdot 10^6\ M^{-1}$), murine eIF4E(28–217) from inclusion bodies, frozen (horizontally striated bars, $\overline{K}_{as} = 102 \pm 4 \cdot 10^6\ M^{-1}$), full-length human eIF4E from inclusion bodies (black bar, $\overline{K}_{as} = 98 \pm 7 \cdot 10^6\ M^{-1}$), full-length human eIF4E eluted from the cap-affinity column with m⁷GTP and purified on the MonoQ column (dark grey bars, $\overline{K}_{as} = 25.0 \pm 1.0 \cdot 10^6\ M^{-1}$, large scatter of the results from batch to batch).

sample from an independent experiment. The active protein concentration ($[P_{act}]$) was introduced as a free parameter in the fitting procedure. Such addition required a novel, very precise method of measuring the course of fluorescence quenching. The attained accuracy, $R^2 = 0.99$ to 0.999, was high enough to extract the lacking information on the actual protein activity from individual titrations.

For samples of the *apo*-protein frozen before the titration the active fraction decreased to <10%. The inactive fraction aggregated and precipitated, despite low concentration, <0.5 mg/ml, the presence of 10% (v/v) glycerol, 0.5 mM EDTA, and flash freezing of the aliquots, from 10 to 50 μl. This was in contrast to the cap-saturated eIF4E that was stabilized by the ligand. The statistics of the results from the titrations by use of various eIF4E samples proved that the K_{as} values were reliable and unaffected by the actual concentration of the active protein in the sample (Fig. 9.5). No statistical difference in the affinity to m⁷GTP was observed between full-length human eIF4E and truncated murine eIF4E(28–217) if purified without any contact with cap (Fig. 9.3). On the contrary, unambiguously decreased and scattered apparent affinities were found for various batches of human eIF4E purified from soluble fraction by cap-affinity chromatography.

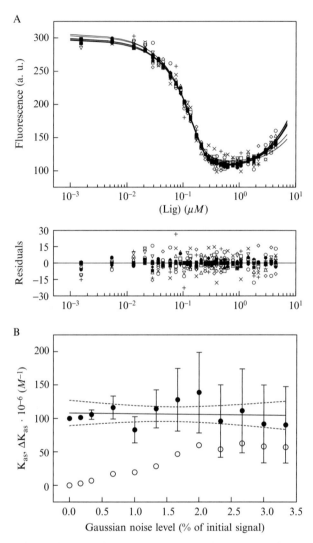

Figure 9.4 (A) Simulations of titration experiments: 12 titration data sets generated according to Eqs. (11) and (12) for $K_{as} = 100 \cdot 10^6 \ M^{-1}$, $[P_{act}] = 0.2 \ \mu M$, $\Delta\phi = 1000$, $\phi_{lig-free} = 10$, and $F(0) = 300$, and different Gaussian noise, from 0 to 3.3% of $F(0)$, and the binding curves fitted to the simulated data; (B) dependence of the fitted K_{as} values (filled circles) and their uncertainties (ΔK_{as}, open circles) on the noise level.

Introduction of the active protein concentration as a free parameter of the fitting was crucial for high-affinity ligands (Fig. 9.6). For the frozen protein (Fig. 9.6A), the fit with the fixed protein concentration determined from absorbance, 0.295 μM, gave an accidental value of $K_{as} = 26 \pm 73 \cdot 10^6$ M^{-1}, $R^2 = 0.38$, $P = 0.0001$. The fit with $[P_{act}]$ as a free parameter showed

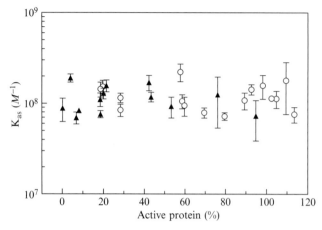

Figure 9.5 Relationship between the association constants, K_{as}, from various titrations of murine eIF4E(28–217) with m^{7}GTP and the fraction of the active protein. The results, K_{as}, obtained by nonlinear fitting of Eqs. (11) and (12) with [P_{act}] as a free parameter are independent from the actual active protein concentration.

an excellent goodness of fit, $R^2 = 0.9998$, $P = 0.63$, and yielded $K_{as} = 83.3 \pm 4.6 \cdot 10^6 \ M^{-1}$ with [P_{act}] = 0.0239 ± 0.0012 μM (i.e., ~8% of the entire protein amount). Moreover, [P_{act}] can significantly influence the results, even if the fitted curve is apparently "nice" (Fig. 9.6B). Two fitted curves, one with the protein concentration fixed as 0.2 μM, and the other with [P_{act}] fitted as 0.1798 ± 0.0014 μM, are apparently almost indistinguishable ($R^2 = 0.9990$ and 0.9999, respectively). However, the former fit the with $K_{as} = 188 \pm 26 \cdot 10^6 \ M^{-1}$ shows a systematic deviation from the experimental data, revealed by the residuals and $P < 0.0001$, whereas the fit with [P_{act}] free yielded Kas = 103.2 ± 5.1 $\cdot \rightarrow 10^6 \ M^{-1}$ and $P = 0.17$. The results obtained from the same data set differ by ~two-fold, even though the concentrations of the active and total protein differ only by 10%.

For the ligands that bind to the protein less strongly ($K_{as} \leq 10^6 \ M^{-1}$), concentration of the active protein looses its numerical importance ($P(v_1, v_2)$ >0.05).

The temperatures at which the activity of eIF4E drops by 50% are in the physiological range (35.4 ± 0.6°), as estimated from the Boltzmann sigmoidal curve (Fig. 9.7A). These results suggest that eIF4E in the living cell is usually complexed with another macromolecule, mRNA 5′ cap or other proteins, to avoid inactivation in the *apo* state. The population of the active protein also depends on the ionic strength (Fig. 9.7B). Relative stabilization is achieved at 150 to 200 mM, a level characteristic for the intracellular fluid.

Knowledge of the actual concentration of an active, nonenzymatic protein is crucial for normalization of every quantity being determined experimentally, analogously to the use of the specific activity of enzymes.

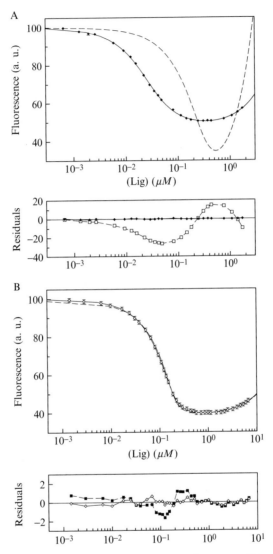

Figure 9.6 Comparison of the binding isotherms for interaction of m^{7}GTP with murine eIF4E(28–217): (A) frozen protein [P$_{act}$] fixed as 0.295 μM, K$_{as}$ = 26 ± 73 · 10^6 M^{-1}, R^2 = 0.38, P = 0.0001 (dashed line, residuals □); [P$_{act}$] as a free parameter, K$_{as}$ = 83.3 ± 4.6 · 10^6 M^{-1}, [P$_{act}$] = 0.0239 ± 0.0012 μM, R^2 = 0.9998, P = 0.63 (solid line, residuals ◆). (B) fresh protein [P$_{act}$] fixed as 0.2 μM, K$_{as}$ = 188 ± 26 · 10^6 M^{-1}, R^2 = 0.9990, P < 0.0001 (dashed line, residuals ■); [P$_{act}$] as a free parameter, K$_{as}$ = 103.2 ± 5.18 · 10^6 M^{-1}, [P$_{act}$] = 0.1798 ± 0.0014 μM, R^2 = 0.9999, P = 0.17 (solid line, residuals ◇).

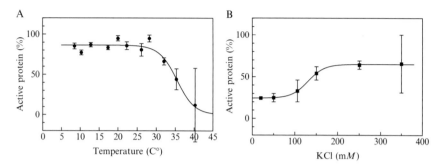

Figure 9.7 (A) Temperature dependence of the active fraction of freshly prepared murine eIF4E(28–217) (●), and (B) KCl dependence of the active protein fraction after storage by ∼6 hours at 25° (■), obtained on titration with m^{7}GTP.

In particular, long-lasting calorimetric and stopped-flow measurements required concurrent control of [P$_{act}$] by fluorescence titrations so that the results are free from systematic errors caused by protein inactivation.

3.9. Incorrectness of data linearization

Most of the previous reports were based on linear representations of the equilibrium equation, which are not suitable for analysis of strong interactions. The commonly used Eadie–Hofstee linear transformation assumes that protein concentration is negligible compared with that of a ligand over the entire course of titration. This is not fulfilled in molecular systems of high affinity. If $K_{as} \sim 10^7 \ M^{-1}$, saturation of $\sim 40\%$ occurs already at [L] = [P$_{act}$], at typical [P$_{act}$] $\sim 0.1 \ \mu M$, hence transformed data become nonlinear at low [L] (Fig. 9.8A). In addition, because of emission of free ligand, transformed data are nonlinear at high [L], too. Moreover, experimental errors are hidden in the x-axis. Inverse representation of the data (Fig. 9.8B) reveals huge and different uncertainty for each data point of the transformed data, usually neglected in the analysis.

   In general, for every kind of interaction the assumptions of linear regression are as follows: scatter of the points around the line follows Gaussian distribution, and the experimental error is the same for each abscissa. As shown in Fig. 9.8, these assumptions are violated. Hence, the values derived from the slope and the intercept of the linear regression are not reliable estimates of the parameters to be determined. On the contrary, the nonlinear fit can be directly applied to the experimental data points with equal errors, and Gaussian distribution of residuals proves that this way is the most appropriate to analyze the data.

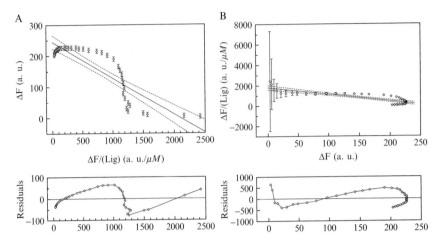

Figure 9.8 Linear representation of the binding data for interaction of murine eIF4E (28–217) with m^{7}GTP: (A) Eadie–Hofstee transformation; (B) inverse Eadie–Hofstee transformation. The experimental data are the same as in Fig. 9.6B.

4. APPLICATION OF MICROCALORIMETRY TO STUDYING EIF4E–CAP INTERACTION

4.1. Principles of microcalorimetry experiments

Isothermal titration calorimetry (ITC) provides a direct insight into thermo-dynamics of bimolecular equilibrium interactions (Ladbury and Chowdhry, 1996). The molar calorimetric enthalpy, ΔH°_{cal}, and the association constant, K_{as}, can be determined in one experiment by a direct measurement of the heat of interaction, when one component is being added to the other. However, two conditions must be simultaneously satisfied. First, a product of the number of interacting molecules and the association constant should be of the order of 100. Second, each injection should give at least 40 μJ of the heat signal (instrument sensitivity). It was impossible to meet these conditions for weakly soluble murine eIF4E(28–217). Hence, a modified "single injection" method that yields accurate ΔH°_{cal} was proposed to confirm a positive value of the molar heat capacity change, ΔC°_p, of m^{7}GpppG-eIF4E association, obtained from van't Hoff analysis (Niedzwiecka *et al.*, 2002b). However, it was possible to run a standard calorimetric titration for more soluble full-length yeast eIF4E with m^{7}GTP to get both K_{as} and ΔH°_{cal} (Kiraga-Motoszko *et al.*, 2003).

4.2. Experimental conditions of calorimetric measurements

The experiments at 15, 20, 25, and 30° were run on OMEGA Ultrasensitive Titration Calorimeter (MicroCal, MA), calibrated by 18-crown-6 titration with $BaCl_2$. The jacket was filled with dry nitrogen to prevent condensation on the outer surface of the cells below room temperature. A refrigerated circulated bath (7°) was connected to keep the surroundings of the cells cooler than the temperature of the experiment. The reference cell was filled with deionized water. Slow stirring at 240 rpm was applied to avoid protein precipitation. Before the automated titration started, the system was subjected to more than 30 min equilibration to get a stable baseline.

The protein sample buffer was exchanged by fourfold centrifugation on 5-kDa Centricon filters (Millipore, MA). The final flow-through buffer was collected to dissolve m^7GpppG, and to run control measurements of the ligand dilution heat. The concentration of the injected ligand was 1.00 ± 0.07 mM, and that of the active eIF4E, $[P_{act}]$ varied from 8.97 to 3.61 μM for measurements, as determined from parallel fluorescence titrations of a part of protein solution at each temperature.

4.3. Modified "single injection" experiment

Low solubility of eIF4E(28–217) hampered direct determination of K_{as} for m^7GpppG by standard ITC measurements, because the first injections resulted in ~35 μJ of the evolved heat, and subsequent signals decreased with the course of the titration becoming indiscernible from the noise. The ligand solution was injected into the calorimetric cell (1386 μl volume) filled with eIF4E solution. Then, the m^7GpppG solution was injected into the buffer to measure the heat of dilution. In a "single injection" method (Fig. 9.9), each experiment consisted of a main, 40-μl, injection, preceded by two 1-μl injections to calculate the correction for the initial leakage from the syringe (a common instrumental artefact), and followed by two 4-μl injections to check the protein saturation with the ligand during the main injection.

4.4. Calorimetric data treatment

After integration of all signals, the corresponding values obtained for injections to the buffer were subtracted from that for injections to the protein solution to yield the total calorimetric enthalpy change. Because the heat exchanged for the two initial 1-μl injections is approximately constant, the integral of the second injection was taken as the proper value for the first injection. The signals were added to yield ΔH°_{cal}. The calorimetric value of $\Delta C^{\circ}_{pcal} = + 1.97 \pm 0.06$ kJ $\cdot$ mol^{-1}K^{-1} was calculated as the slope of the linear dependence of ΔH°_{cal} (Fig. 9.9), in agreement with the van't Hoff result, $\Delta C^{\circ}_{pvH} = + 1.9 \pm 0.9$ kJ $\cdot$ mol^{-1}K^{-1} (Niedzwiecka et al., 2002b).

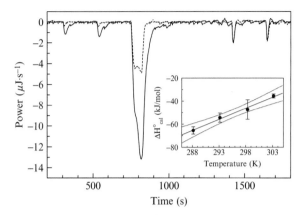

Figure 9.9 Heat signals on binding of m^{7}GpppG to murine eIF4E(28–217) during modified single-injection experiment (solid line) and on dilution of m^{7}GpppG in pure buffer (broken line); 288.1 K, HEPES/KOH 50 mM, 100 mM KCl, 0.5 mM EDTA, 0 mM DTT. Inset shows the calorimetric enthalpy changes determined at various temperatures.

4.5. Calorimetric data analysis

A systematic positive shift between the calorimetric and van't Hoff enthalpies was observed (Niedzwiecka *et al.*, 2002b). Such differences were a subject of various empirical and theoretical analyses (see e.g., Horn *et al.*, 2001). The observed discrepancies were ascribed to contributions from usually unknown coupled processes to ΔH°_{cal}, and/or erroneous apparent values of ΔC°_{p} and ΔH°_{vH} arising from the experimental noise. In case of cap–eIF4E association, the latter obscuring effect is eliminated, as testified by the accordance of the van't Hoff and calorimetric ΔC°_{p} values. The approximately constant difference can be analyzed in terms of the protonation equilibria (Kavanoor and Eftink, 1997). The 7-methylguanine moiety of m^{7}GpppG exists at pH 7.2 as a mixture of the cationic (58%) and zwitterionic (42%), forms, because pK_a(N1-H) = 7.35 ± 0.05 (Wieczorek *et al.*, 1995). The binding studies at various pH (Niedzwiecka *et al.*, 2002a) showed an upward shift of this pK_a value inside the eIF4E-binding site, suggesting a tight binding of the cap in the cationic form. Thus, the association is accompanied by a partial protonation of the cap and must be equilibrated by additional deprotonation of the buffer to keep the cation–zwitterion equilibrium of the free ligand at constant pH. The contribution of the HEPES ionization heat (ΔH°_{ion}) at pH = 7.2 and at a given temperature, Eq. (18),

$$\Delta H^{\circ}_{ion} = \frac{10^{-pK_a}}{10^{-pK_a} + 10^{-pH}} \cdot \Delta H^{\circ}_{H-diss} \tag{18}$$

was calculated from the molar ionization heat, $\Delta H^\circ_{H-diss} = +20.95$ kJ $\cdot$ mol^{-1} at 298.2 K (25°), and its temperature dependence, $\delta/\delta T = -$ $+0.0648$ kJ $\cdot$ $mol^{-1}K^{-1}$ (Christensen *et al.*, 1976), whereas pK_a of m^7GpppG did not change in this temperature range. The values of ΔH°_{ion} were in excellent agreement with the differences between ΔH°_{cal} and ΔH°, thus confirming partial protonation of the cap on binding to eIF4E.

Association of eIF4E with the mRNA 5′ terminus is also directly coupled with self-stacking of the cap, which contributes to the overall thermodynamic parameters of eIF4E–cap binding (see Niedzwiecka *et al.*, 2002b). In general, the binding studies always have to be carefully analyzed in respect to all possible coupled processes, which involve not only the biomolecules being at the focus of the research but also all ingredients of the milieu.

5. STOPPED-FLOW FLUORESCENCE STUDIES OF BINDING CAP ANALOGS TO EIF4E

5.1. Principles of stopped-flow experiments

Stopped-flow spectrometry is a transient kinetic method that allows direct measurement of rate and equilibrium constants governing the dynamics of molecular processes. B. Chance (University of Pennsylvania) invented the technique in 1940. Many companies provide commercial stopped-flow instruments (e.g., Applied Photophysics Ltd. UK used in our measurements). The stopped-flow technique is very useful whenever the reaction of interest is accompanied by a measurable change of an optical signal, UV-Vis absorption, fluorescence, light scattering, CD, IR, NMR, or EPR. After rapid mixing of two reagents in a mixer, the resultant solution flows into an observation cell. The content of the observation cell is flushed into a stopping syringe and can be replaced with freshly mixed reactants. Just before the stopping, a steady-state flow is achieved. A time elapse of several milliseconds or less before the mixture enters the observation cell is a "dead time" of the apparatus. As the content of the observation cell fills the stopping syringe, a plunger hits a block, causing the flow to stop abruptly, triggering the signal measurement as a function of time. In more sophisticated versions, double or even triple reactant mixing are realized, with variable delays (e.g., in BioLogic SFM-400 instrument).

A single-mixing stopped-flow experiment can be performed either in a "binding" or in a "dissociation" mode. In the former, the receptor solution is mixed with the ligand solution, usually in 1:1 volume ratio. Unsymmetrical mixing of unequal volumes is used if a concentration jump of some component(s) is desirable (e.g., in studies of protein unfolding). Binding experiments can be carried out under second-order conditions by mixing the receptor and the ligand of comparable concentrations. Alternately, experiments can be designed to obey pseudo first-order conditions, with the concentration of one component (usually the ligand) much higher than that of the other. In the dissociation experiment, a preequilibrated solution of the receptor and the ligand is mixed with the pure solvent.

5.2. Data analysis of ligand-binding kinetics

An important aspect of stopped-flow experiments is that the initial state of an investigated system (i.e., the moment of the mixing) can be far from the thermodynamic equilibrium, which is subsequently achieved during the course of the reaction. Therefore, kinetic equations describing the time dependence of the transient reactions in the observation cell are usually complex. Two to three decades ago, the data analysis of transient kinetics was performed by simplification of the reaction pathways to provide explicit solution of the kinetic equations. In this regard, the equilibrium perturbation (relaxation) methods that operate close to thermodynamic equilibrium had an advantage over the flow methods. Differential equations describing temporal behavior of a system close to equilibrium can be solved in the form of an exponential function or a sum of such functions. However, the problems of the data analysis in transient experiments were overcome in the last 10 to 15 years by an advance in computational methods used to analyze a reaction time course by numerical integration. The data analysis is no longer restricted to simplified reaction schemes, and the stopped-flow experiments can be performed without restrictions for starting conditions dictated by easy solution of the pseudo first-order rate equations. Here, we describe applications of the two approaches to analysis of the eIF4E–cap association.

Three simplest reaction mechanisms appropriate for a single receptor–single ligand system are described by a one-step reaction scheme:

$$E + L \underset{k_{-1}}{\overset{k_{+1}}{\longleftrightarrow}} EL \tag{19}$$

By a two-step scheme with the binding followed by structural changes of the complex:

$$E + L \underset{k_{-1}}{\overset{k_{+1}}{\longleftrightarrow}} EL^* \underset{k_{-2}}{\overset{k_{+2}}{\longleftrightarrow}} EL \tag{20}$$

By a two-step scheme involving the receptor isomerization before the binding:

$$E^* + L \underset{k_{-1}}{\overset{k_{+1}}{\leftrightarrow}} E + L \underset{k_{-2}}{\overset{k_{+2}}{\leftrightarrow}} EL \qquad (21)$$

It is possible to decide which of these mechanisms is appropriate by analyzing the results of kinetic experiments performed under pseudo first-order conditions (Johnson, 1992; Strickland et al., 1975). The rate equations can be integrated to yield a monoexponential time dependence of the reactants' concentrations. The observed signal is also a monoexponential function of time,

$$F = A\exp(-k_{obs}t) + C \qquad (22)$$

where $A + C$ is the signal at time zero, C is the final signal at equilibrium, and k_{obs} is the apparent pseudo first-order rate constant. For the one-step mechanism, Eq. (19), k_{obs} is a linear function of the total (initial) ligand concentration:

$$k_{obs} = k_{+1}[L] + k_{-1} \qquad (23)$$

For the two-step mechanism, Eq. (20), with the steady-state assumption, and for $[L] >> [E]$, $[EL^*] = [EL] = 0$ at $t = 0$, the observed rate constant reads:

$$k_{obs} = \frac{k_{+1}[L](k_{+2}+k_{-2}) + k_{-1}k_{-2}}{k_{+1}[L] + k_{-1} + k_{+2}} \qquad (24)$$

Finally, for the two-step mechanism described by Eq. (21), with the steady-state assumption, and for $[L] >> [E]$, and $[EL] = 0$ at $t = 0$, the observed rate constant is expressed by:

$$k_{obs} = \frac{k_{+1}(k_{+2}[L] + k_{-2}) + k_{-1}k_{-2}}{k_{+2}[L] + k_{+1} + k_{-1}} \qquad (25)$$

If $k_{-1} >> k_{+2}$, Eq. (24) can be simplified to Eq (26), and Eq (25) to Eq. (27).

$$k_{obs} = \frac{k_{+1}k_{+2}[L]}{k_{+1}[L] + k_{-1}} + k_{-2} \qquad (26)$$

$$k_{obs} = \frac{k_{+1}k_{+2}[L]}{k_{+1} + k_{-1}} + k_{-2} \qquad (27)$$

Subsequently, assuming that k_{-2} is negligibly small, a hyperbolic dependence for the reaction mechanism described by Eq. (20) is obtained:

$$k_{obs} = \frac{k_{+1}k_{+2}[L]}{k_{+1}[L] + k_{-1}} \tag{28}$$

For the mechanism described by Eq. (21), k_{obs} is a linear function of the total ligand concentration irrespective of k_{-2}. Eq (28) can be inverted to give a linear dependence on $1/[L]$:

$$\frac{1}{k_{obs}} = \frac{1}{k_{+2}} + \frac{k_{-1}}{k_{+1}k_{+2}}\frac{1}{[L]} \tag{29}$$

For the association described by Eq. (20), a hyperbolic and a linear dependence of k_{obs} and $1/k_{obs}$, respectively, can be also obtained by neglecting k_{-2} in Eq. (24):

$$k_{obs} = \frac{k_{+1}k_{+2}[L]}{k_{+1}[L] + k_{-1} + k_{+2}}; \quad \frac{1}{k_{obs}} = \frac{1}{k_{+2}} + \frac{k_{-1} + k_{+2}}{k_{+1}k_{+2}}\frac{1}{[L]} \tag{30}$$

Expressions for k_{obs} can also be derived for $[E] \gg [L]$. In the case of the mechanism described by Eq (20), the expressions for k_{obs} are analogous to Eqs (24), (26), (28), (29), and (30), if $[L]$ is substituted by $[E]$. For the mechanism described by Eq. (21), k_{obs} is predicted to be a linear function of $[E]$, if a rapid equilibrium is assumed for the isomerization step, or if the isomerization is rate limiting. In the former case:

$$k_{obs} = \frac{k_{+1}k_{+2}[E]}{k_{-1}} + k_{-2} \tag{31}$$

One can also perform stopped-flow mixing experiments with the protein and ligand concentrations out of the pseudo first-order regimen and try to fit a sum of two or more exponential functions (e.g., for a two-exponential fit):

$$F = A_1 \exp(-k_{obs,1}t) + A_2 \exp(-k_{obs,2}) + C \tag{32}$$

F is the observed signal, A_1 and A_2 are the amplitudes, k_{obs1} and k_{obs2} are the observed rate constants, for the first and the second component of a double-exponential reaction, respectively, and C is the final value of the signal. The amplitudes and the observed rate constants are functions of the rate constants of the elementary reactions and optical properties of all molecular species in the mixed solution. Extraction of these data is usually not simple.

As noted previously, the pseudo first-order conditions require that the concentration of one reagent is much larger than that of the other. It seems that performing the experiments under the second-order conditions combined with the numerical method of the data analysis is a valuable alternative. According to our knowledge, no commercially available stopped-flow instrument is sold with the data analysis package on the basis of numerical integration of the differential kinetic equations for any reaction mechanism. A useful standard tool for this purpose, DynaFit program of Kuzmic (1996), is commercially available from BioKin Ltd. (free for academic users). The program performs nonlinear least-squares regression of chemical kinetics, enzyme kinetics, or ligand-receptor binding data. It runs under control of a script file, which defines tasks to be done, reaction mechanisms, location of the input data and the output files, initial values of parameters, and a character of these parameters (i.e., constants or adjustables). The tasks comprise fitting of the postulated theoretical model to the experimental data and simulation of pseudo-experimental data. A discrimination analysis of various models results in a statistically justified choice of the most appropriate reaction scheme for a given system.

5.3. Practical realization of fluorescence stopped-flow experiments

The eIF4E-cap association is a single receptor–single ligand system, and, therefore, it is not necessary to go beyond a single mixing stopped-flow experiment. Variable factors are as follows: the receptor and the ligand concentrations, temperature, pH, ionic strength, and, if necessary, the presence of some cosolvents. In the association experiments, a solution of eIF4E was mixed with a solution of m^{7}GpppG, and afterward the observed association process proceeded until the equilibrium, in which the rate of the complex formation was equal to the rate of the complex dissociation. In the dilution experiments, an equilibrium mixture of eIF4E and the cap was mixed with the pure buffer, resulting in a shift of the binding equilibrium toward the dissociation. Usually, both kinds of experiments were run with mixing of equal volumes of the solutions, but unsymmetrical mixing may be useful in the dilution experiments. One mixing experiment (i.e., registration of a single kinetic transient) was not sufficient to achieve a satisfactory signal-to-noise ratio, and several transients were added and averaged. Similarly, one averaged transient was insufficient for the data analysis. Several combinations of the reagents at various concentrations were mixed in a given temperature, pH, and ionic strength, leading to a series of the kinetic transients. Simultaneous analysis of all these transients within restrictions imposed by a particular reaction mechanism provided a number of parameters large enough to determine a reliable reaction model.

The eIF4E–cap association is quite fast on the stopped-flow time scale. Therefore, a proper choice of the concentrations of the protein and the ligand is an important issue. By excess of the cap to 1 μM protein solution in the stopped-flow syringe, a nice fluorescence signal at 320 nm (cutoff filter) is observed. However, for relatively low ligand concentration, the association becomes too fast, and it is not possible to work in a pseudo first-order regime (i.e., with the cap concentration at least 10 times greater than that of the protein). As suggested by Eqs. (23) to (25), and their modifications, the k_{obs} values do not depend on the protein concentration. However, decreasing of the protein concentration to (e.g., 0.1 μM) allows us to reach the pseudo first-order requirements, but the signal becomes much weaker. If monoexponential functions are fitted to the transient curves for various protein concentrations, possible differences in the k_{obs} values obtained for the same ligand concentration, in addition to possible uncerntainty of the concentration, can result from the fact that the pseudo first-order conditions are not satisfied for some pairs of the protein and ligand concentrations. On the other hand, one can work with large protein concentration and much lower ligand concentrations. However, this leads to other problems. Each protein concentration may require a new setup of the spectrometer, like photomultiplier voltage. In addition, possible autoassociation of the protein can occur. That is why the numerical procedures for the data analysis implemented in DynaFit program are so important. They allow unrestricted mixing of the solutions without restrictions regarding their concentrations. Even autoassociation of one of the reagents can be easily included into the reaction mechanism. Although, as noted previously, the eIF4E–cap association is fast on the stopped-flow time scale, it is useful to follow the reactions much longer than needed, just to check that nothing wrong happens to the samples. Fluorescence changes caused by photoreactions unnoticed for a short observation time can appear on a longer time-scale and result in some systematic error in data analysis.

5.4. Example of human eIF4E-m⁷GpppG association kinetics

In the last part of this section, we present results of the stopped-flow experiments with m^7GpppG and human eIF4E. Preparation of the samples and the spectral parameters are listed in the section on the steady-state fluorescence experiments. The measurements were run at $20°$ in 50 mM Hepes-KOH buffer, pH 7.2, ionic strength of 150 mM, and the concentrations of 0.2 μM for eIF4E and of 0.05 to 15.0 M for m^7GpppG in the stopped-flow syringes.

Kinetic transients of one of the two series of measurements for various molar concentrations of m^7GpppG are presented in Fig. 9.10. For a low cap

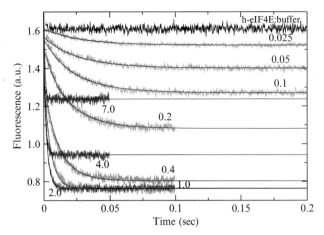

Figure 9.10 Stopped-flow kinetic traces for mixing of human eIF4E, concentration 0.2 μM, with m^{7}GpppG, concentrations in μM indicated by the numbers, and the fitting by DynaFit software assuming the two-step association mechanism; $k_{+1} = 200\ \mu M^{-1}s^{-1}$, $k_{-1} = 17.3\ sec^{-1}$, $k_{+2} = 5.8\ sec^{-1}$, $k_{-2} = 67.5\ sec^{-1}$. The trace marked h-eIF4E buffer corresponds to mixing of the protein solution with equal volume of pure buffer. (See color insert.)

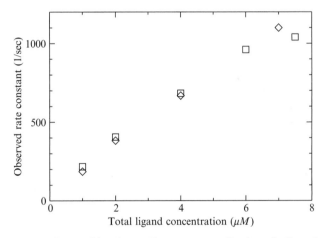

Figure 9.11 Dependence of the observed rate constants (k_{obs}) on the ligand concentration obtained from single-exponent fitting to the kinetic traces registered under pseudo first-order conditions for two independent sets of experiments.

concentration, each transient has a well-marked amplitude and a plateau region. Over the ligand concentration of 2 μM, the plateau regions shift upward because of contribution of the free-ligand fluorescence. Simultaneously, a longer and longer part of the reaction takes place within the dead time. Dependence of k_{obs} vs. total ligand concentration obtained from fitting a monoexponential function to the transients registered under pseudo first-order conditions is shown in Fig. 9.11 for two series of the measurements. One of them exhibits a slightly nonlinear character of the dependence indicating a two-step association reaction. The type of the eIF4E–cap association mechanism (i.e., a one- or a two-step reaction) was an issue in the literature (Blachut-Okrasinska *et al.*, 2000; Dlugosz *et al.*, 2003; Khan and Goss, 2005; Sha *et al.*, 1995; Slepenkov *et al.*, 2006). Simultaneous fittings of all kinetic transients were performed by DynaFit program assuming a one-step, Eq. (19), and a two-step mechanism, Eq. (20). The latter (Fig. 9.10) was indicated by the DynaFit module for statistical discrimination among the reaction models as a more probable.

ACKNOWLEDGMENTS

We are indebted to J. Zuberek, L. Chlebicka, and L. Zhukova for preparation of the protein samples, M. Wszelaka-Rylik for excellent assistance by the ITC measurements, and O. Szklarczyk for running the stopped-flow measurements, as well as to R. E. Rhoads and N. Sonenberg for providing the plasmids. Financial support from the Polish Ministry of Science and Higher Education (2 P04A 033 28, 3 P04A 021 25, 2 P04A 006 28, and BST 1059/BF), and Howard Hughes Medical Institute, grant No.55005604 (to E. D.)

REFERENCES

Adam, A., and Moffat, J. G. (1966). Dismutation reactions of nucleoside polyphosphates. V. Syntheses of P^1,P^4-di(guanosine-5′) tetraphosphate and P^1,P^3-di(guanosine-5′) triphosphate. *J. Am. Chem. Soc.* **88**, 838–842.

Beyer, W. H. (1987). "CRC Standard Mathematical Tables." Vol. 28, p. 536. CRC Press, Boca Raton, FL.

Blachut-Okrasinska, E., Bojarska, E., Niedzwiecka, A., Chlebicka, L., Darzynkiewicz, E., Stolarski, E., Stepinski, J., and Antosiewicz, J. M. (2000). Stopped-flow and Brownian dynamics studies of electrostatic effects in the kinetics of binding of 7-methyl-GpppG to the protein eIF4E, *Eur. Biophys. J.* **29**, 487–498.

Blumenthal, T. (1998). Gene clusters and polycistronic transcription in eukaryotes. *BioEssays* **20**, 480–487.

Cai, A., Jankowska-Anyszka, M., Centers, A., Chlebicka, L., Stepinski, J., Stolarski, R., Darzynkiewicz, E., and Rhoads, R. E. (1999). Quantitative assessment of mRNA cap analogues as inhibitors of *in vitro* translation. *Biochemistry* **38**, 8538–8547.

Christensen, J. J., Hansen, L. D., and Izatt, R. M. (1976). "Handbook of Proton Ionization Heats and Related Thermodynamic Quantities." John Wiley Sons, New York.

Cramer, F., Schatter, H., and Staab, H. A. (1961). Zur Chemie der, energiereichen Phosphate, XI. Darstellung von Imidazoliden der Phosphorsäure. *Chem. Ber.* **94**, 1612–1621.

Darzynkiewicz, E., Ekiel, I., Tahara, S. M., Seliger, L. S., and Shatkin, A. J. (1985). Chemical synthesis and characterization of 7-methylguanosine cap analogues. *Biochemistry* **24**, 1701–1707.

Darzynkiewicz, E., Stepinski, J., Ekiel, I., Jin, Y., Haber, D., Sijuwade, T., and Tahara, S. M. (1988). β-Globin mRNAs capped with m^7G, $m_2^{2,7}G$ or $m_3^{2,2,7}G$ differ in intrinsic translation efficiency. *Nucleic Acids Res.* **16**, 8953–8962.

Darzynkiewicz, E., Stepinski, J., Tahara, S. M., Stolarski, R., Ekiel, I., Haber, D., Neuvonen, K., Lehikoinen, P., Labadi, I., and Lönnberg, H. (1990). Synthesis, conformation and hydrolytic stability of P^1,P^3-dinucleoside triphosphates related to mRNA 5'-cap, and comparative kinetic studies on their nucleoside and nucleoside monophosphate analogs. *Nucleosides Nucleotides* **9,** 599–618.

Edery, L., Altman, M., and Sonenberg, N. (1988). High-level synthesis in *Escherichia coli* of functional cap-binding eukaryotic initiation factor eIF-4E and affinity purification using a simplified cap-analogue resin. *Gene* **74,** 517–525.

Eftink, R. M. (1997). Fluorescence methods for studying equilibrium macromolecule-ligand interactions. *Methods Enzymol.* **278,** 221–257.

Fukuoka, K., Suda, F., Suzuki, R., Ishikawa, M., Takaku, H, and Hata, T. (1994). Large scale synthesis of the cap part in messenger RNA using new type of bifunctional phosphorylating reagent. *Nucleosides Nucleotides* **13,** 1557–1567.

Gingras, A. C., and Raught, B. (1999). eIF4E initiation factors: Effectors of mRNA recruitment to ribosomes and regulators of translation. *Annu. Rev. Biochem.* **68,** 913–963.

Grudzien-Nogalska, E., Stepinski, J., Jemielity, J., Zuberek, J., Stolarski, R., Rhoads, R. E., and Darzynkiewicz, E. (2007). Synthesis of anti-reverse cap analogs (ARCAs) and their application in protein translation and stability. *Methods Enzymol.* **431,** 203–227.

Ha, J. H., Spolar, R. S., and Record, M. T., Jr. (1989). Role of the hydrophobic effect in stability of site-specific protein-DNA complexes. *J. Mol. Biol.* **209,** 801–816.

Hata, T., Nakagawa, I., Shimotohno, K., and Miura, K. (1976). The synthesis of α,γ-dinucleoside triphosphates. The confronted nucleotide structure found at the 5'-terminus of eukaryote messenger ribonucleic acid. *Chemistry Lett.* 987–990.

Hernandez, G., and Vazquez-Pianzola, P. (2005). Functional diversity of the eukaryotic translation initiation factors belonging to eIF4E families. *Mechanisms Dev.* **122,** 865–876.

Hoard, D. E., and Ott, D. G. (1965). Conversion of mono- and oligodeoxyribonucleotides to 5'-triphosphates. *J. Am. Chem. Soc.* **87,** 1785–1788.

Horn, J. R., Russel, D., Lewis, E. A., and Murphy, K. P. (2001). Van't Hoff and calorimetric enthalpies from isothermal titration calorimetry: Are there significant discrepancies. *Biochemistry* **40,** 1774–1778.

Iwase, R., Sekine, M., Tokumoto, Y., Ohshima, Y., and Hata, T. (1989). Synthesis of N^2, N^2,7-trimethylguanosine cap derivatives. *Nucleic Acids Res.* **17,** 8979–8989.

Jankowska, M., Stepinski, J., Stolarski, R., Temeriusz, A., and Darzynkiewicz, E. (1993). Synthesis and properties of new NH_2 and N7 substituted GMP and GTP 5'-mRNA cap analogues. *Collect. Czech. Chem. Commun.* **58**(Special issue), 138–141.

Jemielity, J., Flower, T., Zuberek, J., Stepinski, J., Lewdorowicz, M., Niedzwiecka, A., Stolarski, R., Darzynkiewicz, E., and Rhoads, R. E. (2003). Novel 'anti-reverse' cap analogs with superior translational properties,. *RNA* **9,** 1108–1122.

Johnson, K. A. (1992). Transient-state kinetic analysis of enzyme reaction pathways. *In* "The Enzymes." 3rd ed. XX. Mechanisms of Catalysis (D. S. Sigman, ed.), pp. 1–61. Academic Press, San Diego, CA.

Kadokura, M., Wada, T., Urashima, C., and Sekine, M. (1997). Efficient synthesis of γ-methyl-capped guanosine 5'-triphosphate as a 5'-terminal unique structure of U6 RNA via a new triphosphate bond formation involving activation of methyl phosphorimidazolidate using $ZnCl_2$ as a catalyst in DMF under anhydrous conditions. *Tetrahedron Lett.* **38,** 8359–8362.

Kalek, M., Jemielity, J., Darzynkiewicz, Z. M., Bojarska, E., Stepinski, J., Stolarski, R., Davis, R. E., and Darzynkiewicz, E. (2006). Enzymatically stable 5' mRNA cap analogs: Synthesis and binding studies with human DcpS decapping enzyme. *Bioorg. Med. Chem.* **14,** 3223–3230.

Kalek, M., Jemielity, J., Stepinski, J., Stolarski, R., and Darzynkiewicz, E. (2005). A direct method for the synthesis of nucleoside 5'-methylenebis(phosphonate)s from nucleosides. *Tetrahedron Lett.* **46,** 2417–2421.

Kavanoor, M., and Eftink, M. R. (1997). Characterization of the role of the side-chain interactions in the binding of ligands to apo trp repressor: pH dependence studies. *Biophys. Chem.* **66,** 43–55.

Khan, M. A., and Goss, D. J. (2005). Translation initiation factor (eIF) 4B affects the rates of binding of the mRNA m7G cap analogue to wheat germ eIFiso4F and eIFiso4F·PABP. *Biochemistry* **44,** 4510–4516.

Kiraga-Motoszko, K., Stepinski, J., Niedzwiecka, A., Jemielity, J., Wszelaka-Rychlik, M., Stolarski, R., Zielenkiewicz, W., and Darzynkiewicz, E. (2003). Interaction between yeast eukaryotic initiation factor eIF4E and mRNA 5′ cap analogues differs from that for murine eIF4E. *Nucleosides Nucleotides Nucl. Acids* **22,** 1711–1714.

Kohno, K., Nishiyama, S., Kamimura, T., Sekine, M., Hata, T., Kumagai, I., and Miura, K. (1985). Chemical synthesis of capped RNA fragments and their ability to complex with eukaryotic ribosomes. *Nucleic Acids Res. Symp. Ser.* **16,** 233–236.

Kuzmic, P. (1996). Program DYNAFIT for the analysis of enzyme kinetic data: Application to HIV proteinase. *Anal. Biochem.* **237,** 260–273.

Ladbury, J. E., and Chowdhry, B. Z. (1996). Sensing the heat: The application of isothermal titration calorimetry to thermodynamic studies of biomolecular interactions. *Chem. Biol.* **3,** 791–801.

Mader, S., Lee, H., Pause, A., and Sonenberg, N. (1995). The translation initiation factor eIF-4E binds to a common motif shared by the translation factor eIF-4 gamma and the translational repressors 4E-binding proteins. *Mol. Cell Biol.* **15,** 4990–4997.

Marcotrigiano, J., Gingras, A. C., Sonenberg, N., and Burley, S. (1997). Cocrystal structure of the messenger RNA 5′ cap-binding protein (eIF4E) bound to 7-methyl-GDP. *Cell* **89,** 951–961.

Matsuo, H., Li, H., McGuire, A. M., Fletcher, C. M., Gingras, A. C., Sonenberg, N., and Wagner, G. (1997). Structure of translation factor eIF4E bound to m^7GDP and interaction with 4E-binding protein. *Nature Struct. Biol.* **4,** 717–724.

Morley, S. J., Curtis, P. S., and Pain, V. M. (1997). eIF4G: Translation's mystery factor begins to yield its secrets. *RNA* **3,** 1085–1104.

Nakagawa, I., Konya, S., Ohtani, S., and Hata, T. (1980). A 'capping' agent: P^1-S-phenyl P^2-7-methylguanosine-5′ pyrophosphorothioate. *Synthesis* 556–557.

Niedzwiecka, A., Marcotrigiano, J., Stepinski, J., Jankowska-Anyszka, M., Wyslouch-Cieszynska, A., Dadlez, M., Gingras, A. C., Mak, P., Darzynkiewicz, E., Sonenberg, N., Burley, S., and Stolarski, R. (2002a). Biophysical studies of eIF4E cap-binding protein: Recognition of mRNA 5′ cap structure and synthetic fragments of eIF4G and 4E-BP1 proteins. *J. Mol. Biol.* **319,** 615–635.

Niedzwiecka, A., Stepinski, J., Darzynkiewicz, E., Sonenberg, N., and Stolarski, R. (2002b). Positive heat capacity change upon specific binding of translation initiation factor eIF4E to mRNA 5′ cap. *Biochemistry* **41,** 12140–12148.

Niedzwiecka, A., Darzynkiewicz, E., and Stolarski, R. (2004). Thermodynamics of mRNA 5′ cap binding by eukaryotic translation initiation factor eIF4E. *Biochemistry* **43,** 13305–13317.

Niedzwiecka-Kornas, A., Chlebicka, L., Stepinski, J., Jankowska-Anyszka, M., Wieczorek, Z., Darzynkiewicz, E., Rhoads, R. E., and Stolarski, R. (1999). Spectroscopic studies on association of mRNA cap-analogues with human translation factor eIF4E. From modelling of interactions to inhibitory properties. *Collect. Symp. Ser.* **2,** 214–218.

Parker, C. A. (1968). "Photoluminescence of Solutions." Elsevier Publishing Co., Amsterdam.

Raugh, B., Gingras, A. C., and Sonenberg, N. (2000). Regulation of ribosomal recruitment in eukaryotes. *In* "Translational Control of Gene Expression" (N. Sonenberg,

J. W. Hershey, and M. B. Mathews, eds.), pp. 245–293. Cold Spring Harbor Laboratory Press, New York.

Richter, J. D., and Sonenberg, N. (2005). Regulation of cap-dependent translation by eIF4E inhibitory proteins. *Nature* **433,** 477–480.

Sawai, H., Wakai, H., and Shimazu, M. (1991). Facile synthesis of cap portion of messenger RNA by Mn(II) ion catalyzed pyrophosphate formation in aqueous solution. *Tetrahedron Lett.* **32,** 6905–6906.

Sawai, H., Shimazu, M., Wakai, H., Wakabayashi, H., and Shinozuka, K. (1992). Divalent metal ion-catalyzed pyrophosphate bond formation in aqueous solution. Synthesis of nucleotides containing polyphosphate. *Nucleosides Nucleotides* **11,** 773–785.

Sha, M., Wang, Y., Xiang, T., van Heerden, A., Brownings, K. S., and Goss, D. J. (1995). Interaction of wheat germ protein synthesis initiation factor eIF-(iso)4F and its subunits p28 and p86 with m7GTP and mRNA analogues. *J. Biol. Chem.* **270,** 29904–29909.

Slepenkov, S. V., Darzynkiewicz, E., and Rhoads, R. E. (2006). Stopped-flow kinetic analysis of eIF4E and phosphorylated eIF4E binding to cap analogs and capped oligoribonucleotides. Evidence for a one-step binding mechanism. *J. Biol. Chem.* **281,** 14927–14938.

Stepinski, J., Waddell, C., Stolarski, R., Darzynkiewicz, E., and Rhoads, R. E. (2001). Synthesis and properties of mRNAs containing the novel 'anti-reverse' cap analogues 7-methyl-(3′-O-methyl)GpppG and 7-methyl-(3′-deoxy)GpppG. *RNA* **7,** 1486–1495.

Stepinski, J., Bretner, M., Jankowska, M., Felczak, K., Stolarski, R., Wieczorek, Z., Cai, A., Rhoads, R. E., Temeriusz, A., Haber, D., and Darzynkiewicz, E. (1995). Synthesis and properties of P^1,P^2-, P^1,P^3- and P^1,P^4-dinucleoside di-, tri- and tetraphosphate mRNA 5′-cap analogues. *Nucleosides Nucleotides* **14,** 717–721.

Strickland, S., Palmer, G., and Massey, V. (1975). Determination of dissociation constants and specific rate constants of enzyme-substrate (or protein-ligand) interactions from rapid reaction kinetic data. *J. Biol. Chem.* **250,** 4048–4052.

Stepinski, J., Jemielity, J., Lewdorowicz, M., Jankowska-Anyszka, M., and Darzynkiewicz, E. (2002). Catalytic Efficiency of Divalent Metal Salts in Dinucleoside 5′,5′-Triphosphate Bond Formation. *In* "Collection Symposium Series" (Z. Tocik and M. Hocek, eds.), **5,** 154–158.

Topisirovic, I., and Borden, K. L. B. (2005). Homeodomain proteins and eukaryotic translation initiation factor 4E (eIF4E): An unexpected relationship. *Histol. Histopathol.* **20,** 1275–1284.

Tomoo, K., Shen, X., Okabe, K., Nozoe, Y., Fukuhara, S., Morino, S., Sasaki, M., Taniguchi, T., Miygawa, H., Kitamura, K., Miura, K., and Ishida, T. (2003). Structural features of human initiation factor 4E, studied by X-ray crystal analysis and molecular dynamics simulations. *J. Mol. Biol.* **328,** 365–383.

Volpon, L., Osborne, M. J., Topisirovic, I., Siddiqui, N., and Borden, K. L. B. (2006). Cap-free structure of eIF4E suggests a basis for conformational regulation of its ligands. *EMBO J.* **25,** 5138–5149.

Watkins, S. J., and Norbury, C. J. (2002). Translation initiation and its deregulation during tumorigenesis. *Br. J. Cancer* **86,** 1023–1027.

Webb, N. R., Chari, R. V., DePillis, G., Kozarich, J. W., and Rhoads, R. E. (1984). Purification of the messenger RNA cap-binding protein using a new affinity medium. *Biochemistry* **23,** 177–181.

Wieczorek, Z., Stepinski, J., Jankowska, M., and Lønnberg, H. (1995). Fluorescence and absorption spectroscopic properties of RNA 5′-cap analogues derived from 7-methyl-, N2,7-dimethyl-, and N2,2,7-trimethylguanosines. *J. Photochem. Photobiol. B* **28,** 57–63.

Worch, R., Niedzwiecka, A., Stepinski, J., Mazza, C., Jankowska-Anyszka, M., Darzynkiewicz, E., Cusack, S., and Stolarski, R. (2005). Specificity of recognition of mRNA cap by human nuclear cap-binding complex. *RNA* **11,** 1355–1363.

CHAPTER TEN

BIOPHYSICAL STUDIES OF THE TRANSLATION INITIATION PATHWAY WITH IMMOBILIZED mRNA ANALOGS

John E. G. McCarthy,* Steven Marsden,* *and* Tobias von der Haar[†]

Contents

Abstract

A growing number of biophysical techniques use immobilized reactants for the quantitative study of macromolecular reactions. Examples of such approaches include surface plasmon resonance, atomic force microscopy, total reflection fluorescence microscopy, and others. Some of these methods have already been adapted for work with immobilized RNAs, thus making them available for the study of many reactions relevant to translation. Published examples include the study of kinetic parameters of protein/RNA interactions and the effect of helicases on RNA secondary structure. The common denominator of all of these techniques is the necessity to immobilize RNA molecules in a functional state on solid supports. In this chapter, we describe a number of approaches by which such immobilization can be achieved, followed by two specific examples for applications that use immobilized RNAs.

* Manchester Interdisciplinary Biocentre, University of Manchester, Manchester, United Kingdom
† Protein Science Group, Department of Biosciences, University of Kent, Canterbury, United Kingdom

Methods in Enzymology, Volume 430
ISSN 0076-6879, DOI: 10.1016/S0076-6879(07)30010-4

1. INTRODUCTION

Biochemical and biophysical experiments that use purified or *in vitro*–generated RNA molecules have a long history in translation-related research. Early insights into the workings of the translational apparatus were gained by studying the action of cell extracts or isolated ribosomes either on synthesized homopolymeric nucleic acids such as poly(U) or on easily obtainable, abundant natural transcripts such as globin mRNA. With the identification of the molecular components of the translational apparatus, characterization of individual macromolecular interactions involving mRNAs became an additional focus of attention. The demonstration that RNA could be generated from DNA templates *in vitro* by use of RNA polymerases from bacteriophages (Summers and Jakes, 1971; Summers and Siegel, 1970) greatly extended the range of RNAs available for use in these techniques.

Over recent years, a particular subset of biophysical techniques for the study of macromolecular reactions has developed in which at least one macromolecule is attached to a solid support. Several such techniques have been successfully adapted for work with immobilized RNAs, including surface plasmon resonance (SPR)–based approaches (von der Haar *et al.*, 2000), atomic force microscopy (AFM) (Marsden *et al.*, 2006), total internal reflection fluorescence microscopy (TIRFM) (Blanchard *et al.*, 2004), and laser-trap methods (Liphardt *et al.*, 2001; Tinoco and Bustamante, 2006). In addition, further biophysical methods are currently emerging (Hauck *et al.*, 2002; Lillis *et al.*, 2006) that also rely on immobilization of macromolecules and that should eventually become useful for RNA-related work.

In the following chapter, we will first describe a selection of procedures for the generation and immobilization of RNAs. We will then examine how two biophysical techniques (SPR and AFM) can be used to study these immobilized species.

2. GENERATION OF RNAs

By far the most widely used mode of generating immobilized RNAs involves transcription of a DNA template *in vitro*, although, depending on the application, the purification of endogenous mRNAs or the purely chemical synthesis of oligomeric RNAs may be viable alternatives. In one of the procedures that can be used to facilitate immobilization, biotin moieties can be randomly incorporated into *in vitro* transcripts simply by including biotinylated nucleotide derivatives in the transcription reaction.

Although the biotins can then be efficiently used for immobilization on streptavidin-containing surfaces, the random distribution of biotins throughout the RNA sequence will result in random orientations of the immobilized RNAs. Because it is generally desirable or even necessary to have more control over the position of the attachment site than this approach offers, RNAs are often synthesized in an unmodified state and then modified for immobilization by subsequent reaction steps.

Many suppliers now offer optimized kits for the *in vitro* transcription of DNA templates containing promoter sequences for either the T7 or SP6 bacteriophage RNA polymerases. Suitable templates for such reactions comprise single-stranded or double-stranded synthetic DNA oligomers, PCR products, and linearized plasmid DNA. Moreover, specialized kits are available for the introduction of eukaryotic mRNA end modifications like cap structures. A typical *in vitro* transcription reaction involves the mixing of template DNA, ribonucleotides, purified RNA polymerase, and required buffer components, incubation at 37° for 30 min to several hours, and finally removal of the template DNA by the addition of an RNAse-free DNAse preparation. The reader is referred to the optimized protocols normally accompanying the commercially available RNA polymerases.

An interesting alternative to *in vitro* transcription that can be used for the enzymatic generation of very short capped or uncapped RNA analogs has been developed by Matsuo *et al.* (2000). In principle, their technique relies on a phage T7-derived enzyme called gene 4 primase, which produces short RNA primers beginning with the sequence pppAC on single-stranded DNA templates that contain the internal sequence GT. Further C residues can be appended to the reaction products if the GT recognition sequence is preceded by further Gs (i.e., for example the DNA template N_xGGGTN_y will result in the synthesis of a mixture of pppAC, pppACC, and pppACCC). The longest RNA analog that can be usefully synthesized with this method is $pppAC_5$, because yields drop dramatically with every additional residue.

Although uses for gene 4 primase products are limited because of their shortness, this method is interesting because substituting ATP in the reaction with the dinucleotide cap analog m^7GpppA results in the quantitative production of capped oligoribonucleotides. Although the introduction of cap structures into transcripts can also be achieved in standard T7 RNA polymerase *in vitro* transcription reactions, transcripts generated from templates containing internal Gs display heterogeneous 5′-ends because of the necessity of including both m^7GpppG (for generating the cap structure) and GTP (for incorporation of guanosines during elongation) in the reaction mixture. In contrast, in the absence of ATP but the presence of m^7GpppA, gene 4 primase products are quantitatively capped and are thus particularly useful in investigating cap-binding reactions. We are currently not aware of a commercial source of this enzyme, but protocols for its efficient purification have been published (Frick *et al.*, 1998; Mendelman and Richardson, 1991). One alternative route to achieving 100% capping is to perform *in vitro* synthesis of an RNA that has only one G (i.e., at the 5' end), although this approach can only be used in a restricted number of cases.

2.1. Modification and immobilization of RNAs

Apart from the cotranscriptional incorporation of biotinylated nucleotides mentioned previously, there are three principal approaches available to immobilize RNA sequences on a solid support (Fig. 10.1).

2.1.1. Oligo capture

A convenient way of immobilization requiring relatively little experimental effort is the capture on chemically synthesized DNA oligonucleotides that include 5'- or 3'-biotin modifications. A 3'-biotinylated DNA oligomer can be stably attached to a streptavidin–coated surface and can be used to capture RNA molecules if its sequence is complementary to the 3'-sequence of the RNA to be captured (Fig. 10.1C). Conversely, a DNA oligomer that is 5'-biotinylated can be used to capture a complementary RNA by means of its 5'-terminal sequence. The strength of immobilization with this approach is dependent on the strength of the DNA/RNA duplex being formed (i.e., essentially on the length of the complementary region of RNA and DNA oligomer). An advantage of this approach is that surfaces can often be easily regenerated by dissolving the DNA/RNA duplex. However, results published that use oligo capture onto a BIAcore chip show that the captured RNA is released with significant rates even under normal buffer conditions and in the absence of any external force (see e.g., Lisdat *et al.*, 2001). Therefore, although this method of immobilization is very convenient, the stability of binding needs to be evaluated for every individual application and experimental setup.

2.1.2. End labeling with biotinylated residues

One method for the targeted introduction of biotinylated residues at the 3'-end of an RNA molecule relies on the ability of certain enzymes to append nucleotides to existing RNA sequences. One published method uses poly(A) polymerase to incorporate biotinylated adenine moieties at the 3'-end of a transcript previously generated by standard *in vitro* transcription (Hendy and Cauchi, 1990). In contrast to the oligo-capture method, the biotin moieties are covalently attached to the RNA, and the strength of the bond to the solid support thus corresponds to the strength of the biotin-Streptavidin bond (i.e., it is quasicovalent). The protocol for this procedure is given in detail in Protocol 1.

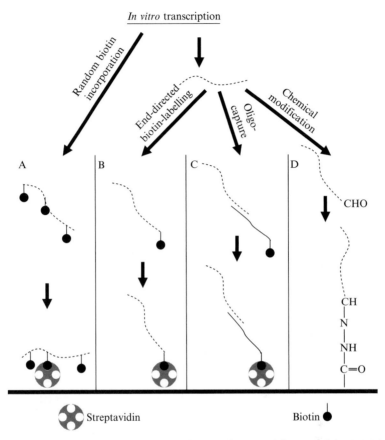

Figure 10.1 Methods for the generation of RNAs for immobilization. (A) Biotinylated nucleotides such as biotin–UTP or biotin–ATP can be incorporated into the transcript by including the modified nucleotides in the transcription reaction. The resulting transcript is randomly labeled, and, hence, the orientation and attachment points of the immobilized RNA will also be of a random nature (giving rise to heterogeneity). (B) Biotinylated nucleotides can be appended to *in vitro* transcripts by means of enzymatic activities (i.e., poly[A] polymerase for the introduction of biotin–ATP at the $3'$-end). (C) Transcripts can be captured by means of complementary, chemically synthesized DNA oligonucleotides that contain $5'$- or $3'$-biotin modifications. (D) Transcript ends can be chemically modified to introduce chemical groups (here an aldehyde group) that are suitable for covalent attachment to reactive groups contained on solid surfaces.

2.1.3. Chemical modification

An alternative strategy to the widely used cotranscriptional introduction of biotins into transcripts is the introduction of chemical groups into the RNA that allow for the establishment of direct, covalent bonds to the surface. Depending on the application, different surface chemistries may be available, although NHS (*N*-hydroxysuccinimide)-ester based chemistries for the coupling of amines have been most widely adopted and are available for most applications. Two published chemical modifications, the introduction of aldehyde groups at the RNA 3'-end and the introduction of a thiol-group at the 5'-end, are described in detail in Protocols 2 and 3. These modifications can be used for covalent binding to NHS-derivatized surfaces by use of bi-active compounds containing an amine group for direct reaction with the surface and a second group for reaction with the modified RNA ends (e.g., cystamine for coupling of the sulfur moiety; hydrazine for coupling of the aldehyde).

2.2. The use of immobilized RNAs in the BIAcore system

In the BIAcore system, published studies that used immobilized RNAs now range from the study of individual protein/RNA and RNA/RNA interactions (Nair et al., 2000; Ptushkina et al., 1999) to the formation of complete E. coli initiation complexes (Karlsson et al., 1999). A problem for the study of RNA/protein interactions is that, on surfaces containing immobilized RNAs, the high density of phosphate moieties creates a relatively high net negative charge. This charge attracts proteins with an isoelectric point above the relevant buffer conditions and can lead to strong, nonspecific ionic interactions. In addition, some nonspecific binding can even be observed if the binding protein has a low isoelectric point but contains localized patches of positive charge. We have found that the inclusion of tRNAs in the eluent buffer to a final concentration of 20 mg/l can greatly reduce nonspecific binding of the cap-binding protein eIF4E to uncapped RNAs, whereas it has no effect on the specific association of this protein with capped RNAs (von der Haar and McCarthy, 2003). This strategy may be generally applicable when studying binding to specific RNA elements, where sequence-independent binding can be out-competed by the soluble tRNA population. Although immobilized RNAs are thus well suited for studying binding of proteins to specific sequence elements, sequence-independent RNA binding is best studied by use of immobilized proteins and soluble RNAs.

We have previously used the principle of studying protein/RNA interactions by means of immobilized RNAs in the BIAcore for investigating the interaction between the cap-binding protein eIF4E and small, capped mRNA analogs that contained single biotin moieties near their 3'-ends (Fig. 10.2A). Although biotin was incorporated cotranscriptionally by means of biotin-UTP in this case, site-directed labeling was achieved by use of a chemically synthesized DNA oligomer as template that introduced only a single U residue near the 3' end of the transcribed sequence. Extra-vidin (Sigma E2511), a streptavidin derivative with favorable nonspecific binding characteristics, was immobilized on CM5 sensor chips by means of a standard amine coupling kit (BIAcore BR 1000–50), and the biotiny-lated RNA was then captured on the immobilized extravidin (Fig. 10.2B).

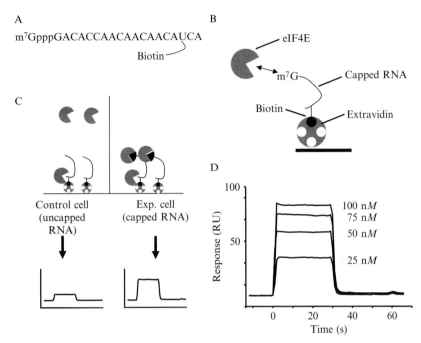

Figure 10.2 Use of immobilized RNA for measuring the eIF4E/cap structure interaction as an example of the usefulness of the SPR techniques. (A) Primary structure of the *in vitro* transcript used for the experiments. Biotin–UTP was inserted cotranscriptionally. (B) Experimental setup for measuring the eIF4E/cap association. Extravidin (a modified streptavidin derivative) was covalently immobilized on a sensor chip surface by standard amine coupling. The biotinylated RNA was then captured on the extravidin, and eIF4E was injected over the RNA-containing surface. (C) Principle of control cell substraction. eIF4E will associate with surfaces containing uncapped RNAs by means of unspecific interactions and with surfaces containing capped RNAs by means of a combination of specific and nonspecific interactions. Subtraction of the control sensorgram on the left from the sensorgram on the right yields a pure sensorgram that reflects only the specific eIF4E/cap interaction. (D) Real-time, control-subtracted sensorgrams from the injection of eIF4E over immobilized, capped RNAs at four different concentrations.

As a control, a second sensor cell was loaded in the same way with a noncapped RNA, to which eIF4E binds with greatly reduced affinity compared with the capped version.

Simultaneous injection of eIF4E over the two surfaces results in a binding signal resulting from nonspecific binding to the RNA in the control cell and a binding signal resulting from nonspecific binding plus specific binding to the cap structure in the experimental cell (schematically shown in Fig. 10.2C). A pure signal for cap-specific binding can then be generated by subtracting the nonspecific signal from that obtained with the capped RNAs.

The extraction of thermodynamic and kinetic data from the curves thus obtained is usually performed in one of two ways. Most frequently, the curve fitting software that is part of the BIAcore package is used to fit the experimentally obtained curves to theoretical binding models. This assumes some prior knowledge of the actual binding model of the interaction, which is usually either a simple Langmuir interaction of the form A + B − AB, where A and B are the interacting components and AB is the complex, or a two-step interaction of the form A + B − AB* − AB, where AB* is an intermediate, unstable complex that can either rapidly decay into its components or undergo a conformational rearrangement to form the final, stable complex AB. Some more complex binding models can also arise from parallel binding reactions caused by impurities or heterogeneities in either of the two binding partners. Although the BIAcore curve-fitting software provides preconfigured models for all of these cases, it is worth noting that the more complicated binding models contain a greater number of free parameters than the Langmuir model and will thus almost always result in apparently better fits. To decide whether the application of more complicated models is appropriate, it is, therefore, essential to test experimentally whether such models are meaningful, for example, by providing evidence for a conformational rearrangement during the interaction that could give rise to a two-step binding mechanism.

Curve fitting will produce detailed kinetic data, including values for the on and off rates of the interaction, but this procedure is limited by the number of data points that are generated before an equilibrium is reached. As can be seen in the example of the eIF4E/cap interaction described later, very fast off rates reach an equilibrium within seconds or less, and in this case, the small number of data points before reaching a plateau in the SPR signal does not permit meaningful curve fitting to be performed. However, in this case equilibrium binding levels (represented by the plateau in the binding curve reached during the injection of the binding partner) can be easily analyzed, and a simple plot of equilibrium binding levels against concentration of the binding partner allows at least the analysis of relevant equilibrium binding constants. In BIAcore experiments, the latter are equivalent to the concentration of injected binding partner at which half-maximal equilibrium binding levels are obtained.

Some typical data obtained with the cap-binding protein eIF4E2 from *S. pombe* (19) are shown in Fig. 10.2D. The resulting sensorgrams are indicative of a reaction where eIF4E cycles extremely fast on and off the cap structures. For the reasons explained previously, the on and off rates of this interaction are too fast to be directly determined by curve fitting. However, the equilibrium affinity could easily be determined by plotting the equilibrium binding levels against the concentration of the injected protein (Ptushkina *et al.*, 2001). The rapid binding/release cycle for the yeast eIF4E/cap interaction was later confirmed in stopped-flow experiments, which revealed on rates of 10^8 to 10^9 M^{-1} sec^{-1} for the interaction with m^7GTP, close to the apparent diffusion limit for this interaction (TvdH, unpublished data). Similar association rates were also obtained for the human protein (Blachut-Okrasinska *et al.*, 2000). In summary, these and other published data show that immobilized RNAs can be a useful tool for studying a variety of protein/RNA interactions.

2.3. The use of immobilized RNAs for atomic force microscopy

In atomic force microscopy (AFM) experiments, immobilized RNAs can be used to probe secondary structure elements, as well as the effect that RNA binding proteins have on the stability of these structural features (Figs. 10.3 and 10.4). Essentially, it is the force required to pull apart secondary structure elements in an RNA that is measured with this particular technique. As an RNA suspended between a surface and an AFM cantilever tip is stretched by slow movement of the tip, the force applied to the latter increases with the distance between tip and surface (Fig. 10.4A). However, the opening of secondary structure elements will, in sudden drops of force applied to the tip (see the transition from step iii to step vi in Fig. 10.4A), result in so-called discontinuity features in the force-distance curves. Analysis of these discontinuity features allows for the interpretation of parameters relevant to the stability of the secondary structures introduced into the immobilized RNA.

We have recently described a series of such experiments designed to probe the effect of RNA helicase activity on artificial hairpin structures introduced into the yeast *GCN4* mRNA leader (Marsden *et al.*, 2006). Suitable RNAs that can be attached to a gold-coated glass slide by means of

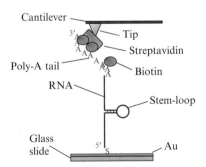

Figure 10.3   A summary of the method used for suspending an RNA molecule between the AFM tip and a gold-coated microscope slide. The RNA molecule was attached to the gold surface by means of a thiol modification at the 5′ end. The 3′ end was modified by the incorporation of biotin-ATP residues, which allowed picking up of this end with a streptavidin-coated cantilever tip. (See color insert.)

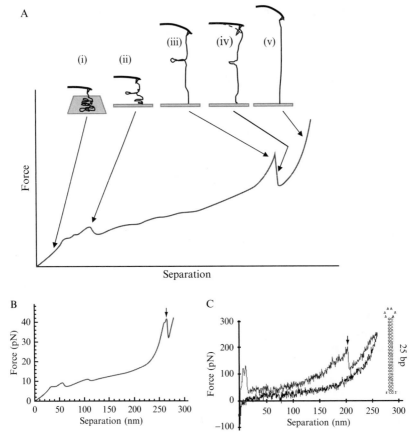

Figure 10.4 (A) A theoretical force-extension curve for structured-RNA stretching carried out by the AFM. The curve is annotated with cartoons showing the state of the RNA molecule and AFM cantilever at different extensions. (i) RNA is uncoiled off the gold surface, causing an entropic force increase; (ii) very weak secondary structure interactions are removed; (iii) enthalpic forces increase as the molecule is pulled taut and force becomes high enough for the hydrogen bonds in the strong specific stem loop to break; (iv) force temporarily decreases as the slack released from stem-loop opening is pulled out; (v) force increases further as the RNA is pulled taut once more. (B) A theoretical retraction force curve for the stretching of the *GCS4* L1-RNA transcript (with a 65-nt Poly[A] tail). This trace is adapted from data generated by use of the online "RNA pulling server" at http://bioserv.mps.ohio-state.edu/rna (Gerland *et al.*, 2001). The model assumes a temperature of 37°, 1 *M* NaCl, and a nucleotide length of 0.334 nm. The predicted ∼40pN GC-rich stem-loop opening feature is labelled with an arrow. (C) Example of an AFM force-curve representing stretching of a *GCN4* RNA molecule containing the GC-rich (25 base pair) stem loop (inset). The discontinuity feature resulting from stem-loop opening is indicated by a small arrow. The approach curve (bottom curve) runs from right to left and the retract curve (top curve) from left to right. (See color insert.)

a 5′ thiol modification, and that have a biotinylated 3′ poly-A tail allowing pickup by a streptavidin coated AFM tip (as shown in Fig. 10.3), can be synthesized by sequential modification of a transcript with the procedures described in Protocols 3 and 1.

For the best results, the RNA should be at least 600-nt long and contain a maximum of two stable stem loops. The *GCN4* leader sequence from yeast is an example of a suitable, relatively "structureless" control construct into which specific stem-loop sequences can be introduced. The RNA construct can be transcribed from this template by use of individually prepared transcription components; however, we have found that very good yields can be achieved with an RNAmaxx transcription kit (Stratagene). A high initial yield of 100 μg is required here, because only 10% of the transcript may remain after the two subsequent modification steps. To minimize RNA degradation, gloves should be worn at all times, and DEPC-treated water containing RNase inhibitors should be used in buffers where possible. Also, the sample should be kept on ice between steps and stored at −80° overnight.

2.3.1. Coating of AFM tips with streptavidin

Si_3N_4 NPS AFM tips (Veeco DNPS-20) were modified by soaking overnight in 50 μl BSA-biotin (1 mg/ml (Sigma A8549-10MG)) at 37°. The tips were then washed with 50 μl DEPC-treated H_2O and fixed in 1% glutaraldehyde (Sigma G7651) for 30 sec before being washed again in 50 μl DEPC-treated H_2O. Immediately before an experiment, 50 μl recombinant streptavidin (1 mg/ml, Roche Diagnostics 11721666001) was added and allowed to bind for 5 min. The tips were then washed with 50 μl DEPC-treated H_2O.

2.3.2. Binding of the RNA construct to a gold-coated AFM slide

To allow primarily single molecules to be picked up, the RNA constructs are best added to the gold surface at a concentration of approximately 500 pM, although this may vary according to conditions. Our purified Th-RNA-biotin constructs were routinely maintained at a higher concentration of 2.7 nM for long-term storage. For application to the gold surface, we diluted 1 μl of this stock into 50 μl of 150 mM NaCl, 1× BSA, 10 mM Tris-HCl, 10 mM $MgCl_2$, 1 U/μl RNasin in DEPC-H_2O. Immediately before experimentation, 10 μl of the diluted RNA were spread evenly over an 11 × 11-mm gold-coated slide (Gold-Arandee™, http://www.Arrandee.com) and left at room temperature for 20 min. The slide was then washed three times with 200 μl of "RNA pulling buffer" (150 mM NaCl, 10 mM Tris-HCl, 10 mM $MgCl_2$, 1 U/μl RNasin) by tilting the slide with tweezers and allowing capillary action to draw the solution off the slide. The slide was then quickly placed on the AFM stage (by means of a mounting disk coated in adhesive), and 50 μl of RNA pulling buffer was dropped onto it. The BSA in the RNA binding buffer is sufficient to reduce nonspecific tip–surface interactions, which are in any case usually of minor importance because the recorded events (the opening of a stem loop) occur only after the backbone of the RNA is stretched out and the tip is physically removed from the surface.

2.3.3. Collecting AFM force spectroscopy data

The following describes the procedure for obtaining RNA force curves with a Digital Instruments AFM (Veeco), with a multimode head, PF scanner, Nanoscope IIIa, and Picoforce controllers with an extender module and Nanoscope software. AFM RNA force spectroscopy was performed by use of the liquid cell and a short (100 μm) V-shaped silicon nitride AFM cantilever with a spring constant of approximately 0.02 N/m, because these produce the least noise of the tips tested. The tip was coated with streptavidin as described previously. Before placing the liquid cell into the AFM head, a 2-μl drop of RNA pulling buffer was placed next to the tip. The buffer drop is then drawn underneath the tip by capillary forces, which prevents the formation of air bubbles when the tip is lowered onto the sample buffer.

The laser was then aligned on the back of the cantilever and the spring constant of the cantilever estimated by use of the thermal tuning method while the tip was in the sample buffer but still greater than 2 μm from the surface. The tip was manually lowered close to the surface, and the software was used to engage the tip in contact mode with a scan size of less than 1 nm and a set point of 1.5 V. Switching to picoforce mode allowed force curves to be generated. A single force curve produced with the trigger force disabled and a ramp length of 1 μm should show a constant compliance region as the tip pushes into the surface. This region was used to determine the cantilever deflection sensitivity.

To obtain RNA force curves, continuous cycles of approach and retract with a trigger force of 200 pN, retract velocity of 1 μm/sec, a surface delay of 400 ms, and a ramp length of just above the contour length of the RNA construct were set up. The tip was moved around the surface in the x and y directions in 100-nm increments until an area was found in which many force-distance curves with long adhesion lengths before pull-off could be measured. Once in this area, the ramp length was reduced to less than 200 pN. Force curves in which both the approach and retract curves bend (rather than the retract curve only) started to appear at this ramp length. The approach curve also bends, because the RNA remains attached to the tip at the end of the retract cycle. Once it was evident that the RNA had been picked up stably in this way, the separation from the surface was increased manually by use of the picoangler, until the contour length of the RNA construct was reached.

By use of this method, up to 100 consecutive full-length, reversible, RNA stretching force curves per RNA molecule should be obtainable. Once the RNA is pulled off the tip, and if it does not reattach within a couple of minutes of pulling, the tip should be moved to a new position on the slide, and the process should be repeated. We found that the RNA preparation can last up to 4 h before RNA force-distance curves can no longer be generated.

To ensure the authenticity of the data produced in these experiments, a series of controls need to be performed. In particular, controls with an unmodified RNA transcript, and with thiolated but not biotinylated RNA, and biotinylated but not thiolated RNA, should show a greatly reduced occurrence of long adhesion length force curves compared with the complete Thiol-RNA-biotin constructs.

At a concentration of 540 pM, control RNA constructs with no specific secondary structure should produce force curves with no discontinuity features within the last 50 nm of the pull in 98% of curves. Strong (>50 kcal/mol) GC-rich stem loops placed within the construct should introduce clear discontinuities (like those shown in Fig. 10.4) in >70% of full-length RNA curves. Weaker structures (50 kcal/mol) show discontinuities that are less visible above background noise and can be identified in less than 50% of curves.

Because the AFM operates at a higher loading rate than is used by other force spectroscopy techniques (such as laser tweezers) and than is assumed by theoretical predictions, measurements should always be taken at different loading rates within the range of the AFM. The modal force should increase linearly as the logarithm of the loading rate is increased. In some cases, this allows extrapolation back to the lower loading rates used by other methods to obtain comparable force estimates. However, if the unfolding event follows a different pathway at the higher range of loading rates, the extrapolation may meet the x-axis before the lower loading rates are reached. An approximate value at lower loading rate may still be obtainable, however, by plotting the mean force of stem-loop opening directly against loading rate and using curve fitting software to extrapolate backwards.

2.3.4. Analysis of AFM force spectroscopy data

Because the process of selecting RNA curves and then identifying secondary structure features on these curves is potentially highly subjective, we have introduced a disciplined regimen designed to help in deciding which curves to include. As a test for subjectivity, a series of "blind" experiments were performed, in which sets of force curves generated with secondary structure-free and stem-loop containing RNAs were mixed and subsequently analyzed. The two types of construct could be reliably distinguished by this analysis, hence proving that any preconceptions about the data do not bias the analysis.

The amount of secondary structure detected by AFM force spectroscopy can be represented in terms of the percentage of force curves containing a discontinuity feature and by the opening force distribution of these features. For reliable analysis, opening forces from many hundreds of curves need to be collected into histograms. When collecting statistics, Th–RNA–biotin force curves are taken to be those reversible curves that display the distinctive initial gradual slope that reach an adhesion length of at least 250 nm and a force of at least 180 pN. The stem–loop opening features are recorded when only a single discontinuity is seen in the last 50 nm of the pull. Only curves where the region between 50 nm from the start and 50 nm from the end of the pull is smooth are counted in the statistical analysis (in practice this includes most curves under our experimental conditions). Curves with a single discontinuity force larger than 250 pN are counted as extraordinarily large forces because of potential stabilization of the stem loop by protein binding and so should not be included in the opening force statistics.

2.3.5. Investigating helicase activity by use of AFM RNA force spectroscopy

RNA helicases can be added to the described AFM RNA force spectroscopy system to measure their effect on RNA secondary structure. AFM RNA force spectroscopy was at first set up to take measurements on RNA alone, to ensure that the tip and working conditions were suitable for producing good RNA force curves. The AFM slide was then removed and washed with 200 μl RNA pulling buffer. Then, 50 μl of RNA pulling buffer containing 1 mM ATP and the RNA helicase were added to the surface. Generation of RNA force curves was then started as quickly as possible. We found that pumping RNA helicase directly into the liquid cell is not advisable, because this may cause removal of RNA from the gold slide and does not allow the concentration of helicase supplied in the solution to be accurately controlled. Moreover, it will not allow instant measurements on the addition of helicase, because pumping in solution invariably detaches RNA from the AFM tip and temporarily destabilizes the system. A potential solution to this problem would be to use a helicase buffer containing ATP that is trapped in a photosensitive cage and that could be released on exposure to UV light (from a flash lamp) to activate the helicase without disturbing the RNA attachment.

3. PROTOCOL 1: 3′-END BIOTINYLATION OF A TRANSCRIPT BY USE OF POLY(A) POLYMERASE

Starting point for this protocol is a T7 RNA polymerase transcript that has been DNAse 1 treated and purified with phenol extraction and ethanol precipitation. The volumes and amounts given are optimal for biotinylation of the product of a 25-μl reaction with the RNAmaxx kit (Strategene #200339).

Additional reagents: Poly-A polymerase with 5× buffer (Amersham Biosciences E74225Y), RNAse inhibitors (e.g., Promega N2111), 1 mM biotin-17-ATP (Enzo labs 42817), 10 mM ATP (Sigma A1852), probequant G-50 microcolumns (Amersham Biosciences 28-9034-08). Also required is DEPC-treated water.

1. Resuspend the precipitated RNA in 17 μl of DEPC-treated water. Add 1 μl RNAse inhibitor, 6 μl 5× polymerase buffer, 4 μl biotin-17-ATP, 1 μl ATP, and 1 μl of the polymerase and mix.
2. Incubate at 37° for 30 min.
3. For removal of the unincorporated nucleotides, equilibrate a G-50 microcolumn by adding 50 μl DEPC-treated water and centrifuging at 2000 rpm for 2 min. Repeat this step twice.
4. Add the polyadenylation reaction to the column, stand at room temperature for 30 sec, then spin at 2000 rpm for 2 min.
5. Analyze a small sample of the eluate on an agarose or polyacrylamide gel suitable for the size of the original RNA. Successful adenylation can be observed by an apparent increase in molecular weight of the transcript, as well as a smeared appearance of the band. If necessary, biotinylation can also be demonstrated with an anti-biotin antibody.

4. PROTOCOL 2: INTRODUCTION OF ALDEHYDE GROUPS BY OXIDATION OF THE mRNA 3'-END

Starting point for this protocol is a T7 RNA polymerase transcript that has been DNAse 1 treated and purified with phenol extraction and ethanol precipitation. The volumes and amounts given are suitable for biotinylation of the product of a 25-μl reaction with the RNAmaxx kit (Strategene #200339).

Additional reagents: 0.1 M Sodium acetate, pH 5.1, sodium meta-periodate (NaIO$_4$, Sigma S1878), NAP 5 column (Pharmacia), DEPC-treated water.

1. Dissolve the pelleted RNA in 225 μl of the sodium acetate solution.
2. Freshly dissolve 20 mg of the meta-periodate in 1 ml of water. Add 25 μl of this stock to the dissolved RNA (to give ~10 mM final concentration of the meta-periodate).
3. Incubate the mixture for 1 h in the dark.
4. Ethanol precipitate the RNA and redissolve the pellet in 250 μl of water.
5. Pass the redissolved RNA through a NAP5 column per the manufacturers instruction to remove all traces of the oxidizing agent.

This procedure essentially destroys the ribose moiety of the 3′-terminal nucleotide by cleaving the ribose ring and introducing two aldehyde groups at the C2 and C3 carbons. Procedures for covalently binding oxidized RNAs for example to BIAcore chips have been described (Karlsson *et al.*, 1999), and can be performed with standard coupling kits available from this company.

5. PROTOCOL 3: THIOLATION AT THE mRNA 5′-END BY USE OF POLYNUCLEOTIDE KINASE

Starting point for this protocol is a T7 RNA polymerase transcript that has been DNAse 1 treated and purified with phenol extraction. For complete removal of unincorporated nucleotides, the RNA pellet should be dissolved in water passed through a desalting column (e.g., Amersham G–50 microcolumns) and then precipitated again with ethanol. The volumes and amounts given are optimal for modification of the product of a 25-μl reaction with the RNAmaxx kit (Strategene #200339).

Additional reagents: 5′ end labeling kit (Vector Labs MB-9001) including shrimp alkaline phosphatase, ATPγS, reaction buffer and polynucleotide kinase; RNAse inhibitors (e.g., Promega N2111), DEPC-treated water.

1. Resuspend the precipitated RNA in 7 μl of DEPC-treated water.
2. Add 1 μl universal reaction buffer, 1 μl RNAse inhibitors, and 1 μl of the shrimp alkaline phosphatase. Incubate at 37° for 30 min.
3. Add 2 μl of universal reaction buffer, 1 μl ATPγS, 6 μl of DEPC-treated water, and 2 μl polynucleotide kinase. Incubate at 37° for 30 min.
4. Make up the volume to 100 μl by adding 79 μl of DEPC-treated water. Purify the labeled RNA by phenol extraction and ethanol precipitation. An alternative purification method involves specific binding of the labeled RNA to thiopropyl sepharose (e.g., Sigma T8387) and subsequent elution that uses reducing agents like 2-mercaptoethanol or DTT (J. Lorsch, personal communication).

Thiol incorporation into intact RNA can be detected by taking a 2-μl aliquot of the labeled sample and incubating with 20 μg biotin-PEG$_3$-maleimide (BP$_3$M) in a final volume of 40 μl DEPC-H$_2$O at 65° for 30 min, running on a 12-cm 2.2% denaturing formaldehyde gel for 2 h at 100V, 400mA, and performing a Northern blot onto a nylon membrane. The membrane is then blocked by soaking for 30 min in 20 ml blocking solution (1% casein, 1× TBS [10 mM Tris, pH 7.5, 100 mM NaCl, 0.1% Tween 20]) and soaked for 4 h in 20 ml of a solution containing 0.5% casein, 1× TBS, 5 μg streptavidin-alkaline phosphatase, before being washed three times in 1× TBS. The blot is then developed by soaking in

30 ml developing solution (50 mM NaHCO$_3$, 5 mM MgCl, 70 μg NBI, 16 μg BCIP) for 10 min. Brown bands on the gel indicated AP activity, hence indirectly showing the presence of the thiol.

Alternately, BP$_3$M-Th-RNA can be run on a G–50 column and then spotted directly onto a nylon membrane that is probed and developed as previously. Comparing the intensity of the brown spot with that produced by use of a commercially available Th-RNA oligo (assumed to have 100% Th-incorporation) allows the percentage of Th-incorporated to be determined. Thiolation of RNA according to the profile described here typically gives a yield of 10 to 15%.

ACKNOWLEDGMENTS

J.E.G.M. thanks the following UK funding bodies for supporting relevant work in his laboratory: BBSRC, EPSRC, The Wellcome Trust, The Royal Society, and the Wolfson Foundation. T.v d H. thanks the Wellcome Trust (UK) for a Research Career Development Fellowship.

REFERENCES

Blachut-Okrasinska, E., Bojarska, E., Niedzwiecka, A., Chlebicka, L., Darzynkiewicz, E., Stolarski, R., Stepinski, J., and Antosiewicz, J. M. (2000). Stopped-flow and Brownian dynamics studies of electrostatic effects in the kinetics of binding of 7-methyl-GpppG to the protein eIF4E. *Eur. Biophys. J.* **29,** 487–498.

Blanchard, S. C., Kim, H. D., Gonzalez, R. L., Jr., Puglisi, J. D., and Chu, S. (2004). tRNA dynamics on the ribosome during translation. *Proc. Natl. Acad. Sci. USA* **101,** 12893–12898.

Frick, D. N., Baradaran, K., and Richardson, C. C. (1998). An N-terminal fragment of the gene 4 helicase/primase of bacteriophage T7 retains primase activity in the absence of helicase activity. *Proc. Natl. Acad. Sci. USA* **95,** 7957–7962.

Gerland, U., Bundschuh, R., and Hwa, T. (2001). Force-induced denaturation of RNA. *Biophys. J.* **81,** 1324–1332.

Hauck, S., Drost, S., Prohaska, E., Wolf, H., and Dübel, S. (2002). Analysis of Protein Interactions Using a Quartz Crystal Microbalance Biosensor. *In* "Protein–Protein Interactions" (E. Golemis, ed.), pp. 273–283. CSHL Press, New York.

Hendy, J. G., and Cauchi, M. N. (1990). Direct detection of beta thalassemic mutations: Use of biotin-labelled allele specific probes. *Am. J. Hematol.* **34,** 151–153.

Karlsson, M., Pavlov, M. Y., Malmqvist, M., Persson, B., and Ehrenberg, M. (1999). Initiation of *Escherichia coli* ribosomes on matrix coupled mRNAs studied by optical biosensor technique. *Biochimie* **81,** 995–1002.

Lillis, B., Manning, M., Berney, H., Hurley, E., Mathewson, A., and Sheehan, M. M. (2006). Dual polarisation interferometry characterisation of DNA immobilisation and hybridisation detection on a silanised support. *Biosens. Bioelectron.* **21,** 1459–1467.

Liphardt, J., Onoa, B., Smith, S. B., Tinoco, I. J., and Bustamante, C. (2001). Reversible unfolding of single RNA molecules by mechanical force. *Science* **292,** 733–737.

Lisdat, F., Utepbergenov, D., Haseloff, R. F., Blasig, I. E., Stocklein, W., Scheller, F. W., and Brigelius-Flohe, R. (2001). An optical method for the detection of oxidative stress using protein-RNA interaction. *Anal. Chem.* **73**, 957–962.

Marsden, S., Nardelli, M., Linder, P., and McCarthy, J. E. (2006). Unwinding single RNA molecules using helicases involved in eukaryotic translation initiation. *J. Mol. Biol.* **361**, 327–335.

Matsuo, H., Moriguchi, T., Takagi, T., Kusakabe, T., Buratowski, S., Sekine, M., Kyogoku, Y., and Wagner, G. (2000). Efficient synthesis of ^{13}C, ^{15}N-labeled RNA containing the cap structure m^{7}GpppA. *J. Am. Chem. Soc.* **122**, 2417–2421.

Mendelman, L. V., and Richardson, C. C. (1991). Requirements for primer synthesis by bacteriophage T7 63-kDa gene 4 protein. Roles of template sequence and T7 56-kDa gene 4 protein. *J. Biol. Chem.* **266**, 23240–23250.

Nair, T. M., Myszka, D. G., and Davis, D. R. (2000). Surface plasmon resonance kinetic studies of the HIV TAR RNA kissing hairpin complex and its stabilization by 2-thiouridine modification. *Nucleic Acids Res.* **28**, 1935–1940.

Ptushkina, M., Berthelot, K., von der Haar, T., Geffers, L., Warwicker, J., and McCarthy, J. E. (2001). A second eIF4E protein in Schizosaccharomyces pombe has distinct eIF4G-binding properties. *Nucleic Acids Res.* **29**, 4561–4569.

Ptushkina, M., von der Haar, T., Karim, M. M., Hughes, J. M., and McCarthy, J. E. (1999). Repressor binding to a dorsal regulatory site traps human eIF4E in a high cap-affinity state. *EMBO J.* **18**, 4068–4075.

Summers, W. C., and Jakes, K. (1971). Phage T7 lysozyme mRNA transcription and translation *in vivo* and *in vitro*. *Biochem. Biophys. Res. Commun.* **45**, 315–320.

Summers, W. C., and Siegel, R. B. (1970). Transcription of late phage RNA by T7 RNA polymerase. *Nature* **228**, 1160–1162.

Tinoco, I. Jr., Li, P. T., and Bustamante, C. (2006). Determination of thermodynamics and kinetics of RNA reactions by force. *Q. Rev. Biophys.* **39**, 325–360.

von der Haar, T., Ball, P. D., and McCarthy, J. E. (2000). Stabilization of eukaryotic initiation factor 4E binding to the mRNA 5'-Cap by domains of eIF4G. *J. Biol. Chem.* **275**, 30551–30555.

von der Haar, T., and McCarthy, J. E. (2003). Studying the assembly of multicomponent protein and ribonucleoprotein complexes using surface plasmon resonance. *Methods* **29**, 167–174.

CHAPTER ELEVEN

PROTECTION-BASED ASSAYS TO MEASURE AMINOACYL-tRNA BINDING TO TRANSLATION INITIATION FACTORS

Yves Mechulam, Laurent Guillon, Laure Yatime, Sylvain Blanquet, *and* Emmanuelle Schmitt

Contents

Abstract

To decipher the mechanisms of translation initiation, the stability of the complexes between tRNA and initiation factors has to be evaluated in a routine manner. A convenient method to measure the parameters of binding of an aminoacyl-tRNA to an initiation factor results from the property that, when specifically complexed to a protein, the aminoacyl-tRNA often resists spontaneous deacylation. This

Laboratoire de Biochimie, CNRS Ecole Polytechnique, Palaiseau Cedex, France

Methods in Enzymology, Volume 430
ISSN 0076-6879, DOI: 10.1016/S0076-6879(07)30011-6

chapter describes the preparation of suitable aminoacyl-tRNA ligands and their use in evaluating the stability of their complexes with various initiation factors, such as e/aIF2 and e/aIF5B. The advantages and the limitations of the method are discussed.

1. INTRODUCTION

Initiation of translation is conserved in all organisms in that selection of a specialized methionine initiator tRNA is always required to identify the proper start codon on mRNA. However, the initiation factors involved in the recruitment of the initiator tRNA by the translation machinery differ in the three domains of life. In bacteria, the monomeric factor IF2 is responsible for channeling initiator formyl-Met-tRNA$_f^{Met}$ toward the small ribosomal subunit (Gualerzi *et al.*, 2000). In Eukarya and in Archaea, initiator Met-tRNA$_i^{Met}$ is carried toward the small ribosomal subunit by the heterotrimeric e/aIF2 factor. Moreover, in these latter organisms, e/aIF5B, a close homolog of bacterial IF2, also participates in initiator tRNA recruitment (Choi *et al.*, 1998). Similarly to IF2, eIF5B promotes joining of the ribosomal subunits during the final steps of translation initiation (Pestova *et al.*, 2000).

Recruitment of an initiator tRNA to the ribosome implies its specific recognition by initiation factors. Therefore, methods to follow the formation of complexes between tRNA and initiation factors are important to unravel the molecular mechanisms of translation initiation. Assays for following the assembly of tRNA/initiation factor complexes have also proved useful for the characterization of inhibitors of protein synthesis (Robert *et al.*, 2006). For many years, the most widely used technique for studying the binding of nucleic acids to proteins has been the nitrocellulose-filter binding assay. This powerful method is based on the preferential retention on a nitrocellulose membrane of nucleic acids complexed to proteins compared with free nucleic acids. The precision and accuracy of this technique for quantitative studies have been improved by the use of an additional DEAE membrane able to trap the free nucleic acid (Wong and Lohman, 1993). Such assays have already proved useful for the study of eukaryotic initiation factors (e.g., Kapp and Lorsch, 2004). However, despite these improvements, the nitrocellulose-binding assay remains a nonequilibrium technique. In particular, the method implies that the complex under study does not dissociate during the filtration step. It can, therefore, be used accurately only in cases where the kinetic dissociation constant of the complex is sufficiently slow, a condition that, in general, corresponds only to high-affinity complexes. The same limitations also apply to another widely used technique, the gel retardation assay. This technique is based on distinct electrophoretic mobilities of the free and complexed nucleic acids, provided that the complex remains stable during the time course of the electrophoresis (Carey, 1991). A third powerful method is the use of the intrinsic protein fluorescence because of the tryptophan residues. In this technique, the fluorescence of the protein (λexc = 295 nm; λem = 340 nm) is monitored during titration by the nucleic acid. Although this is a true equilibrium technique, its setup requires that the protein/nucleic acid complex has a fluorescence yield significantly different from that of the free protein. This method was successfully used for the characterization of several aminoacyl-tRNA synthetase/tRNA complexes (e.g., Avis and Fersht, 1993; Blanquet et al., 1973; Helene et al., 1969).

Interaction of an aminoacyl-tRNA with a translation factor often results in the protection of the esterified nucleic acid against spontaneous deacylation, because the aminoacyl moiety is buried inside the enzyme, thus shielding the ester bond against solvent access (Nissen et al., 1995, 1999; Schmitt et al., 2002). This property enables us to easily assess the stability of a protein/nucleic acid complex, as first established in the case of the elongation factor EF-Tu (Pingoud et al., 1977) or of bacterial IF2 (Petersen et al., 1979). The aim of this chapter is to describe recently used methods taking advantage of protection against spontaneous deacylation to study the interaction of aminoacyl-tRNAs with various initiation factors, such as the archaeal factors aIF2 and aIF5B, and yeast eIF5B (Guillon et al., 2005; Pedulla et al., 2005; Yatime et al., 2005). Indeed, all these translation factors interact with the aminoacyl moiety of the methionylated initiator tRNA. The methods for preparation of the aminoacyl-tRNA substrates, as well as the assays themselves, are detailed. Because these assays are performed under equilibrium conditions, accurate dissociation constants can be derived, provided however that the pre-stationary dynamics leading to the studied complex is fast compared with the deacylation rate. This condition is generally satisfied with medium- to low-affinity complexes. Therefore, the protection-based assays are complementary to the nitrocellulose binding ones, suitable for the study of high-affinity complexes.

2. PURIFICATION OF OVERPRODUCED INITIATOR tRNA

Overproduction of tRNA in E. coli from a cloned gene is a straightforward way to obtain large amounts of material. Various tRNAs, including archaeal initiator tRNAs, can be produced by use of this method. However, the produced tRNAs will carry posttranscriptional modifications specific to the E. coli context, which do not necessarily match the modifications in the authentic tRNA. The desired tRNA gene can be easily assembled from synthetic nucleotides and cloned into an expression vector under the control of the strong constitutive lpp promoter (Meinnel et al., 1988).

Two expression vectors are generally used, pBSTNAV2 (Meinnel *et al.*, 1992) or pBSTNAV3S (Meinnel and Blanquet, 1995). The use of the latter allows accurate 5′-processing by RNase P of tRNA substrates carrying mismatched nucleotides at position 1–72 (such as *E. coli* tRNA$_f^{Met}$). Specificity of RNaseP in this maturation has been described in detail (Meinnel and Blanquet, 1995). Here, we describe the production and purification of *E. coli* initiator tRNA$_f^{Met}$. This tRNA can generally be used as a good model ligand of archaeal aIF2, as well as of eukaryotic/archaeal e/aIF5B. To overproduce *E. coli* tRNA$_f^{Met}$, we use the pBStRNAfmetY2 plasmid (Meinnel and Blanquet, 1995). A recombination deficient strain, such as XL1-Blue (Stratagene) or JM101Tr (Hirel *et al.*, 1988), is preferred to host the pBSTNAV-based tRNA-overproducing plasmids. Total tRNA is extracted by use of the procedure of Zubay (1962). Finally, initiator tRNA is purified by anion exchange chromatography (Guillon *et al.*, 1992; Meinnel *et al.*, 1988).

2.1. Buffers

T1: 1 mM Tris-HCl, pH 7.4; 10 mM magnesium acetate
T2: 20 mM Tris-HCl, pH 7.5; 0.1 mM EDTA; 8 mM MgCl$_2$; 0.2 M NaCl
T3: 20 mM Tris-HCl, pH 7.5; 0.1 mM EDTA; 8 mM MgCl$_2$; 1.0 M NaCl

2.2. Protocols

1. Extraction of tRNAs

- Grow to saturation the cells containing the plasmid at 37° in 1L of 2× TY containing 50 μg/ml of ampicillin.
- Harvest the cells by centrifugation, and resuspend in 8.6 ml of buffer T1.
- Add 10 ml of phenol carefully saturated with buffer T1; mix thoroughly by vortexing.
- Centrifuge 30 min at 15,000g (room temperature) and withdraw the aqueous phase.
- Add 0.1 volume of 5 M NaCl and 2.2 volumes of cold ethanol; mix.
- Centrifuge 30 min at 15,000g (4°) and discard the supernatant.
- Resuspend the pellet in 5 ml of 1 M NaCl.
- Centrifuge 30 min at 15,000g (4°) and discard the pellet containing mRNA.
- Add 2 volumes of cold ethanol to the supernatant, mix, and centrifuge 30 min at 15,000g (4°).
- Discard the supernatant and carefully drain the pellet to remove any fluid.

- Dissolve the pellet in 2 ml of 1.8 M Tris-HCl (pH 8.0).
- Incubate 1 h and 30 min at 37°. This will result in deacylation of tRNAs.
- Precipitate the stripped tRNAs by adding 0.2 ml of 5 M NaCl and 4.4 ml of cold ethanol.
- Dissolve the precipitate in 8 ml of buffer T2 before further purification.

2. Purification of Initiator tRNA

- Load the tRNA preparation at 2.5 ml/min on a Q-Sepharose Fast-Flow column (GE Healthcare; 15 × 16 mm) equilibrated in buffer T2.
- By use of buffers T2 and T3, perform a linear gradient from 0.36 M NaCl to 0.56 M NaCl (2.5 ml/min; 0.1 M/h).
- Measure the methionine acceptance along the profile, pool the desired fractions, raise the NaCl concentration to at least 0.5 M, and precipitate tRNA by adding either 2 volumes of ethanol or 0.6 volume of isopropanol.
- Rinse the precipitate and dissolve in the desired buffer (see note 2 below). Store at −20°.

Notes

1. *E. coli* initiator tRNA is the first tRNA to be eluted. If the overproduction is correct (approximately 20-fold), $tRNA_f^{Met}$ is eluted as a large peak well separated from many other tRNAs. Purity of the recovered material is evaluated by measuring its methionine acceptance. A satisfying preparation has an acceptance of 1300 to 1500 picomoles of methionine per A_{260} unit. This purity is sufficient in most cases. If purity of a recovered tRNA is not satisfactory, for instance in the case in which a tRNA species is less efficiently produced and/or less separated from other tRNAs, a second chromatographic step can be added. In such a case, the most efficient purification can generally be obtained by use of hydrophobic interaction chromatography, for instance on a Phenyl-Sepharose column (GE-Healthcare) eluted by a reverse ammonium sulfate gradient.
2. *E. coli* initiator tRNA can be kept for several months at −20° in water. Some tRNAs, in particular the heterologously expressed ones, tend to unfold when maintained in solution in the absence of magnesium. In this case, it is recommended to store the tRNA in buffer T2. tRNA can nevertheless generally be refolded by heating for 5 min at 65° in buffer T2.

3. Preparation of Methionylated tRNA

Most tRNAs that carry a CAU anticodon can be efficiently aminoacylated with *E. coli* methionyl-tRNA synthetase as the catalyst (Meinnel *et al.*, 1991; Schmitt *et al.*, 1993; Schulman and Pelka, 1988), although a few exceptions occur such as eukaryotic cytoplasmic elongator tRNA^{Met} from mammals, plants, or yeast, all of which are poor substrates of the bacterial synthetase (Guillemaut and Weil, 1975; Meinnel *et al.*, 1992; Petrissant *et al.*, 1970; RajBhandary and Ghosh, 1969). To catalyze aminoacylation, we find it convenient to use a fully active truncated monomeric form of the *E. coli* methionyl-tRNA synthetase. This fragment, which contains the 547 N-terminal residues of the enzyme (M547; Mellot *et al.*, 1989), can be easily overproduced, purified, and stored for years without loss of activity. Overproduction of the M547 enzyme is obtained from the pBSM547 plasmid (Fourmy *et al.*, 1991).

3.1. Buffers

M1: 10 mM potassium phosphate, pH 6.7; 10 mM 2-mercaptoethanol

M2: 10 mM potassium phosphate, pH 6.7; 10 mM 2-mercaptoethanol; 50 mM KCl

M3: 10 mM potassium phosphate, pH 6.7; 10 mM 2-mercaptoethanol; 1 M KCl

10× A1: 200 mM Tris-HCl, pH 7.6; 1 mM EDTA; 100 mM 2-mercaptoethanol

P1: 10 mM sodium acetate, pH 5.0

P2: 10 mM sodium acetate, pH 5.0; 1 M NaCl

3.2. Protocols

1. Purification of M547 (Mellot *et al.*, 1989):

   - Transform JM10Tr cells with pBSM547 and grow the cells at 37° in 1 L of 2× TY medium containing 50 μg/ml of ampicillin.
   - When OD$_{650}$ reaches ~1, add IPTG to a final concentration of 0.3 mM. Continue growth at 37° for at least 3 h.
   - Harvest the cells by centrifugation, and resuspend in 30 ml of Buffer M1.
   - Disrupt cells by ultrasonic disintegration.
   - Remove cell debris by centrifugation (15,000g; 10 min; 4°).
   - To the supernatant, add 0.1 volume of 30% streptomycin sulfate.
   - Remove precipitated nucleic acids by centrifugation (15,000g; 10 min; 4°).

- To the supernatant, add ammonium sulfate to 80% of saturation (i.e., 461 mg/ml of supernatant).
- Recover the precipitated proteins by centrifugation (15,000g; 30 min; 4°).
- Dissolve the pellets in 2 ml of buffer M1. Because of the presence of ammonium sulfate, proteins will not fully dissolve.
- Dialyze 2 h against 2× 1 L of buffer M2. Tighten the dialysis tubing to avoid too large an increase in sample volume. Proteins should redissolve rapidly.
- Load the sample (<8 ml) onto a molecular sieving column (Superose 6, GE Healthcare, 16 mm × 50 cm) equilibrated in buffer M2. Run at 0.2 ml/min.
- Pool the peak corresponding to M547, and load onto an ion exchange column (Q-Hiload; GE Healthcare; 16 × 10 cm) equilibrated in buffer M2.
- Run a 0.05 to 0.4 M KCl gradient by use of buffers M2 and M3 (2.5 ml/min; 0.2 M/h).
- Pool the peak corresponding to M547 and dialyze overnight against 2× 1 L of buffer M1, and then 24 h against 1 L of buffer M1 containing 55% w/v of glycerol.
- Determine the enzyme concentration by measuring A$_{280}$ of an appropriate dilution and by use of an extinction coefficient of 1.72 cm^2/mg (Cassio and Waller, 1971).

The procedure yields 80 to 100 mg of homogeneous M547, which can be stored at −20° for several years.

2. Methionylation of tRNA

Full aminoacylation of tRNA can be obtained by incubating the polynucleotide at 25° in the presence of M547, methionine, and ATP under adequate buffer conditions. Concentration of tRNA should not be raised above 10 to 15 μM, and care has to be taken to maintain the concentration of free methionine above 20 μM. The total concentration of methionine should, therefore, be greater than 20 μM plus the concentration of tRNA in the aminoacylation assay. Analytical assays can be performed to check out for correct aminoacylation before performing the preparative assays. In these analytical assays, the amount of enzyme and/or the incubation time can be varied to ensure that the aminoacylation plateau has been reached.

3.3. Reagents

10× A1 buffer
1.5 M KCl
70 mM MgCl$_2$

200 mM ATP (adjusted to pH 7.6 with Tris) radiolabeled L-methionine (either [^{14}C] or [^{35}S], see notes following) total yeast RNA (4 mg/ml, to be used as carrier)

5% w/v trichloroacetic acid (TCA) containing 0.5 % w/v DL-methionine.

3.4. Analytical assay

- Prepare 100-μl assays containing 150 mM KCl, 7 mM MgCl$_2$, 2 mM ATP, 25 to 40 μM L-Met, 1 to 10 μM tRNA; 20 mM Tris-HCl, pH7.6; 0.1 mM EDTA; 10 mM 2-mercaptoethanol.
- Start the aminoacylation reaction by adding M547 enzyme to a final concentration of 1 μM.
- Incubate 10 min at 25°.
- Precipitate tRNA by adding 2.5 ml of cold TCA containing 0.5 % w/v DL-Met and 20 μl of 4 mg/ml carrier RNA.
- Filter onto Whatman GF-C disks, rinse with cold TCA containing 0.5 % w/v DL-Met, and count the radioactivity retained on the filter.

3.5. Preparative aminoacylation

- Same as the analytical assay, but the volume of the assay can be increased if necessary. After incubation at 25°, withdraw an aliquot for TCA precipitation to check for correct aminoacylation. To the remaining of the reaction mixture, add 0.1 volume of 3 M sodium acetate, pH 5.0, and 2.2 volumes of cold ethanol to precipitate tRNA.
- The precipitated tRNA can be rinsed with ethanol and stored at −20° as small aliquots for several weeks. To use an aliquot, centrifuge, remove the ethanol, and dissolve in water. Determine the level of tRNA aminoacylation by measuring both A$_{260}$ and the amount of TCA-precipitable radioactivity. A tRNA aminoacylation level larger than 90% of the aminoacylation plateau determined in analytical assays can be expected.
- Optionally, the aminoacyl-tRNA can be further purified before use. In this case, proceed rapidly to the purification steps (see below) to avoid spontaneous deacylation.

Notes

1. A purification procedure is described further in the following. This method aims at eliminating the small molecules and the enzyme but not at separating naked tRNA from aminoacyl-tRNA. It is, therefore, important to start with fully aminoacylated tRNA or, at least, to be aware of the fact that nonaminoacylated tRNA can coexist with the aminoacylated one.

2. The specific radioactivity of the radiolabeled methionine has to be adjusted as a function of the concentration of tRNA that will be used in the protection assay. As a rule of thumb, for a concentration of aminoacyl-tRNA of 10 nM in the protection assay, it is recommended to charge tRNA with radioactive methionine having a specific activity of at least 3.7 GBq/mmole (220,000 dpm/picomole). This can be achieved by making an appropriate isotopic dilution of [^{35}S]Met. For a concentration of aminoacyl-tRNA greater than 1 μM, it will generally be enough to use [^{14}C]Met (2 GBq/mmol). Because of some instability of [^{35}S]Met on storage, it is advisable to calibrate the isotopic dilution by comparing the aminoacylation plateaus of two identical tRNA samples, obtained in the presence of either [^{14}C]Met or isotopically diluted [^{35}S]Met.

3. Purification of aminoacyl-tRNA (optional).

To minimize spontaneous deacylation, it is crucial to proceed rapidly to purification after enzymatic aminoacylation and to maintain the sample at an acidic pH (pH 5) and at 4°.

- After ethanol precipitation of the aminoacylation mixture, redissolve the pellet in, for example, 1 ml of buffer P1.
- Load on a Q-Sepharose High-Performance column (GE Healthcare; 1 × 5 cm) previously equilibrated in buffer P1.
- Perform a rapid gradient from 0 M NaCl to 0.4 M NaCl by use of buffer P2 (1 ml/min; 0.04 M/min).
- Rinse with 1.5 column volume of 10 mM sodium acetate pH 5.0, 0.4 M NaCl.
- Elute tRNA by applying a step of 10 mM sodium acetate pH 5.0, 0.8 M NaCl.
- Precipitate the eluted tRNA by adding 0.6 volume of isopropanol. At this stage, it is convenient to separate the material into several aliquots. Centrifuge the aliquots, remove the supernatant, and add, for example, 100 μl of ethanol. The material can be stored at −20° for several weeks.
- To use an aliquot, centrifuge, remove the ethanol, and dissolve in water. Determine the level of tRNA aminoacylation by measuring both A$_{260}$ and the amount of TCA-precipitable radioactivity. An amount of aminoacylated tRNA corresponding to more than 90% of the aminoacylation plateau determined from analytical assays can be expected.

4. PROTECTION ASSAY

The principle of the assay is to incubate aminoacyl-tRNA at a fixed concentration in the presence of various concentrations of the initiation factor. Provided that the contacts between the polynucleotide and the protein involve the aminoacyl moiety esterified to the tRNA, which is

generally the case, bound aminoacyl-tRNA will be protected from spontaneous deacylation compared with free aminoacyl-tRNA. Therefore, by
recording the rate of deacylation of an aminoacyl-tRNA as a function of
the concentration of an initiation factor, information on the stability of the
aminoacyl-tRNA/factor complex can be drawn. Under certain conditions
(see later, a dissociation constant can be rigorously derived from the
experiments. If these conditions are not satisfied, apparent constants, related
to the true dissociation constant, will be obtained. The most important
requirement for a rigorous data processing is that the dynamics of the
equilibrium between the aminoacyl-tRNA and the factor is fast compared
with the deacylation rates. Interestingly, this condition is opposite to that
required for the nitrocellulose binding assay, the validity of which implies
that the studied complexes have a sufficiently slow equilibrium dynamics to
remain stable during the time of the filtration steps. Therefore, the two
techniques may be considered as complementary, the nitrocellulose binding
assay being preferred for high-affinity complexes and the protection assay
being preferred for lower affinity complexes. Equations are shown following to figure the protection assay. These equations suppose that the deacylated tRNA has lost most of its capacity to bind the initiation factor
compared with the aminoacyl-tRNA. This can be easily verified by checking that excess amounts of non-aminoacylated tRNA do not affect the
protection afforded by the factor to aminoacyl-tRNA. Finally, as discussed
later, the processing of the data is simplified if the fixed concentration of
aminoacyl-tRNA in the assay is maintained small with respect to the Kd
value of the aminoacyl-tRNA/factor complex. Indeed, under such a
condition, the fraction of the factor bound to the ligand will be negligible.

Let us define F as the free concentration of the factor, Fo as the total
factor concentration in the assay, A as the free concentration of aminoacyl-
tRNA, T as the concentration of deacylated tRNA, Ao as the concentration
of total aminoacyl-tRNA added at time zero, and C as the stationary
concentration of the aminoacyl-tRNA/factor complex. Under the assumption that the preequilibrium steps enabling the aminoacyl-tRNA and the
factor to form a complex are rapid with respect to the deacylation rate, we
may write at any time:

$$F + A \leftrightarrow C \text{ with a dissociation constant } Kd = \frac{F \times A}{C} \qquad (1)$$

Deacylations of free and bound aminoacyl-tRNA occur at different rates:
$$A \rightarrow T \text{ with a rate constant } k_1$$
$$C \rightarrow F + T \text{ with a rate constant } k_2$$

$$Fo = F + C \qquad (2)$$

$$Ao = A + T + C \tag{3}$$

The rate of production of deacylated tRNA as a function of time is:

$$\frac{dT}{dt} = k_1 A + k_2 C \tag{4}$$

If Ao is small with respect to Kd, then the fraction of bound factor remains small, and we can consider that F is equivalent to Fo. In this case, from Eq. (1), we draw:

$$C = \frac{FoA}{Kd} \tag{5}$$

By substituting the value of C from Eq. (5) in Eqs. (3) and (4), we get:

$$\frac{dT}{dt} = \left[k_1 + (k_2 - k_1) \times \left(\frac{Fo}{Fo + Kd} \right) \right] \times (Ao - T) \tag{6}$$

Therefore, tRNA deacylation follows an exponential law, the rate constant, k, of which is:

$$k = k_1 + (k_2 - k_1) \times \left[\frac{Fo}{Fo + Kd} \right] \tag{7}$$

Kd can then be easily derived from a plot of k as a function of Fo. Note that, in general, deacylation of protected tRNA can be neglected in such a way that k_2 can be approximated to zero.

In cases in which Ao cannot be considered as small compared with Kd, processing of the data remains possible but becomes more complicated, as shown in the following.

For simplification, if we make the assumption that $k_2 = 0$, then Eq. (4) becomes:

$$\frac{dT}{dt} = k_1 A \tag{4'}$$

An expression of A can be obtained by combining Eqs. (1) to (3):

$$A = \frac{-(Fo + Kd + T - Ao) + \sqrt{(Ao - T - Fo + Kd)^2 + 4KdFo}}{2} \tag{8}$$

and therefore:

$$\frac{dT}{dt} = k_1 \times \frac{-(Fo + Kd + T - Ao) + \sqrt{(Ao - T - Fo + Kd)^2 + 4KdFo}}{2}$$

$$(9)$$

Eq. (9) cannot be integrated explicitly. However, a numerical solution of this equation with the Mathematica program (Wolfram Research, Champaign, IL) shows that, at least for values of Kd larger than $0.5 \times Ao$ and values of Fo in the range $0.1 \times Ao$ to $10 \times Ao$, the concentration of deacylated tRNA varies as a function of time following a curve than can satisfyingly be fitted with a single exponential having a rate constant k. From numerical simulations, we observed that a good approximation of such rate constants could be obtained by use of Eq. 10. This equation describes a titration curve in which the concentration of the titrated molecule is not small compared with the Kd value.

$$k = k_1 \times \left[1 - \frac{(Ao + Kd + Fo) - \sqrt{(Ao + Kd - Fo)^2 + 4Kd \times Fo}}{2Ao} \right]$$

$$(10)$$

Eq. 10 can, therefore, be used to derive a Kd value by least square fitting. The entire procedure (i.e., the fitting of the deacylation rates with single exponentials obeying obtained rate constants as defined in Eq. 10) allows the determination of Kd values with slight underestimations ranging from 4.5% (for $Kd = 0.5 \times Ao$) to 1.1% (for $Kd = 10 \times Ao$). Experimental errors generally exceed such systematic errors.

4.1. Buffers and reagents

$5 \times$ D1: 100 mM HEPES-NaOH (pH 8.0); 0.5 M KCl; 25 mM MgCl$_2$; 5 mM DTT; 0.5 mM EDTA; 25% glycerol.
2 mg/ml BSA (bovine serum albumin, Roche) in water.
10 mM GTP or GDPNP adjusted to pH 8.0.
5% w/v trichloroacetic acid (TCA) containing 0.5 % w/v DL–methionine.
4 mg/ml total yeast RNA (carrier).

4.2. Protocols

To measure the Kd of an aminoacyl-tRNA/factor complex, aminoacyl-tRNA at a fixed concentration is incubated in the presence of various concentrations of the factor. For each factor concentration, a rate of deacylation is measured by withdrawing aliquots at various times and by measuring the TCA-precipitable radioactivity in each aliquot. As discussed previously, the concentration of the aminoacyl-tRNA in the assay has to be fixed sufficiently small with respect to the value of the Kd to be measured. Preliminary experiments have, therefore, to be performed to obtain a rough estimate of the Kd value before more extensive measurements are undertaken. For an accurate Kd determination, the factor concentration should be, if possible, varied in the range 0.1 to 10 Kd.

Diol-containing reagents, such as sucrose or glycerol, are known to induce deacylation of aminoacyl-tRNA (Johnson and Adkins, 1984). Therefore, great care must be taken of the buffers in which the protein factor is stored. In particular, glycerol should be avoided, because its uncontrolled presence in the assay will lead to erroneous results. The best way to avoid artifacts is to extensively dialyze the factor to be studied against $1 \times D1$ and to perform dilutions of this factor in the same buffer. Buffer D1 can be modified as desired to match the binding conditions to be tested. For instance, with yeast eIF5B, it seemed necessary to increase salt concentration to ensure full solubilization of the factor (Guillon et $al.$, 2005). It should, however, be checked that, in the presence of the chosen buffer D1, the rate of tRNA deacylation is high enough (a rate of the order of 0.1 min^{-1} is fine) to ensure precise measurements of deacylation rates. Factors acting positively on the deacylation reaction are increasing of the pH, of the temperature, or of the concentration of glycerol. We routinely used a temperature of $51°$ for experiments involving thermophilic archaeal factors (Yatime et $al.$, 2004; 2006), and $30°$ with $S.$ $cerevisiae$ eIF5B (Guillon et $al.$, 2005). Temperatures as high as $65°$ have successfully been used also (Pedulla et $al.$, 2005).

- For each concentration of the factor, prepare a 150-μl incubation mixture as follows:

  $5 \times$ D1: 24 μl
  10 mM GTP or GDPNP: 15 μl (1 mM final concentration)
  2 mg/ml BSA: 15 μl (0.2 mg/ml final concentration)
  Initiation factor in $1 \times$ D1: 30 μl
  Radiolabeled aminoacyl-tRNA: 15 μl
  H$_2$O qsp. 150 μl.

- Incubate at the desired temperature. Withdraw 20-μl aliquots at various times chosen to correctly describe the deacylation curve.
- Pipette each aliquot in a test tube containing 2.5 ml of cold 5% TCA-0.5% DL-Met, and add 20 μl of 4 mg/ml carrier RNA.
- Filter on Whatman GF/C disks, rinse with 10 ml of cold 5% TCA-0.5% DL-Met, and count the radioactivity retained on the filter by liquid scintillation.

4.3. Data processing

- For each concentration of the factor, determine the rate of deacylation by least-square fitting of a single exponential to the experimental points.

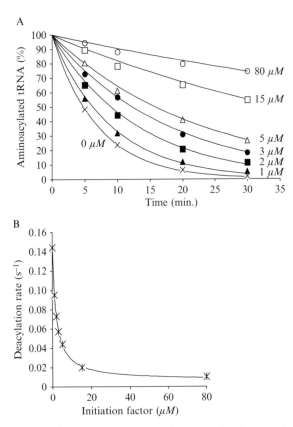

Figure 11.1 Example of dissociation constant determination by use of the protection assay. The binding of *Pyrococcus abyssi* Met-tRNA$_i^{Met}$ produced in *E. coli* to *P. abyssi* initiation factor aIF5B is studied. (A) Time courses of deacylation at 51° of the aminoacyl-tRNA (80 nM) in the presence of the aIF5B concentrations indicated beside each curve. Each set of experimental points is fitted with a single exponential from which the deacylation rate is deduced. (B) Obtained deacylation rates are plotted as a function of aIF5B concentration. The data were fitted with Eq. 7, giving a Kd of 1.9 ± 0.4 μM. Adapted with permission from Guillon *et al.* (2005). Copyright 2005, American Chemical Society.

- Plot the measured deacylation rates as a function of the concentration of the factor in the assay. Determine the value of Kd by least square fitting of Eq. (7) (or of Eq. 10) to the experimental points. We routinely use the MC-Fit program (Dardel, 1994) for least-square fitting. The main advantage of this program is that it uses Monte-Carlo methods to provide accurate confidence limits on the determined parameters. An example of Kd determination is shown in Fig. 11.1.

5. CONCLUSION

Measuring the deacylation rates of an aminoacyl-tRNA in the presence of a variable concentration of an initiation factor represents an easy way to derive the dissociation constant of the protein/nucleic acid complex. Rigorous processing of the data requires that the dynamics of the preequilibrium between the aminoacyl-tRNA and the initiation factor is fast with respect to the deacylation rate. However, even if this condition is not met, the protection method can be used to derive an apparent dissociation constant, related to the stability of the complex, which can be useful to draw some biological conclusions. If the above preequilibrium condition is met, accurate dissociation constants can be derived in a wide range of conditions. Preferably, the concentration of aminoacyl-tRNA in the assay should be chosen small with respect to the dissociation constant to be measured. However, as shown here, the data can be interpreted quantitatively even when the dissociation constant and the aminoacyl-tRNA concentration in the assay are of the same order of magnitude. Finally, provided the aminoacyl-tRNA concentrations in the protection assay are sufficiently larger than the dissociation constant, the polynucleotide can be titrated with the initiation factor. With such experiments, the number of aminoacyl-tRNA binding sites per protein molecule can readily be determined.

ACKNOWLEDGMENTS

We thank Thomas Simonson and Josselin Noirel for helpful discussions about the use of Mathematica.

REFERENCES

Avis, J. M., and Fersht, A. R. (1993). Use of binding energy in catalysis: Optimization of rate in a multistep reaction. *Biochemistry* **32,** 5321–5326.
Blanquet, S., Iwatsubo, M., and Waller, J.-P. (1973). The mechanism of action of methionyl-tRNA synthetase. 1. Fluorescence studies on tRNA^{Met} binding as a function of ligands, ions and pH. *Eur. J. Biochem.* **36,** 213–226.
Carey, J. (1991). Gel retardation. *Methods Enzymol.* **208,** 103–117.
Cassio, D., and Waller, J.-P. (1971). Modification of methionyl-tRNA synthetase by proteolytic cleavage and properties of the trypsin-modified enzyme. *Eur. J. Biochem.* **20,** 283–300.
Choi, S. K., Lee, J. H., Zoll, W. L., Merrick, W. C., and Dever, T. E. (1998). Promotion of met-tRNAiMet binding to ribosomes by yIF2, a bacterial IF2 homolog in yeast. *Science* **280,** 1757–1760.
Dardel, F. (1994). MC-Fit: Using Monte-Carlo methods to get accurate confidence limits on enzyme parameters. *Comput. Applic. Biosci.* **10,** 273–275.

Fourmy, D., Mechulam, Y., Brunie, S., Blanquet, S., and Fayat, G. (1991). Identification of residues involved in the binding of methionine by *Escherichia coli* methionyl-tRNA synthetase. *FEBS Lett.* **292,** 259–263.

Gualerzi, C. O., Brandi, L., Caserta, E., La Teana, A., Spurio, R., Tomsic, J., and Pon, C. L. (2000). Translation initiation in bacteria. *In* "The Ribosome: Structure, Function, Antibiotics and Cellular Interactions" (R. A. Garrett, S. R. Douthwaite, A. Liljas, A. T. Matheson, P. B. Moore, and H. F. Noller, eds.), pp. 477–494. ASM Press, Washington, DC.

Guillemaut, P., and Weil, J. H. (1975). Aminoacylation of *Phaseolus vulgaris* cytoplasmic, chloroplastic and mitochondrial tRNAs^{Met} and of *Escherichia coli* tRNAs^{Met} by homologous and heterologous enzymes. *Biochim. Biophys. Acta* **407,** 240–248.

Guillon, J. M., Meinnel, T., Mechulam, Y., Lazennec, C., Blanquet, S., and Fayat, S. (1992). Nucleotides of tRNA governing the specificity of *Escherichia coli* methionyl-tRNA$^{Met}_f$ formyltransferase. *J. Mol. Biol.* **224,** 359–367.

Guillon, L., Schmitt, E., Blanquet, S., and Mechulam, Y. (2005). Initiator tRNA binding by e/aIF5B, the eukaryotic/archaeal homologue of bacterial initiation factor IF2. *Biochemistry* **44,** 15594–15601.

Helene, C., Brun, F., and Yaniv, M. (1969). Fluorescence study of interactions between valyl-t RNA synthetase and valine-specific tRNA's from *Escherichia coli*. *Biochem. Biophys. Res. Commun.* **37,** 393–398.

Hirel, P.-H., Lévêque, F., Mellot, P., Dardel, F., Panvert, M., Mechulam, Y., and Fayat, G. (1988). Genetic engineering of methionyl-tRNA synthetase: *In vitro* regeneration of an active synthetase by proteolytic cleavage of a methionyl-tRNA synthetase-β-galactosidase chimeric protein. *Biochimie* **70,** 773–782.

Johnson, A. E., and Adkins, H. J. (1984). Glycerol, sucrose, and other diol-containing reagents are not inert components in *in vitro* incubations containing aminoacyl-tRNA. *Anal. Biochem.* **137,** 351–359.

Kapp, L. D., and Lorsch, J. R. (2004). GTP-dependent Recognition of the Methionine Moiety on Initiator tRNA by Translation Factor eIF2. *J. Mol. Biol.* **335,** 923–936.

Meinnel, T., and Blanquet, S. (1995). Maturation of pre-tRNA^{fMet} by *E. coli* RNase P is specified by a guanosine of the 5′ flanking sequence. *J. Biol. Chem.* **270,** 15906–15914.

Meinnel, T., Mechulam, Y., and Fayat, G. (1988). Fast purification of a functional elongator tRNA^{Met} expressed from a synthetic gene *in vivo*. *Nucleic Acids Res.* **16,** 8095–8096.

Meinnel, T., Mechulam, Y., Fayat, G., and Blanquet, S. (1992). Involvement of the size and sequence of the anticodon loop in tRNA recognition by mammalian and *E. coli* methionyl-tRNA synthetases. *Nucleic Acids Res.* **20,** 4741–4746.

Meinnel, T., Mechulam, Y., LeCorre, D., Panvert, M., Blanquet, S., and Fayat, G. (1991). Selection of suppressor methionyl-tRNA synthetases: Mapping the tRNA anticodon binding site. *Proc. Natl. Acad. Sci. USA* **88,** 291–295.

Mellot, P., Mechulam, Y., LeCorre, D., Blanquet, S., and Fayat, G. (1989). Identification of an amino acid region supporting specific methionyl-tRNA synthetase: tRNA recognition. *J. Mol. Biol.* **208,** 429–443.

Nissen, P., Kjeldgaard, M., Thirup, S., Polekhina, G., Reshetnikova, L., Clark, B. F. C., and Nyborg, J. (1995). Crystal structure of the ternary complex of Phe-tRNA^{Phe}, EF-Tu, and a GTP analog. *Science* **270,** 1464–1472.

Nissen, P., Thirup, S., Kjeldgaard, M., and Nyborg, J. (1999). The crystal structure of Cys-tRNACys-EF-Tu-GDPNP reveals general and specific features in the ternary complex and in tRNA. *Structure* **7,** 143–156.

Pedulla, N., Palermo, R., Hasenohrl, D., Blasi, U., Cammarano, P., and Londei, P. (2005). The archaeal eIF2 homologue: Functional properties of an ancient translation initiation factor. *Nucleic Acids Res.* **33,** 1804–1812.

Pestova, T. V., Lomakin, I. B., Lee, J. H., Choi, S. K., Dever, T. E., and Hellen, C. U. (2000). The joining of ribosomal subunits in eukaryotes requires eIF5B. *Nature* **403**, 332–335.

Petersen, H. U., Røll, T., Grunberg-Manago, M., and Clark, B. F. C. (1979). Specific interaction of initiator factor IF$_2$ of *E. coli* with formylmethionyl-tRNA$^{Met}_f$. *Biochem. Byophys. Res. Commun.* **91**, 1068–1074.

Petrissant, G., Boisnard, M., and Puissant, C. (1970). Purification d'un tRNA accepteur de la méthionine dans le foie de lapin. *Biochim. Byophys. Acta* **213**, 223–225.

Pingoud, A., Urbanke, C., Krauss, G., Peters, F., and Maass, G. (1977). Ternary complex formation between elongation factor Tu, GTP and aminoacyl-tRNA: An equilibrium study. *Eur. J. Biochem.* **78**, 403–409.

RajBhandary, U. L., and Ghosh, H. P. (1969). Studies on polynucleotides. XCI. Yeast methionine transfer ribonucleic acid: Purification, properties, and terminal nucleotide sequences. *J. Biol. Chem.* **244**, 1104–1113.

Robert, F., Kapp, L. D., Khan, S. N., Acker, M. G., Kolitz, S., Kazemi, S., Kaufman, R. J., Merrick, W. C., Koromilas, A. E., Lorsch, J. R., and Pelletier, J. (2006). Initiation of protein synthesis by hepatitis C virus is refractory to reduced eIF2.GTP.Met-tRNAiMet ternary complex availability. *Mol. Biol. Cell* **17**, 4632–4644.

Schmitt, E., Blanquet, S., and Mechulam, Y. (2002). The large subunit of initiation factor aIF2 is a close structural homologue of elongation factors. *EMBO J.* **21**, 1821–1832.

Schmitt, E., Meinnel, T., Panvert, M., Mechulam, Y., and Blanquet, S. (1993). Two acidic residues of *Escherichia coli* methionyl-tRNA synthetase are negative discriminants towards the binding of non-cognate tRNA anticodons. *J. Mol. Biol.* **233**, 615–628.

Schulman, L. H., and Pelka, H. (1988). Anticodon switching changes the identity of methionine and valine transfer RNAs. *Science* **242**, 765–768.

Wong, I., and Lohman, T. M. (1993). A double-filter method for nitrocellulose-filter binding: Application to protein-nucleic acid interactions. *Proc. Natl. Acad. Sci. USA* **90**, 5428–5432.

Yatime, L., Mechulam, Y., Blanquet, S., and Schmitt, E. (2006). Structural switch of the gamma subunit in an archaeal aIF2 alpha gamma heterodimer. *Structure* **14**, 119–128.

Yatime, L., Schmitt, E., Blanquet, S., and Mechulam, Y. (2004). Functional molecular mapping of archaeal translation initiation factor 2. *J. Biol. Chem.* **279**, 15984–15993.

Yatime, L., Schmitt, E., Blanquet, S., and Mechulam, Y. (2005). Structure-function relationships of the intact aIF2a subunit from the archaeon *Pyrococcus abyssi*. *Biochemistry* **44**, 8749–8756.

Zubay, G. (1962). The isolation and fractionation of soluble ribonucleic acids. *J. Mol. Biol.* **4**, 347–356.

CHAPTER TWELVE

NMR Methods for Studying Protein–Protein Interactions Involved in Translation Initiation

Assen Marintchev, Dominique Frueh, *and* Gerhard Wagner

Contents

Department of Biological Chemistry and Molecular Pharmacology, Harvard Medical School, Boston, Massachusetts

Methods in Enzymology, Volume 430
ISSN 0076-6879, DOI: 10.1016/S0076-6879(07)30012-8

Abstract

Translation in the cell is carried out by complex molecular machinery involving a dynamic network of protein–protein and protein–RNA interactions. Along the multiple steps of the translation pathway, individual interactions are constantly formed, remodeled, and broken, which presents special challenges when studying this sophisticated system. NMR is a still actively developing technology that has recently been used to solve the structures of several translation factors. However, NMR also has a number of other unique capabilities, of which the broader scientific community may not always be aware. In particular, when studying macromolecular interactions, NMR can be used for a wide range of tasks from testing unambiguously whether two molecules interact to solving the structure of the complex. NMR can also provide insights into the dynamics of the molecules, their folding/unfolding, as well as the effects of interactions with binding partners on these processes.

In this chapter, we have tried to summarize, in a popular format, the various types of information about macromolecular interactions that can be obtained with NMR. Special attention is given to areas where the use of NMR provides unique information that is difficult to obtain with other approaches. Our intent was to help the general scientific audience become more familiar with the power of NMR, the current status of the technological limitations of individual NMR methods, as well as the numerous applications, in particular for studying protein–protein interactions in translation.

1. INTRODUCTION

Translation relies on a complex dynamic network of interactions among RNAs and protein factors, which undergoes structural rearrangements at every step of the process (Marintchev and Wagner, 2004). Remarkable progress has been made in recent years toward understanding the molecular mechanisms of translation initiation, the structures of individual components of the translation network, and their interactions. In this process, NMR methods have had a significant impact: the structures of multiple translation initiation factors have been solved by NMR (Battiste *et al.*, 2000; Cho and Hoffman, 2002; Conte *et al.*, 2006; Elantak *et al.*, 2006; Fleming *et al.*, 2003; Fletcher *et al.*, 1999; Garcia *et al.*, 1995a,b; Gross *et al.*, 2003b; Gutierrez *et al.*, 2004; Ito *et al.*, 2004; Kozlov *et al.*, 2002, 2004; Laursen *et al.*, 2003, 2004; Li and Hoffman, 2001; Matsuo *et al.*, 1997b; Meunier *et al.*, 2000; Moreau *et al.*, 1997; Sette *et al.*, 1997; Siddiqui *et al.*, 2003; Wienk *et al.*, 2005), and an even greater number of interactions have been studied using NMR techniques (the complete list contains more than 50 references, some of which are reviewed in Marintchev and Wagner [2004]). This chapter is intended to introduce the wide range of NMR methods for studying protein–protein interactions to a broad scientific audience. Accordingly, we have striven to present popular descriptions of NMR phenomena and methods, as well as provide ample illustrations and examples, primarily from the field of translation initiation. We have tried to summarize the types of information that can be obtained with individual methods, as well as their limitations. Whereas the discussions are usually centered on protein–protein interaction, most of the approaches apply to interactions with RNA and other ligands as well.

An overview of the various types of interactions observed in translation from a quantitative standpoint is presented in Section 2. Section 3 contains a description of NMR and the variety of information about protein interactions that NMR methods can provide for the different types of interactions, as defined in Section 2. Some frequently used NMR methods for studying protein interactions are then presented, grouped by the type of information they provide. Section 4 contains descriptions of NMR methods for resonance assignments and structure determination, as well as methods for stable isotope labeling, which are also important for studying protein interactions. Because these are general methods, with a broad range of applications, and have been discussed in detail elsewhere, we have tried to provide popular descriptions and focus on recent developments. Some special cases are discussed in Section 5, whereas Section 6 contains, in a table format, a summary of the types of information that can be obtained by NMR, the methods used, and the macromolecular size ranges, for which they are applicable. This chapter is not intended as a detailed review of the literature, and the reader is referred to recent reviews for further information.

2. TYPES OF INTERACTIONS IN TRANSLATION FROM THE PERSPECTIVE OF K_D AND LIFETIME OF THE COMPLEX

2.1. Overview

In translation, as in other multistep processes, as the system progresses along the stages of initiation, the general trend is toward forming more and more stable complexes, which are eventually rearranged/destabilized using energy from NTP hydrolysis (Fig. 12.1). The individual interactions between components of the translation apparatus vary enormously in their strength, roles, and fates:

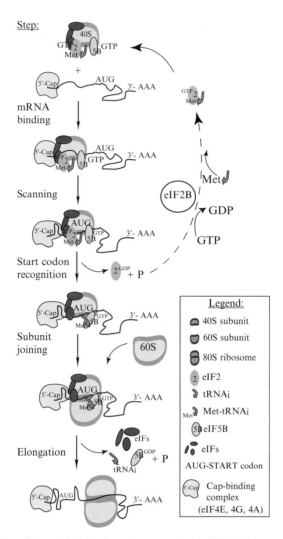

Figure 12.1 Translation initiation in eukaryotes. A simplified diagram of the translation initiation pathway. The legend is shown on the bottom right. (See color insert.)

- Certain complexes are "permanent"—stable for the life of the molecules (e.g., a ribosomal subunit); others need to last for the duration of a process (e.g., a translation initiation factor that has to stay bound to the initiation complex during initiation and then dissociate).
- Some complexes are rigid; others need to undergo rearrangement throughout the process.
- The formation and dissociation of certain complexes need to be regulated.
- Certain interactions need to occur only in the context of specific complexes but not between the free factors in solution.

Stable complexes can be built both from strong interactions and from multiple weaker interactions through cooperativity. The latter alternative often has clear advantages if the complex needs to be subject to regulation; cooperative interactions can be more specific, can form conditionally when/where needed, and can be easily rearranged/broken/regulated. Comparison of the intracellular concentrations of two interacting partners with the equilibrium dissociation constant (K_D) of the complex allows one to estimate whether the equilibrium in the cell is pulled toward complex formation or free species; at concentrations greater than the K_D, there is more complex than free species, whereas at concentrations lower than the K_D, the free species predominate.

For a simple binding reaction of the type

$$A + B \underset{k_{off}}{\overset{k_{on}}{\rightleftarrows}} AB,$$

where k_{on} and k_{off} are the rate constants for complex formation and dissociation, respectively, the equilibrium between complex and free species is obtained from the following equation:

$$K_D = \frac{1}{K_A} = \frac{k_{off}}{k_{on}} = \frac{[A] \cdot [B]}{[AB]} \qquad (1)$$

where K_D and K_A are the equilibrium dissociation and association constants, respectively, [A] and [B] are the concentrations of the free species, and [AB] is the concentration of the complex. The equations become more complicated for multistep reactions (e.g., if the complex is in equilibrium with other species). In such cases, an apparent K_D is often used.

Usually, the equilibrium dissociation constant K_D is used rather than the reciprocal equilibrium association constant, K_A, because K_D has a measure of mol/L and is easier to compare with the concentrations of the interacting species. The intracellular concentrations of translation factors are typically between 0.1 and 100 μM, and most of them are approximately 1 to 10 μM (see e.g., von der Haar and McCarthy [2002]).

Another important aspect is the lifetime of the complex (the reciprocal of the dissociation rate constant k_{off})—by comparing the lifetime of a complex to the length of a process, one can get an idea whether the complex will dissociate appreciably during the process or a given step thereof (in the absence of any additional stabilizing/destabilizing factors). For example, the stability of the scanning translation initiation complex will determine whether scanning along the 5′ untranslated region of a given mRNA is processive or whether a significant portion of the complexes dissociates before reaching the start codon.

The rate of complex formation is usually diffusion controlled, which means that every collision of two molecules leads to binding. For diffusion-controlled protein–protein interactions, k_{on} is typically in the range of 10^6 to 10^8 $M^{-1}sec^{-1}$ and is inversely proportional to the size of the binding partners. However, in certain cases, k_{on} can be much slower, and a complex that seems relatively weak, on the basis of the K_D of the interaction, could have much longer lifetime than expected. In some cases, k_{on} can be faster than expected from diffusion rates. Such deviations usually involve interactions with a strong electrostatic component, because the strength of electrostatic interactions decreases with the first power of the distance between the two molecules and can be significant even when the two molecules are farther apart, whereas the strength of van der Waals interactions, for example, decreases much more steeply—with the sixth power of the distance. Therefore, clusters of charges create a strong electrostatic field that can affect the trajectories of molecules "passing by."

2.2. Types of interactions from the perspective of translation initiation

Individual interactions have different behaviors and fates in the context of the translation process, which depend to a great extent on their equilibrium dissociation constants (K_D), as well as lifetimes. The concentrations of translation initiation factors are typically in the range 0.1 to 10 μM. The requirements for the lifetimes of individual complexes are imposed by the lengths of the corresponding stages of translation initiation, which range from less than 1 sec to more than a min, and the entire process of initiation could be on the order of minutes. Therefore, from a translation standpoint, the interactions can be subdivided into the following groups:

A. K_D is much smaller than the concentrations of the factors and the complex is stable for more than a few minutes (and up to hours). This group includes stable complexes that form spontaneously and do not dissociate for the lifetime of the molecules.

B. K_D is much smaller than the concentrations and the complex is stable for seconds to minutes—interactions occur spontaneously and can be broken down. Such factors can be part of a network of cooperative interactions forming a stable complex, as in A, but easier to regulate. Or, the interactions can be part of a cyclic process, often dissociated on NTP hydrolysis.

C. K_D is comparable to the concentrations. In isolation at physiological concentrations of the factors, equilibrium between free molecules and complexes exists. Because the concentrations of translation factors are often in the low μM range, the lifetimes of the complexes are usually in the millisecond to second range (assuming diffusion-controlled binding rates).

Therefore, in isolation, interactions from this group usually are constantly formed and broken. Two or more such interactions can yield a stable complex through cooperativity, as in B earlier.

D. K_D is much larger than the concentrations. In isolation at physiological concentrations, the equilibrium is strongly shifted toward free molecules (little or no complexes are present present). Such interactions depend on additional interactions for complex formation and can provide stabilization of other interactions through cooperativity.

2.3. Random vs. ordered assembly of multisubunit complexes

Which interaction occurs first is determined by the binding rate constants (k_{on}) and concentrations of the respective factors. Therefore, a stronger interaction need not form first, except in the context of group 2.2.D, where the interaction is too weak and the equilibrium is pulled strongly toward the free species. Certain complexes are assembled in sequential order, where the binding site for the next factor is formed/uncovered in the preceding step. In terms of binding rates, this would translate into slow to infinitely slow rates for binding of the factor to subcomplexes corresponding to earlier steps in the assembly and higher rates of binding to the proper subcomplex.

2.4. Specific vs. nonspecific interactions

When studying interactions *in vitro* at high concentrations, there is often a question whether such interactions occur *in vivo* and whether they are specific or nonspecific. Although no clear distinction usually exists, nonspecific interactions often do not have a fixed stoichiometry and/or a fixed mutual orientation of the binding partners. Note that a nonspecific interaction can be biologically significant. For example many protein–RNA interactions important for translation are not sequence specific. Even very weak interactions (either specific or not) are likely to be significant if the cellular concentration of at least one of the binding partners is comparable to the K_D for the bimolecular interaction or if the two molecules are brought in proximity *in vivo* through other mechanisms.

3. NMR METHODS FOR STUDYING PROTEIN INTERACTIONS IN TRANSLATION

3.1. Overview

NMR can be used for a broad range of tasks from testing unambiguously whether two molecules interact to solving the structure of the complex. NMR also has a number of other applications, such as studying the dynamics of the molecules, their folding/unfolding, as well as the effects of sample conditions (buffer composition, temperature, etc.) and interactions with other molecules on these processes (for recent reviews see Pellecchia [2005] and Vaynberg and Qin [2006]).

NMR is especially useful for weak interactions that are difficult to study by other methods. Often interactions in group 2.2.D and even group 2.2.C earlier cannot be detected by nonequilibrium methods (e.g., affinity chromatography or filter binding) because the binding partners dissociate too quickly and are lost during the washing steps. NMR is an equilibrium method, and the sample concentrations are in the micromolar to millimolar range—appropriate for studying such weak interactions. Furthermore, the studies can be performed on the native proteins, because NMR does not require affinity tags, fluorescent tags, etc., which could affect, or even prevent, interactions.

In NMR, the chemical shifts of individual nuclei can be correlated to the chemical shifts of other nuclei, which results in peaks in the NMR spectra at the intersection of the corresponding chemical shifts (Fig. 12.2). Depending on the experiment, the observed peaks correspond to atoms that are connected to each other through one or more covalent bonds or atoms that are close in space (typically less than 5 Å). Whereas the predominant hydrogen isotope ^{1}H is "visible" by NMR, the most abundant nitrogen (^{14}N) and carbon (^{12}C) isotopes are "invisible." Therefore, it is often necessary to label the proteins with stable isotopes of "better" NMR properties, such as ^{15}N and ^{13}C. This facilitates resonance assignments but also allows us to selectively observe the labeled proteins in complexes with unlabeled macromolecules. It is also sometimes useful to replace ^{1}H with ^{2}H (deuterium, D)—both to make the hydrogens "invisible" and to slow down relaxation (essentially loss of signal) when working with larger proteins. Here, "visible" and "invisible" are used as operational terms. For example, deuterium can be observed by NMR, if desired, but will be absent from a proton-detected spectrum. Labeling with stable isotopes is routinely achieved by growing bacteria in the appropriate media but can also be accomplished with *in vitro* translation. Expression of labeled proteins in eukaryotic cells has also been done but is typically much more expensive.

In the ^{1}H-^{15}N HSQC (Heteronuclear Single Quantum Correlation; Bodenhausen and Ruben [1980]) spectra (Fig. 12.2), the backbone NH group of every residue gives rise to one peak at the intersection of the ^{1}H and ^{15}N chemical shifts (except prolines, which do not have backbone NH groups). Additional peaks correspond to some side-chain NH and NH$_2$ groups. The peaks in the ^{1}H–^{15}N HSQC spectrum of a folded protein are well dispersed, giving a unique "fingerprint" of the protein (Fig. 12.2A, right panel). In contrast, an unfolded protein displays poor signal dispersion (Fig. 12.2A, left panel). The NMR chemical shifts are highly sensitive to the environment and, therefore, when an unlabeled protein is titrated into

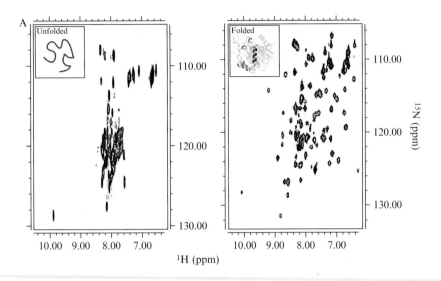

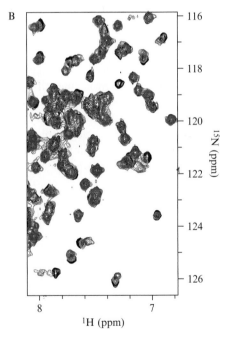

Figure 12.2 Protein–protein interactions visualized by NMR. (A) ^{1}H-^{15}N HSQC spectra of ^{15}N-labeled yeast eIF4G (393–490): free (left) and in complex with unlabeled eIF4E (right). Binding of eIF4E causes folding of eIF4G, which is accompanied by drastic changes in the spectra. Insets in the top left corners of the panels show a cartoon of an unfolded polypeptide (left) and the structure of the folded eIF4G segment (blue) in complex with eIF4E (semitransparent). Adapted from Hershey *et al.*, 1999, reproduced with permission from *J. Biol. Chem.* 1999, 274, 21297–21304. Copyright © 1999

a sample of a labeled protein, the peaks corresponding to residues affected by the interaction "move" on binding. The exploitation of this property is called chemical shift mapping and is one of the most widely used NMR methods for studying protein interactions, because it is both simple and powerful. In this experiment, the NMR spectrum of a protein labeled with an appropriate stable isotope (e.g., ^{15}N) is compared with the spectrum in the presence of an unlabeled (and thus "invisible") binding partner (Fig. 12.2).

3.2. Protein size and NMR

The size of a protein or complex is a major factor in determining what NMR experiments can be performed. Although complexes close to 1 MDa have been observed by NMR, most NMR experiments fail even at much smaller molecular weights. One of the major limitations of NMR is due to relaxation, the process that makes signals decay after excitation by radio-frequency pulses used in the experiments. Relaxation is largely due to the modulation of anisotropic interactions (such as dipolar couplings and chemical shift anisotropies) by the overall tumbling of the molecule; the slower the tumbling, the faster the transverse relaxation, which is most significant for the performance of NMR experiments. The consequence of fast transverse relaxation is a loss of signal intensity and a broadening of the detected signals. Relaxation is slower in smaller molecules that tumble faster, as well as in flexible segments of larger proteins that possess higher mobility. As the size of a protein increases, its relaxation rate increases, and the loss of signal during the NMR experiment is no longer negligible, even to the point that no signal can be detected in a reasonable amount of time. In contrast, small proteins generally give rise to stronger, sharper peaks.

For example, the ^{1}H–^{15}N HSQC spectrum of the large ribosomal subunit of _E. coli_ displays almost exclusively peaks corresponding to the C-terminal domain of L7, which is connected to the rest of L7 through a long flexible linker (Christodoulou _et al._, 2004).

The "regular" ^{1}H–^{15}N HSQC described previously can be used up to at least $\sim$25 kDa. Larger proteins or complexes (30 to 40 kDa) can also be observed but may require longer experiment times. TROSY HSQC (TROSY stands for transverse relaxation optimized spectroscopy; Pervushin _et al._ [1997]) (Fig. 12.2B) would give better sensitivity in this size range and

The American Society for Biochemistry and Molecular Biology. (B) Overlay of a section of the ^{1}H-^{15}N TROSY-HSQC spectra of the second HEAT domain of human eIF4G: free (black) in the presence of substoichiometric amount of unlabeled eIF4A-NTD (blue) and of equimolar amount of unlabeled eIF4A-NTD (red). The interaction is in fast to intermediate exchange, is not accompanied by an unfolded-to-folded transition, and leads to more modest changes in the spectra. (See color insert.)

can also be used for larger complexes. The TROSY effect exploits relaxation interferences to partially compensate for the autorelaxation of a given nucleus. For instance in an H-N spin-system, the interference between the ^{15}N chemical shift anisotropy (CSA) and the HN dipole–dipole interactions counteracts the effect of the two separate relaxation mechanisms. To benefit fully from the TROSY effect, it is necessary to deuterate the ^{15}N-labeled protein at least partially, in particular the Hα position, because the main remaining source of relaxation of the HN nitrogen (and the resulting loss of signal) is the Hα proton (see strategies for stable isotope labeling later) and because the presence of protons reduces the TROSY effect. This becomes essential for complexes larger than $\sim$40 kDa, where the relaxation rates in the protonated protein become too fast compared with the time it takes to transfer magnetization from the ^{15}N to ^{1}H. ^{1}H–^{13}C HMQC, and especially ^{1}H–^{13}C-HMQC of methyl groups (Tugarinov and Kay, 2005) can also be used on larger complexes, provided that the assignments of the methyl groups are available.

3.3. Exchange regimes in NMR

Before describing the various applications of NMR for studying protein interactions, we need to introduce the concept of exchange regimes in NMR.

3.3.1. Equilibrium between alternative conformations

If a molecule is in equilibrium between two conformations: A and A*:

$$A \underset{k_{-1}}{\overset{k_1}{\rightleftarrows}} A^*$$

the individual nuclei experience two different environments. The effect on the chemical shifts, and thus on the peak positions depends on the rates of interconversion, k_1 and k_{-1}, and on the magnitude of the changes in chemical shifts, Δ. Note that because these are conformational changes within the same molecule, the rates of interconversion and the equilibrium are concentration independent.

Slow exchange If the interconversion rate is slow on the time scale of the NMR experiment (k_1 and $k_{-1} \ll \Delta$), where Δ is the difference in chemical shift (in Hz) between states A and A* for a given nucleus, two sets of peaks are observed, corresponding to the two alternative conformations. The ratio between the intensities of the two groups of peaks depends on the ratio between the concentrations of the two conformations at equilibrium (Fig. 12.3A).

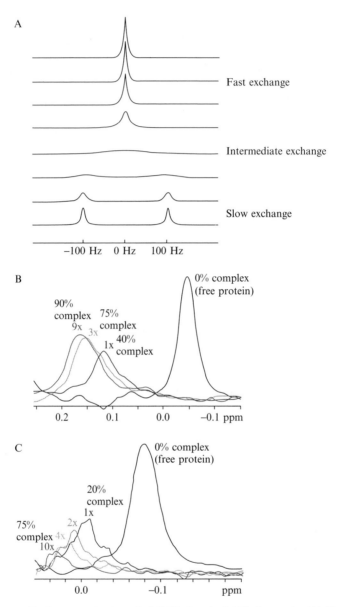

Figure 12.3 Exchange regimes on the NMR chemical shift time scale. (A) Simulation of line shapes of NMR peaks for interconversion between two conformations in fast (top), intermediate, and slow (bottom) exchange regimes, with 200-Hz frequency difference between the peaks corresponding to the two states. The simulation was performed as described in Matsuo *et al.* (1999). (B) 1D slices of a single peak of the protein Bcl2 in the absence and in the presence of increasing concentrations of a small molecule. The binding is in the intermediate exchange regime ($K_D = 20\ \mu M$). The protein concentration is 25 μM, and compound concentration is from 0 to 250 μM. The ratios between

Fast exchange If the interconversion is fast on the NMR time scale (k_1 and $k_{-1} \gg \Delta$), the effects of the two environments are averaged, and one set of peaks is observed. Every peak appears between the positions corresponding to the two conformations, and its exact position depends on the ratio between the populations of the two alternative conformations.

Intermediate exchange If the interconversion rates are intermediate between the fast and slow exchange limits (k_1 and $k_{-1} \sim \Delta$), the effects on the chemical shifts are also intermediate. The averaging between the two chemical shifts is only partial, which leads to a broadening of the peaks and could even make them impossible to detect (Fig. 12.3A).

3.3.2. Equilibrium between free protein and complex

Similar exchange behaviors are observed for protein–protein interactions:

$$A + B \underset{k_{off}}{\overset{k_{on}}{\rightleftarrows}} AB,$$

In this case, the exchange is between the free and bound states, and the exchange regime (slow, intermediate, or fast) depends on the dissociation rate of the complex, k_{off}. Here, the ratios between the free and bound states depend on the concentrations of the two interacting partners (Fig. 12.3B). To facilitate the discussion, we will describe the situation in which a labeled protein A is in presence of an unlabeled molecule B.

- For an interaction in slow exchange ($k_{off} \ll \Delta$), where Δ is the difference in chemical shift (in Hz) between states A and AB for a given nucleus, two signals are observed for the free and the bound state, and the relative intensities of the two signals are proportional to the concentrations of free and bound forms of the labeled protein. Adding unlabeled protein makes the signal of the bound form increase and that of the free form decrease. At equal populations, the broadening δv of the signals

---

compound and protein concentrations and the fraction of the protein that is in complex (in %, calculated from the K_D) are shown above the curves. (C) 1D slices of the same peak as in b on addition of increasing concentration of a different small molecule ($K_D = 80 \, \mu M$) related the one in b and contacting the same residue of Bcl2, with two or more bound conformations in intermediate exchange regime. The protein and small molecule concentrations are the same as in b. The ratios between small molecule and protein concentrations and the fraction of the protein that is in complex (in %, calculated from the K_D) are shown above the curves. Note that, due to the conformational exchange in the bound state, the line broadening is not reversed, and even becomes more severe, as the equilibrium is shifted toward complex formation. Panels b and c are adapted from Reibarkh *et al.* (2006). Reproduced with permission from *J. Am. Chem. Soc.* 2006. 128, 2160–2161. Copyright 2006 American Chemical Society. (See color insert.)

because the exchange is equal to $k_{off}/2\pi$ and independent of Δ. Peaks that are not affected by the interaction remain in the same positions. For peaks whose positions are different between the free and bound states, the intensity of the peak corresponding to the free state gradually decreases when adding the binding partner, whereas the intensity of the peak corresponding to the bound state increases.

- In fast exchange ($k_{off} \gg \Delta$), the position of the peak is the weighted average of the peaks of the free and the bound form, and the peaks "move" gradually from their positions in the free state to their positions in the complex when increasing the concentration of the binding partner. At equal populations of free and bound states, the line broadening δv due to exchange is approximately equal to $(\pi\Delta^2)/4k_{off}$.

- In the intermediate exchange regime, the peaks affected by the interaction become broadened, sometimes beyond detection, at intermediate concentrations of the unlabeled protein, where only a fraction of the labeled protein (A) is in complex; then they "grow" at the position corresponding to the bound state as the equilibrium is shifted toward complex formation (Fig. 12.3B). Note that only peaks whose positions are affected by the interaction become broadened due to intermediate exchange at intermediate concentrations of the unlabeled ligand.

Therefore, irrespective of the type of exchange regime, the peaks corresponding to the bound state can usually be observed in the presence of excess unlabeled protein. Note that the unlabeled protein needs to be in excess with respect to not only the labeled protein but also the K_D of binding for the labeled protein to be predominantly in the bound state (Fig. 12.3B). For example, if the concentration of the labeled protein is $100\ \mu M$ and the K_D is $100\ \mu M$, it can be calculated using Eq. (1) that ~ 2 mM unlabeled protein would be necessary to drive the equilibrium to 95% bound labeled protein, whereas only $115\ \mu M$ unlabeled protein is sufficient for 95% bound labeled protein if the K_D is $1\ \mu M$.

The exchange regime depends on the chemical shift difference Δ (in Hz) between the two states. Therefore, different subsets of the peaks affected by an interaction can be in different exchange regime; peaks that "move" farther would tend to be closer to slow exchange regime than peaks that "move" less. Because Δ is expressed in Hz, its value is proportional to the spectrometer field. Therefore, the use of a spectrometer with higher frequency would shift the exchange regime moderately toward slow exchange (e.g., switching from 400 to 800 MHz doubles the value of Δ). Unfortunately, this does not help much for sharpening up resonances, because changing the value of Δ by only a factor of two has a rather minor effect on the exchange regime. Furthermore, at stoichiometric concentrations, the line broadening of the resonances in slow exchange is independent of Δ, and the only way to sharpen up lines in this exchange regime is to add excess ligand. Varying temperature and/or salt concentrations are the most common approaches when drastic changes in exchange regime are desired (e.g., from intermediate to fast, or to slow exchange).

Line shapes can readily be simulated for all values of k_{off} at given relative concentrations when Δ and relaxation rates are known (see Matsuo *et al.* [1999]). The supplement to that article contains an analytical expression for calculation of line shapes. Note that, whereas Δ and k_{off} have the same units (Hz are by definition equal to sec^{-1}), historically Δ is expressed in Hz and k_{off} in sec^{-1}, which could be confusing.

In NMR of proteins, Δ typically ranges from $\sim$10 to >1000 Hz. Therefore, if k_{off} of a complex is between $10\ sec^{-1}$ and $1000\ sec^{-1}$, a peak with $\Delta \sim 10$ Hz would be approaching fast exchange, whereas a peak with $\Delta \sim 1000$ Hz would be approaching slow exchange. In practice, if Δ is comparable to the peak width, it is difficult to observe slow or intermediate exchange due to peak overlap. Because diffusion-controlled k_{on} is typically in the order of 10^6 to $10^8 M^{-1}sec^{-1}$, a k_{off} of $100\ sec^{-1}$ would correspond to a K_D in the order of 1 to 100 μM.

3.3.3. Additional considerations and special cases

- As described previously, the size of the complex influences the detection of the NMR signals. For example, if a small protein binds to a large protein or complex, it may become difficult or impossible to observe the bound state. In the fast exchange limit, the relaxation of a given protein is partially increased, compared with the free protein. In the slow exchange limit, the relaxation is the one corresponding to the complex, which can be dramatically larger than the one of the free molecule. Therefore, complexes in fast exchange may be easier to study. The ratio between free and bound states has a major impact in this case; as the equilibrium is pulled toward the bound state, the protein spends a greater and greater proportion of the time in the complex, and its relaxation behavior becomes closer to that of a large protein (see Matsuo *et al.* [1999]).
- Whereas changes in peak intensity in slow exchange and "movement" of peaks in fast exchange are directly proportional to the ratio between labeled protein molecules in free and bound state, the intermediate exchange regime appears "nonstoichiometric"; peak broadening and disappearance are evident even at low concentrations of the unlabeled ligand. For example, a 1:3 or even 1:10 ratio between bound and free labeled protein is often sufficient to broaden the affected peaks beyond detection. The reason is that multiple binding and dissociation events occur during the experiment, allowing one molecule of unlabeled protein to bind to (and affect) multiple labeled proteins.
- In a number of cases, part of the peaks corresponding to the bound state cannot be observed. For example, this would be the case if the complex

is in equilibrium between two alternative conformations and the interconversion is in intermediate exchange regime. The peaks whose positions are different between the two alternative conformations would be broadened, irrespective of the ratio between the binding partners (Fig. 12.3C).

- A similar result can be observed if the complexes formed are heterogeneous. For example, when a nonspecific RNA-binding protein binds RNA, individual protein molecules may be in a different environment, depending on what segment of the RNA molecule they interact with. If the resulting chemical shifts for part of the nuclei differ significantly, the corresponding peaks could be broadened beyond detection.
- If the spectrum of the entire labeled protein (or an entire domain) disappears at substoichiometric concentrations of the unlabeled ligand but reappears at near-equimolar ratio, this is most likely not due to intermediate exchange, but to the formation of large complexes (of the type A_nB) when $[A] > [B]$, which are converted to smaller complexes (of the type AB) with increasing the concentration of B ($[A] \sim [B]$). This is typically seen when a protein is titrated with a DNA or RNA fragment long enough to bind multiple protein molecules; even the sequence/structure-specific RNA-binding proteins have a fairly high nonspecific RNA-binding affinity. For example, when GB1-tagged eIF4H is titrated with an RNA oligo, the entire RNA-recognition motif (RRM) domain of eIF4H is broadened at intermediate protein/RNA ratios, and the peaks reappear in the presence of excess RNA. The GB1 tag and the unfolded C-terminal region of eIF4H are unaffected (a representative region of the $^1H-^{15}N$ HSQC spectrum of eIF4H is shown on Fig. 12.4A). A convenient empirical criterion to distinguish between intermediate exchange and large complexes is the comparison of the spectra of the free and bound states; if at the end of the titration, a fraction of the peaks reappears at their original positions from the free state, their disappearance at substoichiometric RNA concentrations is unlikely to be due to intermediate exchange broadening, because the chemical shifts of the corresponding nuclei were not affected by the interaction.

3.4. Applications of NMR for studying protein interactions

1. NMR can be used to test whether two proteins interact. This is particularly useful for weak interactions that are hard to detect with "conventional" biochemical assays.
2. The K_Ds of the interactions can be determined by NMR, when the K_Ds are in the micromolar to millimolar range.
3. The interaction surfaces can be mapped by NMR.

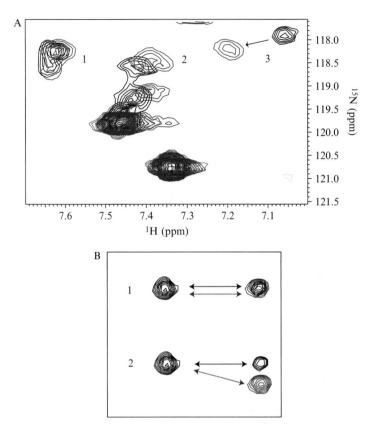

Figure 12.4   Direct vs. indirect effects in NMR titrations. (A) A small region of the ^{15}N-HSQC spectrum of GB1-tagged human eIF4H in the absence of RNA (black) in the presence of substoichiometric amount of a 25-mer RNA oligo (blue) and in the presence of excess RNA (red). The peaks corresponding to the entire RNA-recognition motif (RRM) domain of eIF4H are broadened (most beyond detection) at intermediate RNA concentrations and reappear at excess RNA. Whereas the transient disappearance of peak 3 at substiochiometric concentration of RNA could be due to intermediate exchange broadening, the transient broadening/disappearance of peaks 1 and 2 is unlikely to be due to intermediate exchange, because little or no change exists in chemical shifts. The two strong peaks on the bottom of the panel that are not affected throughout the RNA titration belong to the GB1 tag. (B) Schematic representation of the effects of a ligand on a protein that is in equilibrium between two alternative conformations in slow exchange. Peak 1, ligand binding causes change in equilibrium, but not in peak positions; therefore, there is no change in environment—the corresponding residue is not in direct contact with the ligand, and the effect is indirect. Peak 2, ligand binding causes change in both equilibrium and peak positions—the effect could be either direct or indirect. (See color insert.)

4. Indirect effects of the interaction on the conformation or mobility of other regions of the molecules (outside the contact interface) can also be identified.

5. The overall orientation of the two molecules can be determined.
6. The structure of the complex can be solved.

An important feature of studying interactions with NMR methods is that the process is not all-or-nothing, but stepwise. Thus, the decision how far to take the studies can be taken on the basis of the results obtained from the current stage, depending on how important and/or feasible the next step is.

3.5. Interactions in translation from an NMR perspective

A. Interactions from group 2.2.A (when K_D is much smaller than the factor concentrations and the complex is stable for more than a few minutes): Slow exchange. NMR applications (1), and (3) to (6) from 3.4. The K_D of the interaction (2) is much lower than the protein concentrations in the NMR sample. Therefore, only an upper limit for the K_D can be obtained from NMR. Sometimes the individual binding partners are not soluble in isolation; then only the structure of the complex (6) may be studied, if the size allows it.

B. Group 2.2.B (K_D much smaller than the concentrations and the complex is stable for seconds to minutes): Slow exchange. Same as group A, but the individual factors are more likely to be stable in isolation.

C. Group 2.2.C (K_D comparable to the concentrations, lifetimes of the complexes in the millisecond to second range): Slow, intermediate, or fast exchange. NMR applications (1), and (3) to (6) above. K_Ds (2) can be determined for some of the interactions. NMR can provide an estimate, or upper limit, for dissociation rates and information on how these are affected by other interactions or buffer conditions.

D. Group 2.2.D (K_D much larger than the concentrations): Fast exchange. NMR applications (1) to (6). Solving the structure (6) may be difficult, because the transient character of the complex can make obtaining intermolecular NOEs difficult and may lead to an averaging of residual dipolar couplings (see later for discussions on these methods).

3.6. Applications of NMR for studying protein interactions in translation

3.6.1. Testing for binding

Changes (or absence thereof) in ^1H–^{15}N HSQC spectra on addition of an unlabeled protein (or other molecules) can be used to test whether the two proteins interact. Because the interactions are studied at equilibrium, at high protein concentrations, both strong and weak interactions can be detected. Even very weak binding with K_Ds in the high micromolar, even millimolar, range can be detected, as long as at least one of the interacting partners is soluble in that concentration range. No prior knowledge about the proteins is necessary to test for binding. At the same time, in addition to the "yes/no" answer, the NMR spectra provide valuable information:

- Whether the labeled protein is folded or unfolded, or contains both folded and unfolded segments; whether it contains mostly α-helices or β-strands
- Whether the binding is tight or weak
- The approximate number of residues affected
- Whether the interaction induces significant changes in conformation (e.g., transition from folded to unfolded state of part of the protein) (Fig. 12.2A).

Controls A number of factors could affect the NMR spectra and the observed chemical shifts. Therefore, to avoid false-positive results, the experimental conditions during an NMR titration have to be kept as constant as possible. This includes use of the same spectrometer, at the same temperature, the same experiment, parameters, processing, and calibration. It is also important to keep the buffer conditions (pH, ionic strength, composition) constant. Sometimes for practical reasons, it is impossible to ensure that the buffer conditions are unchanged. For example, a ligand needs to be dissolved in DMSO, or the pH of a stock solution cannot be measured precisely. Control titrations can be used to compare the effects of, for example, DMSO or pH on the spectra of the free protein, to the effects of the putative binding partner.

Another important control is to test for dilution effects over the range of concentrations used in the titration experiment. Sometimes the protein sample is in equilibrium between monomer and dimer or oligomer. Because during the titration the sample is gradually diluted, the monomer/dimer equilibrium would change, which in turn would affect the chemical shifts and lead to false-positive results. In fact, NMR can be used to map homodimerization interfaces by collecting spectra on gradual dilution of a sample, provided the K_D is greater than ~ 10 μM (Pan *et al.*, 2001).

In certain cases, the changes in buffer conditions caused by adding the ligand itself are not negligible. For example, titration with high concentrations of RNA or DNA cause an increase of ionic strength. Therefore, a control titration with increasing salt concentration is warranted. The protein–nucleic acid interactions have a significant electrostatic component, and it is expected that a number of residues will be affected by both RNA binding and high salt. A conservative approach often used in such cases is to exclude all peaks that are affected by both RNA and salt from consideration. However, if binding is unambiguously established (e.g., on the basis of other peaks that are insensitive to salt), residues whose peaks are affected by both RNA and salt could also be considered as candidates for direct contacts.

What chemical shift changes are considered significant? Several factors determine the minimal chemical shift changes that can safely be attributed to interaction. The major ones are:

- Digital resolution of the spectra: chemical shift changes less than half the difference between two adjacent points should not be considered. This translates into 0.01 to 0.02 ppm in the ^{1}H dimension and 0.05 to 0.1 ppm in the ^{15}N dimension of a typical ^{1}H–^{15}N HSQC spectrum.
- Statistical analysis allows us to distinguish significant chemical shift changes from random variation of the measured chemical shifts. Typically, chemical shift changes larger than 1 standard deviation (σ) from the average change for the entire protein are considered significant; 1.5 σ or even 2 σ cutoff have also been used. Because peaks that display significant chemical shift changes are included in the initial calculation of σ, iterative σ calculations, after excluding the outliers from the previous cycle, are sometimes used to better estimate the variation observed among the peaks that are not affected by the interaction. This approach is especially useful when the interaction induces major chemical shift changes.
- In ^{1}H–^{15}N HSQC spectra, the chemical shift changes in the ^{15}N dimension are typically divided by 5 before adding them to the ^{1}H chemical shifts to compensate for the different dynamic range. The chemical shift changes in the ^{1}H (Δ_H) and ^{15}N (Δ_N) dimension are added using either:

$$\Delta = |\Delta_H| + (1/5)|\Delta_N|, \tag{2}$$

adding the absolute values, or

$$\Delta = ((\Delta_H)^2 + (1/5)\Delta_N)^2)^{1/2} \tag{3}$$

The value of 5 for scaling of Δ_N is arbitrary, and values of 6 (on the basis of the gyromagnetic ratio between ^{1}H and ^{15}N) or 10 (if expressing Δ_N and Δ_H in Hz, instead of ppm) have also been used. The approach is empirical, but rather robust; in most cases, the choice of parameters has little, if any, effect on the final results.

The magnitude of the chemical shift changes provides certain qualitative information about the observed interaction; typically larger chemical shift changes correlate with stronger binding. Furthermore, interactions that are accompanied by folding or conformational changes have greater effect on the chemical shifts (Fig. 12.2).

False-negative results, or can NMR miss interactions? It is practically impossible to miss strong interactions, such as those that can be detected by affinity pull-down experiments. However, it is possible to miss weak interactions. Weak interactions often cause smaller chemical shift changes, and it may be impossible to drive the titration to completion, which means that only part of the maximal chemical shift changes can be observed. In the example shown in 3.3.B using Eq. (1), if the labeled protein concentration is 100 μM and the K_D is also 100 μM, then 2 mM unlabeled ligand is necessary to reach 95%:5% bound/free ratio for the labeled protein. Stopping the titration at 100 μM unlabeled protein will reach less than 40%:60% ratio bound/free. Therefore, if the titration is stopped at a 1:1 ratio between A and B, only 40% of the maximum chemical shift would be observed, which may be very small and difficult to observe, especially for a weak interaction. Even if the interaction is 10-fold stronger ($K_D = 10$ μM), the ratio bound/free is ~75%:25%, hence 75% of the maximum chemical shift would be observed. These numbers demonstrate that for weaker interactions, it is important to extend the titration beyond a 1:1 ratio of the two interacting partners, if possible. In practice, this is recommended even for stronger interactions; to account for possible errors in protein concentration, for possible 2:1 stoichiometry, or in case part of the protein molecules are inactive. Modern spectrometers can record simple NMR spectra (1D or HSQC) at concentrations as low as 10 μM. Thus, it is sometimes advisable to perform titrations at low concentrations of the observed (labeled) protein to have a better chance to approach the end point of the titration. This also provides a possibility to see whether the binding is saturable (specific) or not (nonspecific binding).

In case of weak interactions involving only a limited surface, it is theoretically possible that side chains exclusively mediate the interaction; no backbone NH chemical shifts are affected; and no chemical shift changes are observed if ^{1}H–^{15}N HSQC is used for the titration. In practice, this possibility is usually only worth considering if there is uncertainty about the results of ^{1}H–^{15}N HSQC titration or if there are other reasons to expect interaction. The use of a ^{13}C/^{15}N double-labeled sample would allow following the titration in parallel by ^{1}H–^{15}N HSQC and ^{1}H–^{13}C HSQC. The two spectra can be recorded simultaneously (Farmer II, 1991; Sattler et al., 1995). An added benefit of ^{1}H–^{13}C HSQC titrations is that the ^{13}C and ^{1}H chemical shifts observed are usually less sensitive to the buffer conditions than the chemical shifts of the NH groups.

3.6.2. K_D of the interaction

K_D can also be determined in some cases if the interaction is in the fast exchange regime on the NMR time scale (see earlier) and the K_D is comparable to the protein concentrations in the sample. For stronger interactions, only an upper limit for the K_D can be set.

Increasing amounts of unlabeled protein are titrated into the sample of the labeled protein. NMR spectra, typically ^{1}H–^{15}N HSQC, are collected at each step. As the concentration of the unlabeled protein increases, the equilibrium between free and bound labeled protein shifts toward the complex. As explained previously, this causes a set of peaks, corresponding to residues affected by the interaction, to move gradually from their positions in the free protein toward their positions corresponding to the bound species. The titration is continued until the peaks stop moving or until a solubility limit is reached. It is important that there be no significant aggregation of either free proteins or the complex during the titration, and working at low concentrations of the observed protein is advisable (see previously).

The chemical shift changes are related to the fraction of the labeled protein that is in the complex (Y):

$$Y = \Delta/\Delta_{max} = [AB]/[A_T],\qquad(4)$$

where Δ_{max} is the maximum change in chemical shift for the given peak, [AB] is the concentration of the complex, and $[A_T]$ is the total concentration of the labeled protein.

The fraction of the bound labeled protein Y can be expressed as a function of the variable $[B_T]$, the total concentration of unlabeled protein, and the parameters $[A_T]$ and K_D, using Eq. (1) and the following relationships:

$$[A] = [A_T] - [AB],\qquad(5)$$

and

$$[B] = [B_T] - [AB],\qquad(6)$$

The result is a quadratic equation for Y:

$$[A_T]Y^2 - (K_D + [B_T] + [A_T])Y + [B_T] = 0,\qquad(7)$$

with a single meaningful solution $(0 \leq Y \leq 1)$:

$$Y = \frac{((K_D + [B_T] + [A_T]) - \sqrt{(K_D + [B_T] + [A_T])^2 - 4[B_T][A_T]})}{2[A_T]}\qquad(8)$$

K_D can be obtained by fitting Y as a function of $[B_T]$ using appropriate software (e.g., SigmaPlot; SPSS Inc.).

If the titration cannot reach ~100% of bound labeled protein ([A] = [AB]), the exact value of Δ_{max} (and also Y, which is equal to Δ/Δ_{max}) is not available. Therefore, it may be useful to replace Y in the preceding equation with Δ/Δ_{max}, where Δ is the variable and Δ_{max} is a new parameter that can be fitted.

If [A_T] decreases during the titration due to dilution, the preceding equation would have two variables: [B_T] and [A_T]. Therefore, it is important to keep [A_T] constant throughout the experiment. This can be achieved in several ways:

1. If the unlabeled protein (B) stock can be prepared at a concentration much higher than [A_T], the dilution effects on [A_T] during titration would be negligible.
2. If the solubility of the unlabeled protein (B) is limited, A can be added to the stock of B at a final concentration equal to [A_T] in the sample. This approach ensures that [A_T] does not change during titration with increasing amounts of B.
3. The samples for all individual titration steps can be made separately by mixing stocks of A and B at different ratios. However, the previous approach is more convenient in most cases.

To minimize the possibility of errors, it is useful to perform the fitting using several peaks, with good signal-to-noise ratios. For stronger binding, where $K_D < [A_T]$, the contribution of the K_D to the term ($K_D + [B_T] + [A_T]$) is negligible at any [B_T] concentration and, therefore, accurate calculation of K_D is impossible. In such cases, only an upper limit for K_D can be estimated.

3.6.3. Mapping of interaction surfaces

The NMR titration can be used to identify the residues of the proteins that are affected by the interaction if backbone NMR resonance assignments are available (i.e., if it is known which peak in the ^1H–^{15}N HSQC spectrum corresponds to which residue). If the structures of the interaction proteins are known, the contact interface can also be mapped. If the structures of the binding partners have been determined by NMR, no additional information is required to map the binding interface (backbone resonance assignments are one of the prerequisites for structure determination). If the structures have been determined by X-ray crystallography, only backbone NMR resonance assignments are necessary (described later). If no structure is available for one or both of the interaction partners, the structures can be solved by NMR as described later. Alternatively, if the structure of a homologous protein is known, the structures of the binding partners can be modeled, using, for example, SWISS-MODEL (Arnold et al., 2006), because homologous proteins have similar structures. This will allow mapping the binding interface without structure determination.

Chemical shift changes on binding can be due to either direct contacts or indirect effects on the protein. For example, the interaction of eIF4E with eIF4G induces folding of a large ~10 kDa fragment of eIF4G. The transition to a folded state is accompanied by drastic changes in the ^1H–^{15}N HSQC spectrum of eIF4G (Fig. 12.2A) (Gross et al., 2003b). In this case, the changes in chemical shifts are mainly the result of a change in conformation and not only of contacts between the two molecules. The same work also illustrates another type of indirect effect detected by chemical shift changes. Whereas eIF4G does not directly contact the Cap-binding site of eIF4E, its binding to eIF4E stabilizes the eIF4E–cap complex. The results from NMR experiments show chemical shift changes in the Cap-binding site but no structural changes, indicating that eIF4G binding to eIF4E affects the mobility of the Cap-binding site, which leads to an increase in the stability of the eIF4E–Cap interaction (Gross *et al.*, 2003b).

In most cases, chemical shift mapping alone cannot distinguish between direct and indirect effects, and additional experiments are necessary to map direct contacts (see later). One notable exception is when the free protein is in slow exchange equilibrium between two conformations. As described previously, in slow exchange, two sets of peaks are observed, and the ratio in peak intensity reflects the ratio between the two conformations at equilibrium (Fig. 12.4B). If ligand binding shifts the equilibrium toward one conformation, adding increasing amounts of the ligand not only causes chemical shift changes at its own binding site but also changes the ratio in peak intensity between the two sets of peaks corresponding to the two alternative conformations. If only the intensities, but not the positions of a pair of peaks, are affected, the ligand binding does not affect the environment around that nucleus. Therefore, it does not contact the corresponding residue directly (Fig. 12.4B, top). Note, however, that if the conformational exchange in the free protein is in fast exchange (leading to one set of peaks at average chemical shifts), a shift in equilibrium caused by ligand binding would result in peak movement, which would be impossible to distinguish from the chemical shift changes due to direct contacts. Conformational exchange in the fast exchange regime can, however, be studied by relaxation dispersion techniques (Palmer *et al.*, 2005). In short, pulse sequences can be designed to average out the contribution of conformational exchange to transverse relaxation and thus quantify it.

Although chemical shift mapping alone is typically sufficient to identify the main contact surface, it is often necessary to know the precise structure of the complex. The structure of the complex can be solved by NMR or X-ray crystallography. If solving the structure is not possible or necessary, NMR can be used to determine the mutual orientation of the proteins and to distinguish between direct contacts and indirect effects mediated by conformational changes in the labeled protein.

3.6.4. Methods for identification of direct contacts

Intermolecular NOEs   In the NOESY (nuclear Overhauser effect spectroscopy) experiments used for structure determination, magnetization is transferred through space, and NOE (nuclear Overhauser effect) cross-peaks are observed between protons that are close to each other (typically <5 Å). If the peaks can be unambiguously determined to be intermolecular, the corresponding protons are directly at the interface: <5 Å from the other protein. However, it is not always obvious whether a peak is intramolecular or intermolecular. Differential labeling of the two binding partners allows us to observe specifically intermolecular peaks (Gross *et al.*, 2003a; Walters *et al.*, 1997) and is more sensitive than isotope-filtered experiments, especially for larger proteins. One such approach is to form a complex, where one binding partner is deuterated (^{2}H-labeled) and ^{15}N-labeled, and the other one is unlabeled (protonated). If this complex is dissolved in water (^{1}H$_2$O), the only protons on the deuterated protein will be those that exchange with water. Most backbone NH protons of a protein exchange quickly with water, but the side-chain aliphatic and aromatic CH protons do not. Therefore, in ^{15}N-edited NOESY of such a sample, all observed peaks between HN protons and side-chain aliphatic protons will be intermolecular. This approach is commonly used when solving the structures of complexes, but it also allows us to identify NH groups that are at the binding interface, even without assigning the NOE cross-peaks to specific protons, and proceeding with structure determination. Figure 12.5 illustrates another approach using ^{1}H/^{13}C-labeled methyl groups in an otherwise perdeuterated protein in complex with an unlabeled binding partner. In the ^{13}C-edited NOESY spectrum of such complex, all NOEs between methyl and aromatic side-chain protons will be intermolecular (Gross *et al.*, 2003a).

Cross-saturation   In some cases, when it is difficult to observe intermolecular NOEs, saturation transfer experiments can be performed, instead. The labeling scheme is as shown previously. However, instead of collecting a NOESY spectrum, a series of HSQC spectra are collected, where resonances of the aliphatic protons of the unlabeled protein are saturated (their NMR signals are suppressed) for various amounts of time. Saturation can be transferred through space, and after a certain time, nearby protons that were not directly irradiated are partially saturated too. These include NH protons from the ^{2}H-/^{15}N-labeled protein that are at the contact interface. Therefore, NH groups at the binding interface are identified on the basis of decrease of the intensity of the corresponding ^{1}H–^{15}N HSQC peaks (Takahashi *et al.*, 2000). Saturation transfer can be used to study weak interactions between a small to medium size protein and a very large protein or complex (Nakanishi *et al.*, 2002). Cross-saturation of methyl groups has certain advantages, in particular for large complexes (Takahashi *et al.*, 2006). If the off rate of the interaction is fast compared with the relaxation rates of the free protein (fast exchange on the "relaxation time scale"), saturation of resonances in the small protein is accumulated in the free state. This allows us to keep the concentration of the small protein 10- to 100-fold higher than that of the large binding partner and observe intensity changes in the spectrum of the free protein (for recent reviews see Pellecchia [2005] and Shimada [2005]).

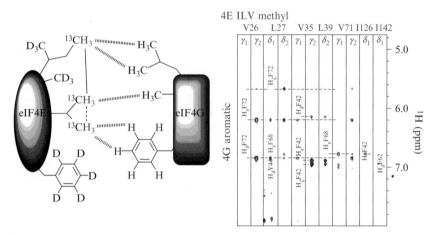

Figure 12.5   Differential labeling strategy for obtaining unambiguous intermolecular NOEs in large complexes. Asymmetrical isotope labeling scheme applied in the NMR analysis of the eIF4E/eIF4G protein complex. One protein is $^{12}C/^{2}H$-labeled, except for the methyl groups of Ile (d_1), Leu, and Val side chains, which are $^{13}C/^{1}H$-labeled. The other protein in the complex is unlabeled. The protons (^1H) attached to ^{13}C methyl carbons are in red; the aromatic protons on the unlabeled eIF4G are in blue. Pairs of protons that can yield NOEs observed in a ^{13}C-edited NOESY spectrum are marked with dashed lines. The right panel shows strips from the aromatic region of the ^{13}C-edited NOESY spectrum of this complex: all observed NOEs between methyl and aromatic side-chain protons are intermolecular. The left panel is adapted from Gross *et al.* (2003). Reproduced with permission from *J. Biomol. NMR* 2003. 25, 235–242. Copyright © 2003 Springer. (See color insert.)

Whereas the intermolecular NOEs can be used in subsequent structure determination, saturation transfer does not provide any information about interproton distances across the binding interface. However, its advantage is that the experiment is more sensitive and shorter than the NOESY experiments.

3.6.5. Determining the mutual orientation of the binding partners

Intermolecular NOEs (see [6]), if assigned to a specific pair of protons, provide direct information about the intersubunit orientation in a complex and are used in structure determination (see following).

Paramagnetic labels   A paramagnetic label can be attached to the side chain of a single surface–exposed cysteine on one of the two proteins. Paramagnetic nuclei affect the NMR resonances over a long range (up to ~20 Å). These effects can be converted into distance restraints and used in structure calculation (Battiste and Wagner, 2000; Battiste *et al.*, 2003; Gross *et al.*, 2003b). If such a cysteine is near the binding interface, the paramagnetic label will affect part of the resonances of the other protein, thus providing intermolecular distance restraints. The advantage of this approach is that it provides a large number of long-range distance restraints (up to ~20 Å, whereas NOE-derived restraints are typically <5.5 Å). A major disadvantage is that it is labor-intensive; often, more than one surface-exposed cysteine is present in a protein and/or the cysteines present are not at the desired position. Therefore, it is usually necessary to mutate all surface cysteines, make sure the protein folding and the interaction are not affected, and then generate several single-cysteine mutants. Mutating buried cysteines is more likely to affect the folding and stability of the protein, but the buried cysteine side chains are unlikely to be modified by the labeling and thus may not need to be mutated, provided proper controls are performed.

Residual dipolar couplings (RDCs)   Dipolar couplings lead to a splitting of lines in an NMR spectrum in the same way that scalar couplings do. The magnitude of the splitting depends on the orientation of the dipole-dipole vector (e.g., an H—N bond) with respect to the magnetic field. In liquids, this effect is averaged out by the reorientation of the molecule, and dipolar couplings only contribute to relaxation. In some cases, it is possible to reintroduce a weak alignment of the molecules, which leads to a residual coupling. This is done either by exploiting the intrinsic susceptibility of a molecule (Tolman *et al.*, 1995) or by means of an alignment medium (Tjandra and Bax, 1997). In particular, it is possible to rapidly determine the orientation of two domains by calculating the alignment tensor for these domains (Fischer *et al.*, 1999). The couplings are measured on simple experiments, derivatives of the HSQC (Ottiger *et al.*, 1998), or the HNCO experiments (Kontaxis *et al.*, 2000). The RDCs can then be used in combination with chemical shift mapping data to guide docking of protein complexes (Clore and Schwieters, 2003). The only prerequisite is to have assigned the resonances of the pairs of nuclei involved in the interaction. One limitation lies in the necessity to find an alignment medium that provides sufficient alignment without having strong interactions with the molecule of interest. Another is due to the line broadening induced by the anisotropic medium.

Relaxation   Relaxation rates can be used to determine domain orientations in a manner similar to residual dipolar couplings. The magnitude of a relaxation process caused by fluctuations of dipolar interactions depends on the orientation of the dipole–dipole vector with respect to the axes of the diffusion tensor (describing the overall tumbling of the molecule). In the case of multiple domains, it is possible to calculate the relative orientations of the axes of each domain from the measured relaxation rates (Bruschweiler *et al.*, 1995; Fushman *et al.*, 2004). These rates typically involve the ^{15}N transverse and longitudinal relaxation rates and the H→N heteronuclear nOe.

Docking algorithms Several methods for docking two proteins to each other or a small molecule to a protein have been developed in recent years. Some of these are entirely driven by shape complementarity and energy. Other approaches, such as Tree-Dock (Fahmy and Wagner, 2002) and HADDOCK (Dominguez *et al.*, 2003), can incorporate experimental data (e.g., from NMR chemical shift mapping or mutations) to limit the search space and/or as part of the docking algorithm. As mentioned previously, docking guided by residual dipolar couplings and chemical-shift mapping has also been pursued with the XPLOR-NIH program (Clore and Schwieters, 2003; Schwieters *et al.*, 2003).

3.6.6. Solving the structures of protein complexes by NMR

The methods for solving the structures of protein complexes by NMR will be described in the next section, together with backbone and side-chain NMR resonance assignments.

4. DESCRIPTIONS OF NMR METHODS

This section is not meant to provide an extensive survey of NMR techniques, which have already been abundantly covered elsewhere (see e.g., Ferentz and Wagner [2000] and Walters *et al.* [2001]). Instead, we would like to mention some basic principles and describe some specific strategies that may be needed for difficult systems.

4.1. Methods for backbone assignments

To benefit from the full range of NMR applications, the resonances in the spectra first need to be related to the atoms in the molecules. For proteins, the signals of the nuclei of the backbone chain (amide protons and nitrogens, alpha carbons, and carbonyl carbons) are assigned first. Standard triple-resonance NMR experiments (Ferentz and Wagner, 2000; Salzmann *et al.*, 1998; Sattler *et al.*, 1999) make use of the network of scalar couplings within a set of nuclei to create specific correlations in a given NMR spectrum. In these experiments, magnetization is transferred between ^1H, ^{15}N, and ^{13}C through covalent bonds (see lower panels in Fig. 12.6 for selected examples of magnetization flowcharts). For instance, the signals of a 3D HNCO[1] experiment (Kay *et al.*, 1990) represent one amide proton attached to its nitrogen, which is itself attached to the carbonyl carbon of the preceding residue. Because one-bond and two-bond $NC\alpha$ scalar couplings have similar amplitudes, it is possible to

---

[1] The common nomenclature for the triple-resonance experiments is to list the nuclei, among which magnetization is being transferred. For example in the HNCO experiment, magnetization is transferred from H to N and to the carbonyl carbon CO.

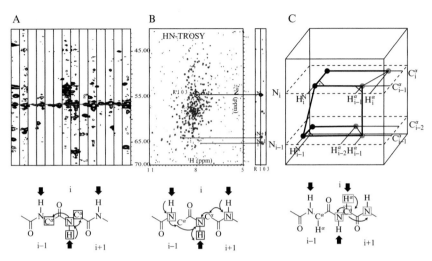

Figure 12.6 Strategies for backbone resonance assignments of proteins. Below each panel, a pictorial representation of the flow of magnetization is represented for the corresponding experiment. The nuclei that are correlated are framed. (A) Strips comparison obtained with the HNCA spectrum of a 37-kDa protein. The experiment leads to a 3D spectrum, depicted by the red dots in c. HNCA correlates amide protons and nitrogens of a given residue (that give rise to the cross-peaks in b) with the Cα carbons of the current and preceding residues (Cα-1), as depicted below panel a. Each strip corresponds to one signal in the HN-TROSY displayed in b, with the vertical axis showing the carbon signals. One residue is selected for the comparison (circled in b) and the shift of its Cα is compared with those of the Cα-1 for all residues. The same procedure is repeated for Cβ and CO nuclei. In this example, the neighbor cannot be identified with only these spectra. (B) "Backbone walk"assignment strategy. A pair of amide nitrogen and proton is correlated with the nitrogens of the neighboring residues, as depicted below panel b. A residue is selected from the HN-TROSY spectrum (circled) and a strip of the double TROSY-hNcaNH displays the nitrogen chemical shifts of the preceding and succeeding residues. The strip is located at the position circled in the HN-TROSY spectrum, with the third dimension (y-axis in the strip) displaying the additional nitrogen correlations. If the resolution is sufficient, the residues can be identified from the hNcaNH and the HN-TROSY. If not, the procedure can be repeated with an HncaNH that would lead to the *proton* chemical shifts of the neighboring residues. Alternatively, the hNcaNH can be combined with the HNCA in a (or any other triple-resonance experiment), which leads to the identification of the successor (red frame in a). Adapted from Frueh *et al.* (2006). Reproduced with permission from *J. Am. Chem. Soc.* 2006. 128, 5757–5763. Copyright 2006 American Chemical Society. (C) "Stairway"assignment procedure. The 3D spectrum is represented schematically. The red peaks correspond to signals obtained with an HNCA (or an HACANH), with the magnetization flow displayed below panel a, and the green peaks to those of an HACAN (or an HNCAHA), with the magnetization flow depicted below panel c. Trapezoidal patterns enable us to identify spin-systems. The chain elongation is easily obtained by linking systems within planes and between planes (bold line). Adapted from Frueh et al. (2005). Reproduced with permission from *J. Biomol. NMR.* 2005. 33, 187–196. Copyright © 2005 Springer. (See color insert.)

correlate a pair of amide proton and nitrogen with both its own alpha carbon (intraresidue signal) and the one of the preceding residue (sequential signal).

Traditionally, experiments are recorded to correlate ^{1}H and ^{15}N chemical shifts of the backbone NH groups to the intraresidue and sequential Cα, Cβ, and CO ^{13}C chemical shifts. Thus, peaks corresponding to the same set of Cα, Cβ, and CO ^{13}C chemical shifts appear as "intraresidue" with respect to the NH group that belongs to the same residue and as "sequential" with respect to the NH group of the next residue in the sequence. Backbone assignments are obtained by matching the intraresidue and sequential ^{13}C chemical shifts correlated to different NH groups, placing the corresponding sets of peaks in sequential order, and then mapping them onto the protein sequence (Fig. 12.6A). This provides connectivities between the signals in a ^{1}H–^{15}N HSQC (such as in Fig. 12.6B). The residue types (Ala, Arg, etc.) can be estimated using the characteristic chemical shifts of the carbon nuclei. By combining the connectivities obtained from CO, Cβ, and Cα nuclei, it is often possible to resolve ambiguities arising from degeneracies in the carbon chemical shifts. This might not be the case for helical proteins that have increased spectral overlap or for proteins with repetitive sequences and obviously for larger proteins with a higher number of signals. One way to overcome those difficulties is to record ultra-high-resolution spectra. Unfortunately, with conventional techniques, this would be accompanied by an increase in precious experimental time. The emergence of various fast-acquisition techniques enables us to achieve such resolutions within an acceptable time (Bruschweiler and Zhang, 2004; Kupce and Freeman, 2003a,b; Rovnyak *et al.*, 2004; Schmieder *et al.*, 1993; Tugarinov *et al.*, 2005). Alternatively, the gain in time can be exploited to record spectra of higher dimensionality (4D and higher).

Another approach to overcome degeneracies in the carbon dimensions is to directly correlate the ^{1}H and ^{15}N chemical shifts of sequential NH groups (Weisemann *et al.*, 1993). Indeed, the ^{1}H–^{15}N HSQC has the best dispersion among the sets of correlations that can be obtained in a protein. In the past, this strategy was not commonly used due to the low sensitivity of the experiments required to establish the correlations and because the resolution required to benefit from the dispersion of HN resonances is rarely achievable in practice.

Recently, such methods with significantly improved sensitivity were developed in our laboratory, suitable for small (Sun *et al.*, 2005a) and larger proteins (Frueh *et al.*, 2006a). These experiments led to a straightforward "backbone walk" assignment strategy, where the neighbors of a given signal are readily identified within the HSQC spectrum (Fig. 12.6B). This circumvents tedious strips comparison. In addition, these 3D experiments upgrade the conventional carbon-dispersed 3D experiments to 4D experiments by providing the ^{15}N or ^{1}H chemical shifts for the successors and predecessors of a given amino acid. The experiments have sensitivities comparable to that of the HNCACB experiment (Wittekind and Mueller, 1993) and critically benefit from perdeuteration.

As previously mentioned, higher dimensions can be recorded to separate signals that overlap. It is also possible to create correlations, not by encoding with the chemical shift of an additional nucleus but by combining data that share common correlations. Thus, for a protonated sample, the appropriate combination of an HNCA (Kay et al., 1990) and an HACAN experiment (Wang et al., 1995) will lead not only to correlations between amide protons, nitrogens, and alpha carbons but also to an additional correlation with alpha protons. Because HNCA correlates a residue with its predecessor whereas HACAN correlates it with its successor, the chain extension can be achieved within a single combination of the two experiments. This results in a "stairway" assignment procedure (Fig. 12.5C), where the chain extension occurs within and between planes of the two combined experiments (Frueh et al., 2005). The strategy is reminiscent of the one used for assigning DNA spectra, where peaks correlating both successors and predecessors with a given residue are observed in a single NOESY spectrum. For small molecules, the HNCA and HACAN spectra can be obtained with a single time-shared experiment (Frueh et al., 2005). In this case, the alpha proton resonances need to be distinct from the water resonance, because the sample needs to be in water to be able to detect amide protons. The advantage of the "stairway technique" is that the additional correlation occurs in the detected dimension, with maximum resolution. The disadvantage is that the sample requires protonation of the alpha carbons, thus limiting the application to relatively small molecules (up to 20 to 25 kDa).

4.2. Methods for side-chain assignments

Once the resonances of the backbone nuclei have been assigned to the residues in the protein, the signals of the side chains need to be identified. In other words, the signals of a ^1H–^{13}C HSQC (correlating carbons to their attached protons) need to be assigned. All atoms of a given residue form a spin system. The assigned amide proton–nitrogen correlations are used as anchors for the assignment of the remaining atoms of the residue. The HCCONH (Grzesiek et al., 1992) experiment is used to identify the protons belonging to the "spin system" by correlating all aliphatic protons to an HN pair, whereas the CCONH (Grzesiek et al., 1992) identifies the corresponding carbons. An HCCH-TOCSY (Kay et al., 1993) or an HCCH-COSY (Ikura et al., 1991) then provides the connectivities between the protons and the carbons, resulting in the assignment of all signals present in a ^1H–^{13}C HSQC spectrum. As for the backbone assignment, spectral overlap leads to ambiguities, and similar strategies are used to overcome this difficulty.

The previously mentioned experiments can be upgraded to higher dimensionality either in a conventional way or using fast acquisition techniques. One example makes use of both the reduced-dimensionality strategy and of nonuniform-sampling acquisition to obtain ultra-high-resolution pseudo-4D spectra for the HCCONH and CCONH experiments (Sun *et al.*, 2005b). For large proteins, the sensitivity of these experiments rapidly deteriorates. Up to 35 kDa, it is possible to obtain good spectra on fractionally deuterated samples. The decrease in the concentration of protons available is largely compensated by the gain in relaxation. In our laboratory, we also assigned successfully a 37-kDa protein by combining a 3D ^{13}C detected MQ–HACACO (Pervushin and Eletsky, 2003) with an HCCH-COSY (Ikura *et al.*, 1991). This bypasses the use of the very insensitive HCCONH and CCONH experiments (Grzesiek *et al.*, 1992). First, the HACACO is used in conjunction with the previously recorded HNCA and HNCO experiments to assign the Hα resonances. The spin systems are then assigned with the HCCH-COSY, starting from the Hα and Cα resonances. The disadvantage is the relatively poor dispersion of the Hα/Cα correlations. Thus, a ^{15}N-dispersed NOESY may need to be used to resolve ambiguities. Compared with the HCCH-TOCSY, the HCCH-COSY has the advantage of simplifying the spectrum and having a shorter length and hence less losses due to relaxation.

4.3. Methods for structure determination

The determination of a structure by NMR consists in establishing various constraints, which will be applied to the molecule studied. These include distances between atoms, dihedral angles, and angles between sets of chemical bonds. Thus, after the assignment of the resonances, structural constraints need to be collected. The secondary structure can readily be determined with the program TALOS (Cornilescu *et al.*, 1999) using the carbon chemical shifts measured during the backbone assignment procedure. In short, these chemical shifts depend on the torsion angles Φ and Ψ, so that it is possible to predict the values of the angles. Torsion angles can be further evaluated by measuring various scalar couplings (Reif *et al.*, 1997) using the Karplus–Bystrov relations (Bystrov, 1976). Distances obtained from NOESY spectra typically constitute the major source of constraints. For small proteins, ^{15}N- and ^{13}C-dispersed ^{1}H-NOESY spectra are recorded. The 3D spectra feature cross-peaks for two nearby protons that are dispersed in a third dimension by the chemical shift of the heteroatom attached to one of the two partners (the target ^{1}H spin in a NOESY-HSQC and the source ^{1}H spin in an HSQC-NOESY). The major difficulty consists in assigning these cross-peaks.

One approach is to first assign all short-range nOes characteristic of the secondary structures in which the residues are located. A few unambiguous long-range nOes are then identified, and an initial fold can be calculated. The remaining nOes are then assigned on the basis of the predicted fold. In short, it is possible to predict the position of the cross-peaks from the initial fold and then check for consistencies and differences with the experimental data. More involved routines are automated and available in programs such as CYANA (Guntert, 2004) or X-PLOR (Brunger, 1992; Schwieters et al., 2003). If a homology model is available for the structure, it can be used as an initial fold, thus significantly speeding up the structure determination.

For large proteins (30 kDa and above), the identification of the cross-peaks rapidly becomes cumbersome due to overlap and ambiguities resulting from proton degeneracies. This is circumvented either by recording 4D spectra or by preparing selectively labeled samples or both. In 4D spectra, in addition to the dispersion with respect to the heteronucleus attached to the target proton, the signals of the source protons are also dispersed with the chemical shift of their attached nuclei. Thus, two HSQC spectra are correlated by means of nOes, making the identification of the signals unambiguous. The major disadvantage is that the experimental time prevents the acquisition of high-resolution spectra with conventional techniques, thus diminishing the advantages of the added dimension. Recently, methods based on nonuniform sampling of the indirect evolution periods allowed to record high-resolution 4D spectra (Tugarinov et al., 2005) within a reasonable time. The signal identification is also facilitated by the preparation of selectively labeled compounds. The most popular sample is uniformly labeled with ^{15}N and ^2H, with the exception of the methyls of Leu, Val, and Ile (at the $\delta 1$ position) that are protonated and ^{13}C labeled. Thus, only nOes involving the protons of these methyls or the amide protons are observed.

A method that allows us to measure the four possible 4D NOESYs in a single experiment, as well as the two 3Ds in another, has recently been developed (Frueh et al., 2006b). The experiments exploit the time-shared technique (Farmer II, 1991; Pascal et al., 1994; Sattler et al., 1995; Xia and Zhu, 2001) to effectuate simultaneous transfers between the protons and their attached heteronuclei. Residual dipolar couplings (Tjandra and Bax, 1997; Tolman et al., 1995), described in 3.6. (5), have gained increased interest due to the high precision of the measurements. Indeed, in theory, it is possible to obtain a structure solely with dipolar couplings (see Bouvignies et al., 2006; Briggman and Tolman, 2003; Tolman, 2002; and references therein). In practice, this is impaired by the necessity to obtain five independent alignment tensors so that the constraints are in general used in conjunction with nOes. Finally, long-range distances can be obtained with paramagnetic labeling (see 3.6. [5]). This provides very valuable constraints. In addition to what was previously described, the drawbacks are the relative low precision of the measurement and the line broadening induced by the paramagnetic center.

4.4. Uniform isotope labeling

Uniform labeling with stable isotopes is commonly done by expressing the protein of interest in bacteria grown on the appropriately labeled medium. For ^{15}N labeling, bacteria are grown on a minimal medium, with ^{15}N-NH_4Cl as the only nitrogen source and glucose as the carbon source. ^{13}C labeling is done in the same medium by replacing unlabeled glucose with ^{13}C-glucose. Partial deuteration is achieved by growing the bacteria in D_2O. During the synthesis of amino acids, most of the Hα protons come from water, whereas a significant part of the protons in the side chains are derived from the glucose. Therefore, if deuteration predominantly at the Hα position is sufficient, there is no need for deuterated glucose. However, for certain applications where complete deuteration is essential, the cells need to be grown on both D_2O and deuterated glucose.

In minimal medium, the bacterial growth rates are slower and the lag phase on adding (inoculation) of the starter culture, grown in LB, to the minimal medium is longer. Replacing H_2O with D_2O slows growth and lengthens the lag phase even further. In addition to making the entire process longer, slow growth often leads to lower expression levels for a number of reasons. It is often possible to achieve protein yields in minimal medium at 40 to 100% the yields in rich medium. The following tips may help improve growth rates and protein yields in minimal medium:

- To achieve maximum expression at slow growth rates, the length of induction often needs to be increased. One empiric approach is to make induction times proportional to the doubling time (e.g., if the doubling time increases from 30 to 60 min between LB and minimal medium, the induction time should be increased from 3 to 6 h). In reality, the results vary depending on the protein being expressed, and even at very slow growth rates, production of new protein often stops in less than a day or even a few hours. Furthermore, some proteins become insoluble if induction is carried on for more than a few hours.
- Whereas adding rare minerals and vitamins is not essential, it increases the growth rates and often improves the yields.
- To shorten the lag time on switching from rich medium (e.g., LB) to minimal medium, it is important to add some fresh LB to the minimal medium: typically 0.5 to 2%. When nearly complete deuteration is necessary and the cells are grown in D_2O with deuterated glucose, adding too much LB could lead to detectable levels of protonation in the protein. In such cases, ~25 mg LB powder per liter minimal medium (corresponding to adding 0.1% LB medium) is sufficient to achieve shorter lag time without affecting the deuteration of the protein.
- Another way to shorten the lag times is to gradually acclimate the cells: starting from rich medium and 100% H_2O and reaching minimal medium and 100% D_2O in two to three steps.

- The use of greater ratio of inoculum to media has a dual effect: it decreases the dilution stress on the cells, leading to shorter lag time, and it also decreases the number of divisions necessary to reach the OD for induction. Starting OD of ~0.1 after inoculation is usually optimal (corresponding to ~2 to 3% inoculum from a typical overnight culture), but for slowly growing cells, OD as high as 0.2 may be appropriate. Resuspending the cultures in fresh medium before inoculation is always recommended but is especially important when expressing a toxic protein in D_2O/minimal medium, using ampicillin resistance for selection. The enzyme β-lactamase is secreted in the medium, and all the antibiotic is degraded long before induction. As a result, there is no longer selection, and cells that have lost the plasmid of interest accumulate. Because bacterial growth causes a drop in pH and ampicillin is inactivated at low pH, use of carbenicillin, or switching to a kanamycin-resistant expression plasmid, is preferable for long growth and induction times.
- As with any expression, it is preferable that the cells used for inoculation have not reached stationary phase. When this is not practical (as is usually the case when acclimating slow-growing cells into medium with D_2O), adding 1% glucose to the medium minimizes leaky expression from lac-based promoters in stationary phase, which often improves yields from toxic proteins. However, extra glucose does not affect the longer lag phase associated with transition of the cells from stationary to log phase on inoculation.
- Whereas ^{15}N–NH_4Cl is relatively inexpensive, the cost of ^{13}C-labeled and especially deuterated glucose is significant; 1 g/L of NH_4Cl and 4 g/L of glucose are sufficient for bacterial growth up to 4 OD units, which is supported by a good minimal medium (Weber *et al.*, 1992). A drop in pH can also cause the bacteria to stop growing before the nutrients are depleted. Therefore, the buffering capacity of the medium used often limits the maximum achievable OD. If, after induction, the cells stop growing at much lower ODs, the amount of glucose can be decreased accordingly.

4.5. Specific labeling schemes (nonuniform labeling)

In certain cases, it is beneficial to label only specific residues or groups in the protein. Accordingly, there is a vast array of labeling schemes for different NMR applications (reviewed in Dotsch and Wagner [1998], Gardner and Kay [1998], Goto and Kay [2000], Kainosho [1997], Koglin *et al.* [2006], Pellecchia [2005], and Tugarinov and Kay [2005]). The most common uses of specific labeling are to decrease crowding in the spectra and help with NMR resonance assignments. Another use, relevant for studying protein–protein interactions, is to aid in unambiguous identification of intermolecular NOEs. One approach that has become popular in recent years is

$^{13}C/^{1}H$-labeling of methyl groups of Ile, Leu, and Val (ILV-labeling) in an otherwise perdeuterated protein (Gardner *et al.*, 1997; Gross *et al.*, 2003a). In a ^{13}C-edited NOESY of such a protein in complex with an unlabeled protein, all NOE cross-peaks to aromatic side chains, Hα, and methylene (CH_2) groups are intermolecular (methyl-methyl NOEs remain ambiguous).

Recently a stereoarray isotope labeling (SAIL) method has been introduced (Kainosho *et al.*, 2006). It exclusively uses chemically and enzymatically synthesized amino acids where a subset of hydrogens are replaced with deuterium. In particular, one of each methylene position is deuterated. This largely reduces line broadening due to $^{1}H-^{1}H$ dipole–dipole interaction. In addition, it dramatically simplifies the spectra without sacrificing the information content. This approach requires cell-free protein expression and the specifically deuterated amino acids, which are not yet easily available. However, SAIL has the potential to significantly facilitate determination of NMR structures of larger proteins at high precision.

4.6. Segmental labeling

Another approach for reducing crowding in the NMR spectra of large proteins is to differentially label individual segments of a single protein. The method makes use of inteins: proteins that are encoded within the polypeptide chain of another protein. After the polypeptide is synthesized, the intein splices out and relegates the two ends of the target polypeptide. The result is two intact proteins from the same polypeptide chain. The individual applications vary greatly. In some cases, a fusion with an intein is used in combination with a synthetic peptide or a native protein that starts with a cysteine. A drawback of this approach is that the ligation is not very efficient.

Another approach uses two polypeptide chains each containing part of the target protein and part of the intein, which are mixed under denaturing conditions, followed by refolding. Folding of the intein allows it to splice out, leaving the desired ligated protein target. The main drawback of this approach is that refolding is successful for only a limited number of protein–intein combinations. Recently, naturally split inteins were discovered; both the intein and the target protein are split between two polypeptide chains, which are translated separately in the cell, after which the two parts of the intein bind to each other. Splicing then yields an intact single–peptide target protein and a folded intein (composed of two polypeptide chains). This system allows us to perform efficient segmental labeling under native conditions *in vitro* by expressing and purifying two polypeptides separately and then mixing them. An alternative approach is to express the two polypeptides sequentially in the same cell culture, with changing the medium (e.g., from unlabeled to labeled) before inducing expression of the second protein. An obvious limitation is that the target protein needs to be split in two segments that remain soluble in the absence of the rest of the protein. Thus, the approach is particularly useful for modular multidomain proteins. Because the two parts of the intein bind tightly to each other, the ligation is much more efficient than chemical ligation (for a recent review see Muralidharan and Muir [2006]).

If a protein cannot be efficiently expressed in minimal medium, commercial rich–labeled media, usually based on algal extracts, can be used, although at a higher cost. Whereas expression in bacteria is the cheapest and most widely used approach for expression of labeled proteins, sometimes *in vitro* translation with labeled amino acids is a viable alternative (Koglin *et al.*, 2006), in particular if special labeling schemes are used.

5. SPECIAL CASES

5.1. Interdomain interactions and orientations

It is often important to know whether individual domains in a protein (or within a complex) contact each other and whether their mutual orientations are fixed or flexible. An intramolecular interaction does not always lead to fixed interdomain orientation; in many cases, a domain or an unfolded tail contacts another domain transiently while tumbling independently. This is often observed when one domain or unfolded tail sterically interferes with an interaction surface on another domain through a weak interaction. Recent studies with high-mobility group (HMG) proteins offer an interesting example: the highly acidic C-terminal region was reported to interact with the central HMG -box DNA-binding domain in rat. These intramolecular interactions were found to modulate the DNA binding activity of the protein (Jung and Lippard, 2003).

- A comparison of the HSQC spectra of the individual domains in isolation with the spectrum of the larger molecule can often provide some clues. For example, changes in peak positions between an isolated domain and the larger construct indicate contacts between the domain and the rest of the molecule, unless the changes are limited to the vicinity of the covalent junction with the rest of the protein (that becomes a terminus in the isolated domain). Clear differences in line widths between peaks belonging to different domains in a large construct are indicative of independent tumbling of the two domains within the molecule. If necessary, these initial observations can be further explored.
- NMR relaxation experiments can determine whether individual domains in the same protein or complex tumble independently or as a rigid body (reviewed in Bruschweiler [2003], Case [2002], Kay [2005], and Palmer [2001]).

- RDC analysis can determine the mutual orientation of individual domains or show that their orientation is not fixed (for recent reviews see Bax [2003] and Bax and Grishaev [2005]).
- NMR titration using isolated domains can be used to show whether they interact and to map the contact interface. It should be noted, however, that if the interactions are very weak, they might be hard to observe when the two domains are not covalently linked (which could keep the effective concentration of one domain in the vicinity of the other much higher).

5.2. Nonspecific vs specific binding

NMR can often help distinguish between specific and nonspecific interactions. In the example shown on Fig. 12.3C, the broadening of HSQC peaks in the complex of a protein with a ligand indicates that the compound may be binding in more than one orientation in the vicinity of the corresponding residue (Reibarkh *et al.*, 2006).

- Another characteristic feature of nonspecific interactions is that they often do not have a fixed stoichiometry. Such behavior would become clear from an NMR titration experiment. NMR can also be used to map the contact surface(s) and identify potential multiple binding sites or binding modes. For example, if a ligand has two binding sites on a protein, NMR titration can be used not only to identify these binding sites but also to determine the two K_Ds; the chemical shift changes will have different concentration dependence for the corresponding sets of peaks.

5.3. Unstable proteins or complexes

Limited solubility of individual proteins and/or their complexes often creates serious challenges for the study of protein–protein interactions by NMR. Two major causes of aggregation are electrostatic interactions and exposed hydrophobic patches on the protein surface. Aggregation caused by electrostatic interactions can usually be avoided using higher salt concentrations. High salt may also help if aggregation is due to hydrophobic interactions, presumably by making the transient exposure of hydrophobic surfaces ("breathing") energetically unfavorable. High salt itself may cause problems: it decreases the sensitivity of the NMR probes, which may at least in part cancel the benefit of increased solubility. Furthermore, many protein–protein interactions have a significant electrostatic component and are weaker/undetectable at high salt. Usually, a balance can be found between salt concentration, protein solubility, and binding affinity. Protein structures have been solved by NMR at salt concentrations as high as $1M$.

If high salt is used and the interaction is significantly weakened, it may not be possible to solve the structure of the complex. However, the mapping of binding interfaces for weak interactions that are in fast exchange regime is often easier, especially for large complexes.

On the other extreme, recent technological advances in NMR, and in particular the use of cold probes, has allowed the detection of proteins at concentrations as low as $\sim 10~\mu M$, which can be more easily achievable even at low salt. However, higher protein concentrations are required if the goal is not only to detect the complex but also to solve its structure by NMR.

Another approach for increasing solubility is the use of solubility-enhancing tags (e.g., the IgG-binding domain 1 [GB1] of protein G) (Zhou et al., 2001).

In the case of aggregation caused by hydrophobic interactions, it may be useful to try adding detergents. Although there have been concerns about the possibility that detergents may change the protein structure, there is currently no evidence that the protein structure is indeed affected. Furthermore, at least when the protein is folded both in the presence and absence of the detergent, a comparison of NMR spectra can be used to rule out any significant conformational changes. Detergents are more likely to help solubility at concentrations higher than their critical micelle concentration (CMC). Above CMC, the individual detergent molecules form micelles, and the proteins with exposed hydrophobic surfaces can become embedded into the micelles. For example, the structure of yeast eIF4E (Matsuo et al., 1997a) and of the complex of yeast eIF4E with a fragment of eIF4G (Gross et al., 2003b) were solved in micelles.

6. Summary

As we have tried to illustrate, NMR has been widely used in the studies of a number of structures and interactions in translation. It also offers a diverse set of applications that have not always been fully appreciated and used. As with any other methods, NMR has its own limitations. Table 12.1 summarizes the scope of information obtainable by NMR, as well as the limits imposed by the size of the macromolecules studied. All numbers and size limits are intended only as general guides. With the ongoing advances in both NMR instrumentation and experimental techniques, the size limits in NMR will continue to be pushed further, allowing the study by NMR of larger macromolecules at lower sample concentrations.

Table 12.1 Questions that can be addressed with protein NMR and size limitations

Question	Experiments	Information obtainable	Size limit[a]	
			Routine[b]	Challenging
Is the protein folded?	1D or 2D NMR spectra	Large dispersion of signals indicates folded structure	100 kDa	1 MDa
Is the protein aggregated?	Inspection of 2D ^{1}H–^{15}N HSQC spectra, counting cross peaks	Broad lines and lack of cross peaks indicate aggregation	100 kDa	1 MDa
What is its secondary structure?	Inspect dispersion of H^N signals in ^{1}H–^{15}N HSQC spectra, are there low-field shifted H$^\alpha$ resonances in 1D ^{1}H spectrum	Large dispersion of H^N resonances and low-field shifted H$^\alpha$ signals indicate presence of β-sheet structure	100 kDa	1 MDa
Does it bind to another protein or a ligand?	Titration of ^{1}H–^{15}N HSQC spectrum with increasing amounts of an unlabeled protein or ligand	Addition of interacting protein or ligand causes changes of resonance positions in 2D spectra	100 kDa	1 MDa
Monitor individual backbone sites	Backbone assignments with triple-resonance experiments	Spatially assigned probes for measuring interactions and mobility, required for the experiments listed below, including structure determination	30–35 kDa[c]	100 kDa[d]

Identify binding epitopes on backbone level	Follow chemical shift changes, selective loss of cross peaks on titration, cross-saturation experiments	Chemical shift mapping identifies all residues affected by interaction, cross-saturation identifies binding epitope directly	The size limits are largely dictated by the ability to obtain backbone assignments, which are a prerequisite[e] (see also footnote c)
Are there mobile regions?	Measure ^{15}N relaxation parameters, primarily transverse relaxation rates or heteronuclear ^{1}H–^{15}N NOEs	Relaxation experiments may indicate flexible tails and suggest trimming of proteins to optimize behavior for NMR and crystallization	
Domain orientation?	Measure residual dipolar couplings (RDCs)	Analysis of RDC values can provide average domain orientation. Comparison of experimental RDCs with those computed from a structural model can provide quantitative measure for quality of structure	
Identify binding epitopes on side-chain level	Multiple-resonance experiments for side chain assignments, particularly Ile, Val, Leu	Sensitive probes for measuring binding of small ligands and other proteins	

Table 12.1 (*continued*)

Question	Experiments	Information obtainable	Size limit[a] Routine[b]	Challenging
Overall fold?	Measure NOEs between H^N, RDCs, chemical shifts and/or NOEs between methyl and aromatic groups in perdeuterated proteins with selectively protonated groups	Low-resolution structure suitable for interaction studies		
High-resolution structure?	Measure and assign NOEs, RDCs, chemical shifts	High-resolution structure	25 kDa	80 kDa

[a] The limits given here are approximate and meant to provide a general idea. They assume that the proteins are soluble (e.g., to >0.5 mM) and stable and do not account for other specific problems, such as difficulty to back-exchange slowly exchanging amide protons in a deuterated protein.

[b] Deuteration of the proteins is assumed for sizes close to the upper limit.

[c] Sometimes assignment of a smaller protein can be transferred to a large complex if the interaction is in fast exchange on the NMR time scale. For interactions in slow exchange, part of the assignments (in particular, those corresponding to residues unaffected by the interaction) can also be transferred to the complex.

[d] Partial assignments up to 800 kDa in special cases (symmetric homooligomers).

[e] For RDCs, depending on the alignment medium and the degree of interaction between the medium and the protein, the size limit can be substantially smaller. The size limits for obtaining nearly complete side-chain resonance assignments necessary for high-resolution structure determination are also lower.

For detailed reviews of size limits of NMR methods and specific NMR experiments see (Kay, 2005; Pellecchia, 2005; Sattler et al., 1999; Tugarinov and Kay, 2005; Tugarinov et al., 2004; Tzakos et al., 2006).

References

Kay, L. E. (2005). NMR studies of protein structure and dynamics. *J. Magn. Reson.* **173,** 193–207.

Pellecchia, M. (2005). Solution nuclear magnetic resonance spectroscopy techniques for probing intermolecular interactions. *Chem. Biol.* **12,** 961–971.

Sattler, M., Schleucher, J., and Griesinger, C. (1999). Heteronuclear multidimentional NMR experiments for the structure determination of proteins in solution employing pulsed field gradients. *Progr. Nucl. Magn. Res. Spectrosc.* **34,** 93–158.

Tugarinov, V., Hwang, P. M. and Kay, L. E. (2004). Nuclear magnetic resonance spectroscopy of high-molecular-weight proteins. *Annu. Rev. Biochem.* **73,** 107–146.

Tugarinov, V., and Kay, L. E. (2005). Methyl groups as probes of structure and dynamics in NMR studies of high-molecular-weight proteins. *Chembiochem.* **6,** 1567–1577.

Tzakos, A. G., Grace, C. R., Lukavsky, P. J., and Riek, R. (2006). NMR techniques for very large proteins and RNas in solution. *Annu. Rev. Biophys. Biomol. Struct.* **35,** 319–342.

ACKNOWLEDGMENTS

This work was supported by grants GM47467 and CA6826 (to G. W.). A. M. was supported by a Howard Temin K01 Career Award from the NCI. We thank members of the Wagner laboratory for valuable comments and discussions.

REFERENCES

Arnold, K., Bordoli, L., Kopp, J., and Schwede, T. (2006). The SWISS-MODEL workspace: A web-based environment for protein structure homology modelling. *Bioinformatics* **22,** 195–201.

Battiste, J. L., Gross, J. D., and Wagner, G. (2003). Global fold determination of large proteins using site-directed spin labeling. *In* "Protein NMR for the Millennium" (N. R. Krishna and L. J. Berliner, eds.), Vol. 20, pp. 79–101. Kluwer Academic/Plenum Publishers, New York.

Battiste, J. L., Pestova, T. V., Hellen, C. U., and Wagner, G. (2000). The eIF1A solution structure reveals a large RNA-binding surface important for scanning function. *Mol. Cell* **5,** 109–119.

Battiste, J. L., and Wagner, G. (2000). Utilization of site-directed spin labeling and high-resolution heteronuclear nuclear magnetic resonance for global fold determination of large proteins with limited nuclear Overhauser effect data. *Biochemistry* **39,** 5355–5365.

Bax, A. (2003). Weak alignment offers new NMR opportunities to study protein structure and dynamics. *Protein Sci.* **12,** 1–16.

Bax, A., and Grishaev, A. (2005). Weak alignment NMR: A hawk-eyed view of biomolecular structure. *Curr. Opin. Struct. Biol.* **15,** 563–570.

Bodenhausen, G., and Ruben, D. J. (1980). Natural abundance nitrogen-15 NMR by enhanced heteronuclear spectroscopy. *Chem. Phys. Lett.* **69,** 185–189.

Bouvignies, G., Markwick, P., Bruschweiler, R., and Blackledge, M. (2006). Simultaneous determination of protein backbone structure and dynamics from residual dipolar couplings. *J. Am. Chem. Soc.* **128,** 15100–15101.

Briggman, K. B., and Tolman, J. R. (2003). *De novo* determination of bond orientations and order parameters from residual dipolar couplings with high accuracy. *J. Am. Chem. Soc.* **125,** 10164–10165.

Brunger, A. T. (1992). X-PLOR, version 3.1. A system for X-ray crystallography and NMR. Yale University Press, New Haven, CT.

Bruschweiler, R. (2003). New approaches to the dynamic interpretation and prediction of NMR relaxation data from proteins. *Curr. Opin. Struct. Biol.* **13,** 175–183.

Bruschweiler, R., Liao, X., and Wright, P. E. (1995). Long-range motional restrictions in a multidomain zinc-finger protein from anisotropic tumbling. *Science* **268,** 886–889.

Bruschweiler, R., and Zhang, F. (2004). Covariance nuclear magnetic resonance spectroscopy. *J. Chem. Phys.* **120,** 5253–5260.

Bystrov, N. S. (1976). Spin-spin couplings and the conformational states of peptide systems. *Progr. Nucl. Magn. Res. Spectrosc.* **10,** 41–81.

Case, D. A. (2002). Molecular dynamics and NMR spin relaxation in proteins. *Acc. Chem. Res.* **35,** 325–331.

Cho, S., and Hoffman, D. W. (2002). Structure of the beta subunit of translation initiation factor 2 from the archaeon *Methanococcus jannaschii*: A representative of the eIF2beta/eIF5 family of proteins. *Biochemistry* **41,** 5730–5742.

Christodoulou, J., Larsson, G., Fucini, P., Connell, S. R., Pertinhez, T. A., Hanson, C. L., Redfield, C., Nierhaus, K. H., Robinson, C. V., Schleucher, J., and Dobson, C. M. (2004). Heteronuclear NMR investigations of dynamic regions of intact *Escherichia coli* ribosomes. *Proc. Natl. Acad. Sci. USA* **101,** 10949–10954.

Clore, G. M., and Schwieters, C. D. (2003). Docking of protein–protein complexes on the basis of highly ambiguous intermolecular distance restraints derived from 1H/15N chemical shift mapping and backbone 15N-1H residual dipolar couplings using conjoined rigid body/torsion angle dynamics. *J. Am. Chem. Soc.* **125,** 2902–2912.

Conte, M. R., Kelly, G., Babon, J., Sanfelice, D., Youell, J., Smerdon, S. J., and Proud, C. G. (2006). Structure of the eukaryotic initiation factor (eIF) 5 reveals a fold common to several translation factors. *Biochemistry* **45,** 4550–4558.

Cornilescu, G., Delaglio, F., and Bax, A. (1999). Protein backbone angle restraints from searching a database for chemical shift and sequence homology. *J. Biomol. NMR* **13,** 289–302.

Dominguez, C., Boelens, R., and Bonvin, A. M. (2003). HADDOCK: A protein–protein docking approach based on biochemical or biophysical information. *J. Am. Chem. Soc.* **125,** 1731–1737.

Dotsch, V., and Wagner, G. (1998). New approaches to structure determination by NMR spectroscopy. *Curr. Opin. Struct. Biol.* **8,** 619–623.

Elantak, L., Tzakos, A. G., Locker, N., and Lukavsky, P. J. (2006). Structure of eIF3b-RRM and its interaction with eIF3j: Structural insights into the recruitment of eIF3b to the 40S ribosomal subunit. *J. Biol. Chem.* **282,** 8165–8174.

Fahmy, A., and Wagner, G. (2002). TreeDock: A tool for protein docking based on minimizing van der Waals energies. *J. Am. Chem. Soc.* **124,** 1241–1250.

Farmer, B., II (1991). Simultaneous [13C,15N]-HMQC, A pseudo-triple resonance experiment. *J. Magn. Reson.* **93,** 635–641.

Ferentz, A. E., and Wagner, G. (2000). NMR spectroscopy: A multifaceted approach to macromolecular structure. *Q. Rev. Biophys.* **33,** 29–65.

Fischer, M. W., Losonczi, J. A., Weaver, J. L., and Prestegard, J. H. (1999). Domain orientation and dynamics in multidomain proteins from residual dipolar couplings. *Biochemistry* **38,** 9013–9022.

Fleming, K., Ghuman, J., Yuan, X., Simpson, P., Szendroi, A., Matthews, S., and Curry, S. (2003). Solution structure and RNA interactions of the RNA recognition motif from eukaryotic translation initiation factor 4B. *Biochemistry* **42,** 8966–8975.

Fletcher, C. M., Pestova, T. V., Hellen, C. U., and Wagner, G. (1999). Structure and interactions of the translation initiation factor eIF1. *EMBO J.* **18,** 2631–2637.

Frueh, D. P., Arthanari, H., and Wagner, G. (2005). Unambiguous assignment of NMR protein backbone signals with a time-shared triple-resonance experiment. *J. Biomol. NMR* **33,** 187–196.

Frueh, D. P., Sun, Z. Y., Vosburg, D. A., Walsh, C. T., Hoch, J. C., and Wagner, G. (2006a). Non-uniformly sampled double-TROSY hNcaNH experiments for NMR sequential assignments of large proteins. *J. Am. Chem. Soc.* **128,** 5757–5763.

Frueh, D. P., Vosburg, D. A., Walsh, C. T., and Wagner, G. (2006b). Determination of all NOES in 1H-13C-Me-ILV-U-2H-15N proteins with two time-shared experiments. *J. Biomol. NMR* **34,** 31–40.

Fushman, D., Varadan, R., Assfalg, M., and Walker, O. (2004). Determining domain orientation in macromolecules by using spin-relaxation and residual dipolar coupling measurements. *Progr. NMR Spectrosc.* **44,** 189–214.

Garcia, C., Fortier, P. L., Blanquet, S., Lallemand, J. Y., and Dardel, F. (1995a). 1H and 15N resonance assignments and structure of the N-terminal domain of *Escherichia coli* initiation factor 3. *Eur. J. Biochem.* **228,** 395–402.

Garcia, C., Fortier, P. L., Blanquet, S., Lallemand, J. Y., and Dardel, F. (1995b). Solution structure of the ribosome-binding domain of *E. coli* translation initiation factor IF3.

Homology with the U1A protein of the eukaryotic spliceosome. *J. Mol. Biol.* **254,** 247–259.

Gardner, K. H., and Kay, L. E. (1998). The use of 2H, 13C, 15N multidimensional NMR to study the structure and dynamics of proteins. *Annu. Rev. Biophys. Biomol. Struct.* **27,** 357–406.

Gardner, K. H., Rosen, M. K., and Kay, L. E. (1997). Global folds of highly deuterated, methyl-protonated proteins by multidimensional NMR. *Biochemistry* **36,** 1389–1401.

Goto, N. K., and Kay, L. E. (2000). New developments in isotope labeling strategies for protein solution NMR spectroscopy. *Curr. Opin. Struct. Biol.* **10,** 585–592.

Gross, J. D., Gelev, V. M., and Wagner, G. (2003a). A sensitive and robust method for obtaining intermolecular NOEs between side chains in large protein complexes. *J. Biomol. NMR* **25,** 235–242.

Gross, J. D., Moerke, N. J., von der Haar, T., Lugovskoy, A. A., Sachs, A. B., McCarthy, J. E., and Wagner, G. (2003b). Ribosome loading onto the mRNA cap is driven by conformational coupling between eIF4G and eIF4E. *Cell* **115,** 739–750.

Grzesiek, S., Anglister, J., and Bax, A. (1992). Correlation of backbone amide and aliphatic side-chain resonances in 13C/15N-enriched proteins by isotropic mixing of 13C magnetization. *JMRB* **101,** 114–119.

Guntert, P. (2004). Automated NMR structure calculation with CYANA. *Methods Mol. Biol.* **278,** 353–378.

Gutierrez, P., Osborne, M. J., Siddiqui, N., Trempe, J. F., Arrowsmith, C., and Gehring, K. (2004). Structure of the archaeal translation initiation factor aIF2 beta from *Methanobacterium thermoautotrophicum*: Implications for translation initiation. *Protein Sci.* **13,** 659–667.

Ikura, M., Kay, L. E., and Bax, A. (1991). Improved three-dimensional ^{1}H-^{13}C-^{1}H correlation spectroscopy of a ^{13}C-labeled protein using constant-time evolution. *J. Biomol. NMR* **1,** 299–304.

Ito, T., Marintchev, A., and Wagner, G. (2004). Solution structure of human initiation factor eIF2alpha reveals homology to the elongation factor eEF1B. *Structure (Camb)* **12,** 1693–1704.

Jung, Y., and Lippard, S. J. (2003). Nature of full-length HMGB1 binding to cisplatin-modified DNA. *Biochemistry* **42,** 2664–2671.

Kainosho, M. (1997). Isotope labelling of macromolecules for structural determinations. *Nat. Struct. Biol.* **4**(Suppl.), 858–861.

Kainosho, M., Torizawa, T., Iwashita, Y., Terauchi, T., Mei Ono, A., and Guntert, P. (2006). Optimal isotope labelling for NMR protein structure determinations. *Nature* **440,** 52–57.

Kay, L. E. (2005). NMR studies of protein structure and dynamics. *J. Magn. Reson.* **173,** 193–207.

Kay, L. E., Ikura, M., Tschudin, R., and Bax, A. (1990). Three-dimensional triple-resonance NMR spectroscopy of isotopically enriched proteins. *J. Magn. Reson.* **89,** 496–514.

Kay, L. E., Xu, G.-Y., Singer, A. U., Muhandiram, D. R., and Rorman-Kay, J. D. (1993). A gradient-enhance HCCH-TOCSY experiment for recording side-chain 1H and 13C correlations in H$_2$O samples of proteins. *J. Magn. Reson.* **101**(Series B), 333–337.

Koglin, A., Klammt, C., Trbovic, N., Schwarz, D., Schneider, B., Schafer, B., Lohr, F., Bernhard, F., and Dotsch, V. (2006). Combination of cell-free expression and NMR spectroscopy as a new approach for structural investigation of membrane proteins. *Magn. Reson. Chem.* **44**(Spec. No.), S17–S23.

Kontaxis, G., Clore, G. M., and Bax, A. (2000). Evaluation of cross-correlation effects and measurement of one-bond couplings in proteins with short transverse relaxation times. *J. Magn. Reson.* **143,** 184–196.

Kozlov, G., De Crescenzo, G., Lim, N. S., Siddiqui, N., Fantus, D., Kahvejian, A., Trempe, J. F., Elias, D., Ekiel, I., Sonenberg, N., O'Connor-McCourt, M., and Gehring, K. (2004). Structural basis of ligand recognition by PABC, a highly specific peptide-binding domain found in poly(A)-binding protein and a HECT ubiquitin ligase. *EMBO J.* **23,** 272–281.

Kozlov, G., Siddiqui, N., Coillet-Matillon, S., Trempe, J. F., Ekiel, I., Sprules, T., and Gehring, K. (2002). Solution structure of the orphan PABC domain from *Saccharomyces cerevisiae* poly(A)-binding protein. *J. Biol. Chem.* **277,** 22822–22828.

Kupce, E., and Freeman, R. (2003a). Projection-reconstruction of three-dimensional NMR spectra. *J. Am. Chem. Soc.* **125,** 13958–13959.

Kupce, E., and Freeman, R. (2003b). Two-dimensional Hadamard spectroscopy. *J. Magn. Reson.* **162,** 300–310.

Laursen, B. S., Kjaergaard, A. C., Mortensen, K. K., Hoffman, D. W., and Sperling-Petersen, H. U. (2004). The N-terminal domain (IF2N) of bacterial translation initiation factor IF2 is connected to the conserved C-terminal domains by a flexible linker. *Protein Sci.* **13,** 230–239.

Laursen, B. S., Mortensen, K. K., Sperling-Petersen, H. U., and Hoffman, D. W. (2003). A conserved structural motif at the N terminus of bacterial translation initiation factor IF2. *J. Biol. Chem.* **278,** 16320–16328.

Li, W., and Hoffman, D. W. (2001). Structure and dynamics of translation initiation factor aIF-1A from the archaeon *Methanococcus jannaschii* determined by NMR spectroscopy. *Protein Sci.* **10,** 2426–2438.

Marintchev, A., and Wagner, G. (2004). Translation initiation: Structures, mechanisms and evolution. *Q. Rev. Biophys.* **37,** 197–284.

Matsuo, H., Li, H., McGuire, A. M., Fletcher, C. M., Gingras, A. C., Sonenberg, N., and Wagner, G. (1997a). Structure of translation factor eIF4E bound to m7GDP and interaction with 4E-binding protein. *Nat. Struct. Biol.* **4,** 717–724.

Matsuo, H., Li, H., McGuire, A. M., Fletcher, C. M., Gingras, A. C., Sonenberg, N., and Wagner, G. (1997b). Structure of translation factor eIF4E bound to m7GDP and interaction with 4E-binding protein. *Nat. Struct. Biol.* **4,** 717–724.

Matsuo, H., Walters, K. J., Teruya, K., Tanaka, T., Gassner, G. T., Lippard, S. J., Kyogoku, Y., and Wagner, G. (1999). Identification by NMR spectroscopy of residues at contact surfaces in large, slowly exchanging macromolecular complexes. *J. Am. Chem. Soc.* **121,** 9903–9904.

Meunier, S., Spurio, R., Czisch, M., Wechselberger, R., Guenneugues, M., Gualerzi, C. O., and Boelens, R. (2000). Structure of the fMet-tRNA(fMet)-binding domain of B. stearothermophilus initiation factor IF2. *EMBO J.* **19,** 1918–1926.

Moreau, M., de Cock, E., Fortier, P. L., Garcia, C., Albaret, C., Blanquet, S., Lallemand, J. Y., and Dardel, F. (1997). Heteronuclear NMR studies of *E. coli* translation initiation factor IF3. Evidence that the inter-domain region is disordered in solution. *J. Mol. Biol.* **266,** 15–22.

Muralidharan, V., and Muir, T. W. (2006). Protein ligation: An enabling technology for the biophysical analysis of proteins. *Nat. Methods* **3,** 429–438.

Nakanishi, T., Miyazawa, M., Sakakura, M., Terasawa, H., Takahashi, H., and Shimada, I. (2002). Determination of the interface of a large protein complex by transferred cross-saturation measurements. *J. Mol. Biol.* **318,** 245–259.

Ottiger, M., Delaglio, F., and Bax, A. (1998). Measurement of J and dipolar couplings from simplified two-dimensional NMR spectra. *J. Magn. Reson.* **131,** 373–378.

Palmer, A. G., 3rd. (2001). Nmr probes of molecular dynamics: Overview and comparison with other techniques. *Annu. Rev. Biophys. Biomol. Struct.* **30,** 129–155.

Palmer, A. G., 3rd, Grey, M. J., and Wang, C. (2005). Solution NMR spin relaxation methods for characterizing chemical exchange in high-molecular-weight systems. *Methods Enzymol.* **394,** 430–465.

Pan, B., Maciejewski, M. W., Marintchev, A., and Mullen, G. P. (2001). Solution structure of the catalytic domain of gamma delta resolvase. Implications for the mechanism of catalysis. *J. Mol. Biol.* **310,** 1089–1107.

Pascal, S., Muhandiram, D., Yamazaki, T., Forman-Kay, J., and Kay, L. (1994). Simultaneous acquisition of 15N and 13C-edited NOE spectra of proteins dissolved in H$_2$O. *J. Magn. Reson. Series B* **103,** 197–201.

Pellecchia, M. (2005). Solution nuclear magnetic resonance spectroscopy techniques for probing intermolecular interactions. *Chem. Biol.* **12,** 961–971.

Pervushin, K., and Eletsky, A. (2003). A new strategy for backbone resonance assignment in large proteins using a MQ-HACACO experiment. *J. Biomol. NMR* **25,** 147–152.

Pervushin, K., Riek, R., Wider, G., and Wuthrich, K. (1997). Attenuated T2 relaxation by mutual cancellation of dipole-dipole coupling and chemical shift anisotropy indicates an avenue to NMR structures of very large biological macromolecules in solution. *Proc. Natl. Acad. Sci. USA* **94,** 12366–12371.

Reibarkh, M., Malia, T. J., and Wagner, G. (2006). NMR distinction of single- and multiple-mode binding of small-molecule protein ligands. *J. Am. Chem. Soc.* **128,** 2160–2161.

Reif, B., Hennig, M., and Griesinger, C. (1997). Direct measurement of angles between bond vectors in high-resolution NMR. *Science* **276,** 1230–1233.

Rovnyak, D., Frueh, D. P., Sastry, M., Sun, Z. Y., Stern, A. S., Hoch, J. C., and Wagner, G. (2004). Accelerated acquisition of high resolution triple-resonance spectra using non-uniform sampling and maximum entropy reconstruction. *J. Magn. Reson.* **170,** 15–21.

Salzmann, M., Pervushin, K., Wider, G., Senn, H., and Wuthrich, K. (1998). TROSY in triple-resonance experiments: New perspectives for sequential NMR assignment of large proteins. *Proc. Natl. Acad. Sci. USA* **95,** 13585–13590.

Sattler, M., Maurer, M., Schleucher, J., and Griesinger, C. (1995). A simultaneous 15N.1H and 13C,1H-HSQC with sensitivity enhancement and a heteronuclear gradient echo. *J. Biomol. NMR* **5,** 97–102.

Sattler, M., Schleucher, J., and Griesinger, C. (1999). Heteronuclear multidimensional NMR experiments for the structure determination of proteins in solution employing pulsed field gradients. *Progr. NMR Spectrosc.* **34,** 93–158.

Schmieder, P., Stern, A. S., Wagner, G., and Hoch, J. C. (1993). Application of nonlinear sampling schemes to COSY-type spectra. *J. Biomol. NMR* **3,** 569–576.

Schwieters, C. D., Kuszewski, J. J., Tjandra, N., and Clore, G. M. (2003). The Xplor-NIH NMR molecular structure determination package. *J. Magn. Reson.* **160,** 65–73.

Sette, M., van Tilborg, P., Spurio, R., Kaptein, R., Paci, M., Gualerzi, C. O., and Boelens, R. (1997). The structure of the translational initiation factor IF1 from *E. coli* contains an oligomer-binding motif. *EMBO J.* **16,** 1436–1443.

Shimada, I. (2005). NMR techniques for identifying the interface of a larger protein–protein complex: Cross-saturation and transferred cross-saturation experiments. *Methods Enzymol.* **394,** 483–506.

Siddiqui, N., Kozlov, G., D'Orso, I., Trempe, J. F., and Gehring, K. (2003). Solution structure of the C-terminal domain from poly(A)-binding protein in *Trypanosoma cruzi*: A vegetal PABC domain. *Protein Sci.* **12**, 1925–1933.

Sun, Z. Y., Frueh, D. P., Selenko, P., Hoch, J. C., and Wagner, G. (2005a). Fast assignment of (15)N-HSQC peaks using high-resolution 3D HNcocaNH experiments with non-uniform sampling. *J. Biomol. NMR* **33**, 43–50.

Sun, Z. Y., Hyberts, S. G., Rovnyak, D., Park, S., Stern, A. S., Hoch, J. C., and Wagner, G. (2005b). High-resolution aliphatic side-chain assignments in 3D HCcoNH experiments with joint H-C evolution and non-uniform sampling. *J. Biomol. NMR* **32**, 55–60.

Takahashi, H., Miyazawa, M., Ina, Y., Fukunishi, Y., Mizukoshi, Y., Nakamura, H., and Shimada, I. (2006). Utilization of methyl proton resonances in cross-saturation measurement for determining the interfaces of large protein-protein complexes. *J. Biomol. NMR* **34**, 167–177.

Takahashi, H., Nakanishi, T., Kami, K., Arata, Y., and Shimada, I. (2000). A novel NMR method for determining the interfaces of large protein–protein complexes. *Nat. Struct. Biol.* **7**, 220–223.

Tjandra, N., and Bax, A. (1997). Direct measurement of distances and angles in biomolecules by NMR in a dilute liquid crystalline medium. *Science* **278**, 1111–1114.

Tolman, J. R. (2002). A novel approach to the retrieval of structural and dynamic information from residual dipolar couplings using several oriented media in biomolecular NMR spectroscopy. *J. Am. Chem. Soc.* **124**, 12020–12030.

Tolman, J. R., Flanagan, J. M., Kennedy, M. A., and Prestegard, J. H. (1995). Nuclear magnetic dipole interactions in field-oriented proteins: Information for structure determination in solution. *Proc. Natl. Acad. Sci. USA* **92**, 9279–9283.

Tugarinov, V., Hwang, P. M., and Kay, L. E. (2004). Nuclear magnetic resonance spectroscopy of high-molecular-weight proteins. *Annu. Rev. Biochem.* **73**, 107–146.

Tugarinov, V., and Kay, L. E. (2005). Methyl groups as probes of structure and dynamics in NMR studies of high-molecular-weight proteins. *Chembiochem* **6**, 1567–1577.

Tugarinov, V., Kay, L. E., Ibraghimov, I., and Orekhov, V. Y. (2005). High-resolution four-dimensional 1H-13C NOE spectroscopy using methyl-TROSY, sparse data acquisition, and multidimensional decomposition. *J. Am. Chem. Soc.* **127**, 2767–2775.

Tzakos, A. G., Grace, C. R., Lukavsky, P. J., and Riek, R. (2006). NMR techniques for very large proteins and RNas in solution. *Annu. Rev. Biophys. Biomal. Struct.* **35**, 319–342.

Vaynberg, J., and Qin, J. (2006). Weak protein–protein interactions as probed by NMR spectroscopy. *Trends Biotechnol.* **24**, 22–27.

von der Haar, T., and McCarthy, J. E. (2002). Intracellular translation initiation factor levels in *Saccharomyces cerevisiae* and their role in cap-complex function. *Mol. Microbiol.* **46**, 531–544.

Walters, K. J., Ferentz, A. E., Hare, B. J., Hidalgo, P., Jasanoff, A., Matsuo, H., and Wagner, G. (2001). Characterizing protein–protein complexes and oligomers by nuclear magnetic resonance spectroscopy. *Methods Enzymol.* **339**, 238–258.

Walters, K. J., Matsuo, H., and Wagner, G. (1997). A simple method to distinguish intermonomer nuclear Overhauser effects in homodimeric proteins with C2 symmetry. *J. Am. Chem. Soc.* **119**, 5958–5959.

Wang, A., Grzesiek, S., Tschudin, R., Lodi, R., and Bax, A. (1995). ...HACAN.... *J. Biomol. NMR* **5**, 376–382.

Weber, D. J., Gittis, A. G., Mullen, G. P., Abeygunawardana, C., Lattman, E. E., and Mildvan, A. S. (1992). NMR docking of a substrate into the X-ray structure of staphylococcal nuclease. *Proteins* **13**, 275–287.

Weisemann, R., Ruterjans, H., and Bermel, W. (1993). 3D triple-resonance NMR techniques for the sequential assignment of NH and 15N resonances in 15N- and 13C-labelled proteins. *J. Biomol. NMR* **3**, 113–120.

Wienk, H., Tomaselli, S., Bernard, C., Spurio, R., Picone, D., Gualerzi, C. O., and Boelens, R. (2005). Solution structure of the C1-subdomain of *Bacillus stearothermophilus* translation initiation factor IF2. *Protein Sci.* **14,** 2461–2468.

Wittekind, M., and Mueller, L. (1993). HNCACB, a high-sensitivity 3D NMR experiment to correlate amide-proton and nitrogen resonances with the alpha- and beta-carbon resonances in proteins. *JMRB* **101,** 201–205.

Xia, Y., and Zhu, G. (2001). 3D Haro-NOESY-CH3NH and Caro-NOESY-CH3NH experiments for double labeled proteins. *J. Biomol. NMR* **19,** 355–360.

Zhou, P., Lugovskoy, A. A., and Wagner, G. (2001). A solubility-enhancement tag (SET) for NMR studies of poorly behaving proteins. *J. Biomol. NMR* **20,** 11–14.

CHAPTER THIRTEEN

STRUCTURAL METHODS FOR STUDYING IRES FUNCTION

Jeffrey S. Kieft, David A. Costantino, Megan E. Filbin, John Hammond, *and* Jennifer S. Pfingsten

Contents

Abstract

Internal ribosome entry sites (IRESs) substitute RNA sequences for some or all of the canonical translation initiation protein factors. Therefore, an important component of understanding IRES function is a description of the three-dimensional structure of the IRES RNA underlying this mechanism. This includes determining the degree to which the RNA folds, the global RNA architecture, and higher resolution information when warranted. Knowledge of the RNA structural features guides ongoing mechanistic and functional

Department of Biochemistry and Molecular Genetics, University of Colorado at Denver and Health Sciences Center, Aurora, Colorado

Methods in Enzymology, Volume 430
ISSN 0076-6879, DOI: 10.1016/S0076-6879(07)30013-X

studies. In this chapter, we present a roadmap to structurally characterize a folded RNA, beginning from initial studies to define the overall architecture and leading to high-resolution structural studies. The experimental strategy presented here is not unique to IRES RNAs but is adaptable to virtually any RNA of interest, although characterization of RNA–protein interactions requires additional methods. Because IRES RNAs have a specific function, we present specific ways in which the data are interpreted to gain insight into that function. We provide protocols for key experiments that are particularly useful for studying IRES RNA structure and that provide a framework onto which additional approaches are integrated. The protocols we present are solution hydroxyl radical probing, RNase T1 probing, native gel electrophoresis, sedimentation velocity analytical ultracentrifugation, and strategies to engineer RNA for crystallization and to obtain initial crystals.

1. INTRODUCTION

Structural analysis of viral internal ribosome entry site (IRES) RNAs uses essentially the same tools used to explore any biologically important RNA. Solution probing, biophysical, and high-resolution structural methods remain the foundation of RNA structural analysis. However, the way these techniques are applied and the data are interpreted is guided by the nature of IRES RNA function. IRESs have evolved to recruit, position, and activate the translation machinery, and, therefore, structural studies are targeted to understand how this function is achieved (Hellen and Sarnow, 2001). For example, an RNA structural biologist focused on understanding ribozyme function is likely most interested in the active site structure and how that active site forms. In the case of IRES RNAs, we are interested in what parts of the molecule are exposed for interaction with the translation machinery and how those recognition surfaces are arrayed in space. Thus, structural techniques successfully used to study IRES RNA function tend to examine the RNA as a large entity of unknown architecture with areas of stable structure, likely areas of dynamic structure, perhaps tightly folded regions, and regions extended into solvent as binding sites for proteins and ribosomes. IRESs may be extended or compact (Costantino and Kieft, 2005; Kieft *et al.*, 1999), and they may interact directly with the ribosome or with transactivating proteins (Bonnal *et al.*, 2003). An IRES RNA that does not possess a higher order fold on its own, but remains flexible and extended in solution, may require protein-binding partners for function. Conversely, a compact fold may lead to a different mechanistic hypothesis. Hence, the overall architecture of the folded IRES RNA, and whether it assumes a stable fold at all, suggests its mode of function, and the experiments presented help provide that structural insight.

1.1. Structural information does not come only from high-resolution techniques

Although high-resolution structure determination methods such as nuclear magnetic resonance and X-ray crystallography are important components of structural biology, a great deal of important three-dimensional (3D) structural information comes from other techniques. In this chapter, we promote a broad view of structural biology in which structural information from a variety of probing, biophysical, etc. methods guides experiments aimed at understanding the mechanism. Although high-resolution information is invaluable, such structures need not be the ultimate goal of the IRES researcher interested in accessing and exploiting structural information. When the mechanistic models motivate high-resolution studies, the insight gained from other probing, biophysical, functional, etc. studies is a solid foundation for successful high-resolution structure determination efforts. In addition, these various methods are invaluable when interpreting and validating structures obtained from crystallography, nuclear magnetic resonance, or cryoelectron microscopy (cryo-EM). New hypotheses gleaned from high-resolution structures are tested by use of these methods, providing the means to develop mechanistic models through the integration of all of these structural techniques. For example, a high-resolution structure of an IRES domain suggests where mutations will affect function, and these are then tested by use of probing and biophysical methods. Hence, the roadmap of Fig. 13.1 is not a linear progression, but a strategy in which the various experimental components build on each other to produce an overall picture of the RNA structure and its relationship to function.

1.2. A roadmap to answer specific questions

When undertaking a structural investigation of a viral IRES RNA, several questions aid in framing the course of experiments. Although these vary on the basis of the nature of the IRES, features of the IRES mechanism, and existing information, the following questions comprise a core set:

1. Does the IRES RNA fold into a higher order structure?
2. What is needed to induce this fold (cations, protein factors)?
3. Is the folded IRES RNA compact or extended? What parts of the RNA are in folded "core" regions and which parts are extended into solution?
4. How many independently folded domains exist? Are they functionally independent?
5. What are the roles for each structural element and roles for each nucleotide in that functional element?
6. Are there areas of structure that are dynamic? How do the dynamic and stable regions respond to interaction with the translation machinery?

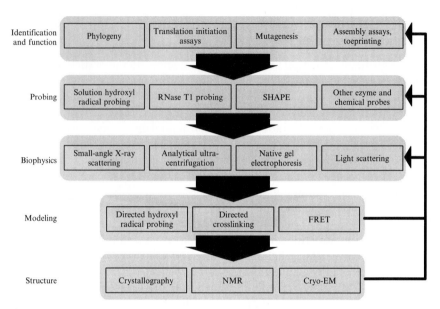

Figure 13.1 Roadmap of experiments to explore IRES RNA structure and relate to function. This chart shows a simple rough roadmap that we have used successfully to explore the structure of IRES RNAs. However, this plan can be used successfully to explore any RNA of biological interest. RNAs that are identified through functional assays are tested for their ability to form specific, higher order, ion-induced folds, and the overall architecture of the fold is explored by use of probing methods. More detailed analysis by biophysical methods establish more details of the fold and are used to explore mutants that affect fold or function. Interesting folding mutants then are assayed by use of functional translation assays, other folding assays, etc., to build a detailed picture linking fold to function. Ultimately, if high-resolution structural studies are warranted, the knowledge gained from the biochemical and biophysical explorations feed into modeling methods and the design of RNAs for solving structures. It is important to realize that this is not a linear flowchart of experiments, but each "family" of experiments integrates with the others. For example, if the RNA structure is solved or a model of the structure is constructed, functional methods, probing methods, and biophysical methods combined with mutagenesis are critical for validating and interpreting that structure or model.

The techniques detailed in this chapter provide a roadmap for exploring these structural questions (Fig. 13.1). We present the experiments in a progression from first experiments designed to detect the presence of a stable fold, through experiments to further understand the characteristics of that fold, and finally to high-resolution structural studies. The information from each experiment guides the next, building a picture of the structure of the IRES "from the ground up." We have used this experimental strategy to successfully characterize two classes of viral IRESs with different architectures, demonstrating its versatility (Costantino and Kieft,

2005; Kieft *et al.*, 1999). The nature of the IRES RNA in question, the interests of the researcher, and the specific questions being asked will drive the sequence of experiments and the priority given each.

We do not provide a comprehensive list of experiments, because the breadth of experimental approaches available to the modern RNA structural biologist is greater than can be represented here. We also do not include protocols for *ab initio* determination of RNA secondary structures, because secondary structure probing methods (Ehresmann *et al.*, 1987; Merino *et al.*, 2005), thermodynamic prediction algorithms (Mathews and Turner, 2006), and phylogenetic comparison methods (James *et al.*, 1989; Pace *et al.*, 1989) for that purpose have been described. Rather, we present experiments aimed at obtaining insight into the 3D, folded structure of the IRES that are, in general, readily available to researchers at virtually any location with very little specialized equipment. Our hope is that readers will use these protocols and build on them with their own expertise and insight.

1.3. Importance of linking structural information to function

The true value of structural information is to lend insight into function and mechanism and to suggest the next set of experiments. Therefore, it is imperative that the experiments described here be linked with functional experiments. For example, when IRES RNA folding mutants are discovered through structural methods, those mutants should be assayed with translation initiation assays, factor binding assays, etc. (Fig. 13.1). Conversely, the approaches outlined here provide a means to analyze mutants identified in functional assays to determine why those mutants fail. This may seem an obvious point, but we mention it because although we do not describe functional assays, they are a critical part of any structural study.

2. Methods

2.1. Solution hydroxyl radical probing

2.1.1. Overview
Hydroxyl radical probing is a straightforward and powerful method to detect a higher order fold within an RNA structure (Tullius and Greenbaum, 2005). The method detects RNA regions that are involved in close backbone packing and regions that extend into solution. The amount of backbone that becomes protected from radicals when the RNA folds indicates the amount that is tightly packed into higher order structure. Hence, this method gives insight into the architecture and global characteristics of the folded RNA. Hydroxyl radical probing is described first, because it is usually the first experiment performed in our laboratory on a new RNA of interest.

Hydroxyl radicals produced in the solvent cleave the RNA backbone in a nonsequence-specific manner and independent of the RNA secondary structure. This makes hydroxyl radical probing fundamentally different from the chemical or enzymatic probing methods that are specific to certain bases or base pairing (Ehresmann *et al.*, 1987). The most useful way to exploit this is by comparing the pattern of cleavage in the absence and presence of tertiary fold–inducing cations (Fig. 13.2). Areas that are cleaved less (protected) in the folded form relative to the unfolded form are involved in close packing of the RNA backbone and are solvent inaccessible (for examples, see Cate *et al.*, 1996; Celander and Cech, 1991; Latham and Cech, 1989; Pan, 1995; Pfingsten *et al.*, 2006; and Takamoto *et al.*, 2002). In general, RNA is protected if it is in areas where the negatively charged backbone is in close proximity to other sugar-phosphate backbones. Hence, the protected areas indicate where the RNA has collapsed to form tight RNA tertiary structure. In addition to protected areas, it is common to observe areas of the RNA that have enhanced cleavage when the RNA

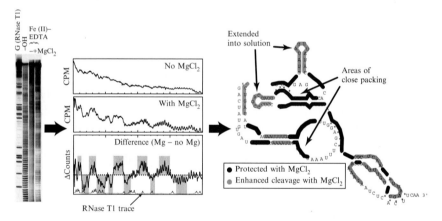

Figure 13.2 Example of a hydroxyl radical probing experiment. In this example, the IRES RNA from the intergenic regions of the *Plautia stali* intestine virus (PSIV) was probed with radicals by use of the protocol described in the text. The gel on the left shows the raw data of the cleavage pattern in the absence and presence of magnesium chloride alongside RNase T1 and alkaline hydrolysis ladders. At the center are traces of the radical-probed lanes and the difference trace obtained by normalizing the two raw traces and calculating the difference. Shaded boxes indicate regions where the degree of cleavage in the (+) magnesium lane is reproducibly higher (positive values) or lower (negative values). By use of the RNase T1 and hydrolysis ladders, these regions were mapped onto the secondary structure shown on the right. This identifies parts of the RNA backbone packed closely in the compact fold and parts extended into solution, defining the architecture. These data are different than secondary structure probing experiments in that the signal depends on the folded conformation of the RNA, not on the base pairing, and hence this technique identifies the RNA as having a compact higher order fold. Data are from Costantino and Kieft (2005).

folds. These areas of the backbone are extended into solution and in IRES RNAs often are found in stem loops that are translation machinery recognition sites (Fig. 13.2). The physical basis for enhanced cleavage is likely that in the unfolded form, the RNA spends equal amounts of time in a protected state and in a solvent-exposed state, but in the folded form it is held in an exposed state. Hence, relative to the unfolded state, these regions are cleaved more.

The folded RNA architecture is revealed by mapping the protected and enhanced areas onto the secondary structure. Parts of the RNA where close backbone packing orders the overall structure are revealed, as are parts that are extended into solution (Fig. 13.2). When much of the RNA is protected from solution, it suggests a compact fold, whereas isolated smaller portions of protection (or no protection) suggest an overall extended fold. The predictive power of this method is illustrated by two examples in which the folded architecture of an IRES RNA was predicted by probing (Costantino and Kieft, 2005; Jan and Sarnow, 2002; Kieft *et al.*, 1999; Nishiyama *et al.*, 2003), and verified by cryo-EM or crystallographic methods (Pfingsten *et al.*, 2006; Spahn *et al.*, 2001, 2004). In the specific cases of IRES RNAs, the locations of regions that are extended into solution are particularly interesting because they may be sites for interaction with the translation machinery or other proteins. Hence, this single experiment readily indicates whether a given IRES RNA is folded, the nature of that fold, what areas are likely to be important for forming the fold, which areas are likely to be important for interaction with the translation machinery, and where mutations can be made to disrupt the fold and function. Thus, this experiment is a foundation on which other structural studies are based.

We describe a chemical-based means to produce hydroxyl radicals used successfully with a large range of RNA targets and that uses readily available reagents. Synchrotron radiation also has been used to produce radicals for kinetic studies (Brenowitz *et al.*, 2002; Ralston *et al.*, 2000) or *in vivo* studies (Adilakshmi *et al.*, 2006), and recent use of stop-flow technology and insight into the radical-producing reactions provide additional tools (Shcherbakova *et al.*, 2006).

2.1.2. Materials

300 mM HEPES-KOH, pH 7.5

100 mM MgCl$_2$

2 to 4 mg/ml tRNA (phenol/chloroform extract, ethanol precipitate, and resuspend in water before use)

Pure ^{32}P-end labeled RNA (>100,000 CPM is best; protocol for end labeling is contained below)

RNase-free water

10 mM sodium ascorbate (can be stored at $-20°$ between uses)

20 mM Na-EDTA (can be stored at $-20°$ between uses)

1.0% hydrogen peroxide (make fresh each time)

100 mM ferrous ammonium sulfate; $(NH_4)_2Fe(SO_4)_2$; (make fresh each time)

100 mM thiourea (can be stored at $-20°$ between uses)

7 M urea in 1× TBE + 0.5 mg/ml each bromophenol blue and xylene cyanol (2× gel loading buffer)

2.1.3. Methods and procedures

1. Begin by annealing and folding the RNA. In two separate reaction tubes, add to each:

   1 μl ^{32}P-end labeled RNA (>100,000 CPM is best)

   3 μl RNase-free water

   Heat to $90°$ for 30 sec, then allow to cool on the bench for 5 min. After this annealing step, add:

   1 μl 300 mM HEPES-KOH pH 7.5

   1 μl 10 mM $MgCl_2$ (folded RNA) or 1 μl 10 mM Na-EDTA (unfolded RNA)

   1 μl 2 to 4 mg/ml tRNA

   Incubate each of the two reactions for 15 to 20 min at the desired temperature (generally $37°$) to reach folding equilibrium.

2. As the RNA reaches folding equilibrium, prepare the following stocks: sodium ascorbate, Na-EDTA, hydrogen peroxide, and ferrous ammonium sulfate.

3. Immediately before performing the experiment, mix 100 μl of the $(NH_4)_2Fe(SO_4)_2$ with 900 μl of the EDTA solution to make the Fe(II)-EDTA solution. Then, to each tube, add (in this order):

   1 μl Fe(II)-EDTA solution

   1 μl H_2O_2 solution

   1 μl sodium ascorbate solution

   Some protocols call for placing these drops on the side of the tube, then vortexing to ensure rapid mixing. We have found this unnecessary. Rather, we remove the reaction tube from the water bath (where it has reached folding equilibrium), rapidly add the three reagents in quick succession, "flick" the tube once or twice to mix, and immediately place the tube at $37°$.

4. Incubate the reaction at $37°$ for 2 min. Quench the reaction with:

   1 μl thiourea solution

   9 μl of denaturing 2× gel loading buffer

   Put the reaction tube on ice.

5. Prepare a 0.4 mm-thick 10% sequencing gel with square wells of ~1 cm width. An excellent comb that produces even wells of ideal size is

manufactured by CBS Scientific (catalog number SG33-0424). Pre-run the gel for 30 min at 65W and then load 10 μl of the reaction per lane (see the *Notes and hints* section for hints on how to ensure equal loading per lane). Alongside the reaction lanes, run an RNase T1 ladder and an alkaline hydrolysis ladder (see protocols below for generating these ladders). Run the gel at 65W for various times depending on the RNA length and the part of the backbone one wishes to visualize. It often requires several repetitions of the experiment with different run times to see the entire backbone. Because only half of each reaction is loaded on the gel, the other half can be loaded after the gel has run several hours to resolve two different parts of the backbone on the same gel. The use of 10% gels improves the resolution of the bands.

6. Remove the gel from the apparatus and transfer to support paper. Cover the gel with plastic wrap and dry the gel by use of a vacuum gel-drying apparatus (with heat). Once dry, expose the gel to a PhosphorImager screen (overnight) and visualize the result by use of a PhosphorImager to obtain an electronic scan of the gel.

2.1.4. Analysis

"By-eye" Usually, RNA regions with diminished or enhanced hydroxyl radical cleavage are readily identified by direct examination of the gel, although the effects are often subtle. This is particularly true of strongly protected regions, several of which are visible in the gel of Fig. 1.2. Nucleo-tide-resolution mapping of the protections is challenging when done manually, but this rough mapping strategy is often sufficient to identify structured domains of the IRES and to determine to what degree the IRES RNA collapses into a folded structure in the presence of cations. In addition, protected regions and enhanced regions that are identified by use of automated or computer-based methods (see next paragraph) should always be confirmed by a careful visual examination of the gel itself.

Quantitative Several quantitative ways exist to analyze the data from the hydroxyl radical probing. In recent years, a program called SAFA (Semi-Automated Footprinting Analysis) has become available (Das *et al.*, 2005), which allows rapid processing of the raw gel data. We encourage readers to explore that resource. Here, we present a step-by-step method that we use in our laboratory with the programs Image Quant, Excel, and Kaleida-Graph. We encourage readers to use this, or a similar method, the first few times they conduct this experiment to gain a feel for the nature of the data, the nature of their specific RNA, the reproducibility of the data, and any potential errors.

1. Open the electronic data file (generally a .gel file) in the program Image Quant (Molecular Dynamics). To produce a trace of intensity versus gel position for a reaction lane, use the **Line** tool (Objects menu) to draw a

line downward over the portion of the lane to be analyzed. Now, select **Create graph** from the Analysis menu and select the line(s) to graph. This displays a two-dimensional graph. At this point, care should be taken to ensure that the same portions of the gel are sampled and the graphs "line up." This is checked by overlaying multiple graphs in Image Quant by use of the **Merge** command in the View menu. To transfer the data to Excel (Microsoft), click on the graph and then press **control-C** (copy) to place the data in the computer's clipboard. Then, open a new Excel spreadsheet, click on the upper left cell, and press **control-V** (paste). The data will be displayed in tabular form. Additional lanes of the gel are processed in this way, leading to a spreadsheet containing data of all the selected lines in Image Quant.

2. Further data processing occurs in Excel in two steps. First, to account for small differences in gel loading or amount of radiation contained in each reaction, the data are normalized. This is accomplished by summing and comparing the total counts in each individual lane to calculate a correction constant. Usually, we consider the lane that contains the reaction without magnesium (unfolded) to be the standard lane, and other lanes are compared and corrected to match this. For example, if the 0 magnesium lane contains $1.1\times$ more total counts than the 10 mM magnesium lane, then each position in the 10 mM magnesium lane must be multiplied by 1.1 to normalize the data. The second step is to calculate the difference between the intensity of the folded lane ($+$ magnesium) and the unfolded ($-$ magnesium). The result contains the signal identifying regions where the RNA is protected from cleavage on folding and where it is enhanced.

3. The data of the difference in intensity versus gel position (and hence backbone position) are now graphed (Fig. 13.2). We prefer to use the program KaleidaGraph, which allows rapid graphing of the data and manipulation of the graphs during analysis. If the data have been carefully quantified and normalized, the plot identifies parts of the backbone protected from cleavage on folding (negative values) and enhanced regions (positive values). It is not unusual for there to be fluctuations around the zero line that do not represent true signal. Authentic changes in the cleavage pattern are parts of the graph where the signal deviates from zero for several nucleotides before returning to zero and where the signal is reproducible between several different experiments (for examples, see: Cate *et al.*, 1996; Costantino and Kieft, 2005; Takamoto *et al.*, 2002). Returning to the actual raw gel image for a visual "reality check" is an excellent way to verify that the protection or enhancement is real.

4. The regions of protection and enhancement then are assigned to specific locations in the backbone by plotting the trace of the RNase T1 and alkaline hydrolysis cleavage ladders alongside the hydroxyl radical probing difference trace. These specific areas of cleavage are transferred to a secondary structure diagram to visualize elements involved in the fold (Fig. 13.2).

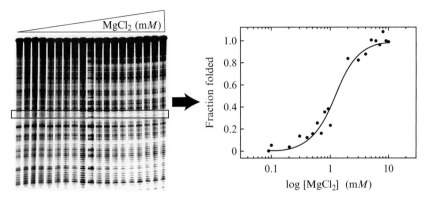

Figure 13.3   The use of hydroxyl radical probing to monitor RNA folding. In this example gel (left), hydroxyl radical probing was performed at increasing [MgCl₂]. As the RNA folds, the protection or enhancement of regions of the backbone was monitored. The degree of protection was measured at the gel location delineated by the box. These raw data were converted to fraction folded by use of knowledge that without MgCl₂ the molecule is unfolded, and at 10 mM MgCl₂ the molecule is fully folded (addition of more MgCl₂ does not change the cleavage pattern). At right, the fraction folded is shown plotted as a function of [MgCl₂]. Fitting of this curve to the Hill equation leads to a value for the [MgCl₂] at which the RNA folds and a measurement of the cooperativity of cation uptake. Data are from Costantino and Kieft (2005).

Notes and hints

1. The protocol previously outlined is for only two magnesium concentrations. Other magnesium concentrations can be used, and titration experiments are a powerful way to assess how much magnesium is needed to fold the RNA and if all parts fold at the same magnesium concentration (Fig. 13.3). Other cations also can be used to examine the specific cation requirements of the fold. Likewise, monovalent salts also can be used; we have used them to concentrations over 300 mM. Note that divalent salts do quench the reaction somewhat by competing with the Fe^{2+} for the EDTA, but up to 80 mM Mg^{2+} has been found to work.

2. For buffers, TRIS or MOPS can also be used, and pH can also be varied. TRIS is a free radical quench, so if it is used, the concentration should be kept below 50 mM.

3. If the reaction seems to be undercutting (cleavage too light for analysis), try increasing the peroxide concentration to 1.5 or 2%. As in all cleavage reactions, the goal is to achieve only "single hit" cleavage, and so most of the RNA should be in the intact, uncleaved band at the top of the gel (Fig. 13.2).

4. The tRNA is optional but seems to improve the look of the gels and prevents RNA from sticking to the tubes.

5. The larger the RNA, the more radioactive counts are needed. Putting 200,000 counts per minute (CPM) *per reaction tube* is not unreasonable for RNAs of 300 bases or more in length.

6. The annealing protocol presented previously is a generic protocol designed to denature the RNA and allow it to return to its native structure. Some other protocols may call for a slower, controlled cool or snap cooling on ice or inclusion of divalent cations during the heat–cool. Different RNAs may require more specialized annealing methods, but we have found that the method outlined here prevents aggregation and promotes proper folding in most RNAs we have tested.

7. Analysis of the gel is much easier if the amount of radioactivity loaded into each lane is identical. This is achieved by being very careful to add identical amounts of labeled RNA to each reaction tube and also by measuring the actual amount of radioactivity contained in each tube before loading the gel. If tubes show different amounts, the loading of the gel is adjusted accordingly. Alternately, if the amount of CPM in each tube is first measured, one can adjust the amount of volume in each tube (with gel loading buffer) to achieve an equal value of CPM/μl in each tube, and then load 10 μl per lane as per norm.

8. Primer extension by reverse transcription also can be used to detect the cleavage. We prefer end labeling, because we can perform experiments with either the 5′ or 3′ end labeled. In our hands, end labeling produces a higher signal-to-noise ratio, requires less manipulation of the RNA, and does not give rise to background "stops" because of reverse transcriptase being interrupted by RNA structure that complicate analysis. However, end labeling requires that the labeled RNA be prepared fresh, because labeled RNA more than a few days old generally gives poor results.

2.2. End labeling protocols (for large RNA molecules, 150 to 400nt)

To obtain homogeneous 5′ and 3′ RNA termini, we routinely engineer our templates with ribozymes on each end that self-cleave during the *in vitro* transcription reaction, leaving a 5′ hydroxyl and a terminal 2′–3′ cyclic phosphate (Ferre'-D'Amare' and Doudna, 1996). RNA transcribed without these ribozymes has a 5′ phosphate and 3′ hydroxyl. End labeling requires a hydroxyl substrate; hence, when a phosphate is present, the end must first be modified. Reactions to accomplish this are included in the following.

2.2.1. 5′ End labeling: Methods and procedures

Step 1 (if necessary). Dephosphorylate the RNA. In the reaction tube, combine:
2 μl of RNA (approximately 5 to 10 μg)
1 μl 10× Buffer (New England Biolabs Buffer #3)

1 μl calf intestinal phosphatase (New England Biolabs)
6 μl RNase-free water

Incubate at 37° for 30 min followed by phenol/chloroform extraction and ethanol precipitation. Suspend in 2 μl RNase-free water and use in 5′ end-labeling reaction.

Step 2. 5′ End label the RNA. In the reaction tube, combine:

2 μl of 5′ dephosphorylated RNA (approximately 5 to 20 μg)
2 μl 10× T4 polynucleotide kinase buffer (New England Biolabs)
3 μl T4 polynucleotide kinase (New England Biolabs)
10 μl RNase-free water
3 μl ^{32}P-γ-ATP (150 mCi/μl)

Incubate at 37° for 30 min, then use a Bio Rad P30 spin column to remove unused ATP. Add 20 μl 7 M denaturing gel loading buffer with dye. Load on a 0.8 mm thick 10% denaturing gel and run at 25W for 1 to 1.5 h, depending on the RNA length. Carefully remove the gel from the plates and wrap in plastic, then expose gel to film for 5 to 15 sec. By use of the film to align the correct lane with the gel, cut the radioactive band and elute into 400 μl elution buffer (0.5 M sodium acetate, pH 5.2 + 0.1% SDS) for at least 2 h at 4° (overnight is acceptable). Remove the buffer from the gel slice and add 1000 μl ice-cold 100% ethanol. Pellet the RNA by centrifugation, carefully wash the pellet with 70% ethanol, then dry the labeled RNA pellet in a speed-vac. This should yield 5 to 15 million counts per minute in a dry pellet.

2.2.2. 3′ End labeling: Methods and procedures

Step 1 (if necessary). Remove the 2′–3′ terminal cyclic phosphate. Combine in the reaction tube:

10 μl RNA solution (50 to 100 μg)
10 μl of 10× dephos. Buffer (400 mM Na-MES, pH 6.0, 100 mM MgCl$_2$, 50 mM DTT)
10 μl T4 polynucleotide kinase (New England Biolabs)
5 μl calf intestinal phosphatase (New England Biolabs)
65 μl RNase-free water

Incubate at 37° for 3 h, followed by phenol/chloroform extraction and ethanol precipitation. Suspend in 10 to 20 μl RNase-free water and use in 3′ end-labeling reaction.

Step 2. Produce radiolabeled p*Cp. Combine in the reaction tube:

5 μl ^{32}P-γ-ATP (150 mCi/μl)
2 μl 1mM Cp (Sigma catalog number C1133-100MG)
1 μl 10× T4 polynucleotide kinase buffer (New England Biolabs)
1 μl T4 polynucleotide kinase (New England Biolabs)
1 μl RNase-free water

Incubate at 37° for 90 min, then at 65° for 11 min (to deactivate the kinase). Use in the 3′ end-labeling reaction without further purification.

Step 3. 3′ End labeled RNA. Combine in the reaction tube:

2 μl (5 to 20 μg) pure RNA (with 3′ -OH)
1 μl dimethyl sulfoxide (DMSO)
1 μl 10× T4 RNA ligase buffer (New England Biolabs)
2 μl T4 RNA ligase (New England Biolabs)
4 μl p*Cp from the preceding reaction

Incubate at 16° overnight (16 h). Purify on a 10% denaturing polyacryl-amide gel as described for 5′ end-labeling.

2.3. Sequencing ladders

2.3.1. RNase T1: Methods and procedures

Combine in the reaction tube:

3 μl denaturing RNase T1 Buffer (20 mM sodium citrate, pH 5.0, 1 mM EDTA, 7 M urea)
1 μl tRNA (2 mg/ml)
1 μl end-labeled RNA (to match the number of counts in experimental lanes)

Heat to 55° for 1 min to denature the RNA, then add 1 μl of 0.3 U/μl RNase T1 (3 μl of catalog number NC9326077 from Fisher diluted into 1 ml water). Incubate the reaction at 55° for 3 min. Add 15 μl denaturing urea gel loading buffer and place on ice. Load 10 μl on the sequencing gel.

Note/hint: The time and/or amount of enzyme may have to be adjusted, depending on the source of RNase T1 and the RNA. A good optimization procedure is to try several reactions with the same amount of enzyme and vary the time from 3 to 20 min at 55°.

2.3.2. RNase U2: Methods and procedures

Combine in the reaction tube:

3 μl denaturing RNase T1 Buffer (20 mM sodium citrate, pH 3.5, 1 mM EDTA, 7 M urea)
1 μl tRNA (2 mg/ml)
1 μl end-labeled RNA

Heat to 55° for 1 min to denature the RNA, then add 1 μl of 0.5 U/μl RNase U2. Incubate the reaction at 55° for 3 min. Add 15 μl denaturing urea gel loading buffer and place on ice. Load 10 μl on the sequencing gel.

Note/hint: At the time of this writing, there is no commercial source of RNase U2, which cuts exclusively after A-bases and thus nicely comple-ments RNase T1. However, the protocol is included here in the hope that a source of this very useful enzyme will reemerge. RNase T2 cuts after all four bases, but preferentially after As, and may be a good substitute for RNase U2, but we have not tested it.

2.3.3. Alkaline hydrolysis ladder: Methods and procedures

Combine in the reaction tube:

1 μl tRNA (2 mg/ml)
1 μl end-labeled RNA
7 μl RNase-free water
1 μl 10× hydrolysis buffer (500 mM sodium bicarbonate, pH 9.2)

Incubate at 85° for 5 min. Add 10 μl denaturing gel loading buffer and place on ice. Load 10 μl on the sequencing gel.

2.4. RNase T1 structure probing

2.4.1. Overview

RNase T1 probing complements hydroxyl radical probing. Probing with this enzyme is fundamentally different from the radical probing for several reasons. First, the enzyme is much larger than the radical and is excluded sterically from more of the folded RNA. Second, the enzyme cleaves only after G-bases in the sequence. Third, the enzyme only cleaves when the G is not base paired and so is not exclusively a tertiary structure probe (Ehresmann et al., 1987). In fact, it is thought of more correctly as a secondary structure probe. The overall information content of RNase T1 probing is, therefore, lower than hydroxyl radical probing, but it provides information not readily gleaned from the hydroxyl radical probing and often verifies predictions made on the basis of those initial probing experiments.

Like other RNA probing techniques, RNase T1 probing is most useful when the unfolded state and folded state (without and with magnesium) cleavage patterns are compared. Locations in the sequence where G-bases are protected, even when structure-inducing cations have not been added, indicate very stable secondary structures or parts of the structure that form a core around which other elements fold. G-bases readily cleaved in the absence of magnesium but protected in the presence of magnesium indicate cation-induced base pairing. Helices or pseudoknots fall into this category, and observation of this behavior verifies their formation in the folded RNA. G-bases that are hypersensitive to the enzyme indicate an area of structure that is extended into solution (Fig. 13.4). In IRES RNAs, we have observed hypersensitive G-bases in the apical loops of hairpins that make contact with the translation machinery (Costantino and Kieft, 2005; Kieft et al., 1999, 2001), predicting a role for these stem loops and confirming the findings of hydroxyl radical probing. In addition, bases that are partially protected when the RNA folds but never completely protected indicate areas where

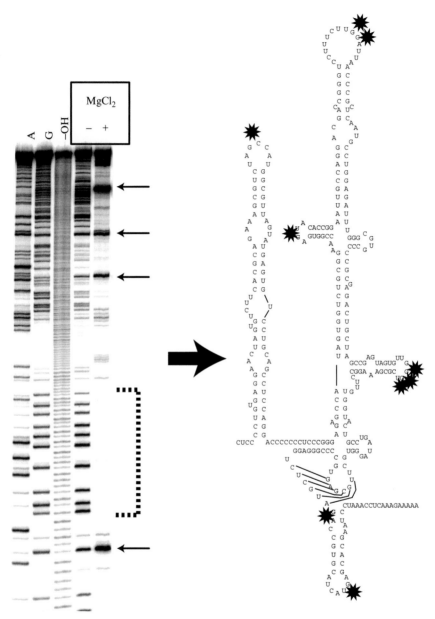

Figure 13.4 Example of RNase T1 probing performed on the hepatitis C virus IRES. At left is an RNase T1 probing gel, showing the difference in the cleavage pattern in the absence and presence of added cations (right two lanes) alongside RNase U2 (A), RNase T1 (G), and alkaline hydrolysis ladders (−OH). Arrows indicate locations in the RNA sequence that become more sensitive to enzyme cleavage when the RNA folds, suggesting they are extended into solution and not base paired. Many other regions of the RNA backbone are protected from the enzyme on folding; an example of this is indicated by the hatched line. At right is a secondary structure with the locations of

the RNA is dynamic or only transiently folded, giving further insight into the RNA structural architecture. When used in conjunction with hydroxyl radical probing, the data from RNase T1 probing help complete a picture of the RNA architecture and guide the next round of mutagenesis, folding, and functional experiments.

2.4.2. Materials

300 mM HEPES-KOH, pH 7.5

100 mM MgCl$_2$

2 mg/ml tRNA (phenol/chloroform extract and ethanol precipitate before use)

Pure ^{32}P-end labeled RNA (>20,000 CPM is best)

RNase-free water

0.3 U/μl RNase T1 (NC9326077 from Fisher diluted into 1 ml water)

7 M urea in 1× TBE+0.5 mg/ml bromophenol blue and 0.5 mg/ml xylene cyanol (2× gel loading buffer)

2.4.3. Methods and procedures

1. Begin by annealing and folding the RNA. In two separate reaction tubes, add to each:

   1 μl ^{32}P-end labeled RNA (>20,000 CPM is best)

   6 μl RNase-free water

   Heat to 90° for 30 sec, then allow to cool on the bench for 5 min. After this annealing step, add:

   1 μl 300 mM HEPES, pH 7.5

   1 μl 10 mM MgCl$_2$ (folded RNA) or 1 μl RNase-free water (unfolded RNA)

   1 μl 2 mg/ml tRNA

   Incubate each of the two reactions for 15 to 20 min at the desired temperature (generally 37°) to reach folding equilibrium.

2. To each tube, add:

   1 μl RNase T1 solution

   Incubate the reaction at 37° for 2 min. Quench the reaction with 10 μl denaturing gel loading buffer and place on ice. Resolve the reactions by running 10 μl of the reaction on a sequencing gel as described in the protocol for **Hydroxyl Radical Probing**. Dry the gel and visualize with a PhosphorImager.

---

enhanced cleavage shown with asterisks. The probing data both supported the secondary structure as drawn and verified that these stem loops are extended in solution and available to interact with the translation machinery. Data are from Kieft *et al.* (1999).

2.4.4. Analysis

The signal from RNase T1 probing reactions is quite robust and usually is observed readily by visual inspection (Fig. 13.4). This is especially true if care is taken to load the gel lanes equally and if the folded (+ magnesium) and unfolded (− magnesium) are run side by side. However, manual analysis is augmented by quantifying the amount of radiation in discrete RNase T1 cleavage bands by use of the Image Quant software (or similar) and directly comparing these measurements. An analysis of this sort would be useful if a magnesium titration were performed, for example. In that case, the degree of cleavage can be plotted as a function of magnesium concentration, demonstrating the formation of a given structural element as a function of cation concentration. In some cases, different parts of the RNA may require different amounts of magnesium to fold.

2.4.5. Notes and hints

1. The cleavage time provided here works for most RNAs, but some optimization may be necessary. We recommend varying the length of the cleavage reaction while keeping the enzyme concentration constant. As in all cleavage reactions, the goal is to achieve only "single-hit" cleavage, and so most of the RNA should be in the intact band at the top of the gel (Fig. 13.4).
2. If end-labeled RNA is used, this experiment can be run simultaneously with the hydroxyl radical probing and the reaction analyzed on the same gel. This allows a clear and direct correlation between the two results. If done this way, one should ensure the same amount of radioactivity is contained in each lane.
3. As with all probing experiments, primer extension by reverse transcription can be used to analyze the results of the reaction. For the reasons stated in the **Hydroxyl Radical Probing** protocol, we prefer end labeling.
4. Other enzymes (such as RNase S1 or V1) and other chemical modifying agents (such as CMCT or DMS) also can be used to probe the RNA. However, as more and more agents are used, one rapidly reaches a point of diminishing returns where the overall 3D architecture of the RNA has been established and additional information does not lend a great deal more insight. We recommend a smaller set of probing experiments that then are augmented by other structural methods.

2.5. Mutagenesis/Native gels

2.5.1. Overview

Native gel electrophoresis takes advantage of the fact that folded RNAs with different shapes travel through a polyacrylamide gel matrix at different rates. Native gels have been used to measure the relative angles between helices emerging from RNA junctions and hence when used in a systematic way, they have the power to provide specific structural information (for examples, see: Bassi *et al.*, 1995; Duckett *et al.*, 1995; Lafontaine *et al.*, 2002). This assay is exquisitely sensitive to the 3D RNA structure, and thus even small structural perturbations are observed. In general, more compact, tightly folded structures migrate through the gel faster than extended or unfolded structures. A mutation that changes the RNA structure by destabilizing a compact fold induces reduction in the migration rate relative to wild-type RNA. Local changes in structure have a minimal effect on the migration rate, whereas global changes in structure produce a substantial change. An example of a local perturbation would be the substitution of one structured tetraloop for another structured tetraloop in the context of a larger IRES RNA, leading to little or no change in the migration rate (Fig. 13.5, mut 1). An example of a global change would be the abrogation of a long-range pseudoknot interaction that completely unfolds the RNA (Fig. 13.5, mut 4). In general, the greater the change in migration rate, the more profound the change in the RNA structure, although one must be careful not to regard this as a hard-and-fast rule. We present the technique here as a simple means to detect whether a given mutation affects the fold of an IRES RNA.

This assay is particularly useful once a hypothesis of the RNA global architecture has been developed from the solution probing experiments outlined earlier in this chapter. Comparing the solvent protection, RNase T1 probing, and RNA sequence conservation supports predictions of what RNA regions stabilize tertiary structure and what parts are extended into solution. Native gel electrophoresis tests these predictions. For example, substitution mutations to RNA positions that are predicted to be extended into solution should not lead to a shift on a native gel. However, mutating sequences within regions protected from the solvent, and therefore predicted to be involved in close backbone packing and tertiary interaction, will lead to a gel-shift if the prediction is correct. Proposed long-range interactions such as pseudoknots or tetraloop–tetraloop receptor interactions are excellent targets for native gel electrophoreses (Fig. 13.5). On native gels many mutants are assayed side by side, generating a data set of which bases are critical for folding the RNA.

One important feature of native gel electrophoresis that must be considered is that the RNA may be undergoing structural transitions between two or more states as it migrates through the gel matrix. Even a folded RNA can change between different states, and an unfolded RNA is rapidly sampling a large ensemble of states. The rate at which an RNA is converting between different structural states affects how it appears on the native gel (Fig. 13.6). We broadly separate the behavior into three kinetic regimes:

- *Fast exchange:* The RNA is exchanging between the states very rapidly compared with the time scale of the experiment. The exchange is fast

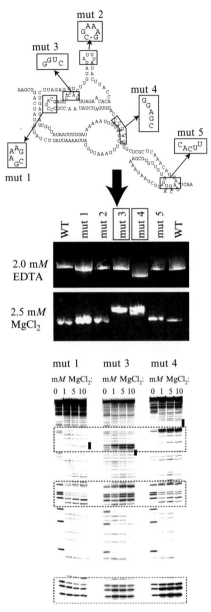

Figure 13.5 Example of native gel analysis coupled with mutagenesis. At the top is the intergenic region of the PSIV IRES that was probed with hydroxyl radicals in Fig. 13.2, here overlaid with several mutations. Below the secondary structure are native gels of these mutant RNAs alongside the wild-type RNA (on either side). Running wild-type RNA on both sides of a set of mutants is recommended to ensure that slight "smiling" of the gel (because of uneven heating) or other effects do not lead to misinterpretation. These gels illustrate two effects described in the text: (1) in the EDTA-containing gel,

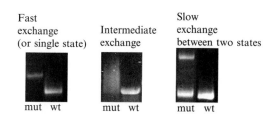

Fast
exchange   Intermediate
(or single state) exchange

Slow
exchange
between two states

mut   wt   mut   wt   mut   wt

Figure 13.6   Types of structural exchange observed on native gels. Shown are examples of the three conformational change regimes described in the text.

enough that the RNA migrates as the populationally-weighted average of the migration rates of the different states, appearing as a single band on the native gel. Note that an RNA that remains in a single folded conformation will also appear as a single band. Because an RNA in fast exchange runs as a single band, changes in the structure or changes in the populations found in different structural states are detected by use of a constant migrating standard.

- *Intermediate exchange:* The exchange between different states is slow enough that sufficient time exists in each state to affect the rate of migration. However, exchange is not slow enough for the different states to resolve themselves on the gel, and the band is smeared.

- *Slow exchange:* The exchange is slow enough that different RNA conformations are resolved on the gel and appear as different bands. The formation of two or more different stable folded conformations, stable dimers, or stable alternate secondary structures falls into this category.

A mutation that affects the RNA fold can have a number of effects, including inducing incorrect but stable structures, a shift in the ensemble

mutant 4 runs slightly faster, likely because of alterations in the RNA secondary structure before it assumes its Mg^{2+}-induced native fold, and (2) in the $MgCl_2$-containing gel, mutant 3 and mutant 4 show a clear retardation, suggesting they do not fold as wild-type RNA (boxed). Hence, the pseudoknots targeted by these mutants are clearly important for the three-dimensional fold. Importantly, mutants 1 and 2 do not affect the RNA migration rate, suggesting they are not involved in the folding. This is consistent with the probing results of Fig. 13.1, which predict these stem loops are extended into solution. Additional mutants (including point mutants) are used to provide more details as to what regions stabilize the ion-induced fold. At the bottom are three RNase T1 probing gels of three of the mutant RNAs: one that folded correctly as assayed by the native gels (mut 1), and two that did not (mut 3 and 4). The location of the mutation is marked with a small vertical bar. Mutant 1's pattern does not deviate from wild-type (not shown), whereas mutants 3 and 4 have several areas where the pattern changes (boxed). By mapping these changes, the specific RNA domains that are altered by each mutation are determined, linking the native gel to specific folding effects. Data are from Costantino and Kieft (2005).

population toward one particular conformation, or a complete destabilization of the structure. Considering these effects when interpreting a native gel helps interpret the effect of a given mutation. Furthermore, mutations that affect the RNA structure as assayed on a native gel can be probed with hydroxyl radical to verify the effect and measure the degree to which the mutation affects the solvent protection of the RNA (Fig. 13.1). By use of these methods hand in hand, one identifies different independently folded domains in the structure and the nucleotides that are important for stabilizing each. For example, in the 342-nucleotide-long HCV IRES RNA, a point mutation in a four-way junction region causes retardation and smearing of the RNA band on a native gel relative to wild-type (Kieft *et al.*, 1999). Hydroxyl radical probing of this mutant shows that the areas around this junction no longer fold, but that another junction region was unaffected. This identifies these two junctions as independently folded regions. Subsequent functional studies and ribosome/factor binding studies lend insight into the roles of these domains, linking sequence to folded structure to factor binding and translation initiation (Kieft *et al.*, 2001).

2.5.2. Materials

$10 \times$ TRIS/HEPES (TH) buffer (660 mM TRIS base, 340 mM HEPES free acid), pH 7.4
1 M MgCl$_2$
0.5 M Na-EDTA, pH 8.0
Pure RNA, at least 2 μg each RNA
40% acrylamide/bisacrylamide 29:1 solution (other crosslinking ratios also can be tried)

2.5.3. Methods and procedures

1. Prepare 500 ml of 10% acrylamide TRIS/HEPES gel stock:

   125 ml 40% 29:1 acrylamide/bisacrylamide 29:1 solution
   50 ml 10$\times$ TH buffer
   1.25 ml 1 M MgCl$_2$ (final concentration of 2.5 mM)
       or 2 ml 0.5 M Na-EDTA, pH 8.0 (final concentration of 2 mM)
   RNase-free water to 500 ml

2. Prepare 10 ml of 2$\times$ sample loading buffer:

   2 ml of 10$\times$ TH buffer, bromophenol blue, and xylene cyanol at 0.5 mg/ml each
   1 ml glycerol
   50 μl 1 M MgCl$_2$ (final concentration of 5 mM)
       or 80 μl 0.5 M EDTA (final concentration of 4 mM)
   RNase-free water to 10 ml

3. Pour two 0.8 to 1.5-mm-thick gels (one with the $+$ MgCl$_2$ gel stock, the other with the +EDTA gel stock) with square wells of $\sim$1 to 1.5 cm width. Treat one side of the smaller plate with a hydrophobic agent (the automobile windshield treatment product Rain-X works well) to aid in removal of the gel after the electrophoresis is completed. In general, we use plates of $\sim$16.5 cm width $\times$ 25 cm height.

4. Add 1$\times$ TH buffer to the buffer chambers (diluted from 10$\times$ stock) with MgCl$_2$ or EDTA added to match the gel and prerun the gels for approximately 30 min at 5W. It is recommended that all native gels be run at 4° (in a cold room) to keep the gels cool.

5. Prepare RNA samples to load onto each gel: add 1 μg of RNA to RNase-free water to a final volume of 10 μl. Heat samples at 65° for 3 min, cool on the bench top for 5 min, then add 10 μl of 2$\times$ sample loading buffer (from step 3 above) for a final volume of 20 μl.

6. Use a syringe filled with buffer to "blow out" any material in the wells on the gel then load each sample by use of gel loading tips to ensure the RNA is in the well in a thin, uniform layer.

7. Run the gels at 5 to 10W in the cold room. The time for the run depends on RNA size and the degree to which the mutations cause retardation. Typically native gels run for 2 to 6 h. On completion of the run, remove the gel from the apparatus and by use of a thin metal spatula, remove the smaller plate from the larger one. The gel should stick to the larger, untreated plate. Apply plastic wrap to the exposed gel, then turn the gel upside down. The gel should stick to the plastic wrap and separate from the glass plate. Again use the metal spatula to assist in this process if needed. Place the plastic wrap (gel facing up) into a large staining dish and cover the gel in a solution of ethidium bromide stain (typically 100 μg/100 ml of water). Note that the gel will be in the proper orientation (i.e., not reversed) after this process. Allow the gel to soak in the stain for 15 min, then remove the gel and visualize on a UV transilluminator.

2.5.4. Notes and hints

1. The protocol above described running a 2.5 mM MgCl$_2$ gel (RNA folded) next to a 2 mM EDTA as a control (unfolded). Mutant RNAs sometimes run slightly differently in the EDTA gel because of changes in secondary structure before the ion-induced fold.

2. To examine the effects of cation concentration, gels are run with varying amounts of MgCl$_2$ (0.1 to 10 mM). Some RNAs will fold at lower salt concentrations, whereas others will require higher concentrations before definitive band shifts are seen.

3. Radiolabeled RNAs also can be used in place of the ethidium bromide staining method, and other stains can be substituted for ethidium bromide.

4. This protocol does not control the temperature of the gel precisely , but maintains the gel at or below room temperature. In-depth analysis of the temperature effects of certain mutations and hence their thermodynamic contribution to the RNA fold is accomplished with gel setups with precise temperature control.

2.6. Sedimentation velocity analytical ultracentrifugation

2.6.1. Overview

Sedimentation velocity analytical ultracentrifugation (AUC) is a biophysical method that directly measures hydrodynamic properties of a particle, in this case a structured macromolecule (for a detailed discussion in regard to the use of this method for protein studies, see Lebowitz *et al.*, 2002). Because the hydrodynamic properties of a folded macromolecule depend on size and shape, this technique reveals global characteristics about the RNA fold and how the overall molecular conformation changes as it folds. This technique has been used to determine the relative magnitude of size and shape change associated with the transition of an RNA domain from an unfolded state to a folded state and to track those structural changes as a function of ionic strength (for examples, see Costantino and Kieft, 2005; Deras *et al.*, 2000; Takamoto *et al.*, 2002).

A detailed description of the physics behind AUC is beyond the scope of this chapter. We limit our discussion to the useful application that provides a physical measurement of global structural changes, complementing probing and native gel electrophoresis studies. We base our use of the technique on the notion that an RNA changing from an extended, unfolded state to a compact, folded state will show a greater change in its hydrodynamic properties than will an RNA going from an extended unfolded state to a folded, but still extended, state. Furthermore, the folding event magnesium dependence is monitored by performing titration experiments, and the presence of multiple steps requiring different magnesium concentrations is detected. This assay also physically measures the degree to which a given mutation affects the overall global RNA conformation when the mutant RNA hydrodynamic properties are compared with the wild-type hydrodynamic properties. Finally, if the IRES RNA contains multiple folded domains, the contribution of each to the overall global RNA conformation is assessed by measuring the effect of mutants that specifically alter one domain.

In AUC, the purified IRES RNA in solution is spun at high speeds so that the molecules sediment to the outside of the sample cell. As they sediment, their distribution is monitored, usually by ultraviolet absorbance. Modern ultracentrifuges allow simultaneous analysis of multiple samples in one 5- to 6-h run. The data obtained for each sample consist of a series of boundary traces measured at different time points (Fig. 13.7). From this set of

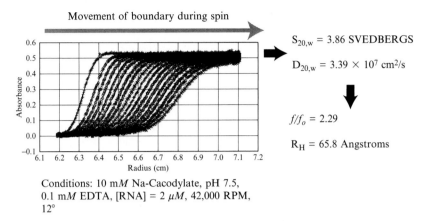

Movement of boundary during spin

$S_{20,w}$ = 3.86 SVEDBERGS

$D_{20,w}$ = 3.39 × 10⁷ cm²/s

f/f_o = 2.29

R_H = 65.8 Angstroms

Conditions: 10 mM Na-Cacodylate, pH 7.5,
0.1 mM EDTA, [RNA] = 2 μM, 42,000 RPM,
12°

Figure 13.7 Example of analytical ultracentrifugation data. The actual data that are gathered in analytical ultracentrifugation are a set of boundaries, collected as the RNA sediments in the sample cell. At the top left is an example of these data, where each trace is a boundary collected at a certain time point. These particular boundaries were collected at 5-min intervals by use of ultraviolet absorbance (at 260 nm wavelength). The boundaries then were fit by use of the program SVEDBERG to yield the experimental values of S and D, which then were converted to standard values. These values are used with the known molecular mass and the partial specific volume (0.53 cm³/g for RNA) to calculate f/fo and R_H, which are model independent values.

boundaries, the hydrodynamic properties sedimentation coefficient (S) and diffusion coefficient (D) are obtained, which depend on the RNA size and shape. If addition of magnesium to an IRES RNA increases the sedimentation coefficient and diffusion coefficient relative to the RNA without magnesium, this indicates an RNA structural change consistent with it becoming more compact. In loose terms, the RNA has become more streamlined and travels through the solvent with less resistance. By measuring these values as a function of [Mg²⁺], we gain insight into the global structural changes that accompany RNA folding.

2.6.2. Materials

100 mM Na–Cacodylate buffer, pH 7.5
100 mM MgCl₂
10 mM Na–EDTA
Pure RNA in RNase-free water (see following)

2.6.3. Methods and procedures

1. Prepare four samples according to the following. The volume needed is set by the cell volume.

SAMPLE	+MgCl₂ Sample	+MgCl₂ Buffer	−MgCl₂ Sample	−MgCl₂ Buffer
BUFFER	10 mM NaCac.	10 mM NaCac.	10 mM NaCac.	10 mM NaCac.
SALT	10 mM MgCl₂	10 mM MgCl₂	1 mM EDTA	1 mM EDTA
RNA	0.01 to 0.025 μg/μl	None	0.01 to 0.025 μg/μl	None

Ideally, the RNA concentration should be set to obtain an absorbance at 260 nm of ∼ 0.7 if UV absorbance detection is used. Lower absorbance results in lower signal to noise, whereas excessively high concentrations can result in aggregation and absorbance measurements out of the linear range. Conducting multiple experiments with various RNA concentrations is an important control (see following and Fig. 13.8).

2. By use of the manufacturer's directions, load and assemble the sample cell. One of the most widely used analytical ultracentrifuges is the Beckman XL-A, and this protocol is based on our experience with that instrument. Place the cells securely into the machine, install the detector, and run the samples at 3000 rpm to check for possible leaks within the cell. Leaks are detected by monitoring the absorbance, and if the overall absorbance in the cell decreases, a leak is likely. If one observes a leak, one must stop the system and reassemble the sample cells before continuing.

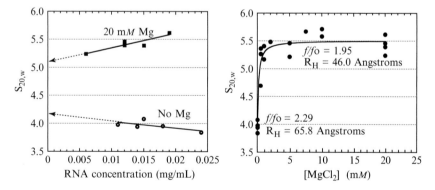

Figure 13.8 Use of AUC data to analyze folding of an RNA. On the left is a graph of the $S_{20,w}$ values measured at several different RNA concentrations, both in the presence and absence of MgCl₂. These data were fit to a linear equation, which is shown. The slight negative slope of the "No Mg" line indicates that the RNA molecules are repelling each other, whereas the upward slope of the "20 mM Mg" line indicates a deviation from ideality because of some RNA self-association. By extrapolating these lines to zero [RNA], one corrects for deviations from ideality. Clearly, addition of magnesium causes a substantial change in the measured sedimentation coefficient (S). On the right is a graph of the $S_{20,w}$ values as a function of [MgCl₂]. A sharp transition occurs as the RNA folds, and the f/f_o value decreases from 2.29 to 1.95. Thus, for this RNA, folding results in a particle that has a shape that is somewhat closer to a sphere than the unfolded state. Data are from Costantino and Kieft (2005).

3. Centrifuge the samples at 20,000 to 35,000 rpm and wait for 10 to 15 min. The smaller the molecule of interest, the faster the spin required. During this initial period, the boundary passes the meniscus. The collection should be set up in the controlling software so that the "buffer-only" absorbance (blank) is subtracted from the corresponding sample absorbance. Program the machine to take a scan of the cell data every 5 min and accumulate a minimum of 60 scans.

2.6.4. Analysis

The data from the AUC consist of a series of boundaries, each a table of absorbance versus position in the cell collected at a different time point (Fig. 13.7). The set of traces can be displayed and analyzed by various software packages. We have had great success with the SVEDBERG/ SEDNTRP package, which is available for free download after online registration (Philo, 1997). This program allows direct selection of the boundaries to be fit and real-time interactive curve fitting, as well as conversion of the experimentally determined S and D values to standard conditions. In addition, the package contains a database of commonly used buffers and salts to determine solvent viscosity and density and easy methods for calculating values such as f/f_o (the ratio of the observed friction coefficient to the coefficient expected from a spherical molecule) and R_H (Stokes radius). A detailed description of the use of this software is beyond the scope of this chapter, and the software is updated frequently, so the reader is referred to other resources for that information (http://www.jphilo. mailway.com/default.htm).

2.6.5. Notes and hints

1. For all sedimentation experiments, it is vital that the samples are at least 95% pure, because contaminating RNAs or proteins will result in boundaries consisting of different species traveling at different rates, which will complicate analysis.
2. Most buffer systems are compatible with AUC. In general, we do not use TRIS, because the pH varies with temperature to a greater degree than other buffers.
3. The speed, time between scans, and length of the run are all variables that require optimization depending on the RNA. The values suggested here are starting points, based on an RNA with MW of 45,000.
4. Straightforward analysis of AUC data requires that the macromolecules behave ideally; that is, there is no macromolecule association or aggregation. In reality, most samples deviate slightly from ideality, and the degree to which this occurs is important to measure. To detect deviation from ideality, measure the S value as a function of RNA concentration and plot the results. In ideal systems, there is no change in the hydrodynamic properties as a function of [RNA]. In reality, there is often slight dependency on the S value as the RNA concentration changes (Fig. 13.8), thus

the sample deviates somewhat from ideality (for examples, see Costantino and Kieft, 2005; Deras *et al.*, 2000).

2.7. Crystal engineering

2.7.1. Overview

The high-resolution RNA structure may be the ultimate goal of a structural investigation, because such a structure yields a large amount of information that explains biochemical and functional observations and leads to new hypotheses and new experimental directions. Crystallography of RNA traditionally has been a specialized technique that has not been readily available to most laboratories, but continued advances in methods and technology are making it accessible to more and more researchers. Because many resources are available to those that wish to understand and use crystallography, we will not present a comprehensive overview here. Rather, we will focus on how to overcome the initial difficulty of obtaining preliminary RNA crystals. In addition to the discussion here, we refer the readers to other excellent resources (Cate and Doudna, 2000; Golden and Kundrot, 2003; Golden *et al.*, 1996, 1997; Holbrook *et al.*, 2001; Kundrot, 1997).

When contemplating a crystallographic effort, the first important questions that must be answered are: What is the information that is sought from solving a structure and what is the crystallization target that will give that information? If one wishes to know the structure of the apical loops of an IRES RNA that make contact with the ribosome, a superior approach is to use NMR to solve the structures of those isolated stem loops (Collier *et al.*, 2002; Kim *et al.*, 2002; Klinck *et al.*, 2000; Lukavsky *et al.*, 2000, 2003). If one wants to see the detailed folded structure within the core of a compact domain of an RNA, a crystal structure of that IRES RNA isolated domain is warranted (Kieft *et al.*, 2002; Pfingsten *et al.*, 2006). Knowledge of the RNA architecture, particularly the presence of independently folded domains and how they interact with one another, is very important because trying to crystallize a multidomained RNA in which the various domains are flexible relative to one another is unlikely to be successful. Furthermore, trying to crystallize an RNA that is unstructured is also unlikely to succeed, and one must wonder if such a structure is useful. If the IRES RNA is known to interact with a protein factor, crystallizing the complex may be necessary, especially if the RNA is only structured when bound to the protein. Careful examination of the data obtained from chemical probing, biophysical analyses, and mutagenesis combined with biochemical and functional assays, must be used to decide (1) whether high-resolution RNA structural studies are useful, (2) what specific part or parts of the IRES should be crystallized, and (3) how to divide the IRES into isolated domain targets if it is a multidomained RNA.

Once a specific structural target is selected, successful crystallization depends on systematic screening of many different variants to find the small number of constructs that yield diffracting crystals. It is generally unfruitful to spend time and energy extensively surveying condition space with a few RNA constructs in the hope of finding a condition in which the RNA will crystallize. Rather, it is far more worthwhile to survey sequence or construct space by screening fewer conditions but far more RNA variants. In some cases, we have found it necessary to screen at least 40 RNAs before a diffracting crystal form is found. A general rule of thumb that we follow is that an RNA that tends to crystallize will tend to do so under many different conditions. Hence, to crystallize an RNA target one must design and construct a library of RNA sequences that model the target, purify those RNAs, and subject them to crystallization trials.

Generating a library of RNA sequences for crystallization trials must be based on knowledge of the overall RNA architecture, because one must be sure that mutations introduced into the RNA to enhance crystallization do not interfere with proper folding. Hence, the chemical and enzymatic probing and biophysical experiments outlined previously provide the foundation for crystallization efforts. For example, IRES RNAs tend to have stem loops that extend into solution to make contact with the translation machinery or transactivating proteins. These stem loops are excellent places in the sequence to mutate in an effort to find a crystallizing construct. Obviously, this means that the detailed structure of the apical loop of the stem loop is lost, but this is a small sacrifice if the structure of the rest of the IRES domain is solved. However, to use stem loops in this way one must be sure that they do, in fact, extend into solution and that they are not critical to forming the fold; this is information gleaned from biochemical and biophysical experiments.

2.7.2. Methods and procedures

The engineering techniques that have proven most useful in producing a library of RNA variants fall into five categories:

- Exploiting phylogenetic variation: The sequence of functional RNA often varies between organisms, virus strains, even different virus isolates. This natural variation is exploited to generate variability within the crystallization library. For certain RNA targets, sequences from thermophilic organisms are useful, because thermostability seems to correlate with crystallizability, although for the IRES RNA researcher, this is not currently an option.
- Introducing GAAA tetraloops: GAAA tetraloops (usually closed with a C-G base pair) are stable RNA tetraloops that have a strong tendency to form intermolecular contacts and hence induce crystallization (for examples, see Adams et al., 2004; Batey et al., 2000; Kieft et al., 2002; Montange and Batey, 2006; Pfingsten et al., 2006). They are used by substituting the tetraloop for wild-type apical stem loops that extend into solution and then varying the length of the corresponding stem (Fig. 13.9). IRES RNAs tend to have stem loops that extend into solution to make contact with the translation machinery, making these prime targets for inserting GAAA tetraloops.

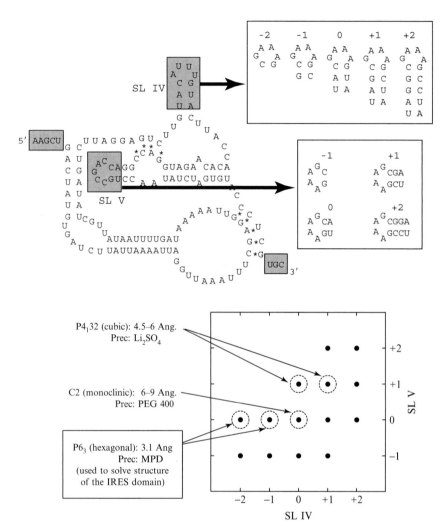

Figure 13.9 Example of crystallization construct engineering applied to the PSIV IGR IRES RNA. The biochemical and biophysical experiments outlined in this chapter led to an understanding of the architecture of this RNA that directed the strategy to engineer the RNA for crystallization. The boxed regions are the regions that were targeted, because they are not involved in forming the compact ion-induced fold but are extended into solution. Hence, whereas mutating these stem loops means that the structures of the apical loops are lost, one gains the structure of the rest of the RNA. Each of the two loops was mutated by replacing them with the GAAA tetraloops shown and by systematically altering the length of the helix on which the GAAA resides. By use of various combinations of these stem loops, a library of sequences was generated. Also, we varied the location of the RNA 5′ and 3′ ends to include more or fewer nucleotides. Combining these variations easily leads to dozens of RNAs to screen. In addition, different species of related viral RNAs were engineered in this way, and U1A-protein binding loops were introduced in place of the GAAA loops (not shown). At bottom is a simple grid

- Introducing U1A binding sites: In this method, an RNA sequence that specifically binds the U1A protein is substituted into a stem-loop structure or internal loop structure and then the RNA is cocrystallized with the protein. The rationale behind this idea is that the bound protein increases the potential for crystal contacts. It has been used successfully by several laboratories (Adams *et al.*, 2004; Ferre-D'Amare' and Doudna, 2000; Ferre'-D'Amare' *et al.*, 1998; Rupert and Ferre-D'Amare', 2001) but has the drawback of requiring that the protein be purified, as well as the additional complications associated with an RNA–protein complex rather than just RNA. As with GAAA tetraloops, these protein-binding loops are placed into the stem loops that extend into solution. In our hands, this has resulted in several crystal forms.
- Variation of helical length: This engineering method is often used in conjunction with the introduction of GAAA tetraloops or U1A-binding loops in that the length of the stem on which these loops reside is systematically altered (Fig. 13.9). This optimizes potential crystal contacts involving these loops and by simultaneously varying several loops, a large number of variants are generated. In addition, if the RNA 5′ and 3′ ends are base-paired within the same helix (often not the case with IRES RNAs), then the length of this helix should be varied.
- Variation of 5′ and 3′ ends: When the 5′ and 3′ ends are base-paired to each other, single-base overhangs can be tried to introduce additional variation in the library (Fig. 13.9). When they are not, the length of the 3′ and 5′ ends are varied. Most viral IRESs for which reliable secondary structure information is available do not have 5′ and 3′ ends that are base-paired, although individual domains may.

When combined, the preceding engineering methods readily give rise to libraries of dozens of RNA constructs, all containing the structural target of interest. Considering all of this, where does one begin? We have found that an initial library of one to two dozen RNAs in which a GAAA tetraloop is introduced in two locations (more if suitable stem loops exist) and the helical length is varied, provides a good start (Fig. 13.9). If any of these RNAs crystallize, one then focuses on constructs similar to the one(s) yielding crystals. It is important to realize that most crystals obtained will not diffract or will

---

representation of the various combinations of the lengths of the two stem loops within the PSIV IRES RNA. For simplicity, we have not included the constructs that could be made by altering the 5′ and 3′ ends or constructs on the basis of other related viruses. Circled constructs generated reproducible crystals; the quality of those crystals is indicated. The fact that these similar constructs yield very different crystals clearly illustrates the need to screen many variations of the RNA target. The addition or subtraction of a single base pair led to crystals of different symmetries, different crystallization requirements (precipitants), and different limits of diffraction. A crystal structure resulting from this is published in Pfingsten *et al.* (2006).

diffract poorly. However, often the difference between a construct that diffracts poorly and one that diffracts very well is a change of a single base pair within one RNA helix; hence, systematic variation and persistence are paramount (Fig 13.10). For example, our recent crystal structure of the ribosome-binding domain of a *Dicistroviridae* IRES required screening more than 50 different RNAs (by use of GAAA tetraloops, phylogenetic variation, and U1A binding sites) (Pfingsten *et al.*, 2006). At least five different crystal forms were obtained and discarded before a suitable crystal form was found.

2.8. A 96-well tray-based method to rapidly screen many RNAs for crystallization

2.8.1. Overview

Having generated a library of RNAs for potential crystallization, the next goal is to identify the RNA construct that is inherently crystallizable and to obtain diffracting crystals. The goal of these initial screens is to find promising initial "hits," or conditions and RNAs that yield preliminary crystals. Successful crystal screening can be accomplished by use of traditional hanging-drop and sitting-drop formats set up by hand (for an overview of methods, see Weber, 1997). However, one would like to screen as many RNAs as quickly as possible, by use of the least amount of reagent, and the traditional methods are fairly labor-intensive and are, therefore, not high-throughput. Many high-throughput methods for crystallization have become available in recent years, including robotic systems. In addition, microfluidics systems, such as those from Fluidigm, allow screening of hundreds of conditions with just a few microliters of macromolecule sample (Frederickson, 2002; Segelke, 2005). These systems, while becoming more widespread, are not available to most laboratories.

Here we describe an easy method for rapidly setting up crystallization screens in 96-well format by use of a small amount of crystallization solution and equipment readily available or purchased. This method was taught to our laboratory by Mair Churchill and her laboratory (Department of Pharmacology, University of Colorado School of Medicine). The procedure takes less than 5 to 10 min per tray and hence is more high throughput than traditional methods. The simplicity and speed of this technique has won over several laboratories on our campus.

We do not describe methods of RNA purification here. Historically, the most widely used method of purifying RNAs for crystallography has been preparative denaturing gel electrophoresis, although other techniques have been described (Anderson *et al.*, 1996; Lukavsky and Puglisi, 2004; Shields *et al.*, 1999). Any method that yields pure, properly folded RNA is acceptable, although obviously higher-throughput methods are preferable, given the number of RNAs that must be prepared. New affinity-based methods offer this possibility (Cheong *et al.*, 2004; Kieft and Batey, 2004).

Although the first crystals obtained out of a screen can be suitable for structure determination, in general these preliminary crystals are small and poorly formed, and subsequent optimization of growth conditions is needed. The details of optimization procedures are not included here. Rather, we limit our discussion to a method to screen rapidly through the library of RNAs to find those first promising crystals.

2.8.2. Materials

These are specific models and supplies that can be used with this method as can equivalent versions.

Pipetter tips: Rainin GPS-L250S and GPS-L10S,
    Gilson DT-125S
Pipetter: Rainin L8-200 and L8-10,
    Gilson D-MAN
Solution blocks: Greiner 780285 (2.2 ml for well solution) and 381080 (covers)
96-well tray: Axygen CP-AXYGEM-96-10
Tape: 4″ CrystalClear tape—Hampton Research HR4-508
100 mM KOH-HEPES, pH 7.5
25 mM MgCl$_2$
Pure RNA at a concentration >5 mg/ml
10 mM spermidine or spermine
RNase-free water

Crystallization trial solutions, commercially available Natrix (Scott *et al.*, 1995), and Crystal Screen #1 (Jancarik and Kim, 1991) (Hampton Research) are good initial screens. In addition, the Nucleic Acid Miniscreen (also available from Hampton Research) (Berger *et al.*, 1996) and other published screens (Doudna *et al.*, 1993; Kundrot, 1997) are also useful.

2.8.3. Methods and procedures

1. Prepare the solution block by transferring up to 2.0 ml of each crystalliza-tion solution into a different well of the solution block. If the block is not going to be used immediately, cover with the block cover and store at 4°.
2. Prepare the 100-μl RNA sample by combining:

    10 μl of 100 mM KOH-HEPES, pH 7.5
    500 μg pure RNA (final concentration will be 5 mg/ml)
    RNase-free water to a volume of 85 μl

    Heat the sample to 65° for 1 min, then place on the bench to cool at room temperature (∼ 5 min). Then add:

    10 μl of 25 mM MgCl$_2$
    5 μl of 10 mM spermidine or spermine

Mix well, then centrifuge at 13,000g for 2 min to pellet any particulate matter. Note that one can also add the $MgCl_2$ before heat cooling as long as this does not induce degradation or RNA aggregation.

3. With an eight-channel pipetter, transfer 100 μl of each well solution from the solution block to the 96-well tray, placing the solution in the lower, well solution chamber. Work quickly to avoid undue evaporation.

4. Load the repeating pipetter with the RNA solution and then deposit 1 μl into the sample well of each of the 96 wells of the crystallization tray. Again, work quickly to minimize evaporation.

5. With the 10-μl capacity eight-channel pipetter, transfer 1 μl of lower well solution from each well into the each sample, combining the 1 μl of well solution with the 1 μl of RNA solution. Change tips after each transfer.

6. When all wells are completed, seal the tray with clear sealing tape.

7. Incubate the tray at the desired temperature. We have had most success at temperatures of 20 to 30°. With a stereomicroscope, check the trays regularly (every few days) for crystal growth.

8. The solution block holds a little more than 2.0 ml of each well solution; therefore, 20 trays are set up before the block must be refilled. Between uses, seal the block and store at 4°.

2.8.4. Notes and hints

1. The initial conditions at which the RNA is screened can be varied from the conditions shown here. We use HEPES buffer, 2.5 to 10 mM $MgCl_2$, and 0.5 mM polyamine as a starting point for our RNA screens, but other conditions are also valid. Adding monovalent cations, higher magnesium concentrations, more or different polyamines, and different buffers of different pH are variable parameters. In general, we find that this condition is sufficient to allow us to find initial "hits" and RNAs that tend to crystallize, at which time we expand condition space.

2. When annealing the RNA (step 2), the $MgCl_2$ can be added before or after the heat–cool step. With some RNAs, adding the $MgCl_2$ before annealing causes aggregation or dimerization. The best conditions to anneal each RNA are determined empirically, but we have found the method described previously to be suitable for most RNAs.

3. One problem associated with this technique is failure to adequately seal the tray with tape. Firmly rubbing the edges of the tray/tape seal with one's fingernail or another object is necessary to provide a seal.

4. It is challenging to extract crystals from these trays for cryocooling or other analysis. A technique is to cut the tape around the specific well of interest, and after removing that section of tape, extract the crystal. We recommend scaling up to larger sitting drop or hanging drop 24-well trays when a crystal form of interest is found and additional crystals are needed.

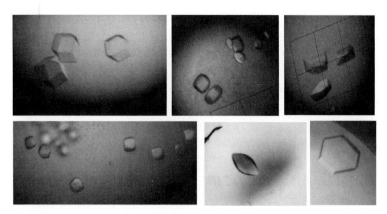

Figure 13.10 Examples of IRES RNA domain crystals obtained by use of the engineering techniques described in the text. (See color insert.)

5. Some 96-well trays allow 3 drops per well, allowing one to screen three different RNAs or three different RNA concentrations in a single 96-well tray. We have not tested these trays extensively, but they should allow even faster screening of multiple RNAs in parallel.

3. CONCLUSIONS

In undertaking structural studies of RNA, we find it useful to think in terms of a roadmap, in which complementary techniques contribute to an overall understanding of the RNA structure and its relationship to function. Hence, structural information does not come solely from solving structures, but through an integrated approach that uses solution probing experiments, biophysical experiments, functional methods, and modeling techniques. In this chapter, we have presented protocols and discussions for methods that we have found particularly useful and that are available to virtually any researcher in any laboratory. This list is far from complete but forms a foundation on which other methods build. For example, methods such as directed hydroxyl radical probing and crosslinking (Harris et al., 1994; Joseph and Noller, 2000), fluorescence resonance energy transfer (FRET), including single molecule studies (Liu and Lu, 2006; Zhuang, 2005), small-angle X-ray scattering (SAXS) (Kieft et al., 1999; Russell et al., 2000), and other methods, have the potential to add greatly to our understanding of IRES RNA structure as they have with other RNAs. As our structural understanding of IRES RNAs grows, so will our ability to propose models for IRES function and to understand IRESs within the context of the world of RNA structure.

REFERENCES

Adams, P. L., Stahley, M. R., Kosek, A. B., Wang, J., and Strobel, S. A. (2004). Crystal structure of a self-splicing group I intron with both exons. *Nature* **430,** 45–50.

Adilakshmi, T., Lease, R. A., and Woodson, S. A. (2006). Hydroxyl radical footprinting *in vivo*: Mapping macromolecular structures with synchrotron radiation. *Nucleic Acids Res.* **34,** e64.

Anderson, A. C., Scaringe, S. A., Earp, B. E., and Frederick, C. A. (1996). HPLC purification of RNA for crystallography and NMR. *RNA* **2,** 110–117.

Bassi, G. S., Mollegaard, N. E., Murchie, A. I., von Kitzing, E., and Lilley, D. M. (1995). Ionic interactions and the global conformations of the hammerhead ribozyme. *Nat. Struct. Biol.* **2,** 45–55.

Batey, R. T., Rambo, R. P., Lucast, L., Rha, B., and Doudna, J. A. (2000). Crystal structure of the ribonucleoprotein core of the signal recognition particle. *Science* **287,** 1232–1239.

Berger, I., Kang, C. H., Sinha, N., Wolters, M., and Rich, A. (1996). A highly efficient 24-condition matrix for the crystallization of nucleic acid fragments. *Acta Crystallogr. D Biol. Crystallogr.* **D52,** 465–468.

Bonnal, S., Boutonnet, C., Prado-Lourenco, L., and Vagner, S. (2003). IRESdb: The Internal Ribosome Entry Site database. *Nucleic Acids Res.* **31,** 427–428.

Brenowitz, M., Chance, M. R., Dhavan, G., and Takamoto, K. (2002). Probing the structural dynamics of nucleic acids by quantitative time-resolved and equilibrium hydroxyl radical "footprinting." *Curr. Opin. Struct. Biol.* **12,** 648–653.

Cate, J. H., and Doudna, J. A. (2000). Solving large RNA structures by X-ray crystallography. *Methods Enzymol.* **317,** 169–180.

Cate, J. H., Gooding, A. R., Podell, E., Zhou, K., Golden, B. L., Kundrot, C. E., Cech, T. R., and Doudna, J. A. (1996). Crystal structure of a group I ribozyme domain: Principles of RNA packing. *Science* **273,** 1678–1685.

Celander, D. W., and Cech, T. R. (1991). Visualizing the higher order folding of a catalytic RNA molecule. *Science* **251,** 401–407.

Cheong, H. K., Hwang, E., Lee, C., Choi, B. S., and Cheong, C. (2004). Rapid preparation of RNA samples for NMR spectroscopy and X-ray crystallography. *Nucleic Acids Res.* **32,** e84.

Collier, A. J., Gallego, J., Klinck, R., Cole, P. T., Harris, S. J., Harrison, G. P., Aboul-Ela, F., Varani, G., and Walker, S. (2002). A conserved RNA structure within the HCV IRES eIF3-binding site. *Nat. Struct. Biol.* **9,** 375–380.

Costantino, D., and Kieft, J. S. (2005). A preformed compact ribosome-binding domain in the cricket paralysis-like virus IRES RNAs. *RNA* **11,** 332–343.

Das, R., Laederach, A., Pearlman, S. M., Herschlag, D., and Altman, R. B. (2005). SAFA: semi-automated footprinting analysis software for high-throughput quantification of nucleic acid footprinting experiments. *RNA* **11,** 344–354.

Deras, M. L., Brenowitz, M., Ralston, C. Y., Chance, M. R., and Woodson, S. A. (2000). Folding mechanism of the tetrahymena ribozyme P4-P6 domain. *Biochemistry* **39,** 10975–10985.

Doudna, J. A., Grosshans, C., Gooding, A., and Kundrot, C. E. (1993). Crystallization of ribozymes and small RNA motifs by a sparse matrix approach. *Proc. Natl. Acad. Sci. USA* **90,** 7829–7833.

Duckett, D. R., Murchie, A. I. H., and Lilley, D. M. J. (1995). The global folding of four-way helical junctions in RNA, including that in U1 snRNA. *Cell* **83,** 1027–1036.

Ehresmann, C., Baudin, F., Mougel, M., Romby, P., Ebel, J. P., and Ehresmann, B. (1987). Probing the structure of RNAs in solution. *Nucleic Acids Res.* **15,** 9109–9128.

Ferre'- D'Amare', A. R., and Doudna, J. A. (1996). Use of cis- and trans-ribozymes to remove 5′ and 3′ heterogeneities from milligrams of *in vitro* transcribed RNA. *Nucleic Acids Res.* **24,** 977–978.

Ferre'-D'Amare', A. R., and Doudna, J. A. (2000). Crystallization and structure determination of a hepatitis delta virus ribozyme: Use of the RNA-binding protein U1A as a crystallization module. *J. Mol. Biol.* **295**, 541–556.

Ferre'-D'Amare', A. R., Zhou, K., and Doudna, J. A. (1998). Crystal structure of a hepatitis delta virus ribozyme. *Nature* **395**, 567–574.

Frederickson, R. M. (2002). Fluidigm. Biochips get indoor plumbing. *Chem. Biol.* **9**, 1161–1162.

Golden, B. L., Gooding, A. R., Podell, E. R., and Cech, T. R. (1996). X-ray crystallography of large RNAs: Heavy-atom derivatives by RNA engineering. *RNA* **2**, 1295–1305.

Golden, B. L., and Kundrot, C. E. (2003). RNA crystallization. *J. Struct. Biol.* **142**, 98–107.

Golden, B. L., Podell, E. R., Gooding, A. R., and Cech, T. R. (1997). Crystals by design: A strategy for crystallization of a ribozyme derived from the tetrahymena group I intron. *J. Mol. Biol.* **270**, 711–723.

Harris, M. E., Nolan, J. M., Malhotra, A., Brown, J. W., Harvey, S. C., and Pace, N. R. (1994). Use of photoaffinity crosslinking and molecular modeling to analyze the global architecture of ribonuclease P RNA. *EMBO J.* **13**, 3953–3963.

Hellen, C. U., and Sarnow, P. (2001). Internal ribosome entry sites in eukaryotic mRNA molecules. *Genes Dev.* **15**, 1593–1612.

Holbrook, S. R., Holbrook, E. L., and Walukiewicz, H. E. (2001). Crystallization of RNA. *Cell. Mol. Life Sci.* **58**, 234–243.

James, B. D., Olsen, G. J., and Pace, N. R. (1989). Phylogenetic comparative analysis of RNA secondary structure. *Methods Enzymol.* **180**, 227–239.

Jan, E., and Sarnow, P. (2002). Factorless ribosome assembly on the internal ribosome entry site of cricket paralysis virus. *J. Mol. Biol.* **324**, 889–902.

Jancarik, J., and Kim, S. H. (1991). Sparse matrix sampling: A screening method for crystallization of proteins. *J. Appl. Cryst.* **24**, 409–411.

Joseph, S., and Noller, H. F. (2000). Directed hydroxyl radical probing using iron(II) tethered to RNA. *Methods Enzymol.* **318**, 175–190.

Kieft, J. S., and Batey, R. T. (2004). A general method for rapid and nondenaturing purification of RNAs. *RNA* **10**, 988–995.

Kieft, J. S., Zhou, K., Grech, A., Jubin, R., and Doudna, J. A. (2002). Crystal structure of an RNA tertiary domain essential to HCV IRES-mediated translation initiation. *Nat. Struct. Biol.* **9**, 370–374.

Kieft, J. S., Zhou, K., Jubin, R., and Doudna, J. A. (2001). Mechanism of ribosome recruitment by hepatitis C IRES RNA. *RNA* **7**, 194–206.

Kieft, J. S., Zhou, K., Jubin, R., Murray, M. G., Lau, J. Y., and Doudna, J. A. (1999). The hepatitis C virus internal ribosome entry site adopts an ion-dependent tertiary fold. *J. Mol. Biol.* **292**, 513–529.

Kim, I., Lukavsky, P. J., and Puglisi, J. D. (2002). NMR study of 100 kDa HCV IRES RNA using segmental isotope labeling. *J. Am. Chem. Soc.* **124**, 9338–9339.

Klinck, R., Westhof, E., Walker, S., Afshar, M., Collier, A., and Aboul-Ela, F. (2000). A potential RNA drug target in the hepatitis C virus internal ribosomal entry site. *RNA* **6**, 1423–1431.

Kundrot, C. E. (1997). Preparation and crystallization of RNA: A sparse matrix approach. *Methods Enzymol.* **276**, 143–157.

Lafontaine, D. A., Norman, D. G., and Lilley, D. M. (2002). The global structure of the VS ribozyme. *EMBO J.* **21**, 2461–2471.

Latham, J. A., and Cech, T. R. (1989). Defining the inside and outside of a catalytic RNA molecule. *Science* **245**, 276–282.

Lebowitz, J., Lewis, M. S., and Schuck, P. (2002). Modern analytical ultracentrifugation in protein science: A tutorial review. *Protein Sci.* **11**, 2067–2079.

Liu, J., and Lu, Y. (2006). Multi-fluorophore fluorescence resonance energy transfer for probing nucleic acids structure and folding. *Methods Mol. Biol.* **335,** 257–271.

Lukavsky, P. J., Kim, I., Otto, G. A., and Puglisi, J. D. (2003). Structure of HCV IRES domain II determined by NMR. *Nat. Struct. Biol.* **10,** 1033–1038.

Lukavsky, P. J., Otto, G. A., Lancaster, A. M., Sarnow, P., and Puglisi, J. D. (2000). Structures of two RNA domains essential for hepatitis C virus internal ribosome entry site function. *Nat. Struct. Biol.* **7,** 1105–1110.

Lukavsky, P. J., and Puglisi, J. D. (2004). Large-scale preparation and purification of polyacrylamide-free RNA oligonucleotides. *RNA* **10,** 889–893.

Mathews, D. H., and Turner, D. H. (2006). Prediction of RNA secondary structure by free energy minimization. *Curr. Opin. Struct. Biol.* **16,** 270–278.

Merino, E. J., Wilkinson, K. A., Coughlan, J. L., and Weeks, K. M. (2005). RNA structure analysis at single nucleotide resolution by selective 2′-hydroxyl acylation and primer extension (SHAPE). *J. Am. Chem. Soc.* **127,** 4223–4231.

Montange, R. K., and Batey, R. T. (2006). Structure of the S-adenosylmethionine riboswitch regulatory mRNA element. *Nature* **441,** 1172–1175.

Nishiyama, T., Yamamoto, H., Shibuya, N., Hatakeyama, Y., Hachimori, A., Uchiumi, T., and Nakashima, N. (2003). Structural elements in the internal ribosome entry site of *Plautia stali* intestine virus responsible for binding with ribosomes. *Nucleic Acids Res.* **31,** 2434–2442.

Pace, N. R., Smith, D. K., Olsen, G. J., and James, B. D. (1989). Phylogenetic comparative analysis and the secondary structure of ribonuclease P RNA—a review. *Gene* **82,** 65–75.

Pan, T. (1995). Higher order folding and domain analysis of the ribozyme from *Bacillus subtilis* ribonuclease P. *Biochemistry* **34,** 902–909.

Pfingsten, J. S., Costantino, D. A., and Kieft, J. S. (2006). Structural basis for cap-independent ribosome recruitment and manipulation by a viral IRES. *Science* **314,** 1450–1454.

Philo, J. S. (1997). An improved function for fitting sedimentation velocity data for low-molecular-weight solutes. *Biophys. J.* **72,** 435–444.

Ralston, C. Y., Sclavi, B., Sullivan, M., Deras, M. L., Woodson, S. A., Chance, M. R., and Brenowitz, M. (2000). Time-resolved synchrotron X-ray footprinting and its application to RNA folding. *Methods Enzymol.* **317,** 353–368.

Rupert, P. B., and Ferre'-D'Amare, A. R. (2001). Crystal structure of a hairpin ribozyme-inhibitor complex with implications for catalysis. *Nature* **410,** 780–786.

Russell, R., Millett, I. S., Doniach, S., and Herschlag, D. (2000). Small angle X-ray scattering reveals a compact intermediate in RNA folding. *Nat. Struct. Biol.* **7,** 367–370.

Scott, W. G., Finch, J. T., Grenfell, R., Fogg, J., Smith, T., Gait, M. J., and Klug, A. (1995). Rapid crystallization of chemically synthesized hammerhead RNAs using a double screening procedure. *J. Mol. Biol.* **250,** 327–332.

Segelke, B. (2005). Macromolecular crystallization with microfluidic free-interface diffusion. *Expert Rev. Proteomics* **2,** 165–172.

Shcherbakova, I., Mitra, S., Beer, R. H., and Brenowitz, M. (2006). Fast Fenton footprinting: A laboratory-based method for the time-resolved analysis of DNA, RNA and proteins. *Nucleic Acids Res.* **34,** e48.

Shields, T. P., Mollova, E., Ste Marie, L., Hansen, M. R., and Pardi, A. (1999). High-performance liquid chromatography purification of homogenous-length RNA produced by trans cleavage with a hammerhead ribozyme. *RNA* **5,** 1259–1267.

Spahn, C. M., Jan, E., Mulder, A., Grassucci, R. A., Sarnow, P., and Frank, J. (2004). Cryo-EM visualization of a viral internal ribosome entry site bound to human ribosomes; The IRES functions as an RNA-based translation factor. *Cell* **118,** 465–475.

Spahn, C. M., Kieft, J. S., Grassucci, R. A., Penczek, P. A., Zhou, K., Doudna, J. A., and Frank, J. (2001). Hepatitis C virus IRES RNA-induced changes in the conformation of the 40s ribosomal subunit. *Science* **291,** 1959–1962.

Takamoto, K., He, Q., Morris, S., Chance, M. R., and Brenowitz, M. (2002). Monovalent cations mediate formation of native tertiary structure of the tetrahymena thermophila ribozyme. *Nat. Struct. Biol.* **9,** 928–933.

Tullius, T. D., and Greenbaum, J. A. (2005). Mapping nucleic acid structure by hydroxyl radical cleavage. *Curr. Opin. Chem. Biol.* **9,** 127–134.

Weber, P. C. (1997). Overview of protein crystallization methods. *Methods Enzymol.* **276,** 13–22.

Zhuang, X. (2005). Single-molecule RNA science. *Annu. Rev. Biophys. Biomol. Struct.* **34,** 399–414.

CHAPTER FOURTEEN

Biophysical and Biochemical Investigations of dsRNA-Activated Kinase PKR

Sean A. McKenna,* Darrin A. Lindhout,* Takashi Shimoike,*,† and Joseph D. Puglisi*,‡

Contents

Abstract

Protein kinase RNA-activated (PKR) is a serine/threonine kinase that contains an N-terminal RNA-binding domain (dsRNA) and a C-terminal kinase domain. On binding viral dsRNA molecules, PKR can become activated and phosphorylate cellular targets, such as eukaryotic translation initiation factor 2α (eIF-2α). Phosphorylation of eIF-2α results in attenuation of protein translation initiation. Therefore, PKR plays an integral role in the antiviral response to cellular infection. Here we provide a methodological framework for probing PKR function by use of assays for phosphorylation, RNA–protein stability, PKR dimerization, and *in vitro* translation. These methods are complemented by nuclear magnetic resonance approaches for probing structural features of PKR activation. Considerations required for both PKR and dsRNA sample preparation are also discussed.

* Department of Structural Biology, Stanford University School of Medicine, Stanford, California
† Department of Virology II, National Institute of Infectious Diseases, Musashi-Murayama, Tokyo, Japan
‡ Stanford Magnetic Resonance Laboratory, Stanford University School of Medicine, Stanford, California

Methods in Enzymology, Volume 430
ISSN 0076-6879, DOI: 10.1016/S0076-6879(07)30014-1

1. INTRODUCTION

A hallmark of viral replication is the dependence on the host–cell protein translation machinery for the production of viral proteins (Gale *et al.*, 2000; Schneider and Mohr, 2003). The ability to suppress translation at the initiation stage represents a crucial host–cell defense against viral attack in eukaryotes. Phosphorylation of the α-subunit of eukaryotic initiation factor 2 (eIF2α) at Ser51 (Chong *et al.*, 1992) inhibits the guanine nucleotide exchange activity of eIF2B (Sonenberg and Dever, 2003) (Fig. 14.1A). The pool of active eIF2 ternary complex is thus reduced, causing a global decrease of both viral and cellular protein synthesis. Phosphorylation of eIF2α is accomplished by an RNA-activated Ser/Thr protein kinase (PKR); PKR is activated through autophosphorylation on association with double-stranded RNA (dsRNA) generated during the replicative cycle of both RNA- and DNA-based viral genomes (Schneider and Mohr, 2003). That PKR is one of many interferon-induced proteins synthesized in response to viral infiltration only highlights its central role in the antiviral response (Malmgaard, 2004). The detailed understanding of the mechanism of activation of PKR by viral dsRNA in eukaryotic cells represents an area of intense interest and will be the focus of this chapter.

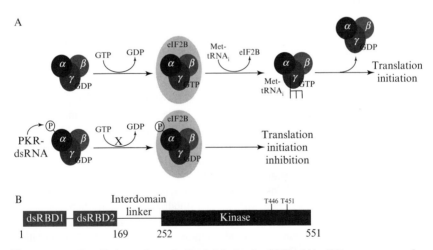

Figure 14.1 Regulation of translation initiation by PKR. (A) eIF2 ternary complex (α, β, γ) is responsible for delivery of initiator Met-tRNA to the translation initiation complex in a GTP-dependent fashion. Activated PKR phosphorylates eukaryotic initiation factor 2α (eIF2α) at Ser51; phosphorylation at this site inhibits the activity of a guanine nucleotide exchange factor, eIF2B. The pool of active eIF2 ternary complex decreases, causing attenuation of translation initiation. (B) Domain organization of PKR. N-terminal dsRBDs, C-terminal kinase domain, and the interdomain linker are shown. Critical autophosphorylation sites (T446, T451) in the kinase domain are indicated.

PKR is a 551-residue enzyme consisting of a N-terminal RNA-binding (dsRBD1/2, 19 kDa) and C-terminal kinase domain (PKR^{KD}, 38 kDa) connected by an 80-residue unstructured interdomain linker (Clemens, 1997; McKenna, 2007) (Fig. 14.1B). The modular domain architecture of PKR has allowed for the high-resolution structure determination and functional characterization of each of these domains individually. The N-terminal domain contains tandem 70-residue dsRNA binding motifs (dsRBDs) that act as an intracellular sensor for dsRNA (Clemens, 1997). PKR is highly specific for dsRNA over other nucleic acid substrates (Gale and Katze, 1998); length and RNA structural features are thought to modulate the interaction (Bevilacqua and Cech, 1996; Kim et al., 2006; McKenna et al., 2006). Each dsRBD adopts a canonical $\alpha-\beta-\beta-\beta-\alpha$ fold in which the α-helices pack against an antiparallel β-sheet (Nanduri et al., 1998a,b). High-affinity PKR–nucleic acid interactions require both dsRBDs; individually expressed dsRBDs do not interact with appreciable affinity (Kim et al., 2006; McCormack et al., 1994; McKenna et al., 2006). Conversely, the C-terminal Ser/Thr kinase domain is responsible for both phosphorylation activity and substrate recognition. The crystal structure of the isolated kinase domain in complex with eIF2α revealed insights into the ATP coordination site, substrate recognition site, and autophosphorylation sites required for activation (Thr446 and T451) (Dar et al., 2005; Dey et al., 2005).

Although the modular nature of PKR has allowed for the detailed biophysical characterization of individual domains, the central mechanistic quandary of how dsRNA leads to kinase domain autophosphorylation remains poorly understood in the context of the full-length protein. Although some groups have proposed an autoinhibitory model in which the latent form of PKR remains inactive through intramolecular association between the N- and C-terminal domains (Gelev et al., 2006; Li et al., 2006; Nanduri et al., 2000; Vattem et al., 2001; Wu and Kaufman, 1997), both ourselves (McKenna, 2007) and others (Lemaire et al., 2006) have observed that in the context of the full-length enzyme, no such interactions are observed. What is well understood is that latent PKR undergoes autophosphorylation on dsRNA binding by means of a bimolecular reaction mechanism in which dimerization between PKR molecules seems important (McKenna, 2007; Romano et al., 1998; Taylor et al., 1996). A potential connection between dsRNA sensing and kinase activity has been observed, because it seems that both dsRNA binding and PKR phosphorylation result in increased self-affinity between PKR monomers.

Here, we present our methods to study PKR from a biophysical perspective to answer central mechanistic questions. The expression and purification of PKR, PKR mutants, and its individual domains will be discussed, as will in vitro transcription of viral RNAs. Various structural and biochemical assays, including kinase phosphorylation kinetics, protein–RNA affinity determinations, dimerization status, low- and high-resolution structural approaches, and *in vitro* translational assays will be discussed. These approaches should prove generally useful for those looking to study dynamic kinases such as PKR or in the study of RNA-binding proteins and their interplay with translation.

 ## 2. Expression and Purification of PKR

Central to the biophysical characterization of any biomolecule is the production of a homogeneous and high-yield sample. The expression and purification of recombinant human PKR in bacteria is particularly challenging because of low yield, heterogeneous phosphorylation state, and non-specific cellular RNA binding. We present here our current protocol for the purification of PKR and various PKR derivatives (Fig. 14.2A).

Full-length human PKR was subcloned into pET-29b(+) vector with NdeI (5′ end) and KpnI (3′ end) such that it could be expressed and purified as C-terminal 6× His fusion protein under control of a T7 promoter. Both this vector and a pET-43.1-based vector containing the sequence for λ-phosphatase were cotransformed into the *Escherichia coli (E. coli)* strain BL21(DE_3)-RIG (Stratagene) for expression. Cultures are grown at 37° to $A_{590} = 0.6$ to 0.8 in LB media supplemented with appropriate antibiotics, and then rapidly cooled on ice for 15 min. After cooling, expression of both PKR and λ-phosphatase is induced with isopropyl β-D-thiogalactopyronoside (IPTG, 1 mM) for 18 h at 20°. We have observed that the combination of cooling and induction at 20° can significantly improve yield by reducing the incorporation of PKR into inclusion bodies.

All subsequent steps were performed at 4° to minimize proteolysis. Cells are harvested by centrifugation at 4000 rpm (Beckman JLA 8.1000 rotor) and resuspended in 10 ml/L culture of disruption buffer (50 mM TRIS/Cl, pH 8.0, 1 M NaCl, 5% glycerol, 5 mM β-mercaptoethanol) and lysed by gentle sonication. After centrifugation at 17,000 rpm for 30 min (Beckman JA 25.50 rotor), the supernatant is applied to a 2.5 ml/L culture Ni-NTA column (Qiagen) preequilibrated with disruption buffer. Bound PKR is washed with 100 column volumes of His-A buffer (50 mM TRIS/Cl, pH 8.0, 1 M NaCl, 5 mM β-mercaptoethanol, 1 mM imidazole), an essential step in the removal nonspecifically bound cellular RNA. We have estimated without the high salt wash, nearly 80% of PKR molecules are contaminated with nucleic acids, whereas with the high-salt wash only 5 to 10% remain bound. The bound fusion protein is then washed with 10 column volumes of His-B buffer (50 mM TRIS/Cl, pH 8.0, 300 mM NaCl, 5% glycerol, 5 mM β-mercaptoethanol, 10 mM imidazole) before elution in 25 ml of His-elution buffer (50 mM TRIS/Cl, pH 8.0, 300 mM NaCl, 100 mM imidazole, 5 mM β-mercaptoethanol). The eluate is then loaded onto a

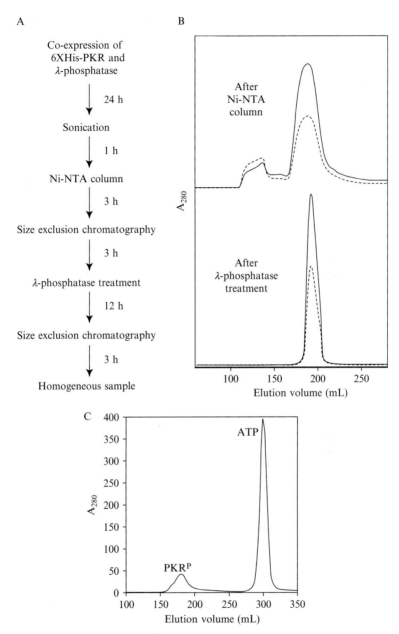

Figure 14.2 Expression and purification of human PKR. (A) Schematic outline of PKR sample preparation. Estimated time for each step is shown. (B) Purification of human PKR by size exclusion chromatography. Elution profile obtained from the size–exclusion chromatography step on the Superdex 200 (26/60) column after either Ni-NTA column (top) or λ-phosphatase treatment (bottom) are shown. Traces corresponding to both 280 nm (solid) and 260 nm (dashed) are shown. (C) Superdex 200 26/60 elution profile of purified PKR^P (15 μM) activated in the absence of dsRNA. PKR^P and ATP peaks are indicated.

Superdex 200 26/60 size exclusion chromatography column (GE Healthcare) with a 50-ml Superloop (GE Healthcare) at 2 ml/min (50 mM TRIS/Cl, pH 8.0, 300 mM NaCl, 5 mM β-mercaptoethanol), and eluted in 5-ml fractions at a flow rate of 2 ml/min. The chromatographic profile typically reveals four distinct peaks that correspond to RNA-bound PKR ($\sim$110 ml), PKR ($\sim$190 ml), and small molecular weight compounds ($\sim$280 ml) (Fig. 14.2B, top). This stage is critical, because RNA-contaminated PKR is completely removed from our sample. The monomeric PKR peak is pooled.

The final purification step involves the removal of phosphorylation heterogeneity from our sample. Although PKR coexpression with λ-phosphatase dephosphorylates most of our PKR population, we have observed that without complete dephosphorylation, constitutive kinase activation in the absence of dsRNA activator can occur. Therefore, the PKR pool is subsequently treated with λPPase (25 kDa, New England Biolabs) overnight at 4° in the presence of 2 mM MnCl$_2$ to ensure complete dephosphorylation. We have confirmed the dephosphorylation state by both activation assays and NMR analysis (McKenna, 2007). The sample is then treated with an excess of ethylenediaminetetraacetic acid (EDTA) to chelate divalent cations and reapplied onto a Superdex 200 26/60 size exclusion chromatography column by means of a 50-ml Superloop at 2 ml/min (50 mM TRIS/Cl, pH 7.5, 100 mM NaCl, 5 mM β-mercaptoethanol), and eluted in 5-ml fractions at a flow rate of 2 ml/min (Fig. 14.2B, bottom). Samples are then concentrated (Vivaspin 20, 10K MWCO, PES membrane) as desired. The 6$\times$ His fusion can be removed, if desired, by incubation overnight at 4° with Thrombin (Novagen), although the tagged and untagged versions appear functionally equivalent in all assays examined. Protein purity and yield were assessed by SDS-PAGE analysis (4 to 20% TRIS HCl gels), Bradford assay (Bio-Rad) and/or A$_{280}$ values by use of a calculated molar extinction coefficient. Typical yields are approximately 5 to 10 mg per initial liter of liquid culture. Full-length PKR is typically stored at 4° at low concentrations ($<$2 μM) to avoid potential RNA-independent activation and is stable if periodically supplemented with β-mercaptoethanol. Freeze/thaw cycles of PKR significantly affect the activity of the enzyme.

PKR offers the advantage that each of its domains can be expressed and purified independently, although activity cannot be restored by adding the domains *in trans*, even in the presence of the interdomain linker (McKenna, 2007). The C-terminal kinase domain (PKR^{KD}, residues 252 to 551) is expressed and purified in an identical manner to the full-length protein, although the extensive high-salt wash can be reduced significantly, because this construct does not associate with bacterial RNA. The N-terminal dsRBD1/2 (residues 1 to 169), dsRBD1 (residues 1 to 99), and dsRBD2 (residues 106 to 169) constructs are expressed and purified in an identical manner to the full-length protein, with the following exceptions: (1) neither rapid cooling before induction nor growth at 20° is necessary; (2) after induction, the dsRBD constructs are grown at 37° for 3 h before harvesting; and (3) a Superdex 75 26/60 size exclusion column is typically used to achieve maximal separation from RNA-bound protein.

Phosphorylated PKR can be generated by incubating purified full-length human PKR at high concentrations (>15 μM) in the presence of ATP (1 mM) and MgCl$_2$ (2 mM) at 30° for 2 h in a buffered solution containing TRIS HCl (pH 7.5), 100 mM NaCl, and 5 mM β-mercaptoethanol. Phosphorylated PKR is then purified in the same buffer on a Superdex 26/60 size exclusion column (Fig. 14.2C). It is important to note that PKR produced in this manner appears homogeneously phosphorylated (McKenna, 2007), whereas bacterially expressed PKR that has not been treated with λ-phosphatase is not.

3. RNA Synthesis

Central to the study of the PKR activation is the production and use of biologically relevant dsRNA ligands for the kinase. Whereas many groups have chosen to use poly I · C (Sigma) as the model RNA ligand, its large size and broad molecular weight distribution are unfavorable for both mechanistic and high-resolution structural studies. In addition to canonical A-form dsRNA, many RNAs, including 3′-UTR of α-tropomyosin, interferon-γ mRNA, human hepatitis δ agent RNA, adenovirus VA$_I$ RNA, Epstein–Barr virus EBER-1 and -2 RNAs, and HIV TAR, regulate PKR activation (Ben-Asouli et al., 2002; Bommer et al., 2002; Maitra et al., 1994; Osman et al., 1999; Spanggord and Beal, 2001; Vuyisich et al., 2002). All known viral activators of PKR are distorted dsRNA helices, possessing structural features beyond simple duplex, such as bulges and hairpin loops. Length and RNA structural features are also thought to modulate the interaction between the dsRBDs and dsRNAs (Bevilacqua and Cech, 1996; Kim et al., 2006; McKenna et al., 2006). Therefore, the validation of a mechanism of PKR activation should include a panel of viral dsRNA molecules. The favored approach for dsRNA ligand production is *in vitro* transcription by use of T7 polymerase. The design and transcription of high-yield RNA samples for biophysical studies have been recently discussed in depth elsewhere (Kim, 2007; Lukavsky and Puglisi, 2004), and, therefore, we will provide only a brief overview of the procedure here (Fig. 14.3A).

A PCR product is generated that contains, in sequential order, the T7 RNA polymerase promoter, desired RNA template sequence to be transcribed, and restriction endonuclease site for vector linearization (Fig. 14.3B).

A

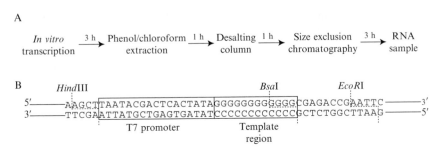

B

Figure 14.3 Design and transcription of RNA ligands. (A) Schematic outline of RNA sample preparation. Estimated time for each step is shown. (B) Synthetic DNA template used for *in vitro* transcription of RNA ligand. Restriction sites for vector ligation (*Hin*dIII, *Eco*RI) and plasmid linearization (*Bsa*I) are indicated. T7 promoter and hypothetical RNA template regions are boxed. Note that the bottom strand (3′–5′) serves as the template for synthesis of the desired RNA.

The PCR product is then ligated into a pUC119 high copy plasmid, which is used as the template for *in vitro* transcription. After restriction enzyme (BstZ17I or BsaI) treatment, the linearized plasmid template is used for *in vitro* transcription by use of T7 polymerase under standard conditions (Kim *et al.*, 2007). After 2 to 3 h of incubation at 37°, pyrophosphate precipitate is pelleted ($3000g \times 10$ min), and the reaction quenched by adding EDTA. An equal volume of phenol/chloroform is added to the supernatant, mixed and centrifuged ($3000g \times 10$ min). In this process, the restriction enzyme, which is carried over with the linearized plasmid DNA, and the T7 RNA polymerase are removed from the reaction mixture, because only the aqueous phase is retained.

After three phenol/chloroform extractions, the aqueous phase (upper) is directly loaded on a desalting column (10DG column, Bio-Rad) to remove phenol and a significant amount of remaining NTPs. After application to a desalting column, the eluted transcription reaction is directly loaded on a size exclusion column without requiring any further treatments (i.e., precipitation or concentration). Milligram quantities of pure RNA can be prepared from transcription in a single day.

4. PHOSPHORYLATION ASSAYS

To probe the mechanism of PKR activation, we developed simple assays to assess kinase activity quantitatively. The general method used ^{32}P-γATP as a probe after the kinase reaction by incorporation of the label into protein. PKR is a Ser/Thr protein kinase that undergoes *trans*-autophosphorylation on productive interaction with dsRNA.

Quantitation of the number of autophosphorylation sites on PKR has varied greatly, indicating that between 1 and 15 sites exist (Lemaire et al., 2005; McKenna et al., 2006). However, there is general agreement that phosphorylation at T446 and T451 seem to be sufficient for ultimate activation of the kinase activity and substrate recognition (Dar et al., 2005; Dey et al., 2005; McKenna et al., 2007). On phosphorylation, PKR is capable of phosphorylating its target substrates; the best character-ized being the α-subunit of eIF2B on Ser51 (Chong et al., 1992; Sonenberg and Dever, 2003). Therefore, phosphorylation assays typically follow the incorporation of radioactivity into either PKR itself, or onto a target substrate such as eIF2α.

Autophosphorylation assays are performed on PKR (100 nM) in 50 mM TRIS/Cl (pH 7.5), 100 mM NaCl, 1 mM ATP, 2 mM MgCl$_2$, and supplemented 1 μCi of ^{32}P-γATP. We typically add activator dsRNA in fourfold molar excess at this protein concentration. Submicromolar concen-trations for both PKR and RNA are chosen to minimize RNA-independent activation that has been observed at higher concentrations. Reactions are incubated at 30° for 0 to 120 min, and quenched with 5× SDS-PAGE load mix. Proteins are separated on a 4 to 20% SDS-PAGE gel (Bio-Rad), dried for 30 min at 80°, and autoradiographed (GE Healthcare) to detect the incorpora-tion of ^{32}P into PKR. Band intensities can be quantitated by use of the Image Quant software program (GE Healthcare) or ImageJ (NIH) (Abramoff et al., 2004). From reaction setup to quantitation typically takes <24 h.

Numerous applications of this assay have been used successfully to probe the mechanism of PKR kinase activity. First, autophosphorylation assays serve as a quality control for PKR preparations; any PKR that remains associated with bacterial RNA or remains phosphorylated will result in dsRNA-independent autophosphorylation and reduce the validity of results. To test our preparations, we perform autophosphorylation assays in the presence and absence of a standard dsRNA activator (HIV-TAR; Kim et al., 2006) and ensure that no incorporation of ^{32}P is observed into PKR unless dsRNA addition has occurred (Fig. 14.4A, lanes 1 and 2). Each preparation of full-length PKR is tested in this manner before use in both biochemical and structural studies to ensure a trustworthy sample.

The dsRNA concentration dependence on PKR autophosphorylation can be examined (Fig. 14.4A). In this variation of the basic autophosphor-ylation assay, the concentration of PKR is fixed (100 nM), whereas the concentration of dsRNA activator is increased from zero to a vast molar excess.

The time-dependence of autophosphorylation yields kinetic parameters for PKR. The standard reaction containing [γ-^{32}P]ATP, PKR, and dsRNA activator is incubated for a series of time points (0 to 120 min), quenched with EDTA, separated by denaturing SDS-PAGE, and quantified by autoradiogra-phy as described previously. A sigmoidal buildup of autophosphorylated PKR

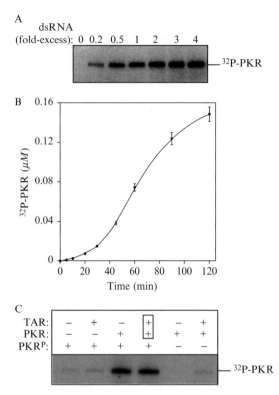

Figure 14.4 PKR phosphorylation assays. (A) Purified PKR (100 nM) was incubated with increasing amounts of HIV-TAR RNA activator in the presence ^{32}P-γATP at 30° for 90 min and resolved on denaturing SDS-PAGE gels. Gels were dried and autoradiographed to quantitate the extent of PKR autophosphorylation. (B) Progress curve of PKR autophosphorylation in the presence of equimolar TAR (100 nM) and ^{32}P-γATP at 30°. Samples were quenched at the time point indicated and quantitated as in (A). (C) *Trans*-autophosphorylation assays in which mixtures of PKR^P (100 nM), TAR (300 nM), PKR (300 nM), or purified PKR-TAR complex (300 nM, boxed) were incubated in the presence of [γ-^{32}P]ATP at 30° for 15 min. The reactions were quenched, resolved, and quantified as in (A).

is typically observed with respect to time, with a lag phase before maximal rates of autophosphorylation (Lemaire *et al.*, 2005, 2006; McKenna *et al.*, 2006) (Fig. 14.4B). These kinetics are characteristic of an autocatalytic process, and as such are well fit by a second order, bimolecular mechanism. By obtaining kinetic data at different PKR–dsRNA complex concentrations, global fitting of these data simultaneously is possible, resulting in extraction of kinetic parameters (including apparent K_M and k_{cat}) (Wang and Wu, 2002).

Phosphorylated PKR (PKR^P) can be used as a kinase capable of *trans*-phosphorylating PKR, PKR mutants, and PKR derivatives (Fig. 14.4C). These experiments remove the need for RNA activators completely, and, therefore, one can dissect the ability of PKR to serve as a substrate for *trans*-phosphorylation in various states that would not otherwise be possible (i.e., mutations, bound to activator or inhibitory dsRNAs). *Trans*-autophosphorylation of PKR by PKR^P occurs at a 20-fold faster rate than dsRNA-mediated autophosphorylation (McKenna *et al.*, 2007) and, therefore, requires only short incubation times (10 min) before quenching. Short reaction times also limit potential side reactions when examining RNA-bound PKR derivatives for their ability to serve as substrates for *trans*-autophosphorylation. We have observed that full-length PKR is equivalently able to serve as a substrate when free or bound to activator dsRNA molecules, but is not competent when bound to viral dsRNA inhibitors. Whereas mutations to the ATP coordination site (PKR^{K296R}) result in wild-type phosphorylation levels, mutations at either Thr446 or Thr451 in the kinase domain attenuate phosphorylation.

Last, substrate phosphorylation can be incorporated into any of the assays previously described to probe the ultimate kinase activity of PKR. Typical substrates have included eIF2α and (histone 2A) H2A. Rates of substrate phosphorylation can be significantly faster than autophosphorylation rates and should be taken into account when selecting time points.

5. MEASURING RNA–PROTEIN STABILITIES

The kinase activity of PKR is directly modulated by the interaction with dsRNA molecules (Ben-Asouli *et al.*, 2002; Bommer *et al.*, 2002; Maitra *et al.*, 1994; Osman *et al.*, 1999; Spanggord and Beal, 2001; Vuyisich *et al.*, 2002), and, therefore, methods to probe RNA–protein stabilities are paramount. Highly quantitative techniques are preferable to differentiate affinities between various RNA ligands and assess the contribution that individual domains in PKR make to the RNA–protein interaction. These stability experiments can then be correlated with activation assays discussed in the previous section to gain insight into the coupling of RNA binding and kinase activation. We have used isothermal titration calorimetry (ITC) and native gel shifts to study RNA–protein thermodynamics.

ITC uses stepwise injections of one reagent into a highly sensitive calorimetric cell containing a second reagent to accurately measure the change in enthalpy, ΔH (Ababou and Ladbury, 2006; Velazquez-Campoy *et al.*, 2004). Because the heat evolved during each injection is proportional to complex formation, the equilibrium binding constant (K_A) is determined. From these results, an entire suite of thermodynamic parameters can be calculated (ΔH, K_D, ΔG, ΔS) as can the stoichiometry (n) of the interaction. Other approaches such as surface plasmon resonance (SPR) or analytical ultracentrifugation are also useful techniques but measure only binding affinity. SPR is further hampered by the requirement for one reagent to be surface anchored, which can introduce experimental artifacts. Therefore, ITC is a quantitaive technique well suited to characterize RNA–protein stabilities in the PKR system.

A VP-ITC microcalorimeter (Microcal) has been used to analyze the thermodynamics of PKR–RNA complexes. The instrument is typically calibrated by injecting a 10% methanol solution (syringe, 320 μl capacity) into water (sample cell, 1.4 ml capacity) as a control to ensure a stable baseline and repetitive evolution of heat. The core of the ITC instrumentation is an extremely sensitive calorimeter, and, therefore, even small sample imperfections can create uninterpretable results. All samples are centrifuged in a tabletop centrifuge extensively and degassed under vacuum immediately before use. The sample cell is also routinely cleaned with detergent (10% Contrad-70) to eliminate solid precipitates. Furthermore, extreme care is taken to eliminate small air bubbles from both the sample cell and syringe. We recommend that the sample cell contain the RNA ligand ($\sim$5 to 10 μM RNA, in 50 mM TRIS/Cl, pH 7.5, 100 mM NaCl) and that the concentrated protein solution ($\sim$50 to 200 μM, identical buffer) is loaded into the syringe. Sample cell and syringe contents are chosen to maintain an excess of RNA over PKR to prevent the formation of nonspecific PKR–RNA aggregates. It should also be noted that the concentrations required in both sample cell and syringe will vary on the basis of the specific interaction being examined. Titrations are performed at 30°, such that the results can be correlated with activation assays, DLS experiments, and NMR studies performed at the same temperature. A standard experiment involves 25 to 30 injections of 10 μl (250 to 300 μl total) into the sample cell containing RNA, and, therefore, significant amounts of both protein and RNA are required. Titration curves are fit by a nonlinear least squares method with a model for one or two binding sites by use of Microcal Origin (version 5.0) to extract thermodynamic parameters. Outlying data points (typically the first two to three acquired) are discarded before analysis. In all cases involving PKR–RNA interactions a single high-affinity site and a second weak affinity nonspecific site were reported, and, therefore, data were fit with a model for two binding sites (Kim *et al.*, 2006; McKenna *et al.*, 2006). Reported results are from the high-affinity binding site only. From these data, the changes in entropy (ΔS) and free energy (ΔG) were calculated by use of the following equations:

$$\Delta G = -RT \ln K$$

$$\Delta G = \Delta H - T\Delta S$$

where R is the universal molar gas constant, and T is the temperature in Kelvin. The accuracy of curve fitting results is highly dependent on the accuracy of the reagent concentrations; accurate concentration determination is likely the largest source of error in the determination of thermodynamic

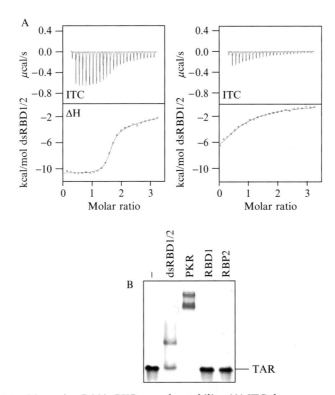

Figure 14.5 Measuring RNA–PKR complex stability. (A) ITC thermograms for titration of a dsRNA ligand (left) or dsRNA negative control (right) (10 μM, sample cell) with dsRBD1/2 (200 μM, syringe), with the heat evolved shown on a per injection basis (top). ITC titration curve for the same reaction is also shown (bottom), where the heat evolved per injection is shown as a function of the molar ratio of dsRBD1/2 to RNA. Experiments were performed at 30°. (B) Native gel mobility shift experiment for HIV TAR dsRNA (1 μM) binding to PKR and domain derivatives (1 μM) on 5% TBE gels.

parameters by ITC. For reliable results, we recommend performing a single ITC measurement in at least triplicate.

ITC has been used successfully to delineate the thermodynamic contributions that both specific structural features in RNA ligands (i.e., loops, bulges, length) and the contributions that each domain of PKR makes to the RNA–protein stability (i.e., dsRB1, dsRBD2, kinase domain, interdomain linker) (Kim *et al.*, 2006; McKenna *et al.*, 2006, 2007). Furthermore, ITC has also been used to distinguish between specific and nonspecific binding events on the basis of measured complex stability. Reported dissociation constants typically fall in the range of 50 to 150 nM for specific PKR–dsRNA complexes and are generally two orders of magnitude higher when nonspecific binding is occurring. A representative example of the use of ITC on a PKR-dsRNA sample is shown in Fig. 14.5A.

Unfortunately, ITC is both sample and time intensive. Native gel shift mobility assays, in which the RNA ligand of interest is mixed with protein under nondenaturing conditions, and RNA-containing complexes are resolved on a native gel, suffer from neither of these limitations and, therefore, provide an extremely useful complementary technique. Highly similar dissociation constants have been measured by use of the gel shift mobility assay and ITC method on PKR–RNA complexes (Kim *et al.*, 2006; McKenna *et al.*, 2006). The negatively charged phosphate backbone of dsRNA makes it the ideal ligand for native gel shift assays; addition of PKR binding partners shifts the dsRNA band to a high molecular weight species, whereas no shift is observed for unstable complexes. Although the inverse experiment is possible, PKR does not migrate significantly into native protein gels unless it is phosphorylated.

Native gel shift mobility assays have proven extremely important in delineating regions in both PKR and dsRNA ligands required for complex formation and maintaining the stability of these complexes (Jammi and Beal, 2001; Kim *et al.*, 2006; McCormack and Samuel, 1995; McKenna *et al.*, 2006, 2007; Schmedt *et al.*, 1995; Zheng and Bevilacqua, 2000) (Fig. 14.5B). Both RNA and protein samples are prepared in a suitable buffer (50 mM TRIS/Cl, pH 7.5, 100 mM NaCl) and incubated at room temperature for 10 min. Samples are then mixed with load mix (0.02% bromophenol blue, 0.01 xylene cyanol FF, 10% glycerol in 1XTBE), and loaded onto a nondenaturing TBE gel (available commercially from Bio-Rad, or can be prepared in house). Sample running buffer contains 0.5× TBE (50 mM TRIS base, 41.5 mM boric acid, and 1 mM EDTA, final pH 8.3). Electrophoresis is performed at 60 V at 4°, so as to minimize sample heating. The band detection method required is dictated primarily by the concentration range used; gels can be stained with 0.1% toluidine blue solution at micromolar concentrations, SybrGreenII fluorescent dye (Molecular Probes Inc.) at submicromolar concentrations, or radiolabeled RNA for subnanomolar complex concentrations. It should be noted that only the latter two approaches are quantitative, whereas toluidine blue is used only for comparative purposes. By varying the protein concentration at a fixed dsRNA concentration, macroscopic dissociation constants can be calculated for the interaction between protein and RNA ligand by fitting data to the following equation: fraction bound = [PKR]/ ([PKR] + K_D) (Jammi and Beal, 2001).

6. MONITORING THE ASSOCIATION STATE OF PKR

The activation of PKR is a bimolecular process, and multiple lines of evidence support the importance of dsRNA in modulating PKR self-affinity, and ultimately kinase activation (Dar *et al.*, 2005; Dey *et al.*, 2005; Gabel *et al.*, 2006; Lemaire *et al.*, 2005; McKenna *et al.*, 2007; Wu and Kaufman, 2004).

In this respect, dynamic light scattering (DLS) has proven an extremely valuable technique to monitor the association state of PKR.

The time dependence of the laser light scattered from a very small region of solution is measured in DLS experiments (Schurr, 1977; Wilson, 2003). The rate of diffusion of molecules in and out of the region being studied is related to their diffusion coefficients and, therefore, hydrodynamic radius of the particles doing the scattering. The polydispersity of the sample is also assessed during fitting of the data, which reflects the uniformity of the sample. These measurements make DLS an ideal technique for the monitoring of association state during PKR activation and as a screening technique before high-resolution structural studies of RNA–protein complexes where monodispersity is favored.

DLS experiments are performed by loading a buffered sample of interest into a small cuvette (<50 μl). We have used a DynaPro-801 molecular sizing instrument (Protein Solutions Co.) equipped with the DYNAMICS (version 6.0) software package to extract the hydrodynamic radius, apparent molecular weight, and polydispersity on the basis of the autocorrelation analysis of scattered light intensity data. It should be noted that under this regimen, average values for species with similar tumbling times are reported (i.e., PKR and PKR-PKR dimer). Because DLS measures scattering phenomena, particulates or bubbles present in the sample cell are problematic. All samples are passed through a 0.22-μM Millex-GV syringe filter (Millipore) to remove particulates, and the sample cuvette is blown with compressed air before use. We recommend a total of 50 data points be collected at 10-sec intervals and that outlying data be excluded from the use of the recommended statistical analyses in the DYNAMICS 6.0 software. We have successfully performed experiments that use PKR concentrations ranging from 0.5 to 10 mg/ml, with the upper limit likely reflecting the solubility limit for PKR. The temperature at which the experiment is conducted can be chosen at will; we typically select a standard temperature of 30° for consistency with other techniques. Samples should be preequilibrated at this temperature before filtration; a significant increase in temperature on placing the sample in the DLS apparatus can introduce bubbling in the cuvette during the DLS experiment.

Two examples of the types of experiments that can be performed by use of DLS are shown in Fig. 14.6. Time-resolved experiments in which a single sample is followed for an extended period time can reveal changes to the hydrodynamic radius on the activation of PKR. Sample preparation and preequilibration are particularly important for these experiments, because they spend a comparatively long period of time in the DLS cuvette. Far simpler is monitoring the dimerization of PKR at various concentrations; individual samples are prepared for a species of interest at each concentration. In this case, DLS measurements are nicely supplemented by performing size exclusion chromatography and native gel shifts on identical complexes.

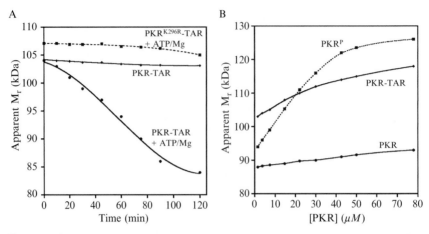

Figure 14.6 Dynamic light scattering of RNA–PKR complexes. (A) Time-dependent molecular weight determination of PKR-TAR (diamonds), PKR-TAR plus ATP/MgCl$_2$ (circles), or K296R plus ATP/MgCl$_2$ (squares, dashed line) by DLS. Equimolar complexes (2 μM) were incubated in the cuvette for the time specified before acquisition. Each data point was repeated in triplicate. Error bars have been omitted for clarity, but reflect less than 5% of the value observed in all cases. (B) Concentration-dependent dimerization of PKR was examined by determining the molecular weight at the specified concentration of PKR, PKR^P, or PKR-TAR. Each data point was repeated in triplicate. Error bars are not shown for clarity, but errors were typically <5% of the value shown.

7. NMR SPECTROSCOPY

The activation of PKR is a highly dynamic process governed primarily by transient protein–RNA interactions (K_D ~0.1 μM) and protein–protein self-association (K_D ~20 to 500 μM). A powerful tool in the characterization of PKR activation would be the high-resolution structure determination of the full-length kinase in various states of activation: free, bound to dsRNA, phosphorylated, and bound to substrate. Structure determination of individual domains has been achieved (Dar *et al.*, 2005; Nanduri *et al.*, 1998b), but as of yet, the full-length protein and RNA–PKR complexes have not been amenable to high-resolution structural studies. Fortunately, nuclear magnetic resonance (NMR) is particularly well suited to the study of dynamic, weakly interacting systems in solution (Bonvin *et al.*, 2005; Gao *et al.*, 2004; Zuiderweg, 2002) and has been extremely usefully in mapping regions of association between of PKR–RNA and PKR–PKR complexes at lower resolution.

The basis of the NMR experiment is that magnetically active nuclei (^{1}H, ^{13}C, ^{15}N) oriented in a strong magnetic field absorb radiation at specific radiofrequencies dictated by their local chemical environment. The resulting spectral lines, commonly referred to as chemical shift, are therefore exquisitely sensitive to changes local environment on perturbations to the system (i.e., ligand binding, dimerization) for ^{15}N, $C\alpha$, and $C\beta$ nuclei (Gao *et al.*, 2004). These observations form the basis for a very useful technique known as chemical shift mapping; one molecule is isotopically enriched with NMR-active nuclei, whereas a binding partner is usually unlabeled. Comparison of chemical shift perturbations before and after interaction allow for the detection of potential binding sites, if resonance assignments are available. These results are enhanced when high-resolution structures of the unbound species are available, because binding sites can be mapped onto the molecular surface.

Mapping the surfaces of interaction on protein (PKR) is accomplished through the ^{1}H-^{15}N heteronuclear single quantum coherence (HSQC) experiment. This experiment essentially provides a two-dimensional finger-print of each backbone amide and nitrogen-containing side chains in a ^{15}N-isotopically enriched protein sample. On addition of unlabeled binding partner, chemical shift perturbations occur and residue-specific information about the interaction is gained (Fig. 14.7). Conversely, the 1D ^{1}H imino

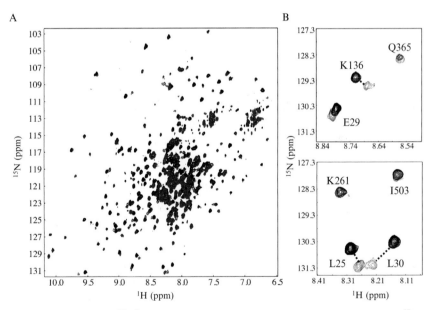

Figure 14.7 NMR of $^{15}N^{2}H$-PKR in the presence and absence of TAR RNA. (A) ^{15}N-TROSYHSQC spectral overlay of unbound PKR1-551 (black) and TAR bound PKR1-551 (grey). NMR spectral data were obtained by use of a Varian Unity INOVA 800-MHz spectrometer at 30° and acquired by use of the sensitivity-enhanced gradient pulse scheme in Biopack (Varian Inc.). (B) Detailed view of the spectra shown in (A), demonstrating chemical shift perturbations only to specific RNA binding resonances within PKR1-551 on TAR binding.

proton resonances (10 to 15 ppm range) are used to map the regions of RNA ligand used for interaction with PKR. Because only the imino protons of G and U nucleotides involved in base-pair interactions can be observed (Furtig *et al.*, 2003), the imino proton spectra provides a fingerprint of each of the base pairs in the RNA stem-loop structures. An additional bonus is that labeled RNA is not required; ^{1}H resonances arising from imino protons are usually nonoverlapping with those from protein (Fig. 14.8).

Isotopically enriched protein necessary for NMR studies is typically expressed in supplemented M9 minimal media containing 1 mM MgSO$_4$, 0.1 mM CaCl$_2$, 90 μM FeSO$_4$, 0.01% (w/v) yeast extract, 1 mg thiamine-HCl, 1mg biotin, 2 g of D-glucose (or ^{13}C-glucose), and 1 g NH$_4$Cl (or ^{15}N-NH$_4$Cl) per liter of liquid media. Deuterated protein, which reduces signal complexity and its increases its intensity in an NMR spectrum, is produced by replacing H$_2$O with the desired D$_2$O/H$_2$O ratio; no preconditioning of the cells in deuterated media is required for optimal growth. Purification is identical to that for unlabeled protein. NMR samples are prepared to a volume of approximately 250 μl (90% H$_2$O, 10% D$_2$O) in a suitable buffered saline solution in 5-mm Shigemi tubes (Shigemi Inc.). Samples are prepared by mixing components at low concentration (<10 μM) and concentrating to the desired concentration with a centrifugal concentration device (Vivascience). When protein–RNA complexes are examined, we also ensure that a slight excess of RNA is present to minimize the formation of nonspecific interactions. Our best samples are typically prepared when these complexes are purified away from unbound components by size exclusion chromatography before NMR analysis.

A necessary precursor to chemical shift mapping is the complete assignment of backbone amide ^{1}H$_N$-^{15}N cross-peaks in the ^{1}H-^{15}N HSQC NMR spectra of protein and imino resonances for RNA samples. Such results are already available for the dsRBDs (Nanduri *et al.*, 1998a) and kinase domain (Gelev *et al.*, 2006) of PKR. Resonance assignment represents the rate-limiting step of this method and its full discussion are beyond the scope of this chapter.

In the context of isolated domains, a weak interaction between dsRBD2 and the kinase domain has been identified by use of ^{1}H-^{15}N HSQCs (Gelev *et al.*, 2006; Li *et al.*, 2006; Nanduri *et al.*, 2000), although in the context of the full-length protein, no such changes in chemical shifts are observed (McKenna *et al.*, 2007). The surfaces of interaction on both PKR and a variety of dsRNA ligands has also been mapped by use of chemical shift perturbation (Kim *et al.*, 2006; McKenna *et al.*, 2006). Studies of this nature in the context of the full-length protein have allowed for novel insights into the communication between the dsRBDs and kinase domain on RNA binding and phosphorylation.

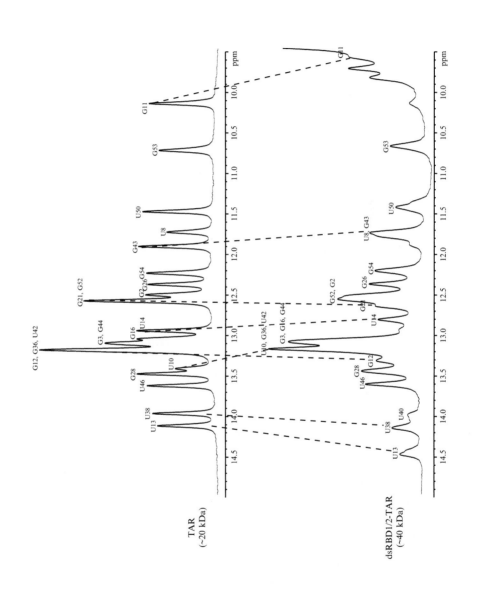

NMR approaches other than chemical shift mapping have also been and will be useful for probing the activity of PKR–RNA complexes. The orientation of dsRNA ligands relative to the dsRBDs has been established by incorporating site-specific spin-labeling into each of the dsRBDs individually (I. Kim, personal communication, May 2006). These approaches use spin labels (commonly TEMPO, a stable free radical nitroxide) incorporated onto cysteine residues in proteins or the termini of RNA molecules and look for the broadening of resonances between the label and NMR nuclei. Because the effectiveness of broadening is proportional to r^{-6} distance, resonances spatially close to the spin-label will preferentially broaden (Zuiderweg, 2002). Backbone and side chain relaxation studies by NMR have also been used effectively to gain insights into the inherent flexibility of PKR, particularly in the dsRBDs (Barnwal *et al.*, 2006; Nanduri *et al.*, 2000). Taken together, it is clear that NMR approaches to study structural characteristics of PKR have provided major insights. With advances in NMR methods and segmental labeling strategies, we are optimistic that further information will be gained on this complex system.

8. *IN VITRO* TRANSLATION ASSAYS

Translation initiation is attenuated by PKR activation through the phosphorylation of eIF2α at Ser51 (Sonenberg and Dever, 2003). Establishing a link between mechanistic/structural observations made and actual effects on the protein synthesis machinery provides an added layer of confidence to reported results. *In vitro* translation assays are a rapid, straightforward method to establish this connection.

The basis for *in vitro* translation assays is a HeLa cell lysate that contains all of the necessary protein translation machinery (Fig. 14.9). The use of a reporter construct that contains a 5′-capped luciferase mRNA (cap-Luc), the activity (intensity of luminescence from substrate) of cap-Luc is determined by the luciferase assay system (Promega) by use of an appropriate luminometer. Preparation of HeLa cell extracts is discussed elsewhere (Otto and Puglisi, 2004). Pellets (approximately 1×10^6 cells) from a 1L HeLa S3 (the National Cell Culture Center) culture are suspended in 1.25 ml of hypotonic buffer in a 5-ml-Dounce homogenizer. After 10 min, 225 μl hypertonic buffer is added,

---

Figure 14.8 Binding interface determined by imino proton titration NMR experiments. Imino proton regions of 1D ^{1}H NMR spectra of free HIV TAR RNA (top) and complex with dsRBD1-dsRBD2 (bottom). Each resonance peak corresponds to a specific base-paired G or U nucleotide and can, therefore, be used to probe regions of interaction with binding partners. Assignments are indicated by numbers above the resonance peaks.

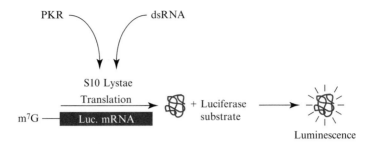

Figure 14.9 *In vitro* translation assay to monitor PKR activity. S10 HeLa cell lysates containing the protein translation machinery are incubated in the presence of 5′-capped luciferase mRNA. If translation proceeds efficiently, addition of luciferase substrates generates luminescence, which can be detected and quantitated. Addition of PKR and dsRNA ligands allows for the monitoring of their effects on translation.

and cells are lysed by 20 rapid strokes of the homogenizer. The lysate is then spun and the supernatant retained.

Inhibition of cap-dependent translation by PKR can be determined by adding purified PKR (70 nM) and RNA ligands (50 nM) to the lysate containing cap-Luc (50 nM). Reporter RNA is preincubated at 65° for 3 min, and then immediately cooled in ice-cold water before addition. The HeLa S3 S10 lysate (50% vol) is supplemented with 0.5 U/μl RNase inhibitor, SUPERase·In (Ambion), 25 μM amino acid mixture, complete (Promega), 60 mM KCl (for cap-dependent translation), and 1 mM MgCl$_2$ (required for activation of PKR). Together the translation reaction is incubated at 30° for 60 min before luciferase activity assay.

The phosphorylation state of PKR and target substrates can also be monitored from the same *in vitro* translation assay to correlate these data with translational competency. An approach that has proven useful is to perform immunoblot analysis by use of site-specific antibodies for detection. Phosphorylated or total (phosphorylated and unphosphorylated) PKR and eIF2α are detected by use of the following antibodies: anti-PKR (prepared from rabbit serum raised against the purified kinase domain of PKR) for detection of total PKR, sc-16565-R (Santa Cruz Biotechnology) for detection of PKR phosphorylation at Thr446, ab5359-50 (abcam) for detection of total eIF2α, and ab4837-50 for detection of eIF2α phosphorylation at Ser51.

9. Conclusions

We have outlined an ensemble of experimental approaches to probe the mechanistic features of PKR, a kinase central to translation initiation regulation on viral infection of cells. PKR is a highly dynamic protein that undergoes ligand binding, dimerization, phosphorylation, and ultimately kinase activation. A panel of different methods was presented to investigate each aspect of PKR activity. Reliable, homogeneous PKR sample preparation is key to the usefulness of these approaches, and methods for expressing and purifying the enzyme were presented. Although high-resolution structures of PKR in various states is an ultimate goal in the field, lower resolution structural approaches should prove extremely useful in delineating the mechanism whereby RNA binding is couple to kinase activation and subsequent translational inhibition.

REFERENCES

Ababou, A., and Ladbury, J. E. (2006). Survey of the year 2004: Literature on applications of isothermal titration calorimetry. *J. Mol. Recognit.* **19,** 79–89.

Abramoff, M. D., Magelhaes, P. J., and Ram, S. J. (2004). Image processing with ImageJ. *Biophotonics Int.* **11,** 36–42.

Barnwal, R. P., Chaudhuri, T. R., Nanduri, S., Qin, J., and Chary, K. V. (2006). Methyl dynamics for understanding hydrophobic core packing of dynamically different motifs of double-stranded RNA binding domain of protein kinase R. *Proteins* **62,** 501–508.

Ben-Asouli, Y., Banai, Y., Pel-Or, Y., Shir, A., and Kaempfer, R. (2002). Human interferon-gamma mRNA autoregulates its translation through a pseudoknot that activates the interferon-inducible protein kinase PKR. *Cell* **108,** 221–232.

Bevilacqua, P. C., and Cech, T. R. (1996). Minor-groove recognition of double-stranded RNA by the double-stranded RNA-binding domain from the RNA-activated protein kinase PKR. *Biochemistry* **35,** 9983–9994.

Bommer, U. A., Borovjagin, A. V., Greagg, M. A., Jeffrey, I. W., Russell, P., Laing, K. G., Lee, M., and Clemens, M. J. (2002). The mRNA of the translationally controlled tumor protein P23/TCTP is a highly structured RNA, which activates the dsRNA-dependent protein kinase PKR. *RNA* **8,** 478–496.

Bonvin, A. M., Boelens, R., and Kaptein, R. (2005). NMR analysis of protein interactions. *Curr. Opin. Chem. Biol.* **9,** 501–508.

Chong, K. L., Feng, L., Schappert, K., Meurs, E., Donahue, T. F., Friesen, J. D., Hovanessian, A. G., and Williams, B. R. (1992). Human p68 kinase exhibits growth suppression in yeast and homology to the translational regulator GCN2. *EMBO J.* **11,** 1553–1562.

Clemens, M. J. (1997). PKR—a protein kinase regulated by double-stranded RNA. *Int. J. Biochem. Cell Biol.* **29,** 945–949.

Dar, A. C., Dever, T. E., and Sicheri, F. (2005). Higher-order substrate recognition of eIF2alpha by the RNA-dependent protein kinase PKR. *Cell* **122,** 887–900.

Dey, M., Cao, C., Dar, A. C., Tamura, T., Ozato, K., Sicheri, F., and Dever, T. E. (2005). Mechanistic link between PKR dimerization, autophosphorylation, and eIF2alpha substrate recognition. *Cell* **122,** 901–913.

Furtig, B., Richter, C., Wohnert, J., and Schwalbe, H. (2003). NMR spectroscopy of RNA. *Chembiochem* **4,** 936–962.

Gabel, F., Wang, D., Madern, D., Sadler, A., Dayie, K., Daryoush, M. Z., Schwahn, D., Zaccai, G., Lee, X., and Williams, B. R. (2006). Dynamic flexibility of double-stranded RNA activated PKR in solution. *J. Mol. Biol.* **359,** 610–623.

Gale, M., Jr., and Katze, M. G. (1998). Molecular mechanisms of interferon resistance mediated by viral-directed inhibition of PKR, the interferon-induced protein kinase. *Pharmacol. Ther.* **78,** 29–46.

Gale, M., Jr., Tan, S. L., and Katze, M. G. (2000). Translational control of viral gene expression in eukaryotes. *Microbiol. Mol. Biol. Rev.* **64,** 239–280.

Gao, G., Williams, J. G., and Campbell, S. L. (2004). Protein-protein interaction analysis by nuclear magnetic resonance spectroscopy. *Methods Mol. Biol.* **261,** 79–92.

Gelev, V., Aktas, H., Marintchev, A., Ito, T., Frueh, D., Hemond, M., Rovnyak, D., Debus, M., Hyberts, S., Usheva, A., *et al.* (2006). Mapping of the auto-inhibitory interactions of protein kinase r by nuclear magnetic resonance. *J. Mol. Biol.* **364,** 352–363.

Jammi, N. V., and Beal, P. A. (2001). Phosphorylation of the RNA-dependent protein kinase regulates its RNA-binding activity. *Nucleic Acids Res.* **29,** 3020–3029.

Kim, I., Liu, C. W., and Puglisi, J. D. (2006). Specific recognition of HIV TAR RNA by the dsRNA binding domains (dsRBD1-dsRBD2) of PKR. *J. Mol. Biol.* **358,** 430–442.

Kim, I., McKenna, S. A., Puglisi, E. V., and Puglisi, J. D. (2007). Rapid purification of RNAs using Fast Performance Liquid Chromatography (FPLC). *RNA* **13,** 1–6.

Lemaire, P. A., Lary, J., and Cole, J. L. (2005). Mechanism of PKR activation: Dimerization and kinase activation in the absence of double-stranded RNA. *J. Mol. Biol.* **345,** 81–90.

Lemaire, P. A., Tessmer, I., Craig, R., Erie, D. A., and Cole, J. L. (2006). Unactivated PKR exists in an open conformation capable of binding nucleotides. *Biochemistry* **45,** 9074–9084.

Li, S., Peters, G. A., Ding, K., Zhang, X., Qin, J., and Sen, G. C. (2006). Molecular basis for PKR activation by PACT or dsRNA. *Proc. Natl. Acad. Sci. USA* **103,** 10005–10010.

Lukavsky, P. J., and Puglisi, J. D. (2004). Large-scale preparation and purification of polyacrylamide-free RNA oligonucleotides. *RNA* **10,** 889–893.

Maitra, R. K., McMillan, N. A., Desai, S., McSwiggen, J., Hovanessian, A. G., Sen, G., Williams, B. R., and Silverman, R. H. (1994). HIV-1 TAR RNA has an intrinsic ability to activate interferon-inducible enzymes. *Virology* **204,** 823–827.

Malmgaard, L. (2004). Induction and regulation of IFNs during viral infections. *J. Interferon Cytokine Res.* **24,** 439–454.

McCormack, S. J., Ortega, L. G., Doohan, J. P., and Samuel, C. E. (1994). Mechanism of interferon action motif I of the interferon-induced, RNA-dependent protein kinase (PKR) is sufficient to mediate RNA-binding activity. *Virology* **198,** 92–99.

McCormack, S. J., and Samuel, C. E. (1995). Mechanism of interferon action: RNA-binding activity of full-length and R-domain forms of the RNA-dependent protein kinase PKR—determination of KD values for VAI and TAR RNAs. *Virology* **206,** 511–519.

McKenna, S. A., Kim, I., Liu, C. W., and Puglisi, J. D. (2006). Uncoupling of RNA binding and PKR kinase activation by viral inhibitor RNAs. *J. Mol. Biol.* **358,** 1270–1285.

McKenna, S. A., Lindhout, D. A., Kim, I., Liu, C. W., Gelev, V. M., Wagner, G., and Puglisi, J. D. (2007). Molecular framework for the activation of PKR. *J. Biol. Chem.* **285,** 11474–11486.

Nanduri, S., Carpick, B., Yang, Y., Williams, B. R., and Qin, J. (1998a). 1H, 13C, 15N resonance assignment of the 20 kDa double stranded RNA binding domain of PKR. *J. Biomol. NMR* **12,** 349–351.

Nanduri, S., Carpick, B. W., Yang, Y., Williams, B. R., and Qin, J. (1998b). Structure of the double-stranded RNA-binding domain of the protein kinase PKR reveals the molecular basis of its dsRNA-mediated activation. *EMBO J.* **17,** 5458–5465.

Nanduri, S., Rahman, F., Williams, B. R., and Qin, J. (2000). A dynamically tuned double-stranded RNA binding mechanism for the activation of antiviral kinase PKR. *EMBO J.* **19,** 5567–5574.

Osman, F., Jarrous, N., Ben-Asouli, Y., and Kaempfer, R. (1999). A cis-acting element in the 3′-untranslated region of human TNF-alpha mRNA renders splicing dependent on the activation of protein kinase PKR. *Genes Dev.* **13,** 3280–3293.

Otto, G. A., and Puglisi, J. D. (2004). The pathway of HCV IRES-mediated translation initiation. *Cell* **119,** 369–380.

Romano, P. R., Garcia-Barrio, M. T., Zhang, X., Wang, Q., Taylor, D. R., Zhang, F., Herring, C., Mathews, M. B., Qin, J., and Hinnebusch, A. G. (1998). Autophosphorylation in the activation loop is required for full kinase activity *in vivo* of human and yeast eukaryotic initiation factor 2alpha kinases PKR and GCN2. *Mol. Cell Biol.* **18,** 2282–2297.

Schmedt, C., Green, S. R., Manche, L., Taylor, D. R., Ma, Y., and Mathews, M. B. (1995). Functional characterization of the RNA-binding domain and motif of the double-stranded RNA-dependent protein kinase DAI (PKR). *J. Mol. Biol.* **249,** 29–44.

Schneider, R. J., and Mohr, I. (2003). Translation initiation and viral tricks. *Trends Biochem. Sci.* **28,** 130–136.

Schurr, J. M. (1977). Dynamic light scattering of biopolymers and biocolloids. *CRC Crit. Rev. Biochem.* **4,** 371–431.

Sonenberg, N., and Dever, T. E. (2003). Eukaryotic translation initiation factors and regulators. *Curr. Opin. Struct. Biol.* **13,** 56–63.

Spanggord, R. J., and Beal, P. A. (2001). Selective binding by the RNA binding domain of PKR revealed by affinity cleavage. *Biochemistry* **40,** 4272–4280.

Taylor, D. R., Lee, S. B., Romano, P. R., Marshak, D. R., Hinnebusch, A. G., Esteban, M., and Mathews, M. B. (1996). Autophosphorylation sites participate in the activation of the double-stranded-RNA-activated protein kinase PKR. *Mol. Cell Biol.* **16,** 6295–6302.

Vattem, K. M., Staschke, K. A., Zhu, S., and Wek, R. C. (2001). Inhibitory sequences in the N-terminus of the double-stranded-RNA-dependent protein kinase, PKR, are important for regulating phosphorylation of eukaryotic initiation factor 2alpha (eIF2alpha). *Eur. J. Biochem.* **268,** 1143–1153.

Velazquez-Campoy, A., Leavitt, S. A., and Freire, E. (2004). Characterization of protein-protein interactions by isothermal titration calorimetry. *Methods Mol. Biol.* **261,** 35–54.

Vuyisich, M., Spanggord, R. J., and Beal, P. A. (2002). The binding site of the RNA-dependent protein kinase (PKR) on EBER1 RNA from Epstein-Barr virus. *EMBO Rep.* **3,** 622–627.

Wang, Z. X., and Wu, J. W. (2002). Autophosphorylation kinetics of protein kinases. *Biochem. J.* **368,** 947–952.

Wilson, W. W. (2003). Light scattering as a diagnostic for protein crystal growth—a practical approach. *J. Struct. Biol.* **142,** 56–65.

Wu, S., and Kaufman, R. J. (1997). A model for the double-stranded RNA (dsRNA)-dependent dimerization and activation of the dsRNA-activated protein kinase PKR. *J. Biol. Chem.* **272,** 1291–1296.

Wu, S., and Kaufman, R. J. (2004). Trans-autophosphorylation by the isolated kinase domain is not sufficient for dimerization or activation of the dsRNA-activated protein kinase PKR. *Biochemistry* **43,** 11027–11034.

Zheng, X., and Bevilacqua, P. C. (2000). Straightening of bulged RNA by the double-stranded RNA-binding domain from the protein kinase PKR. *Proc. Natl. Acad. Sci. USA* **97,** 14162–14167.

Zuiderweg, E. R. (2002). Mapping protein–protein interactions in solution by NMR spectroscopy. *Biochemistry* **41,** 1–7.

CHAPTER FIFTEEN

EXPRESSION AND PURIFICATION OF RECOMBINANT WHEAT TRANSLATION INITIATION FACTORS eIF1, eIF1A, eIF4A, eIF4B, eIF4F, eIF(iso)4F, AND eIF5

Laura K. Mayberry, Michael D. Dennis, M. Leah Allen, Kelley Ruud Nitka, Patricia A. Murphy, Lara Campbell, *and* Karen S. Browning

Contents

Department of Chemistry and Biochemistry and the Institute for Cell and Molecular Biology, University of Texas at Austin, Austin, Texas

Methods in Enzymology, Volume 430

ISSN 0076-6879, DOI: 10.1016/S0076-6879(07)30015-3

Abstract

Protein synthesis initiation factors from wheat germ were cloned into *E. coli* expression vectors for expression and purification. The ability to obtain large amounts of functional initiation factors and mutants of the factors will facilitate the biophysical and biochemical analysis of the process of initiation in plants. The initiation factors, eIF1, eIF1A, eIF4A, eIF4B, eIF4F, eIF(iso)4F, and eIF5, were successfully expressed and purified from *E. coli.* In most cases, the use of 6X-histidine tags was avoided to prevent any possible artifacts of folding or activity because of the presence of the tag. The amounts of highly purified wheat initiation factors obtained ranged from 0.5 to 24 mg of protein per liter of culture, depending on the particular initiation factor. The initiation factors were of very high purity, and the activities of the wheat recombinant factors purified from *E. coli* were found to be comparable to or better than those purified from wheat germ.

1. INTRODUCTION

Wheat germ extracts have been used for more than 30 years for *in vitro* translation of a variety of mRNAs (Marcu and Dudock, 1974; Marcus *et al.*, 1974; Roberts and Paterson, 1973; Walthall *et al.*, 1979). Wheat germ is an excellent source of initiation factors, and an extensive description of the purification of plant initiation factors from wheat germ has been described by this laboratory and others (Lax *et al.*, 1986; Seal *et al.*, 1986). The initiation factors from plants, mammals, and yeast share similarity; however, there are several differences (Browning, 1996, 2004). For example, plants have a unique isozyme form of eIF4F, termed eIF(iso)4F (Browning *et al.*, 1992) and eIF4B is poorly conserved at the amino acid level between plants, mammals, and yeast (Metz *et al.*, 1999). In addition, plants do not seem to regulate the initiation of protein synthesis by the use of phosphorylation of eIF2 and 4E binding proteins (4E-BP) in a manner similar to that of mammals and yeast (Browning, 2004).

In our laboratory, 10 of the wheat translation initiation factors have been cloned and expressed in a functional form in *E. coli*. Expression of wheat translation initiation factors in *E. coli* has many advantages over purification from wheat germ in terms of scale and time required to purify the proteins. The success of the expression of wheat initiation factors in *E. coli* may be because wheat grows over a wide temperature range, and wheat proteins may have evolved to be more tolerant of higher temperatures and, therefore, fold more readily at temperatures used for bacterial expression. For most of the initiation factors, the use of tags has been avoided, although two of the factors, eIF4A and eIF5, do have 6X-histidine tags. The remaining factors were purified by the use of standard ion-exchange resins similar to their native counterparts (Lax *et al.*, 1986). Phosphocellulose is a particularly useful ion-exchange resin for purification of the wheat initiation factors.

eIF1, eIF1A, eIF4B, eIF4F (through the eIF4G subunit), and eIF(iso)4F (through the eIF(iso)4G subunit) all bind to phosphocellulose very well, whereas most *E. coli* proteins do not. Consequently, an almost affinity level purification is obtained by the use of phosphocellulose chromatography for these factors. eIF4A does not bind phosphocellulose and copurifies through several steps with a nonspecific ATPase from *E. coli*. Consequently, a 6X his-tag was added to the C-terminus of eIF4A to facilitate purification and to remove the *E. coli* ATPase activity. eIF5 and eIF1 from wheat germ have not yet been purified to homogeneity by this laboratory. To facilitate purification from *E. coli*, a 6X-his tag was added to the N-terminus of eIF5. However, it has been subsequently observed that eIF5 also binds to phosphocellulose and a his-tag is not necessary for purification. The purification of another cap-binding protein from plants, *novel cap-binding protein* or nCBP, was previously described for *Arabidopsis* (Ruud *et al.*, 1998). The wheat cDNA for nCBP was recently obtained, and it is expected that purification procedures will be similar to that of the *Arabidopsis* form. Purification procedures for *Arabidopsis* initiation factors, eIF(iso)4F complex, eIF(iso)4G, eIF(iso)4E, eIF4E, and eIF4B, are similar to their wheat counterparts (Mayberry, Allen, and Browning, manuscript submitted).

The ability to obtain significant amounts of functional wheat initiation factors or their mutants facilitates the biochemical and biophysical analysis of the process of initiation in plants.

2. MATERIALS

All chemicals were of high quality and were obtained from Sigma, Fisher, VWR or as indicated. Chromatography resins used were phosphocel-lulose P11 (Whatman), Q-Sepharose (Sigma), Ni^{+2}-NTA Superflow (Qiagen), and m^{7}GTP Sepharose (GE Healthcare Systems). The buffers used for chromatography were Buffer B (20 mM HEPES-KOH (pH 7.6), 0.1 mM EDTA, 1 mM DTT, 10% glycerol, and KCl as indicated, Buffer B-100 is Buffer B containing 100 mM KCl), Buffer C (50 mM HEPES-KOH (pH 7.6), 600 mM KCl, 20 mM imidazole), and Buffer D (20 mM HEPES-KOH (pH 7.6), 100 mM KCl, 250 mM imidazole). Buffers were stored up to one week at 4° or at −20° for longer periods of time.

3. PROCEDURES

3.1. Cloning of cDNAs for wheat initiation factors

cDNAs for most of the wheat initiation factors (eIF1A, eIF4A, eIF4B, eIF4E, eIF(iso)4E, eIF(iso)4G) were obtained from a cDNA library prepared in λgt11 (Allen *et al.*, 1992). Table 15.1 lists the sources for the

Table 15.1 Wheat cDNAs used for *E. coli* expression constructs

Wheat initiation factor	GenBank	Source of cDNA
eIF1	EF190328	Oligonucleotide construction
eIF1A	L08060	cDNA library[a]
eIF4A	Z21510	cDNA library[a]
eIF4B	AF021243	cDNA library[a]
eIF4E	Z12616	cDNA library[a]
eIF(iso)4E	M95818	cDNA library[a]
eIF4G	EF190330	Genomic[b] and cDNA library[a]
eIF(iso)4G	M95747	cDNA library[a]
eIF5	EF190329	EST BG313359
nCBP	EF190327	EST BF202792

[a] A wheat cDNA library was prepared in λgt11 (Allen *et al.*, 1992) and was the source for the cDNA for expression clones.
[b] A wheat genomic library was prepared in λZAP (Metz *et al.*, 1999). Construction of the cDNA for eIF4G will be described elsewhere (Allen, Mayberry, Ruud, Campbell, Murphy and Browning, manuscript in preparation).

cDNAs used to prepare the plasmids for expression in *E. coli*. Oligonucleotides were designed to amplify the protein-coding region for insertion into appropriate pET expression vectors (Novagen). Other initiation factor cDNA sequences (eIF1A, eIF5, nCBP) were obtained from wheat expressed sequence tag (EST) libraries (http://compbio.dfci.harvard.edu/tgi/cgi-bin/tgi/gimain.pl?gudb=wheat), amplified, and cloned into pET expression vectors. The cDNA for wheat eIF4G was constructed from a combination of cDNA and genomic sequences. A dicistronic construct containing both eIF4G and eIF4E was prepared for the expression of the eIF4F complex. A similar dicistronic construct was prepared for the subunits of eIF(iso)4F. The details of construction of eIF4G and the dicistronic plasmids will be described elsewhere (Allen, Mayberry, Ruud, Campbell, Murphy, and Browning, manuscript in preparation). The cDNA for wheat eIF1 was prepared *de novo* by the use of overlapping oligonucleotides designed by DNA_Works (http://helixweb.nih.gov/dnaworks/) on the basis of the protein sequence predicted by the TIGR gene indices TC233522 (http://compbio.dfci.harvard.edu/tgi/cgi-bin/tgi/gimain.pl?gudb=wheat).

3.2. Growth of *E. coli* and expression of initiation factors

E. coli cells were transformed with the appropriate expression plasmid by the use of standard methods and selected with the appropriate antibiotic(s) (see Table 15.2). LB media and LB-agar were obtained from USB or Invitrogen and prepared according to the manufacturer's instructions. Cells from a single colony were used to inoculate 50 ml of LB media (in a 250-ml flask) containing the appropriate antibiotics and incubated overnight at 37° with shaking (150 to 170 rpm). The 50-ml overnight culture was transferred to 1 to 1.2 L of LB media in a 6-L flask or 0.8 L in a 4-L flask containing the appropriate antibiotics. The culture was incubated at the indicated temperature and grown to the desired density (A_{600}) as shown in Table 15.2. Cells from a 1-ml aliquot were recovered by centrifugation and resuspended in 100 μl of 1× SDS gel loading buffer (Laemmli, 1970) for the analysis of protein before induction. The expression of the initiation factor was induced by the addition of 5 ml of 100 mM IPTG per L of culture to obtain a final concentration of 0.5 mM IPTG. Incubation of the induced cell cultures was continued for the time and temperature indicated in Table 15.2. A 1-ml aliquot was taken before harvest for the analysis of induction and processed as described previously for the preinduction sample. The before and after induction samples were compared by SDS-PAGE (Laemmli, 1970) with the appropriate initiation factor purified from wheat germ and/or molecular weight markers. The induced cells were harvested by centrifugation (6000g) for 15 min at 4°. The cell pellets were used immediately for purification or quick frozen (powdered dry ice or liquid nitrogen) and stored at −80°.

3.3. Lysis of *E. coli* cells

For non–his-tagged proteins the cells (from up to 6 L of media) were thawed on ice and suspended in 30 ml of Buffer B-500 containing one protease inhibitor tablet (Roche Complete). The cells were disrupted by sonication for 3 × 30 sec at 70% power followed by 2 × 30 sec at 90% power by the use of a Vibra Cell sonicator (Sonics & Materials Inc.). The cells were cooled in ice water for 2 to 3 min between sonication bursts. The lysed cells were centrifuged at 184,000g for 2 h at 4°. The supernatant was removed and either quick frozen (dry ice or liquid nitrogen) and stored at −80° or applied immediately to a column for purification. eIF4F, eIF(iso)4F, eIF4G, or eIF (iso)4G should be applied immediately to a column to minimize proteolysis.

For his-tagged proteins, the cells were suspended in 30 ml of Buffer C containing one protease inhibitor tablet (Roche Complete EDTA-free). The cells were disrupted by sonication and centrifuged as described previously. The supernatant was removed and either quick frozen (dry ice or liquid nitrogen) and stored at −80° or applied immediately to a Ni^{+2}-NTA column for purification.

3.4. General protein purification procedures

All protein purification procedures were carried out at 4°. Protein samples were divided into small aliquots (25 to 100 μl, for assays) or large fractions (>1 ml) for storage or further purification. All protein preparations were fast

Table 15.2 Expression of wheat protein synthesis initiation factors in *E. coli*

Factor	MW	Vector	Antibiotic	*E. coli* strain	Volume[a] (L)	OD[b]	Temp[c] (°C)	Time[d]	Column[e]	Yield	Reference
eIF1	12,642	pET23b (*Nde*I/*Eco*RI)	amp	BL21(DE3)	2	0.45	37/37	3 h	PC	8 mg	
eIF1A	16,291	pET3d (*Nco*I/*Bam*HI)	amp, chlor	BL21(DE3) pLysS	4	0.45	37/37	3 h	PC	6 mg	
eIF4A	46,933	pET23d (*Nco*I/*Xho*I)	amp, chlor	BL21(DE3) pLysS	2	0.5	37/37	3 h	Ni	48 mg	
eIF4B	56,834	pET3d (*Nco*I/*Bam*HI)	amp, chlor, tet	BLR(DE3) pLysS	4	0.5	37/30	3 h	PC, Q	12 mg	Metz *et al.*, 1999
eIF4F[f]		pET3d **4G** (*Nco*I/*Bam*HI) **4E** (*Bam*HI/ *Bam*HI)	amp	BL21(DE3)	2.4[g]	0.8	30/30	3 h	PC, m^{7}G, PC	3.5 mg	
eIF4G	161,747	pET3d (*Nco*/*Bam*HI)	amp	BL21(DE3)	20[g]	0.8	37/30	2 h	PC,DEAE, PC,PC	5 mg	
eIF4E	23,906	pET22b (*Nco*I/*Bam*HI)	amp	BL21(DE3)	2	0.9	37/37	2 h	m^{7}G	2 mg	
eIF(iso) 4F[f]		pET3d **iso4G** (*Nco*I/*Bam*HI) **iso4E** (*Bam*HI/ *Bam*HI)	amp	BL21(DE3)	2.4[g]	0.8	30/30	3 h	PC, m^{7}G, PC	1 mg	
eIF(iso) 4G	86,275	pET3d (*Nco*I/*Bam*HI)	amp, chlor	BL21(DE3) pLysS	8[g]	0.4	37/37	2 h	PC,DEAE, PC,PC	5 mg	Allen *et al.*, 1992
eIF(iso) 4E	23,525	pET3d (*Nco*I/*Bam*HI)	amp	BL21(DE3)	2	0.9	37/37	2 h	M^{7}G	2 mg	Allen *et al.*, 1992
eIF5	49,037	pET15b (*Nde*I/*Bam*HI)	amp	BL21(DE3)	2	0.45	37/30	3 h	Ni, PC	2 mg	

frozen on powdered dry ice or in liquid nitrogen and stored at $-80°$. Protein samples should not go through more than two freeze/thaw cycles. The use of UV monitors (Isco) for protein purification is recommended to determine when the columns are adequately washed and to observe when protein is eluted from the column. Samples were dialyzed for a minimum of 16 h against 50 to 100× the sample volume of the desired buffer, and the buffer was changed once during the dialysis. Dialysis was carried out in a graduated cylinder with stirring at 4°. The dialyzed material was clarified by centrifugation at 4° for 15 min at 16,000g. The purified recombinant initiation factors were analyzed for protein concentration by the Bradford method with bovine serum albumin as a standard (Bradford, 1976) and for purity by SDS-PAGE (Laemmli, 1970). The choice of fractions to pool in each purification was based on the purity (SDS-PAGE) and concentration (Bradford assay). If the purity is satisfactory at any stage, additional purification steps may be omitted if desired. If the protein concentration was lower than desired, the protein was concentrated by the use of an Amicon Ultra centrifugal filter device (10,000 MW cutoff, Millipore).

3.5. Production of rabbit antibodies

Antibodies to recombinant initiation factors were raised in rabbits at the University of Texas M. D. Anderson Cancer Center (Department of Veterinary Sciences, Bastrop, TX).

3.6. Purification of wheat eIF1

The supernatant obtained from 2 L of cells was diluted to 0.1 M KCl with Buffer B-0. The diluted sample was applied to a 5-ml phosphocellulose P11 column (1.5 cm × 2.75 cm) equilibrated in Buffer B-100. The column was washed with Buffer B-100 until the A_{280} returned to baseline. The eIF1 was eluted with a 25-ml gradient (100 to 500 mM KCl) in Buffer B. Fractions of 1 ml were collected. Fractions containing the highest purity and concentration of eIF1 were pooled ($\sim$7 ml), dialyzed against Buffer B-100, and stored as described previously.

---

[a] Liters of cell culture used for a preparation.
[b] Optical density of cells at 600 nm when induced with 0.5 mM IPTG (final concentration).
[c] Temperature of incubation before and after induction.
[d] Length of time cells were grown after IPTG induction until harvest.
[e] Column resin(s) used for purification.
[f] eIF4F consists of two subunits, eIF4G and eIF4E; both subunits are expressed from the same plasmid to make the eIF4F complex. eIF(iso)4F consists of two subunits, eIF(iso)4G and eIF(iso)4E; both subunits are expressed from the same plasmid to make the eIF(iso)4F complex.
[g] The concentration of LB media was increased to 1.5×.

3.7. Purification of wheat eIF1A

The supernatant obtained from 1 L of cells was diluted to 0.25 M KCl with Buffer B-0. The diluted sample was applied to a 5-ml phosphocellulose P11 column (1.5 cm $\times$ 2.75 cm) equilibrated in Buffer B-250. The column was washed with Buffer B-250 until the A_{280} returned to baseline. The eIF1A was eluted with a 40-ml gradient (250 to 750 mM KCl) in Buffer B. Fractions of 1 ml were collected. Fractions containing the highest purity and concentration of eIF1A were pooled ($\sim$7.5 ml), dialyzed against Buffer B-50, and stored as described previously.

3.8. Purification of wheat eIF4A

The supernatant obtained from 2 L of cells was applied to a 2.5-ml Ni^{+2}-NTA column equilibrated in Buffer C. The column was washed with at least 3 column volumes of Buffer C or until the A_{280} returned to baseline. The eIF4A was eluted with Buffer D. Fractions of 1 ml were collected. Because eIF4A is not stable in the presence of the high concentration of imidazole in Buffer D and precipitates, fractions containing the highest amount in protein were immediately identified by combining 5 μl of each fraction to 45 μl Buffer B-0 and adding 150 μl of diluted Bradford reagent (standards not necessary). The fractions with the highest amount of eIF4A (i.e., bluest color) were immediately pooled, dialyzed against 1 L of Buffer B-40, and stored as described previously.

3.9. Purification of wheat eIF4B

The supernatant obtained from 4 L of cells was diluted to 0.1 M KCl with Buffer B-0. The diluted sample was applied to a 5-ml phosphocellulose column (1.5 cm $\times$ 2.75 cm) equilibrated in Buffer B-100. The column was washed with Buffer B-100 until the A_{280} returned to baseline. The eIF4B was eluted with a 50-ml gradient (100 to 400 mM KCl) in Buffer B. Fractions of 1 ml were collected. The fractions containing the highest purity and concentration of eIF4B were pooled. The pooled fractions were stored at $-80°$ until the next column step.

The pooled fractions from the phosphocellulose column were diluted with Buffer B-0 to 40 mM KCl and applied to a 5-ml Q-Sepharose column (1.5 cm $\times$ 2.75 cm) equilibrated in Buffer B-40. The column was washed with 25 ml of Buffer B-40. The eIF4B was eluted with a 50-ml gradient (40 to 400 mM KCl) in Buffer B. Fractions of 1 ml were collected. Fractions containing the highest purity and concentration of eIF4B were pooled and stored as described previously.

3.10. Purification of wheat eIF4F or eIF(iso)4F

To minimize the proteolysis of eIF4G or eIF(iso)4G, the extract was kept packed in ice during the application to the column, and all purification procedures were carried out as quickly as possible. The supernatant obtained from 2.4 L of cells was diluted to 0.1 M KCl with Buffer B-0 and applied in equal portions to two 10-ml phosphocellulose columns (1.5 cm × 7 cm) equilibrated in Buffer B-100. The use of two columns accelerated the application of the extract and minimized degradation of the eIF4G or eIF(iso)4G subunit. The columns were washed with Buffer B-100 until the A_{280} returned to baseline. The eIF4F or eIF(iso)4F was eluted with Buffer B-300. Fractions of 2 ml were collected. The fractions containing the highest concentration of eIF4F or eIF(iso)4F were pooled. The pooled fractions from both phosphocellulose columns were diluted to 0.1 M KCl by the addition of Buffer B-0 and applied to a 4-ml m^7GTP Sepharose column equilibrated in Buffer B-100. The column was washed with Buffer B-100 (20 ml) and with Buffer B-100 containing 100 μM GTP (20 ml) to remove any *E. coli* protein contaminants. The eIF4F or eIF(iso)4F was eluted with Buffer B-100 containing 100 μM m^7GTP. Fractions of 0.5 ml were collected. Fractions containing the highest purity and concentration were pooled. The pooled fractions were applied to a 1-ml phosphocellulose column (1 cm × 1.7 cm) equilibrated in Buffer B-100 and washed with the same buffer (~4 to 5 ml). This purification step removes excess m7GTP and concentrates the protein. The eIF4F or eIF(iso)4F was eluted with Buffer B-300. Fractions of 10 drops were collected by hand (do not use a monitor as it results in a less-concentrated protein preparation). Fractions containing the highest purity and concentration of eIF4F or eIF(iso)4F were pooled and stored as described previously.

3.11. Purification of eIF4G or eIF(iso)4G

eIF4G and eIF(iso)4G are particularly susceptible to degradation. In addition to 3 protease inhibitor tablets, 14 mg soybean trypsin inhibitor (Sigma) and 12 mg PMSF (Sigma, dissolved in 0.5 ml isopropanol) were added to 10 ml of Buffer B-500 used to resuspend the cells before lysis. The purification was as described previously for the dicistronic constructs for the first phosphocellulose column. The pooled fractions from the phosphocellulose columns were diluted to 0.05 M KCl by the addition of Buffer B-0 and applied to two 2.5-ml DEAE-Sepharose columns (1.5 cm × 1.4 cm) equilibrated in Buffer B-50. The use of two columns accelerated application of the protein to the columns and minimized degradation. The matrix from the two DEAE-Sepharose columns was combined into a single column (1.5 cm × 3 cm)and washed with Buffer B-50 until the A_{280} returned to baseline. The eIF4G or eIF(iso)4G was eluted with a 125-ml gradient (50 to 250 mM KCl) in Buffer B. Fractions of 1 ml were collected. Fractions containing the highest purity and concentration of eIF4G or eIF(iso)4G were pooled. The pooled fractions were ~0.1 M KCl, but if they were estimated to be higher, they were diluted to ~0.1 M KCl by the addition of Buffer B-0. The pooled fractions were applied to a 2-ml phosphocellulose column equilibrated in Buffer B-100. The column was washed with Buffer B-100 until the A_{280} returned to baseline. The eIF4G or eIF(iso)4G was eluted with a 30-ml gradient (100 to 250 mM KCl) in Buffer B. Fractions of 1 ml were collected. Fractions containing the highest purity and concentration of eIF4G or eIF(iso)4G were pooled. The eIF4G or eIF(iso)4G was concentrated on a 1-ml phosphocellulose column as described for eIF4F or eIF(iso)4F or by centrifugation as described previously.

3.12. Purification of eIF4E or eIF(iso)4E

The cells for purification of eIF4E or eIF(iso)4E were resuspended in Buffer B-50. The only difference in the purification for eIF(iso)4E was that the HEPES buffer was pH 7 and potassium acetate was substituted for KCl. The supernatant obtained from 2 L of cells was applied directly to a 4-ml m^{7}GTP Sepharose column and eluted as described previously for eIF4F and eIF(iso)4F. No other column purification steps were required. The m^{7}GTP used to elute the column was not removed. NMR analysis of wheat eIF4E indicates that it rapidly unfolds in the absence of m^{7}GTP (Monzingo *et al.*, 2007).

3.13. Purification of wheat eIF5

The supernatant obtained from 2 L of cells was applied to a 2.5-ml Ni^{+2}-NTA column equilibrated in Buffer C. The column was washed with Buffer C until the A_{280} returned to baseline. The eIF5 was eluted from the column with Buffer D. Fractions of 1 ml were collected. The fractions containing the highest concentration of protein were identified as described previously for eIF4A. The fractions containing the highest concentration of eIF5 were immediately applied to a 2-ml phosphocellulose P11 column (1.5 cm × 0.6 cm) equilibrated in Buffer B-100. The column was washed with Buffer B-100 (3 to 5 ml). The eIF5 was eluted with a 10-ml gradient (100 to 500 mM KCl) in Buffer B. Fractions of 0.5 ml were collected. Fractions containing the highest purity and concentration of eIF5 were pooled (~5 ml), dialyzed against Buffer-100, and stored as described previously.

4. Results

SDS-PAGE analysis of the recombinant proteins is shown in Fig. 15.1. The recombinant proteins eIF1A, eIF4A, eIF4B, eIF4F, and eIF(iso)4F show similar mobility to the native factors purified from wheat germ. eIF1 and eIF5 have mobilities consistent with their predicted molecular

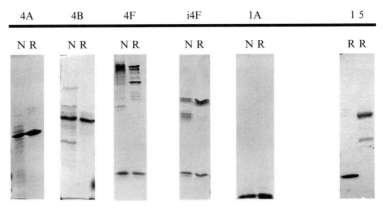

Figure 15.1 SDS-PAGE analysis of expressed wheat initiation factors. eIF1A, 3 μg native (N), 3 μg recombinant (R); eIF4A, 3 μg native (N), 3 μg recombinant (R); eIF4B, 2 μg native (N), 2 μg recombinant (R); eIF4F, 2.5 μg native (N), 2.5 μg recombinant (R); eIF(iso)4F, 2 μg native (N), 2 μg recombinant (R); eIF1, 3 μg recombinant (R); eIF5, 3 μg recombinant (R). The gels were stained with Coomassie brilliant blue.

weights. The antibodies raised to recombinant initiation factors react with the native proteins, as well as the recombinant forms in both ELISA and Western blot analysis (Humphreys *et al.*, 1988).

The abilities of the recombinant factors eIF4A, eIF4B, eIF4F, eIF(iso)4F, and eIF1A to support polypeptide synthesis *in vitro* was measured as previously described (Lax *et al.*, 1986). In all cases, the specific activity was found to be comparable to or higher than the native form. Increased specific activity was probably due to higher purity or lack of inhibitors copurifying with the native preparations. Assays for eIF1 and eIF5 are being developed by the use of antibody depletion of wheat germ S30 (Benkowski *et al.*, 1995), and these recombinant proteins are expected to be fully functional.

5. CONCLUSIONS

The availability of highly purified and active recombinant plant initiation factors provides the amounts and purity of the initiation factors necessary for biophysical analysis, including structure determination. The structure of wheat eIF4E was determined (Monzingo *et al.*, 2007), and an extensive biophysical analysis of wheat eIF4B was carried out (Mayberry, Allen, and Browning, manuscript submitted) using highly purified initiation factors prepared using the procedures described here. In addition, the availability of these purified factors will be very valuable for many other types of *in vitro* biochemical studies.

Further work will be necessary to express and purify eIF4H, eIF5B, eIF6, eIF2 (3 subunits), as well as the subunits of eIF3 (12 subunits), to complete the reconstitution of the wheat translation initiation factors.

ACKNOWLEDGMENTS

This work was supported by grants from DOE (DE-FG02-04ER15575), NSF (MCB0214996), and The Welch Foundation (F1339) to K. S. B. The University of Texas M. D. Anderson Cancer Center, Department of Veterinary Sciences, Bastrop, TX, was supported by grant NIH-NCI CA-16672. The authors thank Joanne M. Ravel for critical reading of the manuscript and Sandra Lax for expert technical assistance.

REFERENCES

Allen, M. L., Metz, A. M., Timmer, R. T., Rhoads, R. E., and Browning, K. S. (1992). Isolation and sequence of the cDNAs encoding the subunits of the isozyme form of wheat protein synthesis initiation factor 4F. *J. Biol. Chem.* **267,** 23232–23236.

Benkowski, L. A., Ravel, J. M., and Browning, K. S. (1995). Development of an *in vitro* translation system from wheat germ that is dependent upon the addition of eukaryotic initiation factor 2. *Anal. Biochem.* **232,** 140–143.

Bradford, M. M. (1976). A rapid and sensitive method for the quantitation of microgram quantities of protein utilizing the principle of protein-dye binding. *Anal. Biochem.* **72,** 248–254.

Browning, K. S. (1996). The plant translational apparatus. *Plant Mol. Biol.* **32,** 107–144.

Browning, K. S. (2004). Plant translation initiation factors: It is not easy to be green. *Biochem. Soc. Trans.* **32,** 589–591.

Browning, K. S., Webster, C., Roberts, J. K. M., and Ravel, J. M. (1992). Identification of an isozyme form of protein synthesis initiation factor 4F in plants. *J. Biol. Chem.* **267,** 10096–10100.

Humphreys, J. S., Browning, K. S., Hobbs, W. R., and Ravel, J. M. (1988). The amounts and molar ratios of the protein synthesis initiation and elongation factors present in wheat germ. *J. Cell Biol.* **107,** 548a.

Laemmli, U. K. (1970). Cleavage of structural proteins during the assembly of the head of bacteriophage T4. *Nature* **227,** 680–685.

Lax, S. R., Lauer, S. J., Browning, K. S., and Ravel, J. M. (1986). Purification and properties of protein synthesis initiation and elongation factors from wheat germ. *Methods Enzymol.* **118,** 109–128.

Marcu, K., and Dudock, B. (1974). Characterization of a highly efficient protein synthesizing system derived from commercial wheat germ. *Nucleic Acids Res.* **1,** 1385–1390.

Marcus, A., Efron, D., and Weeks, D. P. (1974). The wheat embryo cell-free system. *Methods Enzymol.* **30,** 749–754.

Metz, A. M., Wong, K. C. H., Malmström, S. A., and Browning, K. S. (1999). Eukaryotic initiation factor 4B from wheat and *Arabidopsis thaliana* is a member of a multigene family. *Biochem. Biophys. Res. Commun.* **266,** 314–321.

Monzingo, A. F., Dhaliwal, S., Dutt-Chaudhuri, A., Lyon, A., Sadow, J., Hoffman, D. W., Robertus, J. D., and Browning, K. S. (2007). The structure of transaction initiation factor eIF4E from wheat reveals a novel disulfide bond. *Plant Phys.* **143,** 1504–1518.

Roberts, B. E., and Paterson, B. M. (1973). Efficient translation of tobacco mosaic virus RNA and rabbit globin 9S RNA in a cell-free system from commercial wheat germ. *Proc. Natl. Acad. Sci. USA* **70,** 2330–2334.

Ruud, K. A., Kuhlow, C., Goss, D. J., and Browning, K. S. (1998). Identification and characterization of a novel cap-binding protein from *Arabidopsis thaliana*. *J. Biol. Chem.* **273,** 10325–10330.

Seal, S. N., Schmidt, A., and Marcus, A. (1986). The wheat germ protein synthesis system. *Methods Enzymol.* **118,** 128–140.

Walthall, B. J., Spremulli, L. L., Lax, S. R., and Ravel, J. M. (1979). Isolation and purification of protein synthesis initiation factors from wheat germ. *Methods Enzymol.* **60,** 193–204.

CHAPTER SIXTEEN

IN VITRO RECONSTITUTION AND BIOCHEMICAL CHARACTERIZATION OF TRANSLATION INITIATION BY INTERNAL RIBOSOMAL ENTRY

Victoria G. Kolupaeva,* Sylvain de Breyne,* Tatyana V. Pestova,*,† *and* Christopher U. T. Hellen*

Contents

* Department of Microbiology and Immunology, State University of New York Downstate Medical Center, Brooklyn, New York
† A. N. Belozersky Institute of Physicochemical Biology, Moscow State University, Moscow, Russia

Methods in Enzymology, Volume 430
ISSN 0076-6879, DOI: 10.1016/S0076-6879(07)30016-5

Abstract

The internal ribosomal entry sites (IRESs) of encephalomyocarditis virus
(EMCV) and related viruses promote initiation of translation by a noncanonical
end-independent mechanism. To characterize this mechanism at the molecular
level, we have developed biochemical approaches to reconstitute the process
in vitro from individual purified components of the translation apparatus,
developed methods to characterize steps in this process so that the functions
of individual proteins can be characterized, and adapted assays such as
primer extension inhibition ("toe printing") to monitor accurate assembly on
the IRES of ribosomal 48S and 80S complexes. *In vitro* reconstitution of 48S
complex formation offers an approach for the functional identification of IRES
trans-acting factors (ITAFs) that are required for initiation in addition to canonical initiation factors and revealed that despite being related, different EMCV-
like IRESs nevertheless have distinct ITAF requirements. Toe printing revealed
that a common feature of initiation on EMCV-like IRESs is the stable binding of
an eIF4G/eIF4A complex to them near the initiation codon, where it can locally
unwind RNA to facilitate ribosomal attachment. The same toe printing assay
indicated that binding of ITAFs to these IRESs enhances binding of these
two canonical initiation factors. We also describe protocols for chemical and
enzymatic footprinting to determine the interactions of *trans*-acting factors with
the IRES at nucleotide resolution and for directed hydroxyl radical probing to
determine their orientation on the IRES.

1. INTRODUCTION

A number of viral mRNAs are translated after end-independent
initiation on an internal ribosomal entry site (IRES) in the 5′ untranslated
region rather than by the canonical end-dependent scanning mechanism
(Hellen and Sarnow, 2001). IRESs were first identified in the RNA
genomes of the picornaviruses, poliovirus, and encephalomyocarditis virus
(EMCV) (Jang *et al.*, 1988; Pelletier *et al.*, 1988), which exemplify two
major classes of viral IRES. Many of the viral IRESs that have subsequently
been identified belong either to one of these two groups or to two other
structurally distinct groups: Hepatitis C virus (HCV)–like IRESs and the
intergenic region (IGR) IRESs of dicistroviruses such as cricket paralysis

virus (Pisarev *et al.*, 2005). Early observations that translation of poliovirus and EMCV mRNA continues under circumstances that abrogate the canonical initiation mechanism suggested that these IRESs might have distinctive requirements for canonical initiation factors; observations that initiation on several viral IRESs is weak and/or defective in rabbit reticulocyte lysate (which supports efficient initiation by the canonical mechanism) but that it can be enhanced by the ribosomal salt wash fraction from HeLa cells suggests that some IRES also require cell type–specific IRES *trans*-acting factors (ITAFs) that play no part in the canonical mechanism of translation initiation (Dorner *et al.*, 1984; Jackson *et al.*, 1995; Svitkin *et al.*, 1988). Nevertheless, the structural diversity of different IRES groups suggests that they would be unlikely to mediate initiation by a common mechanism. A focus of our research has been to clarify understanding of internal initiation on viral IRESs that are representative of these different groups by use of biochemical approaches to define the molecular mechanism of each process. As a result, we have characterized the outlines of distinct mechanisms of initiation on three unrelated groups of IRESs (Pestova and Hellen, 2003; Pestova *et al.*, 1996a,b, 1998a,b, 1999; Pilipenko *et al.*, 2000; Wilson *et al.*, 2000). Here we describe the approaches that we have used to explain the mechanism of initiation on the EMCV and related Theiler's murine encephalomyelitis virus (TMEV) and foot-and-mouth disease virus (FMDV) IRESs.

2. CHEMICALS, ENZYMES, AND BIOLOGICAL MATERIALS

Standard chemicals are from Fisher Scientific (Pittsburgh, PA) and Sigma Aldrich (St. Louis, MO). The protease inhibitor cocktail set III and puromycin are from Calbiochem (La Jolla, CA), GMP-PNP and Pefabloc are from Roche (Indianapolis, IN), and isopropyl-β-D-thiogalactopyranoside (IPTG) is from Bioworld (Dublin, OH). 1-cyclohexyl-3-(2-morpholinoethyl)carbodiimide metho-p-toluenesulfonate (CMCT) and dime;thylsulfate (DMS) are from Sigma Aldrich. Fe(II)-1-(p-bromoacetamidobenzyl)-EDTA (Fe-BABE) is from Dojindo Molecular Technologies (Gaithersburg, MD). 7-diethylamino-3-(4-(iodacetyl)phenyl)-4-methyl coumarin (DCIA) is from Molecular Probes/Invitrogen (Carlsbad, CA). Ni^{2+}-nitrilotriacetic acid (Ni-NTA) and imidazole are from Qiagen (Valencia, CA). Unlabeled NTPs and dNTPs, poly(U) sepharose, and 7-Methyl-GTP Sepharose 4B are from GE Healthcare (Piscataway, NJ). Oligonucleotides are from Invitrogen. Whatman 3MM paper, Spectrum Spectra/Por 3RC 25-mm dialysis tubing, Kontes chromatography columns, Whatman DE52 DEAE cellulose, and Whatman P11 phosphocellulose are from Fisher Scientific. Polyprep chromatography columns are from BioRad (Hercules, CA). Centricon YM100 and YM10 centrifugal filter units are from Millipore (Billerica, MA). Sephadex G-50 Spin-50 mini-columns were from USA Scientific (Ocala, FL). Polyallomer centrifuge tubes are from Beckman Coulter (Fullerton, CA). BD Precison-Glide 20G1 1/2 syringe needles are from Becton Dickinson (Franklin Lakes, NJ). Radiochemicals [^{35}S]methionine (37 TBq/mmol), [α-^{32}P]dATP (111 TBq/mmol), and [α- ^{32}P]UTP (111 Tbq/mmol) are from MP Biomedicals (Irvine, CA).

DNA modifying enzymes and restriction endonucleases are from New England BioLabs (Beverley, MA), Fermentas (Hanover, MD), and Roche. RQ1 RNase-free DNase, avian myeloblastosis virus (AMV) reverse transcriptase and RNase ONE are from Promega Corp. (Madison, WI). T7 RNA polymerase is from Fermentas or Ambion (Austin, TX); the T7 RNA polymerase mutant DEL172–3 was a gift from D. Lyakhov and W. McAllister. Ribonuclease inhibitors are from Promega, GE Healthcare, and Invitrogen. Sequenase T7 DNA polymerase is from USB Corp. (Cleveland, OH). RNase T1 is from Ambion, and RNase V1 from Pierce (Milwaukee, WI). Polymerase chain reactions (PCR) are done by use of the Expand High Fidelity (Plus) PCR System (Roche) or Vent DNA polymerase (New England BioLabs). Rabbit reticulocyte lysate (RRL) for purification of ribosomes and factors is from Green Hectares (Oregon, WI) and for *in vitro* translation is from Promega and Ambion. Native calf liver tRNA is from EMD Biosciences.

Escherichia coli strain MRE 600 (ATCC No 29417) was from the American Type Culture Collection (Manassas, VA). *E. coli* BL21 Star (DE3): F- *ompT hsd*SB (rB-mB-) *gal dcm rne131* (DE3) and *E. coli* DH5α-T1^R: F- ϕ80*lac*ZΔM15 Δ(*lac*ZYA-*arg*F) U169 *deo*R *rec*A1 *end*A1 *hsd*R17 (rk-, mk+) *pho*A *sup*E44 λ- *thi*-1 *gyr*A96 *rel*A1 *ton*A were from Invitrogen.

Items of apparatus used in the procedures described in the following include an Eppendorf thermomixer and a 5417C microcentrifuge (Eppendorf North America, Westbury, NY), a Branson S-450D digital sonifier and a Bransonic 1510 ultrasonic cleaner (Branson, Danbury CT), Aktapurifier chromatography systems (GE Healthcare), a Stratalinker UV cross linker with a 365-nm bulb (Stratagene, La Jolla, CA), and a GeneAmp 2400 PCR system (Applied Biosystems, Foster City, CA). For electrophoresis, we use BioRad Mini-PROTEAN II or Mini-PROTEAN III electrophoresis cells in conjunction with an EPS 301 power supply and a vertical sequencing apparatus (Owl Separation Systems, Rochester, NY) with an EPS 3501 power supply. Gels are dried by use of a model 583 gel dryer (Bio-Rad). Centrifugation is done by use of a Sorvall RC-5C Plus centrifuge (Thermo Electron Corp., Asheville, NC) and Beckman L8-M ultracentrifuges (Beckman Coulter), together with Sorvall GSA and SS34 rotors and Beckman SW41 and SW55 rotors, respectively.

3. PLASMIDS

Plasmids for transcription of the EMCV IRES are based on pTE1 (Evstafieva *et al.*, 1991), which contains EMCV nt. 378 to 1155 cloned downstream of a T7 promoter. Plasmids for transcription of the FMDV IRES are based on pSP449 (Luz and Beck, 1990), which contains nt. 362 to 831 of the FMDV O_1K strain cloned downstream of a T7 promoter (numbering of FMDV nucleotides starts after the poly(C) tract rather than from the 5′ end). Plasmids for transcription of the TMEV IRES were made with TMEV (GDVII strain) cDNA (Pilipenko *et al.*, 2000). pHP5′UTR/VP contains human parechovirus type 1 (HPeV1) nt. 1 to 1653 (which include the IRES) cloned downstream of a T7 promoter (Nateri *et al.*, 2000). pBS$^-$β-globin (Hellen *et al.*, 1993) contains the complete β-globin cDNA cloned downstream of a T3 promoter, and pTRNA1 (Pestova and Hellen, 2001) contains the mammalian $tRNA^{Met}_i$ gene cloned into pBR322 so that it is flanked by an upstream T7 RNA polymerase promoter and a downstream BstN1 restriction site, which permits the 3′-terminal -CCA sequence of tRNA to be transcribed precisely.

Vectors for expression in *E. coli* of all single subunit initiation factors are described by Pisarev *et al.* (2007) and include pQE(His_6-eIF1), pET(His_6-eIF1), pET(His_6-eIF1A), pET(His_6-eIF4A), and pET(His_6-eIF4B) (Pestova *et al.*, 1996a, 1998a). pET(His_6-eIF4A(R362Q)) was derived from a vector for expression of this eIF4A mutant lacking a Hexa histidine tag (Pause *et al.*, 1993). pET15b-based vectors for expression of fragments of eIF4GI and mutants thereof and pET28-based vectors for expression of $eIF4GII_{745-1003}$ and derivatives thereof have been described (Kolupaeva *et al.*, 2003; Lomakin *et al.*, 2000; Marcotrigiano *et al.*, 2001; Pestova *et al.*, 1996b). The numbering of amino acid residues in eIF4GI has been revised several times since the work described here was initiated: the nomenclature used here describes the longest isoform (Genbank accession No. AY082886). pET28b-p97(NAT1)$_{62-330}$ was constructed by inserting a PCR fragment corresponding to amino acid residues 62 to 330 of NAT1 (Genbank accession No. U76111) into pET28b. The vector for expression of full-length PTB-1 was made by inserting an *Xho*1 fragment containing full-length PTB-1 cDNA into pET3a (Hellen *et al.*, 1993) and is renamed pET3a-PTB1 here. pET(His_6-ITAF$_{45}$) for expression of full-length ITAF$_{45}$ was made by insertion of PCR-amplified cDNA of the ITAF$_{45}$

coding sequence from mouse embryo cDNA into pET28b (Pilipenko *et al.*, 2000).

4. PURIFICATION OF FACTORS AND RIBOSOMES

4.1. Buffers

Buffer A 20 mM Tris-HCl, pH 7.5; 10% glycerol
Buffer B 20 mM HEPES, pH 7.5, 2 mM DTT, 0.1 mM EDTA, 5% glycerol
Buffer C 20 mM Tris-HCl, pH 7.5, 2 mM DTT, 0.1 mM EDTA, 5% glycerol
Buffer D 20 mM Tris-HCl, pH 7.5, 2 mM DTT, 0.2 mM EDTA, 10% glycerol

For the sake of brevity, methods for the purification of ribosomal subunits, native initiation factors, and most recombinant factors are not described here, and interested readers should instead consult the chapter by Pisarev *et al.* (2007) in this volume.

Vectors for expression of eIF4GI$_{653-1600}$ and derivatives thereof, of eIF4GII$_{745-1003}$ and derivatives thereof, and of p97(NAT1)$_{62-330}$ are used to transform *E. coli* BL21 Star (DE3), and after overnight growth on LB/kanamycin plates, cells are washed off into 5 ml LB/kanamycin medium. *E. coli* BL21 Star (DE3) is similarly transformed with pET(His$_6$-ITAF$_{45}$) or pET3a-PTB1, and after overnight growth on LB/ampicillin plates, cells are washed off into 5 ml LB/ampicillin medium. Cultures grown for expression of ITAF$_{45}$, PTB, and short fragments of eIF4G1 (such as eIF4GI$_{737-1116}$), of eIF4GII and of p97 are added to 1 L LB with the appropriate antibiotic, incubated with shaking at 37° until 0.4 <OD$_{600}$ <0.5, at which stage expression of the recombinant protein is induced with 1 mM IPTG, and the bacterial suspension is incubated with shaking for 4 h at 37°. eIF4GI$_{653-1600}$ is significantly less well expressed than shorter eIF4G fragments, and cultures grown to express it are therefore transferred to 4 L LB/kanamycin, incubated with shaking at 30° until 0.3 <OD$_{600}$ <0.4, and then for another 2 h at 30° after induction of expression of the recombinant protein with 1 mM IPTG.

After centrifugation of bacterial cultures for 20 min at 5000 rpm at 4° in an GSA rotor in a Sorvall RC5B centrifuge, the pellet is resuspended in 20 ml buffer A + 300 mM KCl and mixed with 50 μl protease inhibitor cocktail set III in 30 ml Corex (or similar heavy duty) glass tubes (except from the 4 L bacterial cultures, which are resuspended in 60 ml buffer A + 300 mM KCl and mixed with 100 μl protease inhibitor). Bacterial suspensions are then sonicated on ice for a total of 10 min with a cycle of 10 sec sonication and 55 sec off.

To purify ITAF$_{45}$ and fragments of eIF4G, supernatants derived from the bacterial lysate by centrifugation in a Sorvall SS34 rotor at 15,000 rpm for 25 min at 4° are added to a Ni^{2+}-NTA column that has been prepared by adding 800 μl of a 50% Ni^{2+}-NTA suspension to a BioRad Polyprep chromatography column, allowing ethanol to flow out and then washing with 10 ml deionized water followed by buffer A + 300 ml KCl. After adding the supernatant, the bound material is washed sequentially with 10-ml aliquots of buffer A + 300 mM KCl, buffer A + 800 mM KCl, buffer A + 100 mM KCl, and buffer A + 100 mM KCl + 20 mM imidazole. The high salt and low (20 mM) imidazole washes remove proteins and nucleic acids that bind nonspecifically to the Ni^{2+}-NTA matrix before elution of the factor of interest with six 1-ml aliquots of buffer A + 100 mM KCl + 300 mM imidazole. DTT is then added to each 500-μl fraction to 2 mM final concentration and the partially purified proteins are assayed by SDS-PAGE.

Fractions containing ΔeIF4G are dialyzed overnight at 4° against 1 L buffer B + 100 mM KCl and are then applied to a FPLC MonoS HR 5/5 column. Fractions are collected across a 100 to 500 mM KCl gradient: apparently homogenous ΔeIF4G elutes with ~200 mM KCl. The ITAF$_{45}$ eluate from the Ni^{2+}-NTA matrix is dialyzed overnight at 4° against 1 L buffer C + 50 mM KCl, and applied to a FPLC MonoQ HR 5/5 column. Fractions are collected across a 50 to 500 mM KCl gradient: apparently homogenous ITAF$_{45}$ elutes with ~100 mM KCl.

To purify PTB, the supernatant prepared from the bacterial lysate by centrifugation in a Sorvall SS34 rotor at 15,000 r.p.m. for 25 min at 4° is dialyzed against 2 L buffer D + 100 mM KCl and applied to a pre-swollen DEAE column. The flow-through fraction, which contains PTB, is applied to a Kontes column containing 500 μl of poly(U)-sepharose that has been washed once with 10 ml buffer D + 100 mM KCl. The loaded column is washed with 10 ml buffer D + 250 mM KCl; bound material is then eluted with 10 1-ml aliquots of buffer D + 500 mM KCl, and the column is then cleaned by use of 10 column washes of buffer D + 1000 mM KCl and regenerated. Fractions containing PTB1 are dialyzed against 1 L buffer B + 100 mM KCl and applied to a FPLC MonoS HR 5/5 column. Fractions are collected across a 100 to 500 mM KCl gradient: apparently homogenous PTB elutes with ~175 mM KCl.

5. PREPARATION OF IRES-CONTAINING mRNA AND AMINOACYLATED INITIATOR tRNA

Protocols for *in vitro* transcription of human tRNA$^{Met}_i$, expression and purification of recombinant His$_6$-tagged *E. coli* methionyl-tRNA synthetase, and aminoacylation of [^{35}S]Met-tRNA$^{Met}_i$ (Lomakin *et al.*, 2006;

Pestova and Hellen, 2001) are described in the chapter by Pisarev *et al.* (2007) in this volume.

IRES-containing mRNAs are transcribed *in vitro* from linearized plasmids by use of wild-type T7 RNA polymerase, except for Pst1-linearized EMCV plasmids, which have 3′-overhanging ends that cause wild-type but not DEL172–3 mutant T7 RNA polymerase to produce aberrant transcription products (Lyakhov *et al.*, 1992). We, therefore, either use this mutant RNA polymerase or fill in the protruding ends after Pst1 digestion by use of T4 DNA polymerase before transcription with wild-type T7 RNA polymerase. For *in vitro* transcription of unlabeled RNAs, we use a 100-μl transcription mixture that contains 10 μg linearized plasmid DNA, 10 U/μl T7 RNA polymerase, 4 mM each of the four nucleotide triphosphates, 80 mg/ml PEG8000, 1 U/μl RNase inhibitor, 20 mM MgCl$_2$, and 10 mM DTT. α-[^{32}P]UTP-labeled RNAs used in sucrose density gradient centrifugation analysis of 48S complexes are transcribed in reaction mixtures that contain 0.1 mM unlabeled UTP and α-[^{32}P]UTP (111 Tbq/mmol) instead of 4 mM unlabeled UTP but are otherwise identical. After incubation at 37° for 1 h, RNA is extracted with acidic phenol/chloroform (pH 4.7), and ethanol-precipitated. RNA transcripts are separated from DNA restriction fragments and unincorporated nucleotides by gel filtration on FPLC Superdex G75 or a microSpin G-50 column, and their integrity is confirmed by gel electrophoresis. Radiolabeled RNAs had specific activities of $\sim$300,000 to 500,000 cpm/μg.

6. Strategies for Identification of Viral IRESs

IRESs are identified by their ability to mediate translation of the downstream cistron irrespective of translation of the upstream cistron when inserted between cistrons in a synthetic dicistronic mRNA (Kaminski *et al.*, 1994). Translation of the upstream cistron is abolished by insertion of a stable hairpin at the extreme 5′ end of the dicistronic mRNA; in most cases, translation of the upstream cistron is also impaired without affecting IRES-mediated translation if the dicistronic mRNA lacks a 5′-terminal m⁷GTP cap, by cleavage of eIF4G by picornavirus proteases, or by sequestration of eIF4E by 4E-binding protein. Stringent controls (van Eden *et al.*, 2004) are required to rule out possible trivial explanations for translation of the downstream cistron in transfected cells, such as aberrant splicing or the presence of a cryptic promoter. The activity of many IRESs in mediating translation of a downstream cistron is detectable only by use of highly sensitive reporter assays (such as luciferase). It is essential to note that IRESs for which this is necessary are not appropriate candidates for biochemical analysis of the initiation mechanism by the methods described in the following section. The IRESs that have been characterized by use of these approaches are all derived from RNA viruses that replicate in the cytoplasm and are sufficiently active in cell-free translation systems to mediate *in vitro* translation at a level that permits ready detection of [^{35}S] methionine-labeled translation products. The ability of an IRES to exhibit this level of activity in a cell-free lysate indicates that the lysate can potentially be fractionated, allowing the essential factors that are in it to be isolated and characterized. The assays described here are done exclusively in cell-free extracts or *in vitro* reconstituted systems, obviating any need to assay for aberrant splicing. However, the integrity of mRNAs is verified both after transcription and after initiation complex formation by direct analysis of input mRNA and of mRNA after phenol-extraction from 48S and ribonucleoprotein (RNP) complexes, as well as by reverse transcription (e.g., Pestova *et al.*, 1996a).

Translation initiation on various IRESs is impaired after replacement of homologous by heterologous reporter sequences, either because IRES folding is altered or because inhibitory secondary structures are introduced that occlude the ribosomal binding site (e.g., Fletcher *et al.*, 2002; Shibuya *et al.*, 2003). The ribosomal mRNA binding cleft extends ~15nt. downstream of the initiation codon, and we assume that the rapidly evolving genomes of RNA viruses are under selective pressure to maintain IRES-proximal coding sequences in an unstructured conformation that is compatible with efficient ribosomal binding. We, therefore, endeavor to include at least 50 nt. of homologous coding sequence downstream of the initiation codon in IRES constructs; the experiments described in the following that relate to the EMCV IRES (~nt.400 to 834 of the genomic RNA) were done with mRNAs that included 300+ nt. of homologous coding sequence.

The accuracy of IRES-mediated 48S complex formation is monitored by "toe printing," a technique in which the position of 48S complexes on mRNA is detected by primer extension inhibition (see later). Stops caused by bound complexes can potentially be obscured if reverse transcription on naked mRNA yields "strong stops" immediately upstream of the binding site; this problem can be overcome by reducing local secondary structure, for example by substituting residues in this region of the viral coding sequence by CAA triplets.

7. ASSAYS FOR 48S COMPLEX FORMATION IN CELL-FREE EXTRACTS: SUCROSE DENSITY GRADIENT CENTRIFUGATION

7.1. Buffers

Buffer E 20 mM Tris-HCl, pH 7.5, 2 mM DTT, 100 mM potassium acetate

We use two methods to assess the ability of cell-free extracts to support formation of 48S complexes. Nonhydrolyzable analogs of GTP (such as guanylimidodiphosphate (GMP-PNP)) can substitute for GTP in the eIF2·GTP·Met–tRNA$^{Met}_i$ ternary complex and allow normal formation of 48S complexes that accumulate because they cannot progress further to the subunit joining stage of initiation. Similarly, inclusion of elongation inhibitors such as cycloheximide or anisomycin in cell-free translation extracts leads to accumulation of 80S ribosomes (Anthony and Merrick, 1992). Ribosomal 48S and 80S complexes are stable and can be resolved from ribonucleoprotein (RNP) complexes and from each other by sucrose density gradient centrifugation.

48S complexes are formed by incubating 0.5 µg [^{32}P]-labeled EMCV nt. 315 to 1155 RNA in 25 µl of RRL for 5 min at 30° under translation conditions in the presence of 1 mM GMPPNP (Pestova *et al.*, 1996a). In this and all reactions described in the following, an equimolar amount of Mg^{2+} must be added to a reaction mixture whenever GMPPNP, GTP, or ATP is included to compensate for their chelation of Mg^{2+}. Reaction mixtures are then layered on top of 10 to 30% linear sucrose gradients (in buffer E with 6 mM magnesium acetate) prepared in Beckman 14 × 89-mm polyallomer centrifuge tubes (11 ml total volume) with a Hoefer SG15 gradient maker. The elevated magnesium concentration serves to stabilize the 48S complex on the IRES. Sucrose density gradients are prepared at least 2 h in advance and are kept at 4° until needed. Gradients are centrifuged for 3.5 h at 4° at 40,000 rpm by use of a Beckman SW41 rotor or for 1.5 h at 53,000 rpm by use of a Beckman SW55 rotor to resolve ribosomal complexes and RNP complexes. Gradients are then fractionated; we clamp tubes on top of a thin, lightly greased rubber bung, pierce both the bung and the tube very carefully with a BD 20G1 $\frac{1}{2}$ syringe needle, collect fractions across the gradient, and measure their optical density and radioactivity. In analogous experiments, 80S ribosomes accumulate in RRL programmed with EMCV IRES-containing mRNA and preincubated with cycloheximide (Wilson *et al.*, 2000). The sedimentation coefficient of RNP complexes assembled on long dicistronic mRNAs containing conventional reporter genes can potentially lead them to obscure 48S complexes; thus, to maximize resolution of 48S and RNP complexes, it is advisable to use short mRNA in these experiments (i.e., with truncated open reading frames).

8. Assays for 48S Complex Formation in Cell-Free Extracts: Toe Printing

8.1. Buffers

Buffer E 20 mM Tris–HCl, pH 7.5, 2 mM DTT, 100 mM potassium acetate

Sucrose density gradient centrifugation can be used to determine the efficiency of 48S complex formation on an mRNA but does not yield data concerning the position of these arrested ribosomal complexes on it. IRESs frequently contain "silent" AUG triplets upstream of the actual initiation codon, and it is, therefore, important to distinguish between the possibilities that 48S complexes detected by sucrose density gradient centrifugation assembled by an IRES-mediated mechanism at the correct initiation codon or by the canonical end-dependent mechanism at an upstream AUG triplet. Primer extension inhibition ("toe printing") can be used to map the position of stably arrested ribosomal 48S and 80S complexes on mRNAs at nucleotide resolution and thus to determine whether ribosomal complexes assembled on the presumed internal initiation codon or at an upstream location. Toe printing involves cDNA synthesis by reverse transcriptase (RT) on a template RNA to which a ribosome or protein is bound (Fig. 16.1). cDNA synthesis is arrested if the complex is stable, yielding stops ("toe prints") at its leading edge. If complex formation is efficient and sequesters a high proportion of the mRNA, the yield of full-length cDNA will be reduced significantly. Eukaryotic 48S complexes usually arrest primer extension 15 to 17 nt. 3′ to the A of the initiation codon; this can vary slightly depending on the mRNA (Anthony and Merrick, 1992; Pestova et al., 1996a, 1998a,b; Pilipenko et al., 2000). 80S ribosomes yield toe prints at the same position as 48S complexes (Alkalaeva et al., 2006; Pestova and Hellen, 2003, 2005), indicating that the toe print is caused by the leading edge of the 40S subunit and that joining of a 60S subunit to the 48S complex does not alter its relative position.

The position on an mRNA of ribosomal complexes assembled in RRL can be determined by toe printing, but in our experience direct toe printing of complexes assembled in cell-free translation extracts of HeLa and murine Krebs 2 ascites cells is not possible, likely because of RNAse H in these extracts that cleaves the mRNA after annealing to the oligonucleotide primer (e.g., Pestova et al., 1989). However, toe printing can be used to map the position of ribosomal complexes assembled in such extracts if they are first purified by sucrose density gradient centrifugation. For toe printing analysis of 48S complexes assembled in RRL, 0.5 μg of RNA is incubated in 15 μl of RRL in the presence of 1 mM GMPPNP for 5 min at 30°. The reaction mixture is diluted with buffer E with 2.5 mM magnesium acetate to 100 μl final volume before the addition of 1 μl of a 17 to 25 nt-long primer (1 μg/μl), designed to hybridize to mRNA $\sim$100 nt. downstream of the anticipated ribosome binding site. For example, in experiments on the TMEV GDVII IRES, in which the initiation codon is $AUG_{1068-1070}$, we used a primer complementary to TMEV nt. 1195 to 1214 (Pilipenko et al., 2000). Avian myeloblastosis virus (AMV) RT (15 U), 1 μl magnesium acetate (320 mM), 4 μl dNTPs (5 mM dCTP, dGTP and dTTP; 1 mM dATP), 1 μl [α-^{32}P]dATP (111 TBq/mmol) are added to the reaction mixture, which is then incubated for 45 min at 37°. The efficiency of phenol extraction of cDNA products is increased by supplementing this

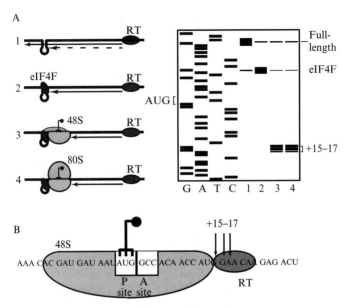

Figure 16.1 The primer extension inhibition ("toe printing") assay to map the position of ribonucleoprotein (RNP) and ribosomal complexes on mRNA. After assembly of RNP or ribosomal complexes on an mRNA, an oligonucleotide primer is annealed to it and is extended by use of reverse transcriptase (RT) by incorporation of unlabeled dUTP, dCTP, dGTP, and [^{32}P]-labeled dATP into cDNA. On naked mRNA (1) primer extension mostly proceeds to the 5' end of the mRNA, generating a full-length cDNA product, but on structured mRNAs such as IRESs, may be arrested at the 3' border of stable helical elements, yielding shorter cDNAs. Proteins that bind to the 5'UTR to form stable RNP complexes (2), 48S initiation complexes (3), and 80S ribosomes (4) that assemble at the initiation codon arrest cDNA synthesis at intermediate positions on the mRNA that can be identified precisely by resolving radiolabeled cDNA products on denaturing sequencing gels in parallel with dideoxynucleotide sequencing products generated by use of appropriate cloned cDNA and the same oligonucleotide primer. The 40S subunit forms the leading edge in both 48S initiation complexes and 80S ribosomes and characteristically arrests primer extension in both $^+$15 and $^+$17 nt from the first nucleotide in the P site (by definition, +1).

reaction mixture with 100 μl H$_2$O, 2.5 μl 10% SDS, and 2.5 μl 0.5M EDTA. After extraction with phenol (pH 8.0), cDNA products are ethanol-precipitated, resuspended in 8 μl loading buffer (0.05% Bromophenol blue, 0.05% xylene cyanol FF, 20 mM EDTA (pH.8.0), 91% formamide), and heated at 95° for 3 min. Aliquots (2 μl) are then loaded on 6% polyacrylamide/7 M urea sequencing gels that have been prerun for 45 min (to avoid band compression). cDNA products are compared with dideoxynucleotide sequence ladders obtained by use of the same primer and appropriate plasmid DNA in reactions with Sequenase, done according to the manufacturer's instructions. The aliquots from toe printing and sequencing reactions loaded into each well should contain equal amounts of buffer (particularly of formamide) to ensure that cDNAs of identical length have identical mobility on these gels. After electrophoresis, urea is removed by fixing the gel for 20 min in a 10% acetic acid/10% methanol solution. It is then transferred to prewetted Whatman 3MM paper, dried, and exposed to X-ray film.

9. IDENTIFICATION OF THE MINIMUM SET OF FACTORS REQUIRED FOR 48S COMPLEX FORMATION

The ability of a cell-free extract to support IRES-mediated initiation allows the minimum set of factors needed for initiation to be determined by biochemical fractionation of that extract, which can theoretically be done *ab initio*. However, because protocols have been developed for purification of all native eIFs required for cap-dependent initiation, and because all single subunit eIFs can be expressed in *E. coli* and purified as active recombinant proteins [see the chapter by Pisarev *et al.* (2007) in this volume], a more convenient strategy is first to determine whether this full set of eIFs is sufficient to support, for example, 48S complex formation on the IRES in the presence of all other necessary components of the translation apparatus ([^{32}P]-labeled mRNA, [^{35}S]Met-tRNA$^{Met}_i$, ribosomal subunits, ATP, and GTP/GMPPNP). If it is, factors can systematically be omitted to determine the minimum necessary set (e.g., Pestova *et al.*, 1996a; Pilipenko *et al.*, 2000). Although some IRESs require only a subset of canonical initiation factors, many IRESs additionally require one or more noncanonical initiation factors, otherwise known as IRES *trans*-acting factors (ITAFs), that are usually specific RNA-binding proteins with functions unrelated to translation that have been co-opted by IRESs. One of the first ITAFs to be identified was the pyrimidine tract–binding protein (PTB), a regulator of alternative splicing that binds to all picornavirus IRESs and enhances initiation on a subset of them (see later) (Borovjagin *et al.*, 1994; Hellen *et al.*, 1993, Kaminski and Jackson, 1994; Luz and Beck, 1990; Pestova *et al.*, 1991, 1996a; Pilipenko *et al.*, 2000). Novel factors can be purified on the basis of activity in complementing the function of a complete set of canonical eIFs in supporting, for example, 48S complex formation on the IRES, and identified by N-terminal sequencing and sequencing of tryptic peptides, followed by database searches and cDNA cloning. Once a set of factors has been identified that mediates initiation with a similar efficiency to that of the starting material, individual factors can be systematically omitted to identify the minimum set required for 48S complex formation. In our analysis of initiation on EMCV and related picornavirus IRESs, we focused on factors required for 48S complex formation, because this stage in IRES-mediated initiation seemed likely to differ most significantly from the canonical mechanism. Details of the methods that were used to identify the factors required for 48S complex formation on the EMCV, TMEV, and FMDV IRESs, including ITAF$_{45}$, a novel ITAF, have been described (Pestova *et al.*, 1996a,b; Pilipenko *et al.*, 2000) and will not be repeated here because details of the strategy necessary for purification of the essential set of factors for initiation on other IRESs will undoubtedly be different. However, some general comments may be useful.

1. Initiation in cell-free extracts occurs efficiently on EMCV and related IRESs at relatively high magnesium concentrations, but the salt optima for other IRESs can be significantly different (e.g., Borman *et al.*, 1995). Conditions for translation initiation on novel IRESs must, therefore, first be optimized.

2. The availability of recombinant factors to replace native purified factors eliminates the possibility that cross-contamination could lead to errors in characterizing factors as being nonessential and can allow the requirement for individual subunits of a factor and even domains of a subunit to be determined (e.g., Lomakin *et al.*, 2000; Marcotrigiano *et al.*, 2001; Pestova *et al.*, 1996b, 1998a, 2000).

3. Although initiation factors are conventionally purified from the RSW from fractions that precipitate between 0 and 70% A.S., ITAF$_{45}$, which is required for initiation on the FMDV IRES, was purified from the 70 to 95% A.S. fraction (Pilipenko *et al.*, 2000). It is, therefore, wise to retain and store at $-80°$ all fractions left after purification, including the post-ribosomal fraction that remains after centrifugation of polysomes and the supernatant left after precipitation of the 50 to 70% A.S. fraction.

4. Replacement of a RSW subfraction by individual purified components may result in a higher level of activity than that obtained with the starting material, because purification may eliminate specific inhibitory proteins or general RNA-binding proteins that compete with initiation factors for binding sites on the IRES (e.g., Merrill and Gromeier, 2006).

5. The efficiency of 48S complex formation can be assayed by sucrose density gradient centrifugation, but toe printing assays should be used to verify the fidelity of initiation codon selection on the IRES. First, as noted previously, it is important to verify that 48S complexes identified on the basis of sedimentation in sucrose density gradients did not assemble by a conventional scanning mechanism on upstream AUG triplets in the IRES, particularly near its 5′ end. This can be assayed by use of one or more oligonucleotide primers whose binding sites are chosen to enable potential initiation complex formation at upstream AUG triplets to be monitored. Second, several viral mRNAs contain AUG triplets in the immediate vicinity of the initiation codon, which is, nevertheless, selected by ribosomes with high specificity. It is important

that initiation codon selection in reconstituted reactions and either *in vivo* or in cell-free translation systems are comparable. The factors eIF1 and eIF1A play important roles in ensuring the fidelity of initiation codon selection during the scanning stage of cap-mediated initiation, and they can also influence this process during IRES-mediated initiation, enhancing the fidelity of initiation codon selection on the EMCV IRES (Pestova *et al.*, 1998a).

 ## 10. RECONSTITUTION OF 48S COMPLEX FORMATION ON TYPE 2 PICORNAVIRUS IRESS

10.1. Buffer

Buffer E 20 mM Tris-HCl, pH 7.5, 2 mM DTT, 100 mM potassium acetate

The strategy of successively replacing subfractions of the 0.5 M Kcl RSW by individual initiation factors enabled 48S complex formation on the EMCV IRES to be reconstituted in a reaction containing canonical eIFs, PTB, ribosomal subunits, mRNA, and Met-tRNA$^{Met}_i$. Subsequent systematic omission experiments allowed the contribution of each factor to initiation on the IRES to be assessed.

For toe printing analysis, ribosomal complexes are assembled by incubating 0.5 μg of EMCV RNA for 10 min at 30° in a 40-μl reaction mixture that contains buffer E with 2.5 mM magnesium acetate, 100 U RNase inhibitor, 1 mM ATP, and 0.1 mM GMPPNP (and, as noted previously, an equivalent amount of Mg^{2+}), 0.25 mM spermidine, 6 pmol [^{35}S]Met-tRNA$^{Met}_i$, 3 pmol 40S subunits and combinations of PTB (1.5 μg), and the initiation factors eIF1 (0.5 μg). eIF1A (0.5 μg), eIF2 (3 μg), eIF3 (7 μg), eIF4A (2 μg), eIF4B (2 μg), and eIF4F (2 μg) (e.g., Pestova *et al.*, 1996a, b, 1998a). Incubation is continued for 3 min at 30° after addition of 1 μl of an appropriate primer (1 μg/μ1), before primer extension done by use of AMV RT as described previously. By systematic omission from reactions assayed by sucrose density gradient centrifugation and toe printing, the PTB (an ITAF) was found to contribute marginally to initiation. eIF4F could, in the presence of 2 μg eIF4A, be replaced by 1 μg eIF4G$_{653-1600}$ or 2 μg eIF4G$_{653-1130}$; these eIF4G fragments lack the domain that binds the eIF4E subunit of eIF4F, so that neither eIF4E nor its eIF4G binding domain are required for initiation on this IRES. Toe printing showed that eIF1 strongly increased 48S complex formation at the authentic initiation codon (AUG$_{834}$) and reduced 48S complex formation at AUG$_{826}$ (which is barely used *in vivo*, but accumulates a significant proportion of the 48S complexes *in vitro* in the absence of eIF1) (Pestova *et al.*, 1998a). Similar reaction conditions were sufficient for initiation on the TMEV IRES, which, however, required PTB, whereas initiation on the FMDV IRES required PTB and a novel factor, ITAF$_{45}$, in addition to the same set of canonical eIFs as the EMCV IRES (Pilipenko *et al.*, 2000).

11. TOE PRINTING TO MAP STABLE INTERACTIONS OF THE IRES WITH COMPONENTS OF THE TRANSLATION APPARATUS

Early models for IRES-mediated initiation suggested that conserved structural IRES domains or motifs might be recognized directly by the initiating ribosome and/or associated factors or by ITAFs that might either be recognized by the ribosomal 43S complex or might stabilize an active three-dimensional conformation of the IRES (Jackson and Kaminski, 1995). The appearance of toe prints in addition to those attributable to 48S complexes assembled on various IRES during factor omission experiments (Pestova *et al.*, 1996a,b, 1998b; Pilipenko *et al.*, 2000) was suggestive of such IRES-factor interactions. The processivity of RT is such that many proteins that bind IRESs specifically, such as PTB, are, nevertheless, displaced from them and do not yield toe prints. However, eIF3 and eIF4F bind specifically to large, highly structured elements of HCV-like and EMCV-like IRES, respectively, and the specificity and stability of the resulting complexes is sufficiently great to arrest RT, yielding prominent toe prints (e.g., Pestova *et al.*, 1996a,b, 1998b; Pilipenko *et al.*, 2000).

To analyze potential IRES-factor interactions, we allow RNP complexes to form by incubating eIFs and IRES mRNA together. For example, toe printing assays were done on complexes formed by incubating 1 μg EMCV IRES-containing mRNA (4 pmol) and either 3 μg eIF4F (13 pmol) or combinations of 1 μg (22 pmol) eIF4A, 1.5 μg (21 pmol) eIF4B, 1 μg (10 pmol) eIF4G$_{653-1600}$, or 2 μg (10 pmol) eIF4G$_{653-1130}$ for 5 min at 37° in buffer E with 2.5 mM magnesium acetate and 0.25 mM spermidine in a 20-μl reaction volume. Toe print analysis of the resulting IRES/factor complexes was done by use of a primer complementary to EMCV nt 957 to 974 and AMV RT, essentially as described previously. These complexes yielded a specific toe print at C$_{786}$ in the J-K domain of the IRES, which indicated that eIF4F, eIF4G (or even just its central domain) bound this domain specifically, and that eIF4A enhanced binding of eIF4G (Lomakin *et al.*, 2000; Pestova *et al.*, 1996a,b). This observation is important because eIF4A and eIF4G normally interact with each other as part of the eIF4F heterotrimer. Cooperativity in the binding of factor to an IRES can be confirmed and quantified by use of an electrophoretic mobility shift assay (e.g., Lomakin *et al.*, 2000). The toe printing assay in not quantitative, but it can, nevertheless, also be used to characterize the mutual influence of proteins that do not bind to each other on their binding to an IRES. For example, inclusion of $ITAF_{45}$ and PTB in RNP assembly reactions significantly enhanced the toe print attributable to bound eIF4G on the FMDV IRES (Pilipenko et al., 2000; Fig. 16.2C,D). This result supports the hypothesis that ITAFs function as chaperones that stabilize the IRES in an active conformation, which was based on observations that PTB binds to multiple sites on EMCV and related IRESs (see later) and that its requirement for initiation on the EMCV IRES is conditional and depends on slight variations in the structure of the J-K domain (Kaminski and Jackson, 1998).

The specificity of interaction of a protein with the IRES can be confirmed in several ways.

A. Binding of the factor to related IRESs can be assayed to determine whether an interaction is a general characteristic of a class of IRESs and what sequence/structural variation in the binding site is tolerated. For example, in addition to binding to the EMCV IRES, eIF4G binds FMDV, TMEV, and HPeV1 IRESs sufficiently stably to be detectable by toe printing (Pilipenko et al., 2000; Fig. 16.2).

B. Toe printing can be used to determine whether a factor binds to an individual domain of an IRES inserted into a heterologous RNA; thus, eIF4G binds to the EMCV J-K domain inserted into a heterologous background (Kolupaeva et al., 2003).

C. The effects of mutations in the IRES domain on interactions with the cognate factor can be assayed in toe printing experiments and compared with their effects on the activity of the natural IRES in supporting initiation in in vitro translation reactions and in reconstituted initiation reactions (e.g., Kolupaeva et al., 2003; Pestova et al., 1996b).

D. The ability of closely related factors to bind to the IRES can be used to characterize the specificity of binding. Thus, neither wheat eIF4F nor the mammalian eIF4G-like protein p97 bound to the EMCV IRES, whereas mammalian eIF4GI and eIF4GII bind specifically (Lomakin et al., 2000; Marcotrigiano et al., 2001; Pestova et al., 1996a,b).

E. The influence of deletions and substitutions in the IRES-binding protein on its ability to bind to the IRES and to function in initiation of translation can be used to characterize IRES-binding determinants in a factor and to determine whether this activity is sufficient to account for the factor's involvement in initiation or whether it plays other key roles (such as recruiting other factors or even the ribosome to an IRES). Analysis of this type indicated that binding of eIF4G to the IRES and recruitment by it of eIF4A were both essential aspects of this mechanism of initiation (Lomakin et al., 2000; Marcotrigiano et al., 2001; Pestova et al., 1996b).

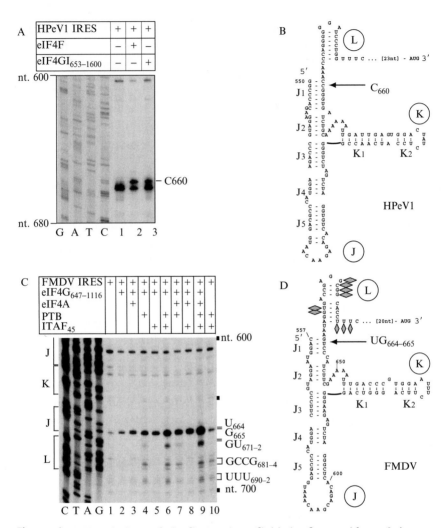

Figure 16.2 Toe printing analysis of interactions of initiation factors with type 2 picorna-virus IRESs. (A, C) Toe printing analysis of RNP complexes assembled on (A) the HPeV1 IRES and (C) the FMDV O₁K IRES by incubation under conditions as described in the text with eIF4F, eIF4GI$_{653-1600}$, eIF4GI$_{647-1116}$, eIF4A, PTB, and ITAF$_{45}$ in combinations as indicated. Toe printing was done by use of oligonucleotides complementary to HPeV1 nt. 857 to 840 and to FMDV nt. 730 to 713, respectively. The labels to the right of each panel indi-cate the nucleotides at which primer extension was arrested, and the reference lanes C, T, A, and G depict the appropriate negative strand cDNA sequence. (B, D) Secondary struc-tures of segments of (B) Human parechovirus type 1 (HPeV)1 and (D) foot-and-mouth disease virus (FMDV) O₁K IRESs, on which the prominent toe prints caused by eIF4F/eIF4G are indicated by labeled arrows. Toe prints that appear on the FMDV IRES in the presence of eIF4F/eIF4G and additional factors (PTB, ITAF4₅) are indicated by lozenges. Panel C is reproduced from Pilipenko *et al.* (2000) with permission (Copyright © Cold Spring Harbor Laboratory Press, *Genes and Development* 14, 2028–2045 [2000]).

Taken together, these approaches are sufficient to establish the specificity and functional importance of binding of a factor to the IRES.

 12. MAPPING INTERACTIONS OF THE IRES WITH COMPONENTS OF THE TRANSLATION APPARATUS BY CHEMICAL AND ENZYMATIC FOOTPRINTING

12.1. Buffers

Buffer E 20 mM Tris-HCl, pH 7.5, 2 mM DTT, 100 mM potassium acetate

Buffer F 1.5 M Sodium acetate, pH 5.0, 1 M β-mercaptoethanol, 0.1 mM EDTA

Buffer G 20 mM HEPES, pH 7.5, 100 mM potassium acetate, 2.5 mM magnesium acetate

Buffer H 50 mM Tris-HCl, pH 7.5, 1 mM EDTA, 100 mM KCl

Buffer I 150 mM Tris-HCl, pH 7.4, 40 mM DTT, 40 mM magnesium chloride

EMCV-like IRESs have a conserved structure that is thought to orient binding sites in individual IRES domains to permit binding by components of the translation apparatus and ultimately by the 43S complex, leading to 48S complex formation. The use of chemical and enzymatic footprinting techniques has enabled the interactions of ITAFs and canonical eIFs with EMCV and FMDV IRESs to be mapped and the consequent conformational changes in these IRESs to be characterized at nucleotide resolution (Fig. 16.3; Kolupaeva *et al.*, 1996, 1998, 2003; Pilipenko *et al.*, 2000). These chemical and enzymatic footprinting approaches are generally applicable, and have also been used to characterize interactions of HCV-like IRESs with eIF3 and of HCV-like and dicistrovirus IGR IRESs with the 40S subunit (Kieft *et al.*, 2001; Kolupaeva *et al.*, 2000a,b; Nishiyama *et al.*, 2003; Sizova *et al.*, 1998).

CMCT reacts with N-3 of uridine and N-1 of guanine residues, DMS reacts with N-1 of adenine residues, with N-7 of guanosine residues and to a lesser extent with N-3 of cytosine residues, and hydroxyl radicals generated from free Fe(II)-EDTA attack hydrogens at C-1' and C-4' positions of the ribose (Ehresmann *et al.*, 1987). In addition to these chemical probes, we use the enzymatic probes RNAse V1, which cleaves double-stranded and other base paired or stacked RNA, RNAse ONE, which cleaves single-stranded RNA without base specificity, and RNAse T1, which cleaves after unpaired G residues. The combined use of chemical and enzymatic probes overcomes some of their shortcomings (enzymatic probes are sensitive to steric hindrance whereas modification by DMS and CMCT is inhibited by base-pairing in RNA, which precludes their use for analysis of helical elements). The choice of reagent is governed in part by the structure (if known) of the RNA; for example, for highly structured RNAs, RNAse ONE can be omitted, which, because of its size, cuts at very few sites on such RNAs. Reverse transcription is arrested by enzymatic cleavage and by chemical modification of RNA at most positions, which can, therefore, be detected by primer extension inhibition. Arrest occurs at the base immediately 3′ to the modified base or the cleaved bond, and numbering of residues accordingly refers to the modified base or the nucleotide to the 3′ side of the cleaved bond.

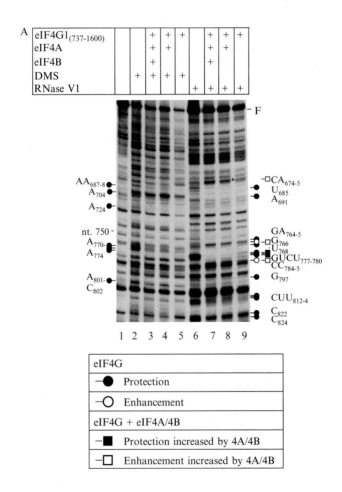

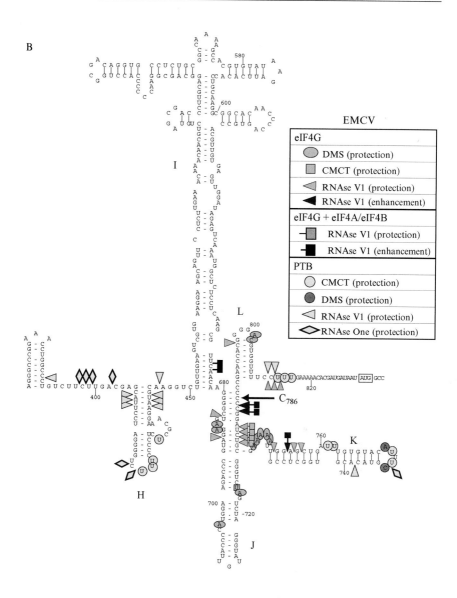

Figure 16.3 Chemical and enzymatic footprinting analyses of ribonucleoprotein complexes assembled on the EMCV IRES. (A) Chemical and enzymatic footprinting of the EMCV IRES in complexes formed with eIF4GI$_{737-1600}$, eIF4A, and eIF4B, as indicated. DMS modification or RNase V1 cleavage of the IRES, either alone or complexed with factors, was assayed by primer extension inhibition followed by polyacrylamide-urea gel fractionation of cDNA products. cDNAs obtained after primer extension of untreated EMCV RNA are shown in lane 1. The full-length cDNA extension product is marked F, protected residues are numbered, and their positions are indicated by the

RNP complexes are assembled by incubating IRES-containing mRNA transcripts with a twofold to threefold molar excess of IRES-binding proteins in buffer E + 2 mM magnesium acetate (20 μl total reaction volume) for 5 min at 30°. Typical reaction mixtures include 1 μg EMCV nt. 378 to 1155 RNA (4 pmol), and either 0.6 μg PTB1 (10 pmol), 3 μg eIF4F (13 pmol), or 1 μg eIF4GI$_{653-1600}$ (10 pmol) alone or with combinations of 1 μg eIF4A (22 pmol), 1.5 μg eIF4B (21 pmol), and Mg^{2+}-ATP (1 mM). Chemical modification is initiated by addition to reaction mixtures containing free RNA (as a control) and assembled RNP complexes of 2 μl of a fresh 5% solution of DMS in ethanol and incubation for 5 min at 30° or by addition of 10 μl of a solution of CMCT (42 g/ml) in buffer E + 2 mM magnesium acetate and incubation for 10 min at 30°. The conditions for probing each RNA should be experimentally optimized; for EMCV-like IRESs, we found it is easier to vary the concentration of the DMS solution (3 to 10%) keeping the duration of incubation constant, whereas for CMCT, we varied the period of incubation (10 to 30 min) and kept the concentration of CMCT constant. Complex formation and modification and/or cleavage can be also done at 37°. Modification of RNA by CMCT is optimal at pH 8.0, but remains significant at pH 7.5, which is optimal for 48S complex formation on the IRES. Complexes were, therefore, assembled at pH 7.5 for probing. Both reactions are quenched by addition of $\frac{1}{4}$ volume of buffer F, after which RNAs are phenol-extracted and precipitated with 3 volumes of ethanol. The RNA precipitation step is crucial, and to minimize loss of samples that by definition can be heterogeneous in length, we suggest precipitation overnight at −20°. Enzymatic cleavage is initiated by addition to reaction mixtures containing free RNA (as a control) and assembled RNP complexes of freshly diluted RNAse V1 (final concentration of 0.0007 U/ml), RNAse ONE (0.0005 U/ml final concentration), or RNAse T1 (0.015 U/ml final concentration), followed by incubation at 30° for 15 min, extraction twice with phenol-chloroform, and precipitation with three volumes of ethanol. Hydroxyl radical cleavage

---

symbols shown in the key under the gel. A dideoxynucleotide sequence generated with the same primer was run in parallel (lanes C, T, A, and G in panel A). This panel is reproduced from Kolupaeva *et al.* (2003) with permission (Copyright © American Society for Microbiology, *Molecular and Cellular Biology* 23, 687–698 [2003]). (B) Sites at which chemical modification by DMS and CMCT and enzymatic cleavage by RNase V1 of the IRES were altered by eIF4G$_{737-1600}$ alone and in the presence of eIFA/eIF4B are mapped onto a secondary structure model of nt 382 to 815 of the EMCV IRES and are indicated by the symbols shown in the key on the right. These footprinting data are from reports by Kolupaeva *et al.* (1998, 2003). Domains of the IRES domain are designated H to L (13), the initiation codon AUG$_{834}$ is boxed and the position of the toe print because of RT arrest in the J-K domain by bound eIF4G is indicated by a black arrow.

is done essentially as described (Huttenhofer *et al.*, 1992), by preparing RNP complexes in buffer G (36 μl final volume) and placing them on ice. To initiate cleavage, 5 μl H$_2$O, 5 μl 50 mM ferrous ammonium sulfate, and 5 μl 100 mM sodium EDTA, 5 μl 2.5% (vol/vol) H$_2$O$_2$, and 5 μl 250 mM sodium ascorbate are first mixed sequentially, and 4 μl of this mixture is immediately added to the RNP mixture and to the "free RNA" control. The reaction is done at 0° for 10 min and quenched by adding thiourea (10 mM final concentration). Samples are then ethanol-precipitated, resuspended, extracted twice with phenol-chloroform, and used for primer extension. Reactions can be scaled up proportionately if a long RNA is to be analyzed by use of a panel of primers.

Primer annealing conditions are designed both to maximize unfolding of the RNA so that the primer anneals efficiently, and to minimize degradation of RNA. Cleaved or modified IRES-containing RNAs are dissolved in 9 μl (hybridization) buffer H, incubated with a threefold excess of an appropriate DNA primer, heated to 65° and allowed to cool slowly to room temperature (20°). Some primers hybridize so readily that annealing can be done at 37° with a fivefold excess of primer. As in toe printing assays, we use primers that are 17 to 25 nt-long and are designed to hybridize to mRNA at ~200-nt. intervals (which corresponds to the limit of resolution of sequencing gels for such analysis). The choice of primers is dictated by the length of an mRNA and by the location of a putative binding site on it. The optimum distance from a primer to the site of interest is 50 to 150 nt; it may be helpful to verify data by use of additional primers that bind elsewhere within this region.

Primer extension is done by adding 4 μl (reverse transcriptase) buffer I, 1 μl dNTPs (5 mM dCTP, dGTP and dTTP; 1 mM dATP), 1 μl [α-^{32}P]dATP (111 TBq/mmol), 10 U RNase inhibitor, and 2 U AMV RT. Reaction mixtures are incubated for 30 min at 37° and then for 2 min at 100° before degrading the RNA template by incubation with RNAse A (0.1 mg/ml final concentration) for 20 min at 37°. cDNA products are phenol-extracted (pH 8.0), ethanol-precipitated, resuspended in 8 μl loading buffer, and heated at 95° for 3 min. Aliquots (2 μl) are then analyzed by electrophoresis in a prerun 8 to 10% polyacrylamide/7 M urea sequencing gel. cDNA products are compared with dideoxynucleotide sequence ladders obtained by use of appropriate plasmid DNA and the same primer in reactions with Sequenase, exactly as for toe printing. If modifications to the RNA occur very close to the primer, toe printing can instead be done by use of an oligodeoxynucleotide primer that has been [^{32}P]-end-labeled by use of T4 polynucleotide kinase in primer extension reactions that contain 1 μl of dNTPs (20 mM total concentration) and that lack [α-^{32}P]dATP.

Analysis of cleavage/modification patterns obtained by use of free RNA should be compared with the pattern of reverse transcriptase stops on untreated RNA to identify endogenous "strong stops" or breaks in the template and can then be used to evaluate the relevant IRES secondary structure model. Comparison of patterns of cleavage/modification of IRES RNA in RNP complexes with corresponding patterns of free RNA can then be used to identify binding sites for proteins and to monitor conformational changes in the RNA. In evaluating data, the fact that rates of modification at different positions do vary should be borne in mind; DMS reacts more slowly with N-3 of cytosines than with N-1 of adenines, and CMCT reacts more slowly with N-1 of guanosine than with N-3 of uridine. We have used the same methods to map binding sites for eIF3 and the 40S subunit on HCV and related IRESs; probing of mRNA/ribosome complexes is complicated by the fact that 18S rRNA of the 40S subunit can act as a "sink" for these probes, In such cases, the concentration of chemical or enzymatic probe in reaction mixtures is raised slightly (e.g., Kolupaeva *et al.*, 2000a,b).

The approaches described here have been used to confirm that the eIF4G/4A complex binds primarily to the J-K domains of EMCV and FMDV IRESs and that eIF4B enhanced protection of the IRES by this complex, consistent with its increased stability that is apparent from toe printing data (Kolupaeva *et al.*, 1998, 2003; Pilipenko *et al.*, 2000). Binding of the eIF4G/eIF4A–eIF4B complex to the IRES also induced conformational changes in it that are apparent as enhanced RNAse cleavage at a few defined sites. PTB (which contains four RNA-binding domains) binds to three noncontiguous sites on the EMCV and FMDV IRESs (in and around domain H, in domain K, and downstream of domain L), consistent with its proposed role as a chaperone that maintains the IRES in an active conformation (Fig. 16.3B).

13. Directed Hydroxyl Radical Cleavage

13.1. Buffers

Buffer J 80 mM HEPES, pH 7.5, 300 mM KCl, 10% glycerol
Buffer K 80 mM HEPES, pH 7.5, 100 mM KCl, 2.5 mM MgAc, 10% glycerol

Directed hydroxyl radical cleavage constitutes a valuable complement to footprinting in defining the interactions of initiation factors with an IRES. Whereas footprinting yields data concerning the elements of an IRES that are bound by a factor, directed hydroxyl radical cleavage can, depending on whether the structure of only the factor or also of its binding site are known, define the orientation of the factor on the element of this IRES to which it binds or provide data that allows the two to be docked together. In this approach, iron (II) tethered to different unique surface-exposed cysteine residues on a protein by means of the linker 1-(p-bromoacetamidobenzyl)-EDTA (BABE) is used to catalyze formation of hydroxyl radicals by Fenton chemistry that cleave nearby RNA (Culver and Noller, 2000). Correlation of the cleavage sites with the positions of the modified cysteine residues on the mutant factors yields a set of constraints that can be used to dock the protein onto the RNA if the structure of both is known, or as in the example described in the following, allows their relative orientation to be determined (Fig. 16.4).

First, mutant forms of the protein of interest are prepared that lack all cysteine residues; mutants can also be prepared that retain single native cysteine residues. The cysteine-less mutant protein is then used as the basis for synthesis by conventional PCR mutagenesis of a panel of mutants with single cysteine residues at well-distributed, surface-exposed positions. The first important control is to determine that such mutant factors have comparable activity to the wild-type factor both before and after derivatization with Fe(II)-BABE. In the present example, proteins are assayed for their activity in binding to the IRES and supporting 48S complex formation; only those that had comparable activity to the wild-type protein can be expected to yield data that leads to a valid model for the interaction of the protein with the IRES.

Single cysteine mutant proteins are derivatized with Fe(II)-BABE by use of a modified version of an established procedure (Culver and Noller, 2000). Twenty to 50 μg of single cysteine mutant protein and (as a negative control) cysteine-less mutant protein in 60 μl buffer J are incubated with 5 μl 20 mM Fe(II)-BABE at 37° for 30 min. For consistency, we modify all single-cysteine mutant proteins simultaneously by use of the same batch of reagents. Fe(II)-BABE–derivatized protein is then loaded onto Microcon YM-10 microconcentrators and separated from unincorporated Fe(II)-BABE by four cycles of concentration and dilution with buffer J. Modified proteins are stored at a concentration of 0.3 to 0.4 mg/ml at −80°. Protein can be adsorbed to these microconcentrators, and we, therefore, verify protein recovery by SDS-PAGE. Comparison of the reactivity of the fluorescent reagent DCIA with cysteine sulfhydryl groups before derivitization and after (when reactive groups should be blocked by bound Fe(II)-BABE derivatization) is used to estimate the accessibility of cysteine residues in mutant proteins to Fe(II)-BABE derivatization and the efficiency of this reaction as described (Culver and Noller, 2000). The cysteine-less mutant form of the protein can be used as a control that does not bind DCIA and thus did not fluoresce after incubation with it.

To analyze the proximity of residues of the protein of interest to elements of the IRES, complexes of the IRES (0.2 μg) with ~1 μg [Fe(II)-BABE]–derivatized protein alone and with similar amounts of any potential cofactors are formed in 20 μl buffer K by incubating at 37° for

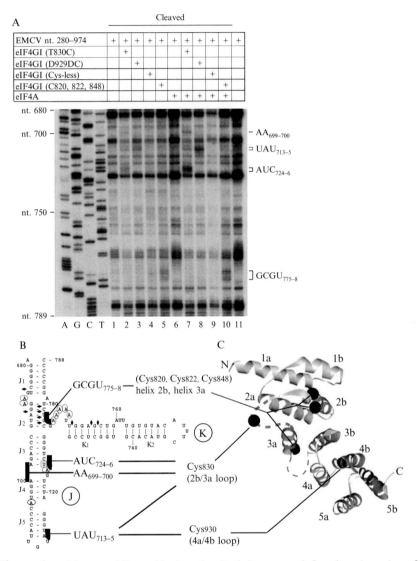

Figure 16.4 The use of directed hydroxyl radical cleavage to define the orientation of the central HEAT-repeat domain of eIF4G on the J-K domain of the EMCV IRES. The activity of unmodified and Fe(II)-BABE–derivatized mutant forms of eIF4GI$_{737-1116}$ in binding to the EMCV IRES and in supporting 48S complex formation on it in *in vitro* reconstituted initiation reactions were comparable to that of the analogous wild-type eIF4GI fragment. (A) Directed hydroxyl radical probing. Probing reactions contained EMCV nt. 280 to 974 RNA, Fe(II)-BABE–derivatized mutant forms of eIF4GI737–1116 as indicated, and eIF4A (lanes 6 to 10). Specific hydroxyl radical cleavages are seen as additional bands, as indicated to the right of the panel. The dideoxynucleotide sequencing lanes used to identify sites of cleavage are indicated (A, G, C, and T). (B) Sites of directed hydroxyl radical cleavages in the EMCV IRES from Fe(II) tethered to specific

5 min and are then chilled on ice. A 0.1 mM ascorbic acid/0.05% H_2O_2 mixture is prepared freshly and 2-μl aliquots of it are added to the factor/ IRES complexes to generate hydroxyl radicals in the vicinity of the tethered Fe(II). After incubation on ice for 5 min, reactions are quenched by addition of 20 μl 20 mM thiourea, and the IRES RNA is then phenol-extracted and ethanol precipitated. Sites of hydroxyl radical cleavage are identified by primer extension, done essentially as described previously by use of AMV RT and primers complementary to different regions of the RNA. As in toe printing and footprinting experiments, the pattern of stops in these experiments should be compared with the pattern of RT stops on mock-treated RNA (i.e., incubated with cysteine-less parental protein that had been treated with Fe(II)-BABE) to identify endogenous "strong stops." Stops that are due to cleavage by hydroxyl radicals are categorized as being strong, medium, or weak and can be correlated with the distance between the position of Fe(II)BABE and the site of cleavage: strong cleavage occurs within 0 to 22Å, medium cleavage from 12 to 36 Å, and weak cleavage from 20 to 44 Å (Culver and Noller, 2000).

The utility of the directed hydroxyl radical probing is illustrated here by data derived by use of [Fe(II)-BABE]–derivatized eIF4G$_{737-1116}$ bound to the EMCV IRES in the presence and absence of eIF4A (Kolupaeva et al., 2003). This eIF4G fragment consists of five stacked pairs of α-helices ("HEAT repeats") (Marcotrigiano et al., 2001), binds to both eIF4A and the IRES, and is the essential central domain of eIF4G that is required for initiation on this IRES (Lomakin et al., 2000). The directed hydroxyl radical probing data (Fig. 16.4A) were used to determine the orientation of this domain of eIF4G on the J-K domain of the EMCV IRES (Fig. 16.4B, C), revealing that it binds to it in a manner that would enable it to direct the helicase activity of eIF4A to the vicinity of the initiation codon. In addition to being used to determine the orientation of an RNA-binding protein when bound to its target, this approach can also be used to map the location of initiation factors on the ribosome and by extrapolation to deduce the relative location of IRESs bound to ribosome/factor complexes (Lomakin et al., 2003; Pestova et al., 2004).

positions on eIF4G shown on a model of the IRES' J-K domain which, for comparison, is annotated to show residues in this domain that are protected from chemical modification or enzymatic cleavage by eIF4G. (C) Positions of cysteine residues used as tethering sites for Fe(II)-BABE are indicated as black-filled circles on a ribbon drawing of the central region of eIF4G, viewed along the central axis of the α-helices, which are labeled 1a, 1b, etc. Crystallographically disordered loops are indicated by dashes. This figure is adapted from Kolupaeva et al. (2003) and is reproduced with permission (Copyright © American Society for Microbiology, Molecular and Cellular Biology 23, 687–698 [2003]).

ACKNOWLEDGMENTS

This work was supported in part by Grant AI-51340 from the National Institutes of Health.

REFERENCES

Alkalaeva, E. Z., Pisarev, A. V., Frolova, L. Y., Kisselev, L. L., and Pestova, T. V. (2006). *In vitro* reconstitution of eukaryotic translation reveals cooperativity between release factors eRF1 and eRF3. *Cell* **125,** 1125–1136.
Anthony, D. D., and Merrick, W. C. (1992). Analysis of 40 S and 80 S complexes with mRNA as measured by sucrose density gradients and primer extension inhibition. *J. Biol. Chem.* **267,** 1554–1562.
Borman, A. M., Bailly, J. L., Girard, M., and Kean, K. M. (1995). Picornavirus internal ribosome entry segments: Comparison of translation efficiency and the requirements for optimal internal initiation of translation *in vitro. Nucleic Acids Res.* **23,** 3656–3663.
Borovjagin, A., Pestova, T., and Shatsky, I. (1994). Pyrimidine tract binding protein strongly stimulates *in vitro* encephalomyocarditis virus RNA translation at the level of preinitiation complex formation. *FEBS Lett.* **351,** 299–302.
Culver, G. M., and Noller, H. F. (2000). Directed hydroxyl radical probing of RNA from iron(II) tethered to proteins in ribonucleoprotein complexes. *Methods Enzymol.* **318,** 461–475.
Dorner, A. J., Semler, B. L., Jackson, R. J., Hanecak, R., Duprey, E., and Wimmer, E. (1984). *In vitro* translation of poliovirus RNA: Utilization of internal initiation sites in reticulocyte lysate. *J. Virol.* **50,** 507–514.
Ehresmann, C., Baudin, F., Mougel, M., Romby, P., Ebel, J. P., and Ehresmann, B. (1987). Probing the structure of RNAs in solution. *Nucleic Acids Res.* **15,** 9109–9128.
Evstafieva, A. G., Ugarova, T. Y., Chernov, B. K., and Shatsky, I. N. (1991). A complex RNA sequence determines the internal initiation of encephalomyocarditis virus RNA translation. *Nucleic Acids Res.* **19,** 665–671.
Fletcher, S. P., Ali, I. K., Kaminski, A., Digard, P., and Jackson, R. J. (2002). The influence of viral coding sequences on pestivirus IRES activity reveals further parallels with translation initiation in prokaryotes. *RNA* **8,** 1558–1571.
Hellen, C. U. T., and Sarnow, P. (2001). Internal ribosome entry sites in eukaryotic mRNA molecules. *Genes Dev.* **15,** 1593–1612.
Hellen, C. U. T., Witherell, G. W., Schmid, M., Shin, S. H., Pestova, T. V., Gil, A., and Wimmer, E. (1993). A cytoplasmic 57-kDa protein that is required for translation of picornavirus RNA by internal ribosomal entry is identical to the nuclear pyrimidine tract-binding protein. *Proc. Natl. Acad. Sci. USA* **90,** 7642–7646.
Huttenhofer, A., and Noller, H. F. (1992). Hydroxyl radical cleavage of tRNA in the ribosomal P site. *Proc. Natl. Acad. Sci. USA* **89,** 7851–7855.
Jackson, R. J., and Kaminski, A. (1995). Internal initiation of translation in eukaryotes: The picornavirus paradigm and beyond. *RNA* **1,** 985–1000.
Jang, S. K., Krausslich, H. G., Nicklin, M. J., Duke, G. M., Palmenberg, A. C., and Wimmer, E. (1988). A segment of the 5′ nontranslated region of encephalomyocarditis virus RNA directs internal entry of ribosomes during *in vitro* translation. *J. Virol.* **62,** 2636–2643.
Kaminski, A., Hunt, S. L., Gibbs, C. L., and Jackson, R. J. (1994). Internal initiation of mRNA translation in eukaryotes. *Genet. Eng.* **16,** 115–155.

Kaminski, A., and Jackson, R. J. (1998). The polypyrimidine tract binding protein (PTB) requirement for internal initiation of translation of cardiovirus RNAs is conditional rather than absolute. *RNA* **4,** 626–638.

Kieft, J. S., Zhou, K., Jubin, R., and Doudna, J. A. (2001). Mechanism of ribosome recruitment by hepatitis C IRES RNA. *RNA* **7,** 194–206.

Kolupaeva, V. G., Hellen, C. U. T., and Shatsky, I. N. (1996). Structural analysis of the interaction of the pyrimidine tract–binding protein with the internal ribosomal entry site of encephalomyocarditis virus and foot-and-mouth disease virus RNAs. *RNA* **2,** 1199–1212.

Kolupaeva, V. G., Lomakin, I. B., Pestova, T. V., and Hellen, C. U. T. (2003). Eukaryotic initiation factors 4G and 4A mediate conformational changes downstream of the initiation codon of the encephalomyocarditis virus internal ribosomal entry site. *Mol. Cell. Biol.* **23,** 687–698.

Kolupaeva, V. G., Pestova, T. V., and Hellen, C. U. T. (2000a). An enzymatic footprinting analysis of the interaction of 40S ribosomal subunits with the internal ribosomal entry site of hepatitis C virus. *J. Virol.* **74,** 6242–6250.

Kolupaeva, V. G., Pestova, T. V., and Hellen, C. U. T. (2000b). Ribosomal binding to the internal ribosomal entry site of classical swine fever virus. *RNA* **6,** 1791–1807.

Kolupaeva, V. G., Pestova, T. V., Hellen, C. U. T., and Shatsky, I. N. (1998). Translation eukaryotic initiation factor 4G recognizes a specific structural element within the internal ribosome entry site of encephalomyocarditis virus RNA. *J. Biol. Chem.* **273,** 18599–18604.

Lomakin, I. B., Hellen, C. U. T., and Pestova, T. V. (2000). Physical association of eukaryotic initiation factor 4G (eIF4G) with eIF4A strongly enhances binding of eIF4G to the internal ribosomal entry site of encephalomyocarditis virus and is required for internal initiation of translation. *Mol. Cell. Biol.* **20,** 6019–6029.

Lomakin, I. B., Kolupaeva, V. G., Marintchev, A., Wagner, G., and Pestova, T. V. (2003). Position of eukaryotic initiation factor eIF1 on the 40S ribosomal subunit determined by directed hydroxyl radical probing. *Genes Dev.* **17,** 2786–2797.

Lomakin, I. B., Shirokikh, N. E., Yusupov, M. M., Hellen, C. U. T., and Pestova, T. V. (2006). The fidelity of translation initiation: Reciprocal activities of eIF1, IF3 and YciH. *EMBO J.* **25,** 196–210.

Luz, N., and Beck, E. (1990). A cellular 57 kDa protein binds to two regions of the internal translation initiation site of foot-and-mouth disease virus. *FEBS Lett.* **269,** 311–314.

Lyakhov, D. L., Ilgenfritz, H., Chernov, B. K., Dragan, S. M., Rechinsky, V. O., Pokholok, D. K., Tunitskaya, V. L., and Kochetkov, S. N. (1992). Site-specific mutagenesis of the Lys-172 residue in phage T7 RNA polymerase: Characterization of the transcriptional properties of mutant proteins. *Mol. Biol.* **26,** 679–687.

Marcotrigiano, J., Lomakin, I. B., Sonenberg, N., Pestova, T. V., Hellen, C. U., and Burley, S. K. (2001). A conserved HEAT domain within eIF4G directs assembly of the translation initiation machinery. *Mol. Cell* **7,** 193–203.

Merrill, M. K., and Gromeier, M. (2006). The double-stranded RNA binding protein 76:NF45 heterodimer inhibits translation initiation at the rhinovirus type 2 internal ribosome entry site. *J. Virol.* **80,** 6936–6942.

Nateri, A.S, Highes, P. J., and Stanway, G. (2000). *In vivo* and *in vitro* identification of structural and sequence elements of the human percehovirus 5′ untranslated region required for internal initiation. *J. Virol.* **74,** 6269–6277.

Nishiyama, T., Yamamoto, H., Shibuya, N., Hatakeyama, Y., Hachimori, A., Uchiumi, T., and Nakashima, N. (2003). Structural elements in the internal ribosome entry site of *Plautia stali* intestine virus responsible for binding with ribosomes. *Nucleic Acids Res.* **31,** 2434–2442.

Pause, A., Methot, N., and Sonenberg, N. (1993). The HRIGRXXR region of the DEAD box RNA helicase eukaryotic translation initiation factor 4A is required for RNA binding and ATP hydrolysis. *Mol. Cell. Biol.* **13,** 6789–6798.

Pelletier, J, and Sonenberg, N. (1988). Internal initiation of translation of eukaryotic mRNA directed by a sequence derived from poliovirus RNA. *Nature* **334,** 320–325.

Pestova, T. V., Borukhov, S. I., and Hellen, C. U. T. (1998a). Eukaryotic ribosomes require initiation factors 1 and 1A to locate initiation codons. *Nature* **394,** 854–859.

Pestova, T. V., and Hellen, C. U. T. (1999). Internal initiation of translation of bovine viral diarrhea virus RNA. *Virology* **258,** 249–256.

Pestova, T. V., and Hellen, C. U. T. (2001). Preparation and activity of synthetic unmodified mammalian tRNAi(Met) in initiation of translation *in vitro*. *RNA* **7,** 1496–1505.

Pestova, T. V., and Hellen, C. U. T. (2003). Translation elongation after assembly of ribosomes on the Cricket paralysis virus internal ribosomal entry site without initiation factors or initiator tRNA. *Genes Dev.* **17,** 181–186.

Pestova, T. V., and Hellen, C. U. T. (2005). Reconstitution of eukaryotic translation elongation *in vitro* following initiation by internal ribosomal entry. *Methods* **36,** 261–269.

Pestova, T. V., Hellen, C. U. T., and Shatsky, I. N. (1996a). Canonical eukaryotic initiation factors determine initiation of translation by internal ribosomal entry. *Mol. Cell. Biol.* **16,** 6859–6869.

Pestova, T. V., Hellen, C. U. T., and Wimmer, E. (1991). Translation of poliovirus RNA: Role of an essential cis-acting oligopyrimidine element within the 5′ nontranslated region and involvement of a cellular 57-kilodalton protein. *J. Virol.* **65,** 6194–6204.

Pestova, T. V., Lomakin, I. B., and Hellen, C. U. T. (2004). Position of the CrPV IRES on the 40S subunit and factor dependence of IRES/80S ribosome assembly. *EMBO Rep.* **5,** 906–913.

Pestova, T. V., Maslova, S. V., Potapov, V. K., and Agol, V. I. (1989). Distinct modes of poliovirus polyprotein initiation *in vitro*. *Virus Res.* **14,** 107–118.

Pestova, T. V., Shatsky, I. N., Fletcher, S. P., Jackson, R. J., and Hellen, C. U. T. (1998b). A prokaryotic-like mode of cytoplasmic eukaryotic ribosome binding to the initiation codon during internal translation initiation of hepatitis C and classical swine fever virus RNAs. *Genes Dev.* **12,** 67–83.

Pestova, T. V., Shatsky, I. N., and Hellen, C. U. T. (1996b). Functional dissection of eukaryotic initiation factor 4F: The 4A subunit and the central domain of the 4G subunit are sufficient to mediate internal entry of 43S preinitiation complexes. *Mol. Cell. Biol.* **16,** 6870–6878.

Pilipenko, E. V., Pestova, T. V., Kolupaeva, V. G., Khitrina, E. V., Poperechnaya, A. N., Agol, V. I., and Hellen, C. U. T. (2000). A cell cycle-dependent protein serves as a template-specific translation initiation factor. *Genes Dev.* **14,** 2028–2045.

Pisarev, A. V., Shirokikh, N. E., and Hellen, C. U. T. (2005). Translation initiation by factor-independent binding of eukaryotic ribosomes to internal ribosomal entry sites. *C. R. Biol.* **328,** 589–605.

Pisarev, A. V., Unbehaun, A., Hellen, C. U. T., and Pestova, T. V. (2007). Assembly and analysis of eukaryotic translation initiation complexs. *Methods Enzymol.* **430,** 147–177.

Shibuya, N., Nishiyama, T., Kanamori, Y., Saito, H., and Nakashima, N. (2003). Conditional rather than absolute requirements of the capsid coding sequence for initiation of methionine-independent translation in *Plautia stali* intestine virus. *J. Virol.* **77,** 12002–12010.

Sizova, D. V., Kolupaeva, V. G., Pestova, T. V., Shatsky, I. N., and Hellen, C. U. T. (1998). Specific interaction of eukaryotic translation initiation factor 3 with the 5′ nontranslated regions of hepatitis C virus and classical swine fever virus RNAs. *J. Virol.* **72,** 4775–4782.

Svitkin, Y., Pestova, T. V., Maslova, S. V., and Agol, V. I. (1988). Point mutations modify the response of poliovirus RNA to a translation initiation factor: A comparison of neurovirulent and attenuated strains. *Virology* **166,** 394–404.

Van Eden, M. E., Byrd, M. P., Sherrill, K. W., and Lloyd, R. E. (2004). Demonstrating internal ribosome entry sites in eukaryotic mRNAs using stringent RNA test procedures. *RNA* **10,** 720–730.

Wilson, J. E., Pestova, T. V., Hellen, C. U. T., and Sarnow, P. (2000). Initiation of protein synthesis from the A site of the ribosome. *Cell* **102,** 511–520.

Author Index

Subject Index

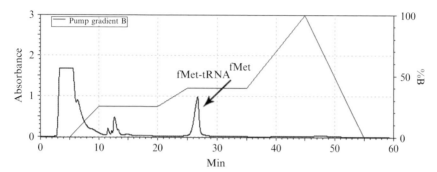

Sean M. Studer and Simpson Joseph, Figure 2.2 Purification of fMet-tRNA^{fMet}: Shown is an HPLC profile of fMet-tRNA^{fMet} eluting from a C18 column. The sample was eluted with a gradient of 0 to 100% buffer B over 45 min. Buffer A was 20-mM Tris-acetate (pH 5.0), 400-mM NaCl, and 10-mM magnesium acetate. Buffer B was the same as buffer A with 60% methanol. Aminoacylated tRNA elutes from the column at 40% buffer B.

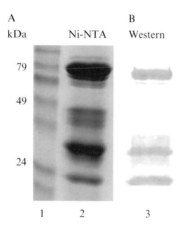

Domenick G. Grasso et al., Figure 4.1 Spectrum of proteins present in Ni-NTA–purified preparations of IF2$_{mt}$. (A) A sample (10 μg) of the Ni-NTA–purified IF2$_{mt}$ was run on a 10% SDS-polyacrylamide gel and stained with Coomassie Blue (lane 2). Lane 1 contains the Benchmark^{TM} Protein Ladder. (B) Western blot (lane 3) of the partially purified IF2$_{mt}$ preparation showing that many of the contaminating bands arise from proteolytic degradation of the full-length product.

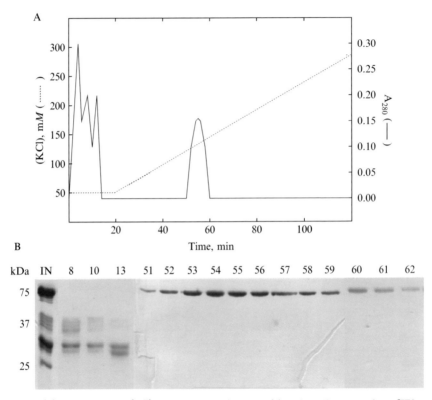

Domenick G. Grasso *et al.*, Figure 4.2   Purification of the Ni–NTA preparation of IF2$_{mt}$ on TSK gel DEAE-5PW HPLC. A sample of partially purified IF2$_{mt}$ was subjected to chromatography on HPLC as described in the text. (A) The absorbance pattern at 280 nm was monitored during chromatography with an ISCO UA-6 Absorbance detector on a 0.5 absorbance scale using a 5-mm flow cell. (B) Aliquots (16 μl) of the indicated fractions were examined by electrophoresis on 10% SDS-polyacrylamide gels for the presence of the 74-kDa mature form of the expressed IF2$_{mt}$. The designations kDa and IN indicate the Benchmark^{TM} Protein Ladder and the starting sample that had been passed through the Ni–NTA column and that was applied to the HPLC column. Fractions 52 to 59 were pooled and used as the source of purified IF2$_{mt}$. Three separate gels were required for the analysis of the full-column pattern. The figure shown is a composite of the relevant regions of these three gels.

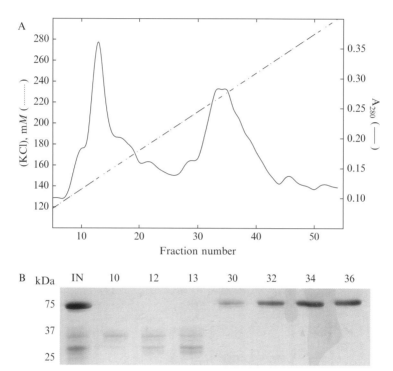

Domenick G. Grasso *et al.*, Figure 4.3 Purification of IF2$_{mt}$ on a gravity DEAE-Sepharose column. (A) The absorbance pattern at 280 nm was monitored during chromatography with an ISCO UA-6 Absorbance detector on a 0.5 absorbance scale with a 5 mm flow cell. (B) Analysis of the purity of the IF2$_{mt}$ in the column fractions. Aliquots (10 μl) of the indicated fractions were subjected to electrophoresis on 10% SDS-polyacrylamide gels and stained with Coomassie Blue. IN represents the protein pattern in the input to the DEAE-Sepharose column.

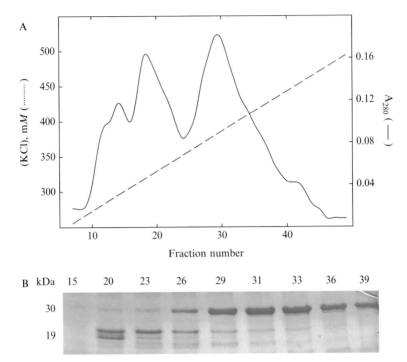

Domenick G. Grasso *et al.*, Figure 4.5 Elution profile of IF3$_{mt}$ from the first gravity S-Sepharose column using a 250 to 550 mM KCl gradient in Buffer I. (A) The absorbance pattern at 280 nm was monitored during chromatography with an ISCO UA-6 Absorbance detector and chart recorder on a 0.5 absorbance scale and a 5-mm flow cell. (B) Aliquots (5 μl) of the indicated fractions were run on a 15% SDS-polyacrylamide gel and the protein bands located by staining with Coomassie Blue. Fractions 26 to 39 were pooled and dialyzed against 100 vol of ice cold Buffer I for 2 h with a change of buffer after 1 h. This sample (∼10.5 ml, ∼2.5 mg) was applied to a second S-Sepharose column and subjected to chromatography as shown in Fig. 6.

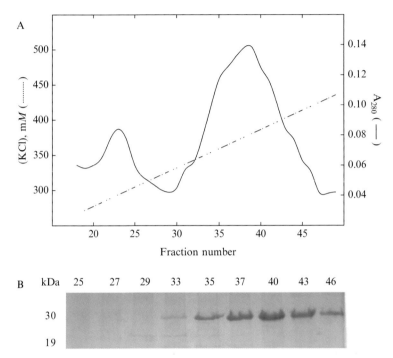

Domenick G. Grasso *et al.*, Figure 4.6 Passage of the partially purified IF3$_{mt}$ on a second S-Sepharose column using the same gradient conditions. A sample (~10.5 ml, ~2.5 mg) from the first S-Sepharose column was applied to an identical second S-Sepharose column and subjected to chromatography as described in the text. (A) The absorbance pattern at 280 nm was monitored during chromatography with an ISCO UA-6 Absorbance detector and chart recorder on a 0.5 absorbance scale. (B) Analysis of fractions for the presence of intact IF3$_{mt}$. Aliquots (10 μl) of the indicated fractions were applied to a 15% SDS-polyacrylamide gel and the protein bands were located by staining with Coomassie Blue. The material in fractions 35 to 46 was pooled and dialyzed against 100 vol of ice cold Buffer I for 2 h with a change of buffer after 1 h.

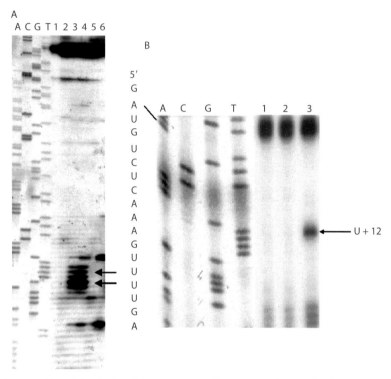

Dario Benelli and Paola Londei, Figure 5.3 "Toeprinting" assays for determining the position of the 30S subunit on leadered and leaderless mRNAs. (A) Samples containing 30S subunits and a leadered mRNA were incubated for 15 min before primer extension. 1 and 2, wild-type mRNA 104, at 37° and 65°, without ribosomes. 3 and 4, wild-type mRNA with 30S subunits at 37° and 65°. 5 and 6, SD-less leadered mRNA with 30S subunits at 37° and 65°. (B) 1, leaderless mRNA alone; 2, mRNA and 30S subunits; 3, mRNA, 30S subunits and Met-tRNAi. Lanes ACGT, sequencing reaction. The arrows indicate the position of the toeprint signals. (Modified from Benelli *et al.*, 2003; Condo *et al.*, 1999.)

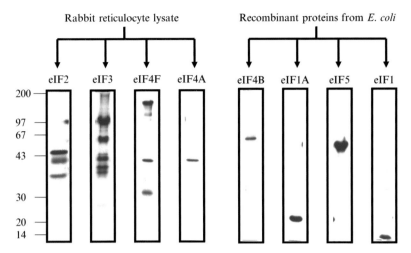

Romit Majumdar et al., Figure 8.4 SDS-polyacrylamide gel electrophoresis of purified translation initiation factors.

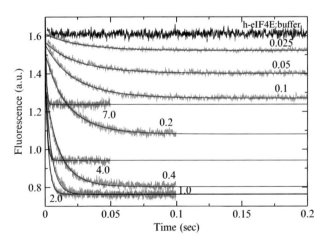

Anna Niedzwiecka et al., Figure 9.10 Stopped-flow kinetic traces for mixing of human eIF4E, concentration 0.2 μM, with m[7]GpppG, concentrations in μM indicated by the numbers, and the fitting by DynaFit software assuming the two-step association mechanism; $k_{+1} = 200\ \mu M^{-1}s^{-1}$, $k_{-1} = 17.3\ sec^{-1}$, $k_{+2} = 5.8\ sec^{-1}$, $k_{-2} = 67.5\ sec^{-1}$. The trace marked h-eIF4E buffer corresponds to mixing of the protein solution with equal volume of pure buffer.

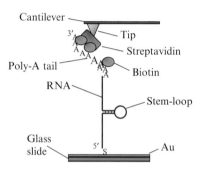

John E. G. McCarthy *et al.*, Figure 10.3 A summary of the method used for suspending an RNA molecule between the AFM tip and a gold-coated microscope slide. The RNA molecule was attached to the gold surface by means of a thiol modification at the 5′ end. The 3′ end was modified by the incorporation of biotin–ATP residues, which allowed picking up of this end with a streptavidin–coated cantilever tip.

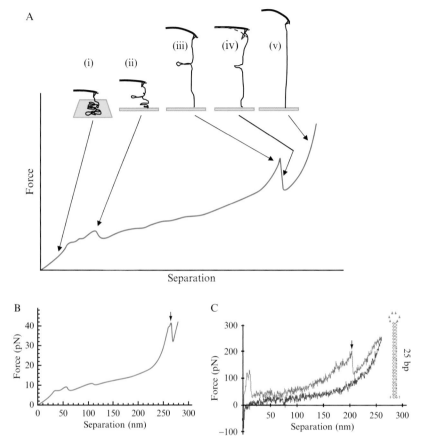

John E. G. McCarthy *et al.*, Figure 10.4 (A) A theoretical force-extension curve for structured-RNA stretching carried out by the AFM. The curve is annotated with cartoons showing the state of the RNA molecule and AFM cantilever at different extensions. (i) RNA is uncoiled off the gold surface, causing an entropic force increase; (ii) very weak secondary structure interactions are removed; (iii) enthalpic forces increase as the molecule is pulled taut and force becomes high enough for the hydrogen bonds in the strong specific stem loop to break; (iv) force temporarily decreases as the slack released from stem-loop opening is pulled out; (v) force increases further once the RNA is pulled taut once more. (B) A theoretical retraction force curve for the stretching of the *GCS4* L1-RNA transcript (with a 65-nt Poly[A] tail). This trace is adapted from data generated by use of the online "RNA pulling server" at http://bioserv.mps.ohio-state.edu/rna (Gerland *et al.*, 2001). The model assumes a temperature of 37°, 1 M NaCl, and a nucleotide length of 0.334 nm. The predicted ~40pN GC-rich stem-loop opening feature is labelled with an arrow. (C) Example of an AFM force-curve representing stretching of a *GCN4* RNA molecule containing the GC-rich (25 base pair) stem loop (inset). The discontinuity feature resulting from stem-loop opening is indicated by a small arrow. The approach curve (bottom curve) runs from right to left and the retract curve (top curve) from left to right.

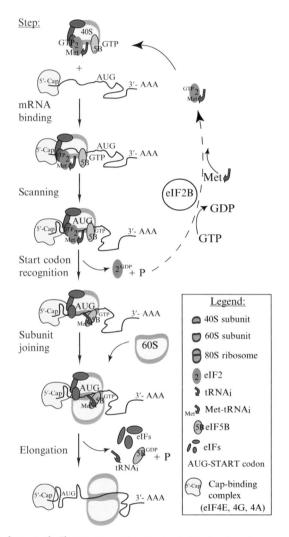

Assen Marintchev *et al.*, Figure 12.1 Translation initiation in eukaryotes. A simplified diagram of the translation initiation pathway. The legend is shown on the bottom right.

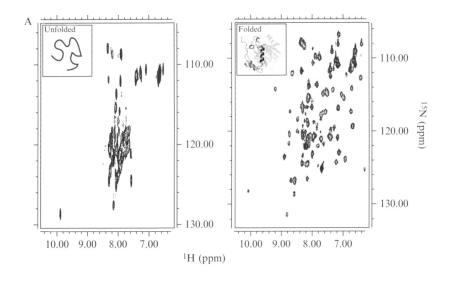

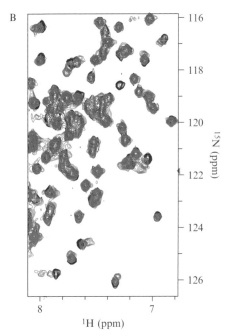

Assen Marintchev *et al.*, Figure 12.2 Protein–protein interactions visualized by NMR. (A) ^{1}H-^{15}N HSQC spectra of ^{15}N-labeled yeast eIF4G (393–490): free (left) and in complex with unlabeled eIF4E (right). Binding of eIF4E causes folding of eIF4G, which is accompanied by drastic changes in the spectra. Insets in the top left corners of the panels show a cartoon of an unfolded polypeptide (left) and the structure of the folded eIF4G segment (blue) in complex with eIF4E (semitransparent). Adapted from Hershey *et al.*, 1999, reproduced with permission from *J. Biol. Chem.* 1999, 274, 21297–21304. Copyright © 1999 The American Society for Biochemistry and Molecular Biology. (B) Overlay of a section of the ^{1}H-^{15}N TROSY-HSQC spectra of the second HEAT domain of human eIF4G: free (black) in the presence of substoichiometric amount of unlabeled eIF4A-NTD (blue) and of equimolar amount of unlabeled eIF4A-NTD (red). The interaction is in fast to intermediate exchange, is not accompanied by an unfolded-to-folded transition, and leads to more modest changes in the spectra.

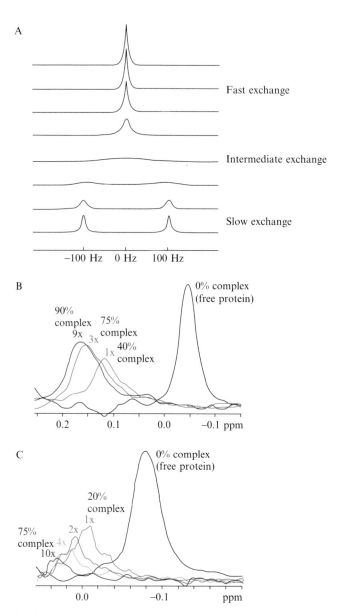

Assen Marintchev et al., Figure 12.3 Exchange regimens on the NMR chemical shift time scale. (A) Simulation of line shapes of NMR peaks for interconversion between two conformations in fast (top), intermediate, and slow (bottom) exchange regimens, with 200-Hz frequency difference between the peaks corresponding to the two states. The simulation was performed as described in Matsuo *et al.* (1999). (B) 1D slices of a single peak of the protein Bcl2 in the absence and in the presence of increasing concentrations of a small molecule. The binding is in the intermediate exchange regime ($K_D = 20\ \mu M$). The protein concentration is 25 μM, and compound concentration is from 0 to 250 μM. The ratios between compound and protein concentrations and the fraction of the protein that is in complex (in %, calculated from the K_D) are shown above the curves. (C) 1D slices of the same peak as in b on addition of increasing concentration of a different small molecule ($K_D = 80\ \mu M$) related the one in b and contacting the same residue of Bcl2, with two or more bound conformations in intermediate exchange regimen. The

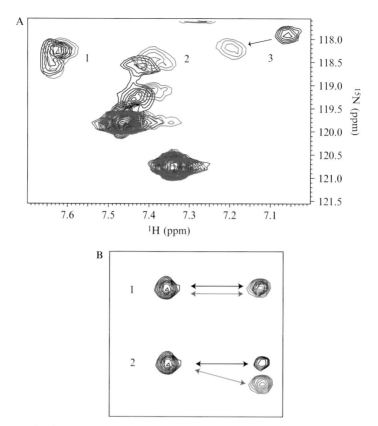

Assen Marintchev et al., Figure 12.4 Direct vs. indirect effects in NMR titrations. (A) A small region of the ^{15}N-HSQC spectrum of GB1-tagged human eIF4H in the absence of RNA (black) in the presence of substoichiometric amount of a 25-mer RNA oligo (blue) and in the presence of excess RNA (red). The peaks corresponding to the entire RNA-recognition motif (RRM) domain of eIF4H are broadened (most beyond detection) at intermediate RNA concentrations and reappear at excess RNA. Whereas the transient disappearance of peak 3 at substoichiometric concentration of RNA could be due to intermediate exchange broadening, the transient broadening/disappearance of peaks 1 and 2 is unlikely to be due to intermediate exchange, because little or no change exists in chemical shifts. The two strong peaks on the bottom of the panel that are not affected throughout the RNA titration belong to the GB1 tag. (B) Schematic representation of the effects of a ligand on a protein that is in equilibrium between two alternative conformations in slow exchange. Peak 1, ligand binding causes change in equilibrium, but not in peak positions; therefore, there is no change in environment—the corresponding residue is not in direct contact with the ligand, and the effect is indirect. Peak 2, ligand binding causes change in both equilibrium and peak positions—the effect could be either direct or indirect.

protein and small molecule concentrations are the same as in b. The ratios between small molecule and protein concentrations and the fraction of the protein that is in complex (in %, calculated from the K_D) are shown above the curves. Note that, because of the conformational exchange in the bound state, the line broadening is not reversed, and even becomes more severe, as the equilibrium is shifted toward complex formation. Panels b and c are adapted from Reibarkh et al. (2006). Reproduced with permission from *J. Am. Chem. Soc.* 2006. 128, 2160–2161. Copyright 2006 American Chemical Society.

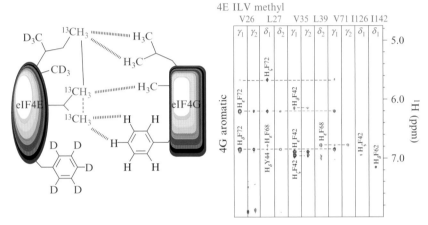

Assen Marintchev et al., Figure 12.5 Differential labeling strategy for obtaining unambiguous intermolecular NOEs in large complexes. Asymmetrical isotope labeling scheme applied in the NMR analysis of the eIF4E/eIF4G protein complex. One protein is ^{12}C/^{2}H-labeled, except for the methyl groups of Ile (d$_1$), Leu, and Val side chains, which are ^{13}C/^{1}H-labeled. The other protein in the complex is unlabeled. The protons (^{1}H) attached to ^{13}C methyl carbons are in red; the aromatic protons on the unlabeled eIF4G are in blue. Pairs of protons that can yield NOEs observed in a ^{13}C-edited NOESY spectrum are marked with dashed lines. The right panel shows strips from the aromatic region of the ^{13}C-edited NOESY spectrum of this complex: all observed NOEs between methyl and aromatic side-chain protons are intermolecular. The left panel is adapted from Gross et al. (2003). Reproduced with permission from *J. Biomol. NMR* 2003. 25, 235–242. Copyright © 2003 Springer.

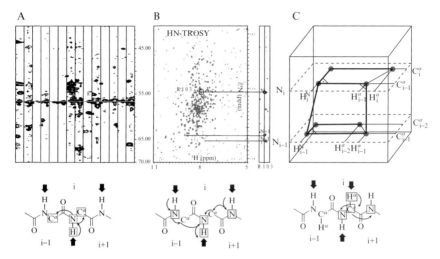

Assen Marintchev *et al.*, Figure 12.6 Strategies for backbone resonance assignments of proteins. Below each panel, a pictorial representation of the flow of magnetization is represented for the corresponding experiment. The nuclei that are correlated are framed. (A) Strips comparison obtained with the HNCA spectrum of a 37-kDa protein. The experiment leads to a 3D spectrum, depicted by the red dots in c. HNCA correlates amide protons and nitrogens of a given residue (that give rise to the cross-peaks in b) with the Cα carbons of the current and preceding residues (Cα-1), as depicted below panel a. Each strip corresponds to one signal in the HN-TROSY displayed in b, with the vertical axis showing the carbon signals. One residue is selected for the comparison (circled in b) and the shift of its Cα is compared with those of the Cα-1 for all residues. The same procedure is repeated for Cβ and CO nuclei. In this example, the neighbor cannot be identified with only these spectra. (B) "Backbone walk" assignment strategy. A pair of amide nitrogen and proton is correlated with the nitrogens of the neighboring residues, as depicted below panel b. A residue is selected from the HN-TROSY spectrum (circled) and a strip of the double TROSY-hNcaNH displays the nitrogen chemical shifts of the preceding and succeeding residues. The strip is located at the position circled in the HN-TROSY spectrum, with the third dimension (y-axis in the strip) displaying the additional nitrogen correlations. If the resolution is sufficient, the residues can be identified from the hNcaNH and the HN-TROSY. If not, the procedure can be repeated with an HncaNH that would lead to the *proton* chemical shifts of the neighboring residues. Alternately, the hNcaNH can be combined with the HNCA in a (or any other triple-resonance experiment), which leads to the identification of the successor (red frame in a). Adapted from Frueh *et al.* (2006). Reproduced with permission from *J. Am. Chem. Soc.* 2006. 128, 5757–5763. Copyright 2006 American Chemical Society. (C) "Stairway" assignment procedure. The 3D spectrum is represented schematically. The red peaks correspond to signals obtained with an HNCA (or an HACANH), with the magnetization flow displayed below panel a, and the green peaks to those of an HACAN (or an HNCAHA), with the magnetization flow depicted below panel c. Trapezoidal patterns enable us to identify spin-systems. The chain elongation is easily obtained by linking systems within planes and between planes (bold line). Adapted from Frueh *et al.* (2005). Reproduced with permission from *J. Biomol. NMR.* 2005. 33, 187–196. Copyright © 2005 Springer.

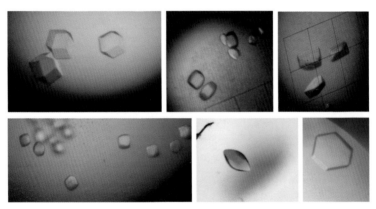

Jeffrey S. Kieft *et al.*, Figure 13.10   Examples of IRES RNA domain crystals obtained by use of the engineering techniques described in the text.